PRECISION FARMING : A NEW APPROACH

About the Editors

Dr. Tulasa Ram, presently working as Sr. Scientist (Agronomy), Project Directorate for Farming Systems Research (PDFSR), Meerut – was born in 1971 in Rajasthan. Dr. Ram obtained Ph.D in Agronomy from CCS HAU, Haryana. He has the experience of working at SKUAST-K (J&K) aend RARSS, Kargil, Ladakh (J&K). He has also been Assistant Commissioner (Crops) in the Department of Agriculture & Cooperation, Ministry of Agriculture, Krishi Bhawan, New Delhi during 2004-10. He has been responsible for monitoring, management, policy formulation, quarantine and boost through new technologies in crop production. His research works have been published in national and international journals and several book chapters.

Er. Shiv Kumar Lohan did his M.Tech (Farm Power & Machinery) from CCS Haryana Agricultural University, Hisar in 1999. He has authored nearly 20 research papers in national/ international journals of repute, besides he has published practical manuals, technical bulletins and popular articles. He is also the member of editorial board of Science Publishing group, International Journal of Energy and Power Engineering, and Journal of Technology Innovations in Renewable Energy. He has served CCS Haryana Agricultural University, Hisar, Project Directorate for Farming System Research, Meerut and SKUAST, Srinagar. Presently, he is working as Assistant Research Engineer, Department of Farm Machinery & Power Engineering, PAU, Ludhiana.

Dr. Ranveer Singh, Assistant Agriculture Officer, Government of Rajasthan was born in 1980 in Pilani (Rajasthan). He has studied at SK Rajasthan Agricultural University, Bikaner and Allahabad Agricultural Institute and holds B.Sc.Ag.(Hons), M.Sc. (Agronomy) and Ph.D degrees. Dr. Ranveer Singh is actively involved in implementation of various Agriculture Schemes of the State Government and the NGOs.

Dr. Purshotam Singh, Assistant Professor, Agronomy in SKUAST-Kashmir, J&K. Born in 1967 Srinagar (Kashmir), Dr. Purshotam Singh holds two B.Sc degrees (each from University of Kashmir and Guru Nanak Dev University, Amritsar). After doing M.Sc. (Agronomy) from Dr. BRA University, Agra, he attained Ph.D. from G.B. Pant University and Technology, Pantnagar in the field of weed management and crop production. Recently, he has worked on various university sponsored projects on weed management at Division of Agronomy SKUAST-Kashmir.

PRECISION FARMING : A NEW APPROACH

Editors:-

Tulasa Ram
PDFSR (ICAR), Modipuram (U.P.) - 250110

Shiv Kumar Lohan
*Deptt. of Farm Machinery & Power Engineering,
Punjab Agricultural University,
Ludhiana (Punjab) - 141001*

Ranveer Singh
State Agril. Deptt., Rajasthan

Purshotam Singh
*Division of Agronomy,
SKUAST-K, Srinagar (J&K) - 191121*

2014

Daya Publishing House®

A Division of

Astral International Pvt. Ltd.

New Delhi – 110 002

Published by   : **Daya Publishing House®**
                       A Division of
                       Astral International Pvt. Ltd.
                       – ISO 9001:2008 Certified Company –
                       4760-61/23, Ansari Road, Darya Ganj
                       New Delhi-110 002
                       Ph. 011-43549197, 23278134
                       E-mail: info@astralint.com
                       Website: www.astralint.com

Laser Typesetting : **Classic Computer Services**, Delhi - 110 035

Printed at   : **Replika Press Pvt. Ltd.**

PRINTED IN INDIA

PREFACE

The present agriculture scenario in India has led to certain vital points of concern for the planners and agricultural scientists to feed its growing population. The technological break through of "Green revolution" led to increased productivity of existing land by increasing number of crops or improving the input utilization like fertilizer, herbicides, pesticides and water. But it also yielded several negative ecological consequences such as decline in soil fertility, soil salinization, soil erosion, deterioration of environment, health hazards, degradation of biodiversity and poor sustainability of agriculture. Since the sustainable agriculture in the context of development efforts has to meet production efficiency, sensitivity to ecosystems, appropriate technology, maintenance of the environment, cultural diversity and satisfaction of the basic needs. Hence, modern agricultural management practices are changing from assuming homogenous fields to attempting to address field variability by dividing the field into smaller zones and managing these zones separately. A vision for an innovative route of development in agriculture with the backdrop of WTO regime and ecological crises that threaten to bring down productivity could truly be derived from the convergence of biotechnology with space and informatics known as 'Precision Farming'.

Precision farming is one of the most scientific and modern approaches to sustainable agriculture that has gained momentum in 21st century. The potential of precision farming for economical and environmental benefits could be visualized through reduced use of water, fertilizers, herbicides and pesticides besides the farm equipments. A precision farming approach recognizes site-specific differences within fields and adjusts management actions accordingly. Precision Agriculture offers the potential to automate and simplify the collection and analysis of information. In Precision agriculture, based on soil pH, nutritional status, pest infestation, yield rates, and other factors that affect crop production the field is broken into "management zones". Based on the requirements of each zone, the management decisions are taken with application of precision agriculture technologies such as field mapping, Global Positioning System (GPS) receivers, yield monitoring and mapping, grid soil sampling, variable-rate fertilizer (VRT) application, remote sensing and Geographic information systems (GIS) etc. But opinion about Precision Farming means to the large growers of the US or European countries. But, this is far from the truth as this approach has a large potential for improving the agricultural production in developing world too. Precision Farming in India could be unique in nature; it would be primarily based more upon knowledge and less upon sophisticated techniques.

The present book is an attempt to collect and comprehend the theoretical and practical information on various Precision Farming techniques by exploring their applicability and affordability in Indian agriculture. This book encompasses 24 chapters on use of Spectral Sensors, Variable Rate Technology, Remote Sensing, Precision Irrigation, Laser Guided Land Leveling, Precision Farming driven Crop Protection, Site Specific Nutrient Management and application of Information and Communication Technology in agricultural production invited from researchers and scholars. Besides, some chapters on management of Plant Genetic Resources (PGR's) for Precision Farming, System of Rice Intensification and Geoinformatics have also included.

The editors are grateful to all the contributors to the book for providing latest information on diverse aspects of Precision Farming and the information based on the experimental data generated from different research programmes. We also feel obliged to Dr. K.N. Singh, Head, Department of Agronomy for his encouragement and inspiration. Editorial assistance provided by Dr. N. K. Jat and Dr. Sudhir Kumar are gratefully acknowledged.

We are sure that this book will prove to be a valuable source of information to all those involved in Precision Farming including scientists, developmental personnel, policy makers, NGOs and farmers. It is hoped that this comprehensive treatise will stimulate and motivate more intensified research, accelerated developmental efforts, favourable policy initiatives and spread of Precision Farming at the grass roots for the production of more food with higher input use efficiencies. Once again we thank all those directly or indirectly associated with compilation of the information for this manuscript.

☞ **Tulasa Ram**
☞ **Shiv Kumar Lohan**
☞ **Ranveer Singh**
☞ **Purshotam Singh**

Contents

PRECISION FARMING : A NEW APPROACH

1. Precision Farming: An Indian Perspective.
2. Importance, Concept and Approaches for Precision Farming in India.
3. Engineering Interventions in Precision Farming.
4. Use of Specral Sensors in Crop Production.
5. Management of Plant Genetic Resources (PGR) for Precision Farming.
6. Scope for Application of Variable Rate Technology in India.
7. Remote Sensing and Image Processing
8. Precision Irrigation Systems in Agriculture: A Perspective.
9. Laser Guided Land leveling and Grading for Precision Farming.
10. Effective Water Resources Management Using Efficient Irrigation Scheduling.
11. Site Specific Varietal Improvement Strategies for Rice Breeding in Temperate Regions.
12. Expanding Horizons of Precision Farming Driven Crop Protection.
13. Storage Fungi Infestation, their Detection and Management.
14. Utility, Scope and Implementation of system of Rice Intensification (SRI) Under Temperate Conditions.
15. Site Specific Nutrient Management for Increasing Crop Productivity.
16. An Introduction of Plant Nutrients and Foliar Fetilization.
17. Accountability of Information and Communication Technology into Precision Farming in Climate Change Era.
18. Precision Farming of Seed Spices.
19. Remote Sensing Application for Crop Acreage and Production Estimation.
20. Yield Monitoring System for Grain Combine Harvester.
21. Database Requirements for Precision Farming.
22. Precision Farming in Vegetables.
23. Rangelands: Concepts and Management.
24. Geoinformatics in Forest Mapping.

1

CHAPTER

Precision Farming: An Indian Perspective

☞ Vijay Pooniya[1] and ☞ Ummed Singh[2]

INTRODUCTION : AN OVERVIEW

The present agriculture scenario in India has led to certain vital points of concern for the planners and agricultural scientists to feed its growing population. Land is precious natural resource for agriculture and per capita availability of the land has decrease drastically to nearly one third from 0.46 ha in 1951 to 0.15 ha in 1996-97. This lead us with the need to increase productivity of existing land by increasing number of crops or improving the input utilization like fertilizer, herbicides, pesticides and irrigation etc. or both. However, green revolution technology has helped to raise crop yields during mid-1960s to mid-1980s. The green revolution has not only increased productivity, but also it has several negative ecological consequences such as decline in soil fertility, soil salinization, soil erosion, deterioration of environment, health hazards, poor sustainability of agricultural lands and degradation of biodiversity. Indiscriminate use of pesticides, irrigation and imbalanced fertilization has threatened sustainability. On the other hand, issues like declining use efficiency of inputs and dwindling output–input ratio have rendered crop production less remunerative. According to CGIAR, sustainable agriculture is the successful management of resources to satisfy the changing human needs, while maintaining or enhancing the quality of environmental and conserving natural resources. Nehru said, everything can wait but not the agriculture. Therefore, agricultural research seeks the generation of new technologies to reorient the current and future needs and constraints. The new technology should be highly productive, cost-effective and ecologically sustainable. In the present context, maintenance of ecological balances through precise and site-specific management is most desirable. Planners have long recognized that an accurate and timely crop production forecasting system is essential for strengthening the food security. The concept of precision farming may be appropriate to solve these problems and it is not impossible to adopt in India. Research efforts are needed to find out its applicability in the Indian agricultural scenario. The M. S. Swaminathan Research Foundation, Chennai, India has joined hands with Israel to initiate precision farming on an experimental basis, and also conducting training programmes.

[1] Scientist, Division of Agronomy, Indian Agricultural Research Institute, New Delhi 110 012
[2] Division of Crop Production, IIPR, Kanpur-208 024

Management technologies which are considered to be applied uniformly over larger areas may not help us to increase yield because of variation in micro-climate, socio-economics of farmers and soil variability at micro level to apply input resources accordingly. It may also be recognized that the requirement of a crop for a particular inputs also vary spatially within field scale. In order to tackle these new challenges appropriate new technologies must be utilized with present day agronomic practices.

PRECISION FARMING

Precision farming or satellite farming is a farming management concept based on observing and responding to intra field variations. Today, precision farming is about, whole farm management with the goal of optimizing returns on input while preserving resources.

Precision farming involves site specific management practices paying due considerations to the spatial variability of land in order to maximize crop production and minimize cost of production and environmental damage (Robert, 1999). In precision agriculture, after assessing the in-field variability, same is taken care of through precision land leveling to manage landscape variability, variable rate technology, site specific planting, site specific nutrient and other input management (Jat *et al.,* 2004). The crop inputs are distributed on a spatially selective basis through gird sampling or management zone approach. India, being a land of geo-physical, agro-climatic and greater socio-economic variability and simultaneously making indiscriminate use of irrigation and pesticides and imbalanced use of fertilizers during last four decades, badly needs precision farming for increasing use efficiency of crop inputs and also maintaining the sustainability with enhanced productivity and reduced environmental damage. Precision farming in India could be unique in nature, different from what is being practiced in developed nations. It would be primarily based more upon knowledge and less upon sophisticated techniques (Sharma *et al.,* 2005).

Literary, precision farming is made up of two words *i.e.* "Precision" and "Farming". The term precision refers to the quality or state of being precise. Where precise means minutely exact or term synonymous with correct. Therefore, precision farming refers to exactness and implies correctness or accuracy in any aspect of production. According to Stafford (1996) "precision farming involves the targeting of inputs to arable crop production according to crop requirements on the localized basis". Thus the intent of precision farming is to match agricultural inputs and practices to localized conditions within a field to do the right things, in the right place, at right time and in the right ways (Pierce *et al.,* 1994). On the basis of discussion, the definition of precision farming may be defined as "the application of technologies and principles to manage spatial and temporal variability associated with all aspects of agricultural production for the purpose of improving crop performance and environmental quality". The objectives of precision farming are to increase production efficiency, improve product quality, use of chemicals more efficiently, energy conservation and soil and ground water protection.

Components of Precsision Farming

    (*i*)   Computers

    (*ii*)   Remote sensing

    (*iii*)   Geographic information system (GIS)

    (*iv*)   Differential global positioning system (DGPS)

    (*v*)   Variable rate applicator

Computers: Many technologies support precision farming, but none is important than computers in making precision agriculture possible. Also, it is not computers along that are important but their ability to communicate that is powerful for agriculture. The fusion of computers and communication is defined as the age of process (Taylor and Wacker, 1997). Precision farming requires the acquisition, management analysis and output of large amount of spatial and temporal data. Mobile computing systems were needed to function on the go in farming operations because desk top systems in the farm office were not efficient. These mobile systems needed micro-processors that could operate at speeds of millions of instructions per seconds (MIPS), had expansive memory and could store massive amount of data.

Remote Sensing: Remote sensing is collection of data from a distance and have been used for several years to distinguish crop species and locate stress conditions in the field. Remote sensing involves the detection and measurement of photons of differing energies emanating from distant materials. These photons may be identified and categorized by class/ type, substance and spatial distribution with most designed to monitor reflected radiation (Frazier *et al.*, 1997). Soil nutrient sensors are used to detecting the soil fertility status. With the help of these soil nutrients sensors the fertilizer rate is determined for accurate place where particular nutrient is deficient and other applications include yield monitors or moisture sensors. Remote sensing holds great promise for precision farming because of its potential for monitoring spatial variability over time at high resolution (Moran *et al.*, 1997).

Geographical Information System: It is a software application that is designed to provide the tools to manipulate and display spatial data (Fig. 1.1). It is an effective way of computerizing maps. An important function of an agricultural GIS is to store layers of information such as yields, soil survey maps, remoledly sensed data, crop scouting reports and soil nutrients levels. GIS technology allows the manager to store field inputs and output data as separate map layers in digital maps and to retrieve and utilize these data for future input allocation decision. Clark and McGucken (1996) refers to GIS as the brain of a precision farming system because precision farming is concerned with spatial and temporal variability and it is information based and decision focused. GIS has the capabilities to anlyze spatial variability.

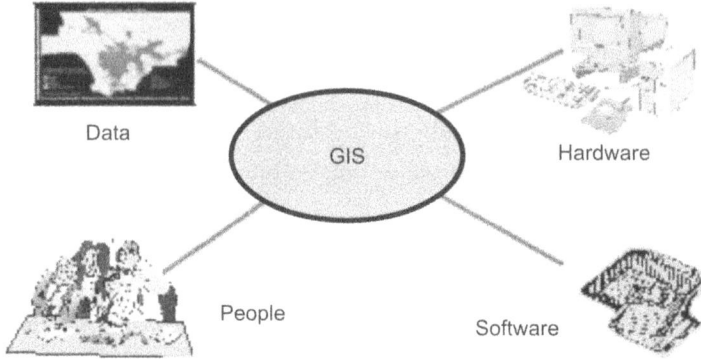

Fig. 1.1. *Components of GIS*

Differential Global Positioning System (DGPS): The DGPS was developed by American Military for accurate positioning of military personnel. It is a navigation system based network of earth-orbiting satellites that lets user's record near-instantaneous positional information (latitude, longitude and elevation) with accuracy ranging from 100 m to 0.01 m (Lang, 1992). The GPS technology enables precision farming because all phases of precision farming require positioning information. GPS is able to provide the positioning in a practical and efficient manner for a field locations so that input can be applied to individual field segments based on performance criteria and previous input application.

Variable Rate Applicator: The variable rate applicator has three components *viz.* control computer, locator and, actuator. The control computer co-ordinates the field operation, It has a map of desired activity as a function of geographic location. It receives the equipment's current location from the locator, which has a GPS in it, and decides what to do base upon the map in its memory or data storage. Then it issues the command to the actuator, which does the input application (Ravi and Jagadeesha, 2002). In variable rate technology, crop production input rate changed with a field in response to variable factors that affect the optimum rate of application. Uniform application of crop production inputs does not allow optimum efficiency or profitability because factors that affect crop production are not always uniform within fields. It has the potential to improve or ideally maximize efficiency of inputs and profitability of individual fields by targeting application where needed and at optimum rate. Variable rate technology is not a new concept. The University of Illionis published a circular in 1929 outlining practices to intensively soil sample fields for the purpose of mapping soil pH variation and determining treatable areas for variable limestone application.

PRACTICAL PROBLEMS IN ADOPTION OF PRECISION FARMING IN INDIA

Precision Agriculture has been mostly confined to developed countries. Reasons of limitations of its implementation in developing countries like India are small land holdings, heterogeneity of cropping systems and market imperfections, high cost, lack of technical, expertise knowledge and technology etc. In India, major problem is the small field size. More than 58 per cent of operational holdings in the country have size less than 1 ha. Only in the states of Punjab, Rajasthan, Haryana and Gujarat more than 20 per cent of agricultural lands have operational holding size of more than four hectare. There is a scope of implementing precision agriculture for crops like, rice and wheat especially in states of Punjab and Haryana. Commercial as well as horticultural crops show a wider scope for precision agriculture. In India, broadly two types of agriculture *viz.* high input agriculture characterized by the provision of assured irrigation and other agricultural inputs, and subsistence farming, which are confined mostly to rainfed or dry land regions. Nearly two-third arable land in India is rainfed. The crop yields are very low (≈ 1 t/ ha) and very good potential exists for increasing productivity of rainfed cropping systems.

Steps to be taken for Implementing Precision Farming in India

In the present existent situation, the potential of precision agriculture in India is limited by the lack of appropriate measurement and analysis techniques for agronomical important factors (National Research Council, 1997). High accuracy sensing and data management tools must be developed and validated to support both research and production. The limitation in data quality/availability has become a major obstacle in the demonstration and adoption of the precision technologies. The adoption of precision agriculture needs combined efforts on behalf of scientists, farmers and the government. The following methodology could be adopted in order to operationalize precision farming in the country.

- Creation of multidisciplinary teams involving agricultural scientists in various fields, engineers, manufacturers and economists to study the overall scope of precision agriculture.
- Formation of farmer's co-operatives since many of the precision agriculture tools are costly (GIS, GPS, remote sensing etc.).
- Government legislation restraining farmers using indiscriminate farm inputs and thereby causing ecological/environmental imbalance would induce the farmer to go for alternative approach.
- Pilot study should be conducted on farmer's field to show the results of precision agriculture implementation.
- Creating awareness amongst farmers about consequences of applying imbalanced doses of farm inputs like irrigation, fertilizers, insecticides and pesticides.

Realizing the potential of space technology in precision farming, the Department of Space, Government of India has initiated eight pilot studies in well-managed agricultural farms of the ICRISAT, the Indian Council of Agricultural Research and the Agricultural Universities, as well as at farmer's fields. The pilot studies aim at delineating homogeneous zones with respect to soil fertility and crop yield, estimation of potential yield, yield gap analysis, monitoring seasonally variable soil and crop conditions using optical and microwave sensor data, and matching the farm inputs to bridge the gap between potential and actual yield through spatial decision support systems (SDSS). The test sites are spread over a fairly large area across a cross section of agro-climatic zones of the Indian sub-continent, and cover some of the important crops like wheat, rice, sorghum, pigeon pea, chickpea, soybean and groundnut. The next step would be to generate detailed-level information on soil resources addressing potentials and limitations of individual fields since except for states like Punjab, Haryana, Madhya Pradesh and Maharashtra where fields size is quite large, practically individual field could be treated as a homogenous management unit for the purpose of precision farming. Prior operational work in remote sensing with respect to agriculture has been undertaken by the space community which has been summarized in the following table 1.1.

Table 1.1 : Major Indian remote sensing missions for agriculture (Current & immediate future)

Missions	Year of launch	Sensors
IRS-IA, IB	1988	LISS-1 (72.5 m resolution,148 km swath)
	1991	LISS-II (36.25 m resolution,142 km swath)
IRS-P2	1994	LISS-II (36 m resolution, 142 km swath
IRS-IC, ID	1995	PAN(5.8 m resolution,70 m swath)
	1997	LISS-III (23.5m, 70.5 m resolution, 141 km,148 km swath) WiFS (188.3 m resolution; 774 km swath)
IRS-P3	1996	WiFS (188.3 m resolution,774 km swath)
TES	2001	PAN (1 m resolution,14 km swath)
RESOURCESAT-I	2001	LISS-IV (6 m resolution,25 km swath / LISS-III/ 23 m resolution,140 km swath/AWiFS /80 m resolution, 800 km swath)
CARTOSAT-1	2002	PAN Stereo (2.5 m resolution,30 km swath)
CARTOSAT-2	2002/03	PAN Stereo (1 m resolution,12 km swath)

(Source: Gowrisankar and Adiga, 2001)

IDENTIFICATION AND ASSESSMENT OF VARIABILITY

Gird soil sampling: It uses the same principle as soil sampling but increased the intensity of soil sampling compared to traditional sampling. Soil samples collected in a systematic gird also have location information that allowed the data to be mapped. The goal of gird sampling is to generate the map of nutrient/water requirement, called an application map.

Crop scouting: In-season observations of crop conditions like weed patches, insect and fungal infestation, crop tissue nutrient status are also helpful when explaining variations in yield map. Precision farming techniques for assess variability. In real time assessment of variability is possible only through advanced tools of precision agriculture.

MANAGEMENT OF VARIABILITY

Variable rate application: Gird soil samples are analysed in laboratory and an interpretation of crop inputs (nutrient/water) needs is made for each sample. Then the input map is plotted using the entire set of soil samples. The input application map is loaded into a computer on variable rate input applicator. The computer used the input application map and geographical information system receiver to direct a product deliver controller that changes the amount and kind of input according to the application map.

Yield monitoring and mapping: Yield information provides important feedback in determining the effects of managed inputs such as fertilizer amendments, seed, pesticides, cultural practices including tillage and irrigation. Since yield measurements of a single year may be influenced by climate, it is always advisable to examine the data several years to find out the correct value.

ECONOMIC ANALYSIS OF DGPS NETWORKS

As a first step towards operationalization of precision farming, it is necessary to establish a DGPS infrastructure for the country. It would enable the farmers' to get an accuracy of few centimeters in the various unit processes (sowing, fertilizer application, herbicide and pesticide application etc.) involved in precision farming. A DGPS network would cater to the needs of multitude of applications like meteorology, transportation, geodetic survey, crustal deformation studies, disaster management and mitigation, etc. Each DGPS master reference is capable of providing the services within a radius of 100-kilometer radius. Hence, the number of master stations required to establish the DGPS infrastructure, which would cover the entire country, is calculated. This can be calculated as follows (only a random calculation).

Geographical area of India	= 329 million hectares
Area of GPS (circular area, PI = 3.14)	= PI * (100)² sq. km
	= 31400 sq. km
	= 3.14 million hectares
Total no. of GPS reference stations required for the country	= 329/3.14
Total GPS reference stations required	= 105
Cost of a single DGPS set	= ₹ 3 lakhs
Total cost of the entire infrastructure	= ₹ 3.15 crores
Solution:	
Area of GPS (circular area, PI = 3.14)	= PI * (100)² sq. km
	= 3.14 × 100 × 100 sq.km
	= 31400 sq.km

APPROACHES OF PRECISION FARMING IN INDIAN PERSPECTIVES

Crop management over the past four decades in India was driven by the increasing use of external inputs and blanket recommendations for fertilizer use over wide area. However, future gains in productivity and input use efficiency will require soil and crop management technologies that are more knowledge intensive and tailored to the specific characteristics of individual farm and fields.

Random sampling: State farming of western countries, field size is very small in India as the majority being small and marginal farmers with continuous sub-division and fragmentation of land. Therefore, the gird sampling and production level management zone approach may not be feasible in Indian agriculture. The different small fields in possession of a farmer or farmer may consider as management zones and the recommendations could be on random sampling from each zone. If the field size is found very small to study the in-field variability and recommend variable rate application in an area.

Crop establishment: Despite much progress in mechanization, use of country ploughs for tillage is not obsolete in India. Germination, crop stand, vigor and yield are dependent on proper tillage and crop establishment methods. Simply, mechanization does not ensure precision; therefore, low cost precision planters with précised seed metering devices are to be popularized to ensure optimum plant stand with less seed rates. The recent approaches of conservation agriculture with the development of precision planters *viz.* no-till multi-crop planter with new generation seed metering systems; reduced till raised bed planter with multi-crop planting systems in one of the right direction of precision farming in Indian perspective and can be properly followed when the soil physico-chemical variability and constraints are well known.

Nutrient management: Nitrogenous fertilizer being a basic and widely applied nutrient in Indian agriculture needs special attention for precision management practices. For improving nitrogen use efficiency, reducing nitrogen losses and environmental pollutions, through the use of simple precision tools *viz.* leaf colour charts (Fig. 1.2) and chlorophyll meter (SPAD) (Fig. 1.3) has offered wide scope for farm level adoption of these technologies. The leaf colour chart (LCC) based on real time nitrogen management can be used to optimize or synchronize nitrogen application with crop demand or improve existing fixed split nitrogen recommendations (Shukla *et al.*, 2004a). They further reported that the net returns in rice-wheat cropping system were increased by 19-31 per cent in LCC based N-management then in fixed time nitrogen application. The site specific nutrient management practice responds significantly not only to the primary nutrients but also to the secondary and micro-nutrients due to emerging deficiencies for the targeted yields. The site specific nutrient management package (N_{170}, P_{30}, K_{120}, S_{20}, Zn_7, Mn_{17}, $B_{0.6}$ in rice and N_{150}, P_{30}, K_{120} in wheat) gave significant yield advantages both in rice and wheat compared to soil test lab response, local adhoc recommendation and farmer's practice (Shukla *et al.*, 2004b). The yield and yield attributes of rice were significantly higher in the LCC based nitrogen application compared to other treatments (Budhar, 2005).

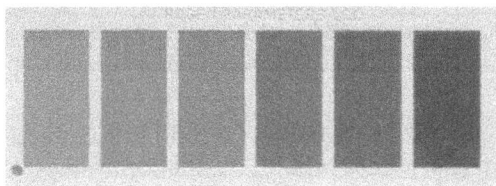

Fig. 1.2: *Leaf Colour Chart*   **Fig. 1.3:** *SPAD Meter*

Precision water management: Water is a critical and most limiting input for crop production. Crop yield generally shows linear correlation with amount of water transpired (Howell, 1990). When water is a limiting factor, crop yield is much lower than the potential yield. The excess watering may induce stresses with respect to aeration and nutrient availability. Therefore, the only option is precision water management which has three approaches *viz.* variable rate irrigation soil-landscape management and drainage.

Site-specific weed management: In general farmer's uniformly broadcast or spray herbicides to control weeds. However, it is well documented that a degree of spatial variability exists in weed distributions (Johnson *et al.*, 1997). Therefore, over application and under application of herbicides occurs obviously. Over application of herbicides causes environmental contamination, increased cost and injury to the crops. Likewise, under application causes poor weed control with substantial yield losses. The spatial variability in weed practice of managing landscape variability in India is leveling through animal or tractor drawn scrappers. The precision weed management involves three main components *viz.* preventing weeds through adapted crop management, improvised decision making on weed control, and developing precision control technology.

Precision pest and disease management: Application of chemicals for management of pest and diseases increasing rapidly in India leading the problem of chemical residues and resistance in pest. Although, commendable research has been made on integrated pest management but in many cases it is not been practically feasible and the farmer's are not aware of the components of IPM. Much research is required on IPM for different crops and cropping systems at different locations. The pests could not manage without proper monitoring which is essence of precision pest management. Assessing pest incidence at regional level in predominant crops concentrated is specific locations could be made through the precision farming technologies (remote sensing and geographical information system) and the pesticides to manage the specific pest can be suggested to the farmer at regional scale. This type of information based technologies help in reducing pesticide loads as well as better management with less investment.

Keys to Success of Precision Farming

Information: Information is most valuable resource for modern farmers. Timely and accurate information is essential in all phases of production from planning through post-harvest. Information available to the farmer includes crop characteristics, soil properties, fertility requirements, weed populations, insect populations, plant growth response, harvest data and post-harvest processing data. The precision farmer must seek out and use the information available at each step in the system.

Technology: Modern technology in agriculture is the second key to success. Computer software, spreadsheets, databases, geographic information systems and other types of application software are readily available. The global positioning system has given the farmer means to locate position in the field to within a few feet.

Management: Management is the third key to success, combines information obtained and available technology into a comprehensive system. Without proper management, precision crop production would not be effective. Farmers must know how to interpret the information available, how to utilize the technology and how to make sound production decisions.

CONCLUSION

Agriculture has continued to be the cornerstone of Indian economy. In the years to come, precision agriculture may help Indian farmers to harvest the fruits of frontier technologies without compromising quality of land and produce. The adoption of such a novel technique would trigger a techno-green revolution in India which is the need of hour.

- Precision farming incorporates modern technology and past experiences to practice crop production
- Precision farming has potential to improve both the profitability of farm and reduce environmental damage of agriculture
- Precision farming can improve the fertilizer use efficiency and other input use efficiency so that farmers can save considerable amount of fertilizer
- Precision farming improving the management efficiency which helps in raising the economic condition of farmers

REFERENCES

Budhar, M N. (2005). Leaf colour chart based N -management in direct seeded puddle rice. *Fertilizer News* 50(3) 41-42.

Clark, RL and McGucken RL. (1996). Variable rate application technology: An overview. In "Proceeding of the third international conference on precision agriculture, MN, 23-26 June 1996" (P.C. Robert., R.H. Rust and W.E. Larson, Eds.). ASA Miscellaneous publication, pp. 855-862, CSSA and SSSA, Madison, WI.

Frazier, BN, Walters CS and Perry EM. (1997). Role of remote sensing in site- specific management. In "The state of site-specific management for agriculture". ASA Miscellaneous publications, pp. 149-160.

Gowrisankar, D and Adiga S. (2001). Remote sensing in agricultural applications: An overview. pp: 9-14, In: Proc, the First National Conference on Agro-Informatics (NCAI), INSAIT. Dharwad.

Howell, T. (1990). In: Irrigation of agricultural crops (E. A. Stewart and Nilseon, Eds.), Agronomy monograph No. 30, ASA-CSSA-SSSA, Madision, WI, pp. 391-434.

Jat, ML, Pal SS, Subba Rao AVM, Sirohi K, Sharma SK and Gupta RK. (2004). In: Proceeding of national conference on conservation agriculture: Conserving resources enhancing productivity. September, 22-23, NASC complex, Pusa, New Delhi. pp. 9-10.

Johnson, GA, Cardina J and Mortensen DA. (1997). In: The state of site specific management of agriculture, ASA-CSSA-SSSA, Madision,WI, pp. 131-147.

Lang, L. 1992. GPS + GIS + Remote Sensing: An overview. Earth observation magazine. April, 23-26.

Moran, M S, Inoue Y and Barnes E M. (1997). Opportunities and limitations for image-based remote sensing in precision crop management. *Remote Sensing and the Environment*. **61** 319-396.

National research council, (1997). Precision Agriculture in the 21st Century: Geospatial and Information Technologies in Crop Management. Washington, DC: National Academy Press, pp. 31.

Pierce, FJ, Robert PC and Mangold G. (1994). Site-specific management. In "Proceedings of the international crop management conference, Iowa State University". pp. 17-21.

Ravi, N and Jagadeesha CJ. (2002). *Precision Agriculture*, Training course on Remote Sensing and GIS Applications in Agriculture, May 27th –7th June, 2002, RRSSC- Bangalore, pp: 225-228.

Robert, PC 1999. Precision agriculture: research needs and status in the USA. *Precision Agriculture*, **1** 19-33.

Sharma, SK, Jat ML and Biswas C. (2005). Precision farming and its relevance in Indian agriculture. *Indian Journal of Fertilizer* 1 (4) 13-18 & 21-26.

Shukla, AK, Ladha JK, Singh, VK, Dwivedi BS, Sharma SK , Balasubramanian V, Gupta R K, Singh Y, Pathak H, Pandey P S and Padre A T. (2004a). Calibrating the leaf colour chart for nitrogen management in different genotypes of rice and wheat in a systems perspective. *Agronomy Journal*, **96** 31-46.

Shukla, AK, Singh VK, Dwivedi BS, Sharma SK and Tiwari KN (2004b). Site-specific nutrient management for maximum economic yield of the rice-wheat cropping system. *Better Crops*, **88** 18-21.

Stafford, JV. 1996. Essential technology for precision agriculture. In "Proceedings of the third international conference on precision agriculture. Minneapolis, MN, 23-26 June (1996)" (P.C. Robert., R. H. Rust and W. E. Larson, Eds.), ASA Miscellaneous publication, pp. 595-604.

Taylor, J and Wacker W. (1997). "The 500 years delta: What happens after what comes next". Harper Business, New York.

2

Importance, Concept and Approaches for Precision Farming in India

☞ Jagvir Dixit[1], ☞ Anoop Kumar Dixit[2], ☞ Shiv Kumar Lohan[3] and ☞ Dinesh Kumar[4]

INTRODUCTION

"India lives in the villages and Agriculture is the backbone of the Indian Economy" said Mahatma Gandhi five decades ago. Even today, as we enter the new millennium, situation is still the same with almost entire economy being sustained by agriculture which is the mainstay of the villages. Not only the economy but also every one of us looks up to agriculture for our sustenance too. Despite all the natural advantages, India's productivity of food grains per hectare is no more than three fourths of the world average and less than half of that in agriculturally advanced countries. Per capita food grain availability even after green revolution, has been less than two -thirds of the world average. Only five states in India namely Himachal Pradesh, Punjab, Haryana, Uttar Pradesh and Madhya Pradesh produce more food grain than their consumption. The combined population of the above five states is less than one third of the total countries population. More than two thirds of the population lives in states that are still food deficit. This requires transport of lakhs of tonnes of food grain, involving high costs and pilferage. Our effort should have been to make all the states self-sufficient with respect to food grains and if some disturbances occurred due to unnoticed natural calamities, Nation must be in an ever ready position to mitigate such challenging tasks.

The Indian green revolution is also associated with negative ecological and environmental consequences. The status of Indian environment shows that in India, about 182 million ha of the country's total geographical area (328.7 million ha) is affected by land degradation, of this 141.33 million ha are due to water erosion, 11.50 million ha due to wind erosion, 12.63 and 13.24 million ha due to water logging and chemical deterioration (salinization and loss of nutrients), respectively (Ray *et al.*, 2001). On the other end, India shares 17 per cent of world's population with 2.5 per cent of geographical area, 1 per cent of gross world production, 4 per cent of world carbon emission and hardly 2 per cent of world forest area. The Indian status on environment is though not alarming when compared to developed countries, it gives an early warning to take appropriate precautionary measures. For decades now, the farmers have been applying fertilizers based on recommendations

[1] Division of Agricultural Engineering, Sher-e-Kashmir University of Agricultural Sciences & Technology of Kashmir, Srinagar-191121
[2, 3&4] Deptt. of Farm Machinery & Power Engineering, PAU, Ludhiana - 141001

emanating from research and field trials under specific agro-climatic conditions, since soil nutrients and characteristics vary not only from one region to another but also from field to field (Ladha *et al.*, 2000).

The growth rate of grain production during ninth plan has been less than the population growth rate. *The poor agricultural performance has not been because of vagaries of monsoon* from 1997-98 to 2002-03, as rainfall was between 92 and 106 per cent of the normal. Per capita availability of grain and per capita calorie intake, which were less than the minimum required for adequate nutrition have declined further. According to human development report 2003, the percentage of undernourished in India which was 21 per cent a few years ago, has now reached 24 per cent. The Government claims that India has emerged as the seventh largest exporter of food grains in the world. There is nothing to be proud of if we take into account that total Indian grain export in 2002-03 did not add up to even 4 per cent of total world exports and the value of our food grain exports did not add up to even the value of our imports of vegetable oils and pulses. The more crucial question, however, is whether it is morally justifiable to export grain when 24 per cent of population remains under nourished. The decline in agricultural growth and increase in rural poverty have been due to the long persisting government indifference towards the farm sector, which is evident from plan outlays on agricultural and allied activities. Rural development and irrigation, which added up to 37.1 per cent of the total during first plan were brought down to only 19.4 per cent during the ninth plan. However, main reason for poor performance of farm sector has been long persisting adverse terms of trade policies for agriculturists in addition to mismanagement of natural resources leading to ever ending crisis.

Ever since the man appeared on the earth, he has been harnessing the natural resources to meet his basic requirements. Reference to soil, water and air as basic resources, their management and means to keep them pure are mentioned in the Vedas, Upanishads and in ancient Hindu literature. The phenomenal increase in population of both man and animal in last century and fast growing industrialization and urbanization in last few decades have overstrained the natural resource base, which are getting degraded much faster than ever before. Thus, the attention of whole world is focused on how to increase production to feed the burgeoning population and the question uppermost in every one's mind is "Can we produce enough food in a sustainable manner without demaging natural resource-base." Be lying all predictions made to the contrary, India could achieve unprecedented increase in the food grain production as a result of expansion of irrigation and technological advancement in agriculture. While it has been a satisfying experience, Indian agriculture would need a new vision to make rapid progress in ensuing millennium. To achieve the required growth will not be easy as some of the existing production systems are based on unsustainable use of the resources. The signs of fatigue in the natural resources have already appeared which is a cause of serious concern to the planners, decision-makers and researchers alike.

In recent times, the researchers in the field have been busy formulating methodologies and fabricating new implements for precision farming. It is here that the challenge arises considering the implementation of technology at various levels in the global community. The need of the hour is not application of technology but adoption of appropriate technology which would suit the particular level of the global community. In India, farming practices are too haphazard and non-scientific, that they need some forethought before implementing any new technology.

Applications of agricultural inputs at uniform rates across the field without due regard to in-field variations in soil fertility and crop conditions does not yield desirable results in terms of crop yield. The management of in field variability in soil fertility and crop conditions for improving the crop production and minimizing the environmental impact is crux of precision farming. Thus, the information on spatial variability in soil fertility status and crop conditions is a prerequisite for adoption of precision farming. Space technology including global positioning system and geographical information system holds good promise in deriving information on soil attributes and crop yield and allows monitoring seasonally variable soil and crop characteristics, namely soil moisture, crop phonology, growth, evapo-transpiration, nutrient deficiency, crop disease, weed and insect infestation etc. which, in turn to help in optimizing inputs and maximizing crop yield and income.

THE NEED FOR PRECISION FARMING

The 'Green revolution' of 1960's has made our country self sufficient in food production. In 1947, the country produced a little over six million tonnes of wheat, in 1999 our farmers harvested over 72 million tonnes, taking the country to second position in wheat production in the world. The production of food grains in five decades has increased more than threefold, yield during this period has increased more than two folds. All this has been possible due to high input application like increase in fertilization, irrigation, pesticides, use of high yielding varieties, increase in cropping intensity and increase in mechanization of agriculture.

(*i*) Fatigue of Green Revolution

Green revolution of course contributed a lot. However, even with spectacular growth in agriculture, productivity levels of many major crops are far below than expectation. We have not achieved even lowest level of potential productivity of Indian high yielding varieties, whereas the world's highest productive countries have crop yield levels significantly higher than the upper limit of potential of Indian HYV's. Even the crop yields of India's agriculturally rich state like Punjab is far below than the average yield of many high productive countries (Ray *et al.*, 2001).

(*ii*) Natural Resource Depletion

One of the major reasons for environment degradation is the population growth of 2.2 per cent in 1970–2000. The Indian status on environment is though not alarming when compared to developed countries gives an early warning.

In this context, there is need to convert this green revolution into an evergreen revolution, which will be triggered by farming systems approach that can help to produce more from available land, water and labour resources without either ecological or social harm (Shanwad *et al.*, 2002). Since precision farming proposes to prescribe tailor made management practices, it can help to serve this purpose.

- Precision Farming is the title given to a method of crop management by which areas of land or crop within a field may be managed with different levels of input depending upon the yield potential of crop in that particular area of land.
- It is also defined as the application of technologies and principles to manage spatial and temporal variability associated with all aspects of agricultural production (Pierce and Nowak, 1999).

In other words, precision farming is matching of resource application and agronomic practices with soil attributes and crop requirements as they vary across a field. Precision farming is useful in many situations in developing countries. Rice, wheat, sugar beet, onion, potato and cotton among the field crops and apple, grape, tea, coffee and oil palm among horticultural crops are perhaps the most relevant. Some have a very high value per hectare making excellent cases for site-specific management. For all these crops, yield mapping is the first step to determine the precise locations of highest and lowest yield areas of the field. Researchers at Kyoto University recently developed a two row rice harvester for determining yields on a micro plot basis (Iida *et al., 1998*).

NUTRIENT STRESS MANAGEMENT

Nutrient stress management is another area where precision farming can help Indian farmers. Most cultivated soils in India are acidic and spatial variation in pH is high. Detecting nutrient stresses using remote sensing and combining data in GIS can help in site-specific applications of fertilizers and soil amendments such as lime, manure, compost, gypsum and sulphur. This in turn would increase fertilizer use efficiency and reduce nutrient losses. In semi arid and arid tropics, precision technologies can help growers in scheduling irrigation more profitably by varying the timing, amounts and placement of water. *For example*, drip irrigation, coupled with information from remotely sensed stress conditions (*e.g.* canopy temperature) can increase the effective use of applied water from 60 to 95 per cent there by, reducing runoff from 23 to 1 per cent and deep percolation from 18 to 4 per cent.

PESTS AND DISEASES

Pests and diseases cause huge losses to Indian crops. If remote sensing can help in detecting small problem areas caused by pathogens, timing of applications of fungicides can be optimized. Recent studies in Japan show that pre-visual crop stress or incipient crop damage can be detected by using radio-controlled aircraft and near-infrared narrow-band sensors. Likewise, airborne video data and GIS have been shown to effectively detect and map black fly infestations in citrus orchards, making it possible to achieve precision in pest control. Perennial weeds, which are usually position-specific (Wilson and Scott, 1982) and grow in concentrated areas, are also a major problem in developing countries. Remote sensing combined with GIS and GPS can help in site-specific weed management. Although thorough cost benefit analysis has not been done yet, the possible use of precision technologies in managing the environmental side effects of farming and reducing pollution is appealing. The success in precision agriculture depends on the accurate assessment of variability, its management and evaluation in space-time continuum in crop production. The agronomic feasibility of precision agriculture has been intuitive, depending largely on the application of traditional recommendations at finer scales. The agronomic success of precision agriculture has been quite convincing in crops like sugar beet, sugarcane, tea and coffee. The potential for economic, environmental and social benefits of precision agriculture is largely unrealized because the space-time continuum of crop production has not been adequately addressed. Precision agriculture must fit the needs and capabilities of farmer and must be profitable.

TRADITIONAL PRACTICE VERSUS PRECISION FARMING

Precision farming technologies may be relatively new concept in India but precision management is not. Indian farmers have long known that soil conditions, fertility, moisture etc. vary widely across a single field and that various parts within fields responded differently to different types of inputs and cultural practices. The small size of their farms often permitted effective monitoring of spatial and temporal yield variation and variable application of inputs by simple observation and manual means. The advent of green revolution in the country pushed up production levels by leveraging the advantages of high yielding varieties, fertilizer application and liberal use of pesticides. The trend has continued and generalized regional recommendations based on crops grown are followed. *For example*, pesticide application rate is on a per hectare basis irrespective of what percentage of crop is really affected. Providing subsidies has encouraged fertilizer use. Fertilizers application rate have been developed on a regional scale by agricultural experiment stations irrespective of whether the variety planted can make full use of it or not. Withdrawal of subsidies contemplated by government would lead to decline in fertilizer consumption. Judicious use of fertilizer at precise point of need can facilitate higher yields and also reduce the degradation of soil and pollution of ground water (Khosla *et al.*, 2001). The right time for implementing precision farming has arrived.

Many precision farming technologies are multifunctional and their adoption should result in favorable changes in various aspects of farming. Indian farms have traditionally worked by many generations of farmers, who have accumulated substantial location specific knowledge and skills. Precision farming technologies allow integration of local and traditional knowledge into farm resource management to create a historical spatial database, which would be useful for posterity. Further, the collection and analysis of geo-referenced data from all fields in a village provides a unique opportunity to gain new insights into the functioning of Indian agricultural systems. The data will have added value, especially when integrated into regional databases. It is therefore abundantly clear that precision farming technologies are and will be increasingly relevant to the development of sustainable and profitable cropping systems in India.

THE PRECISION FARMING SYSTEM CONCEPT

Precision farming sometimes called site-specific management (SSM) is an emerging technology that allows for adjustments to address with-in field variability in characteristics such as soil fertility, soil moisture, weed intensity and insect-pest infestation. As per Khosla *et al.,* 2002, precision farming can help today's farmer meet these new challenges by applying the 5 "R's" *i.e.,* **Right** input, in the **Right** amount, to the **Right** place, at the **Right** time, and in the **Right** manner. The importance and success of precision farming lies in these five "R's". Precision farming is also defined as the application of technologies and principles to manage spatial and temporal variability associated with all aspects of agricultural production. (Pierce and Nowak, 1999). Precision farming may be used to improve a field or a farm management from several perspectives.

- Agronomical perspective: adjustment of cultural practices to take into account the real needs of the crop (*e.g.* better fertilization management).

- Technical perspective: better time management at the farm level (*e.g.* planning of agricultural activities).

- Environmental perspective: reduction of agricultural impacts (better estimation of crop nitrogen needs implying limitation of nitrogen run-off).

- Economical perspective: increase of the output and/or reduction of the input, increase of efficiency (*e.g.* lower cost of nitrogen fertilization practice).

The technology has the potential to reduce production costs through more efficient and effective application of crop inputs. It can also reduce environmental degradation by allowing farmers to apply agricultural inputs at appropriate rates at places where these are needed. Spatial, temporal and predictive aspects of soil and crop variability are the vital elements of precision farming. It involves the sampling, mapping, analysis and management of specific areas within a field in recognition with spatial and temporal variability with respect to soil fertility, moisture availability, crop characteristics and insect-pest population.

The philosophy behind adoption of precision farming is to find ways to reduce cost of stable supply of food and fibre, enable the enterprising farmer to obtain the ability to handle variations in productivity within a field and to maximize financial returns, reduce wastes and minimize negative impacts on environment. This concept is not new. What is new is the ability to automate data collection, documentation and utilization of this information for strategic farm management decisions in field operations through mechanization, sensing and communication technology. Site specific crop management refers to developing agricultural management system that promotes variable management practices with-in a field according to site or soil conditions. Current whole field management approaches ignore variability in soil related characteristics and seek to apply crop production inputs in uniform manner. With such approaches, the likelihood of over application and/ or under application of inputs in a single field cannot be avoided. Precision farming employs a systems engineering approach to crop production where inputs are applied on "as need basis" and is achievable by recent innovations in information technology such as microcomputers, geographic information systems, positioning technologies and automatic control of farm machinery. It is a holistic approach to micro-manage spatial and temporal variability in agricultural landscapes based on integrated soil, plant information, and engineering management technologies as well as economics.

ASPECT OF PRECISION FARMING

Aspect of precision farming encompass a broad array of topics including variability of the soil resource base, weather, plant genetics, crop diversity, machinery performance and physical, chemical and biological inputs used in crop production. It is usually presumed that precision farming is only applicable to large holdings as seen in developed countries. Flexibility is an inherent feature of precision farming and hence type and size of farms are no hindrances in adoption of a well designed precision farming system. It offers the opportunity to improve agricultural productivity and product quality. Timeliness is one of the built in advantage of precision farming and helps to maintain punctuality despite local and farm level variability in sowing, application of fertilizers and pesticides and harvesting.

Precision farming system is based on the recognition of spatial and temporal variability in crop production. Variability is accounted for in-farm management with aim of increasing productivity and reducing environmental risks. In developed countries, farms are often large (sometimes 1000 ha or more) and comprise several fields. The spatial variability in large farms, therefore, two components like within field variability and between-field variability. The precision farming system within a field is also referred to as site-specific crop management. According to the second international conference on site specific management for agricultural systems, held in Minneapolis, Minnesota, in March 1994, SSCM refers to a developing agricultural management system that promotes variable management practices within a field according to site or soil conditions (National Research Council, 1997).

BASIC COMPONENTS OF PRECISION FARMING

Precision farming is characterized by a number of sophisticated tools that assist in monitoring variation and managing inputs. These include:

- Remote Sensing
- Global Positioning System (GPS) – a referencing device capable of identifying sites within a field
- Sensors and data loggers – crop, soil and climate information can be monitored at a high frequency using these technologies
- Geographic Information Systems (GIS) – maps of these attributes can be generated and analyzed using simple browsers or complex models
- Variable rate applicator

REMOTE SENSING

Precision farming needs information about mean characteristics of small, relatively homogeneous management zones. These mean characteristics may be obtained from soil tests for nutrient availability, yield monitors for crop yield, soil samples for organic matter content, information on soil maps, or ground conductivity meters for soil moisture. Generally, the fields are manually sampled along a regular grid and the analyzed results of samples are interpolated using geo-statistical techniques. Geo-statistical modeling of soil, water and crop variability requires that large numbers of samples at close intervals are collected throughout the agricultural landscape. Such samplings are costly and time consuming. Various workers have shown the advantages of using remote sensing technology to obtain spatially and temporally variable information for precision farming. Remote sensing imagery for PF can be obtained either through satellite-based sensors or CIR video digital cameras on board small aircraft. Moran *et al.* (1997) in their review paper summarized the applications of remote sensing for precision farming. There are basically three approaches for use of remote sensing for precision farming (Barnes *et al.*, 1996).

In first approach, the multi-spectral images can be used for anomaly detection. These anomalies can be in the forms of disease or pest, weed growth, water stress, etc. Using the reflectance measurements in visible part of the spectrum, it has been possible to detect diseases and identify weeds from crops. The difference between remotely sensed surface temperature and ground based measurements of air temperature has been established as a method to detect water stress in plants. However, such type of anomaly detection needs regular observation of crop through remote sensing sensor. This calls for use of a remote sensing system with high temporal resolution. It can provide atleast 5-6 observation in a season. Hence, the temporal resolution needed is of the order of a fortnight. The next approach is based on correlating variations in spectral response to specific variables such as soil properties or crop yield. Soil physical properties such as soil water, organic matter, soil texture can be correlated to spectral reflectance. Vegetation spectral response has also been used to infer other soil conditions. Crop yields for many crops like, rice, wheat etc. have been found to be highly correlated with spectral vegetation index during maximum vegetative cover. Thus, the yield map generated from spectral images can be used to form management units.

To find out within field variability, the remote sensing data should have high spatial resolution. Typically to analyze the variability one is looking for about 750 to 1,500 data points per hectare. With current satellites, one can see areas that are 30 metres × 30 metres (11.1 measurements/ha) 23 × 23 metres (18 measurements/ha) 10 × 10 metres (100 measurements/ha) and 5 × 5 metres (400 measurements/ha). With future satellites, we will be receiving data that have a variety of spatial resolutions (Table 2.1) that in some cases will be as detailed as 1 × 1 metre or over 10000 data points per hectare.

Table 2.1. Near future high resolution earth observation satellites

Mission/ Agency	Major Specifications
IRS 1-A, 1-B	PAN (Resolution 3m, 5m, Swath 120 km), MSS (Resolution 10m, 20m, Swath 120 km) VEGETATION payload (Resolution: 1 km, Swath: 2200 km)
ORBVIEW-3, Orbital Science Inc., USA	PAN (Resolution 1m, 2 m, Swath: 8 km) MSS (Resolution 8 m, Swath 8 km)

Contd.

QUICK BIRD, Earthwatch Inc., USA	PAN (Resolution 1m, 2 m, Swath 36 km) MSS (Resolution 4 m, Swath 36 km)
RESOURCESAT-1ISRO, India	LISS-IV (Resolution 6m, Swath 25 km) LISS-III (Resolution 23m, Swath 140 km) AWiFS (Resolution 60m, Swath 740 km)
CARTOSAT-1ISRO, India	PAN Stereo (Resolution 2.5 m, Swath 30 km)
CARTOSAT-2ISRO, India	Panchromatic (Resolution 1m, Swath 12 km)

The third approach is to integrate biophysical parameters (such as Leaf Area Index or Temperature) derived from high resolution satellite based remote sensing data, with physical crop growth modeling towards an operational decision support system for precision farming. *For example*, Moran *et al.* (1997) utilized remotely sensed estimates of LAI and evapo-transpiration as inputs to a simple alfalfa growth model. To derive biophysical parameter, the remote sensing system need to have high spectral resolution, covering the whole range of optical and thermal region.

However, use of RS data for mapping has many inherent limitations, which includes, requirements for instrument calibration, atmospheric correction, normalization of off nadir effects on optical data, cloud screening for data especially during monsoon period, processing images from airborne video and digital cameras (Moran *et al.*, 1997). Various research workers (Hanson *et al.*, 1995, Taylor *et al.*, 1997, Moran *et al.*, 1997) have shown the advantages of using remote sensing technology to obtain spatially and temporally variable information for precision farming.

DGPS ROLE IN PRECISION AGRICULTURE

'The right thing are in the right place and in right time'- This is where GPS comes into picture. In addition the accuracy is important factor in PF. So demands of DGPS comes (Differential Global Positioning System). GPS makes use of a series of military satellites that identify the location of farm equipment with-in a metre of an actual site in the field. Knowing right thing to do may involve all kinds of high tech equipments and fancy statistics or other analysis. Doing the right thing however starts with good managers and good operators doing a good job of using common tools such as planters, fertilizer applicators, harvesters and whatever else might be needed. The value of knowing a precise location within inches is that

- Locations of soil samples and laboratory results can be compared to a soil map
- Fertilizer and pesticides can be prescribed to fit soil properties (clay and organic matter content) and soil conditions (relief and drainage).
- Tillage adjustments can be made as one finds various conditions across the fields
- One can monitor and record yield data as one goes across the field.

The GPS technology provides accurate positioning system necessary for field implementation of variable rate technology. The use of GPS in agriculture is limited but it is fair to expect wide spread use of GPS in future. Recently a GPS based crop duster (precision GPS Helicopter), which can spray an area as small as 4×4 meter is attracting great attention. Some progressive farmer's are now beginning to use GPS for recording observations such as weed growth, unusual plant stress, colouring and growth conditions, which can then be mapped with a GIS programmes.

Fig. 2.1: *Global Positioning System and GPS Survey Equipment*
(Source: http://en.wikipedia.org/wiki/File:GPS_Satellite_NASA_art-iif.jpg)

In coming years to come, the role of GPS system in precision agriculture will help Indian farmers to harvest the fruits of frontier technologies without compromising the quality of land and produce. As a first step towards operationalization of PF, it is necessary to establish a DGPS infrastructure for the country. It would enable the farmers to get an accuracy of few centimeters in various unit processes (sowing, fertilizer application, herbicide/pesticide application etc.) involved in PF. A DGPS network would cater the needs of multiple applications (meteorology, transportation, geodetic survey, crustal deformation studies, disaster management and mitigation etc.) of which PF is one. Each DGPS master reference is capable of providing the services within a radius of 100 kilometre radius. Hence, the number of master stations required to establish DGPS infrastructure which would cover the entire country is calculated and given in Table 2.1.

Table 2.1: Requirement of master stations

Area of India	= 329 million hectares
Area of GPS (circular area, π =3.14)	= π × 100 × 100 sq. km
	= 3.14 million hectare
Total no. GPS reference station required for the country	= 329/3.14
Total GPS reference station required	= 105
Cost of a single DGPS set	= Rs. 3.0 lakh
Total cost of the entire infrastructure	= Rs. 3.15 crores

(*Source:* Singh, 2010)

This is affordable by any means for a country as a whole considering the innumerable applications it can cater.

GEOGRAPHICAL INFORMATION SYSTEM (GIS)

The GIS contributes significantly to precision farming by allowing presentation of spatial data in the form of a map. In addition, it forms an ideal platform for storage and management of model input data and presentation of model results, which the process model provides. GIS have evolved rapidly within production agriculture especially in the area of precision agriculture. GIS can provide the producer valuable insight into field variability, soil and plant interactions and yield results. GIS is most effective information tool the producer has to store, retrieve, analyze map and manage agricultural data. In many ways agricultural producers have always been GIS users. Most producers use some type of map for planning strategies for the coming years. Usually the maps have farm and field boundaries, along with any additional information that the producer might record for helping to make decisions. GIS is the link between the field and the office. GIS allows the producer to (i) Compare different types of agricultural data, (ii) Query to find relationships within and between data sets and (iii) Produce maps and charts to visualize, interpret and present the analysis results.

The degree of spatial variability present in a field will determine whether unique treatments are warranted for certain areas. Post-harvest analysis of the variation in crop yield and the measured factors influencing crop yield will provide useful information for next growing season. Each year, the producer is able to look back and benefit from previous year's management decisions to help and guide in making current decisions. The farm map is still the foundation of farm GIS, but now it is a digital map instead of a sketch or lines drawn on aerial photograph. Now the digital map includes other information and features, all associated with coordinates and time. All of this is linked in a computer database, able to be queried and analyzed.

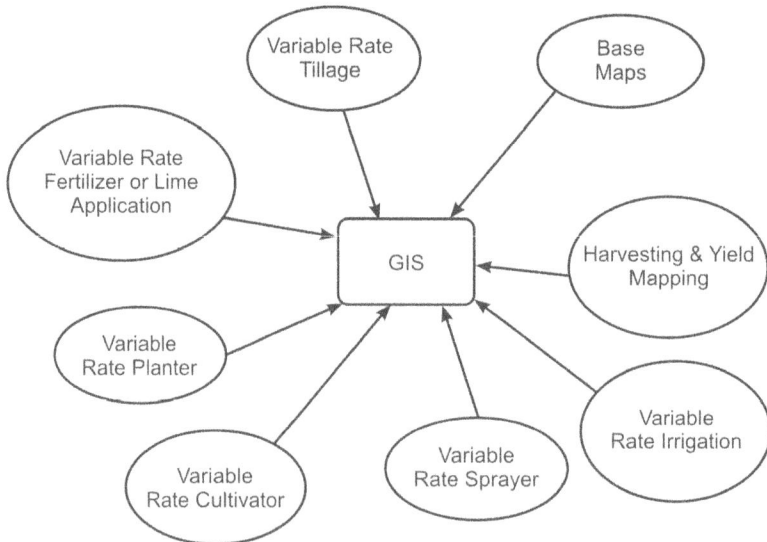

Fig. 2.2: *Precision farming overview*

The real reward to GIS as an information technology for making precision agriculture useful, is the way GIS helps producer to visualize the entire agricultural production system. GIS provides producer a holistic view from beginning to end. A farming GIS database can provide information on: field topography, soil types, surface drainage, subsurface drainage, soil testing, irrigation, chemical application rates and crop yield. Once analyzed, this information is used to understand the relationships between the various elements affecting a crop on a specific site (Trimble, 2005).

Precision farming basically depends on measurement and understanding of variability. The main components of precision farming system must address the variability. The components include (the enabling technologies), remote sensing, geographical information system, global positioning system, soil testing, yield monitors and variable rate technology.

Precision farming requires the requisition, management, analysis and output of large amount of spatial and temporal data. Mobile computing systems were needed to function on the go in farming operations because desktop systems in the farm office were not sufficient. Precision farming is concerned with spatial and temporal variability and it is information based and decision focused. It is the spatial analysis capabilities of GIS that enable precision agriculture. GPS, DGPS has greatly enabled precision farming and is of great importance to precision farming, particularly for guidance and digital evaluation modeling position accuracies at the centimeter level are possible in DGPS receivers. Accurate guidance and navigation systems will allow for farming operations at height and even under unfavourable weather conditions. In India, all these technologies are available and can be implemented through agricultural training centers by giving training to agriculture officers regarding these technologies.

VARIABLE RATE TECHNOLOGIES (URT)

VRT describes machines that can automatically change their application rates in response to their position. The Variable rate applicator has three components, (i) Control computer, (ii) Locator and (iii) Actuator.

The control computer coordinates the field operation. It has a map of desired activity as a function of geographic location. It receives the equipment's current location from the locator which has a GPS in it and decides what to do, based upon the map in its memory or data storage. It then issues the command to actuator which does the input application (Ravi and Jagadeesha, 2002). VRT systems are available for applying a variety of substances including granular and liquid fertilizers, pesticides, seed and irrigation water. The most widely recognized VRT machines are large chemical applicators that control up to 11 different materials at once. VRT applicators consist of a controller that adjusts actual material flow rate, a positioning system and a map of the desired application rates for the field. The controllers are very similar to those used on many sprayers, spreaders and agricultural machines. On conventional machines, operator controls the application.

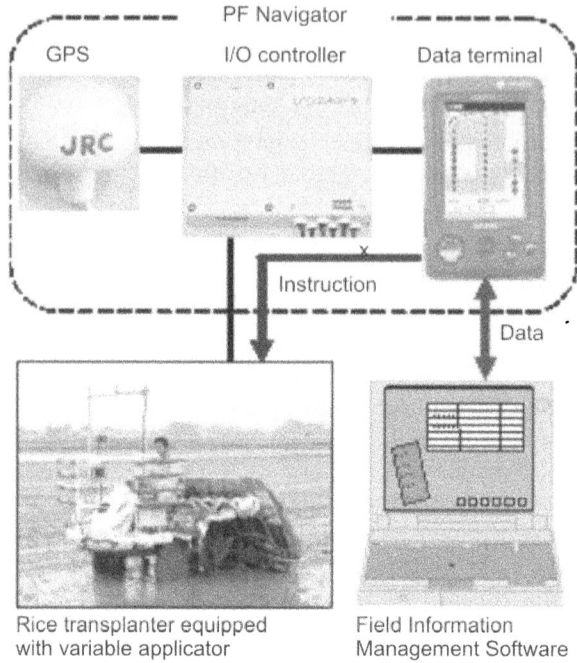

Fig. 2.3: *Variable Rate Transplanter in operation*

(Source: http://www.naro.affrc.go.jp/org/brain/PF-E/machin/vra/vra.htm)

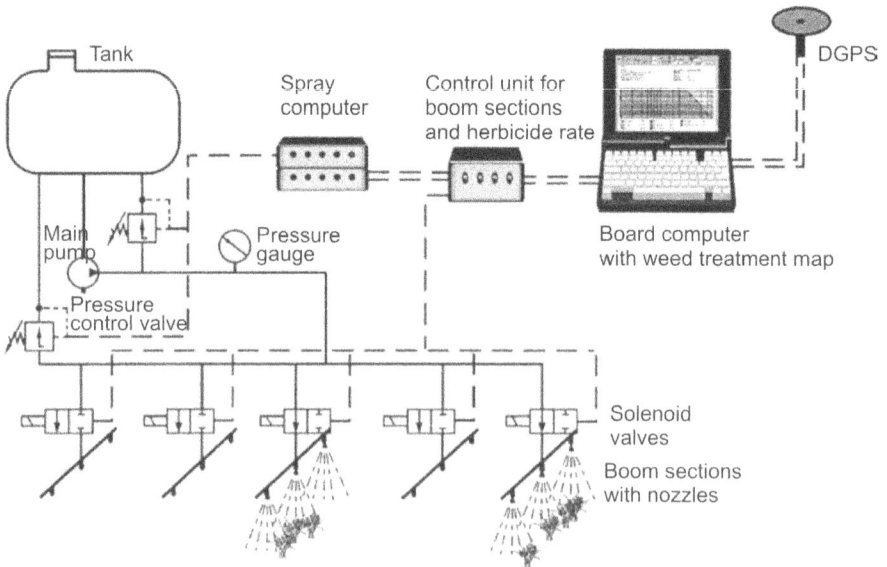

Fig. 2.4: *Sprayer control system for site-specific herbicide application (Gerhards et al., 2000)*

Yield Monitor

A yield monitor, combined with GPS technology, is an electronic tool that collects data on crop performance for a given year. The yield monitor for grain measures and records information such as grain flow, grain moisture, area covered and location. Yields are automatically calculated. Yield monitors are also available for commodities such as peanuts, cotton, forage silage and sugar beets. These monitors have some elements in common with grain yield monitors. While cost of yield monitor is reasonable, the commitment of time and resources required to effectively use this technology can be significant. Yield monitors come with various technical designs and features. However, yield monitors alone do not generate yield maps. A yield monitor is most useful with a DGPS receiver.

Yield Monitor Systems

Yield monitors are combination of several components (Fig. 2.5). They typically include a data storage device, user interface (display and key pad) and a console located in the combine cab which controls the integration and interaction of these components. The sensors measure the mass or volume of grain flow (grain flow sensors), separator speed, ground speed, grain moisture and header height. Yield is determined as a product of the various parameters being sensed.

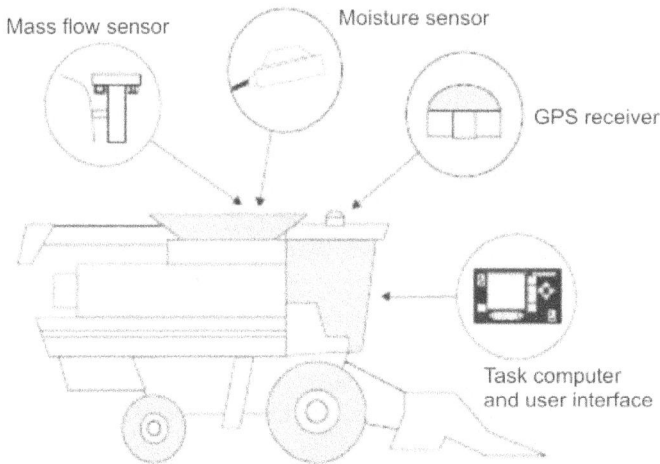

Fig. 2.5: *Components of a yield-monitoring system*

(Source: *http://www.bae.uky.edu/precag/PrecisionAg/Exten_pubs/pa1.pdf*)

Yield monitors typically provide a periodic yield report. The operator can usually select the amount of data that is collected (*e.g.* 1, 2, or 3 second intervals). Consider a combine harvesting six 30-inch rows, operating at 5 mph and harvesting corn with an average yield of 9.37 ton/ha. It requires about 15 seconds for grain entering the header of combine to reach grain tank. The combine using a yield monitor to collect data at 2-second intervals will collect about 200 yield measurements per acre, many more data points than any other precision agriculture tool.

Yield and Moisture Sensors

A sensor in the stream of clean grain measures the mass flow. Yield sensors can measure one or more of the following: (*i*) the force of the grain hitting a plate, (*ii*) the attenuation of light passing through the grain stream, (*iii*) the weight of the grain collected for a period of time, or (*iv*) the volume of grain on an elevator paddle. The most common method is to measure the force of grain striking a plate located at the top of clean grain elevator (Fig. 2.6). The calibration of these units will depend on the elevator speed, the type of crop, and the moisture of the grain.

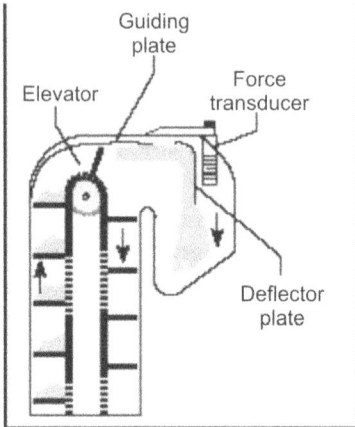

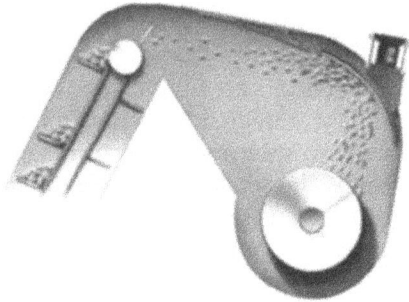

Fig 2.6: *Mass flow sensor on yield-monitoring system*

(Source: *http://www.bae.uky.edu/precag/PrecisionAg/Exten_pubs/pa1.pdf*)

Yield and moisture data are collected simultaneously to obtain accurate yields. Moisture sensors are often located in the clean grain elevator or the clean grain auger. Grain passing over the moisture sensing plate can leave deposits that can affect moisture readings. Buildup can introduce bias into the moisture measurements. The moisture sensor is essentially a conductive shell or a series of metal plates with an electrically isolated internal metal fin. As grain rises in the clean grain elevator, a small amount enters the top of the moisture sensor and moves between the metal plates. A small paddle wheel located in the bottom of sensor housing ensures that grain always covers the plates. The paddle wheel also controls the rate at which grain re-enters the clean grain elevator. Periodically take manual measurements to check the performance of moisture sensor, especially when operating in severe conditions that can coat the sensor with soil or plant sap. Over-estimated moisture readings from a malfunctioning sensor will underestimate yield. Clean the plates often when combine is operated in weedy or moist grains. These conditions can cause a buildup of dirt or plant residue on the sensing elements, which interferes with grain moisture measurements.

Yield Monitor Console

The yield monitor console is a data collection unit and computer that records data from yield sensor, moisture sensor and DGPS receiver. The console is also used to enter field names, grain type, calibration numbers, correction factors and other user-specified data. The console may also monitor or record elevation, elevator speed, ground speed, swath width, header height and electronic flags manually set by the operator. Electronic flags are often used to record the location of weeds (known to be highly correlated with yield reductions). These flags may locate and identify other problems or obstacles such as rocks, terrace failures, standing water etc. A yield monitor equipped with a DGPS receiver stores data in a format that includes position information. These spatially indexed data are later used to produce maps of yield, moisture, elevation or any other information collected during harvest. Most yield monitors can display instantaneous readings of yield and moisture and provide statistics for loads or batches of grain from field or within an area of field.

Data Collection and Storage

Data are often recorded on removable memory cartridges, such as a personal computer memory. Data from these cards can be downloaded to a computer. Download data daily to ensure that yield monitor is working properly and to protect against accidental data loss. Memory cards may store several megabytes of data. The card capacity is sometimes stated in hours of operation since data are typically stored on periodic basis. One megabyte of memory can store 15 to 45 hours of information for yield data collection intervals of 1 to 3 seconds.

Yield Monitor Calibration and Accuracy

A yield monitor must be calibrated to provide accurate yield data. Calibration must be performed for each type of grain harvested at the beginning of the harvest season. Accuracy usually improves when several loads are used to perform the calibration. Recalibration should be performed as necessary, especially later in season as average moisture content drops or when there is a significant change in crop conditions. Calibration is usually as simple as weighing and recording the moisture of first several loads collected under a variety of conditions, such as various operating speeds or grain flow rates. Consult the operator's manual for specific instructions. The accuracy of a yield monitor depends not only on its design but on how carefully the calibration procedure is followed. Some companies offer a training session or videotapes to teach calibration. These procedures vary considerably among manufacturers but all require carefully weighing several loads of grain, which can become a logistical problem on some farms.

Factors Involved in Calibrating a Yield Monitor

Yield monitor manufacturers make every effort to build accuracy into their systems, however every combine and installation may have different errors. You must consider four elements when calibrating a new yield monitor *i.e.* (*i*) distance, (*ii*) header height, (*iii*) mass-flow rate of grain and (*iv*) grain moisture content.

(*i*) **Distance:** A speed sensor in transmission is used to determine ground speed. The calibration procedure relates the actual distance traveled to a specific number of pulses from sensor. The sensor is calibrated by operating combine over a known distance (*e.g.* 40 m or length specified by manufacturer) at typical harvest speed and field condition. The yield monitor then generates a scale factor to calibrated travel speed. When calibrating the speed sensor, match actual operating conditions of combine at harvest as closely as possible. *For example*, operate a loaded combine on soft ground or a hillside if this is typical of field conditions at harvest. Ground speed radar must be calibrated in the same manner. Yield monitors that rely on the use of GPS for ground speed determination do not require calibration for speed.

(*ii*) **Header Height:** Header height determination is important as it establishes the beginning and ending of data logging and area accumulation. There are principally three different methods for sensing header height. One method is a magnetic sensor that opens a contact when the header reaches a predetermined position. A second method uses a rotary potentiometer for sensing the angle or elevation of the header. At the option of the combine operator, the start and stop positions determined by the potentiometer can be adjusted on a control panel located in the combine cab. The third method involves tracking the length of time the header height control switch is in the "up" or "down" position. Once the actuation time exceeds a predetermined value (*e.g.* 1.5 seconds), area accumulation and data logging is either turned on or off, depending on whether the header is being lowered or raised. Regardless of the methodology use to start and stop data logging and area accumulation, should read the operator's manual and thoroughly understand the operation of this feature because quality and integrity of yield data depend heavily on its use.

(*iii*) **Mass flow Rate of Grain:** To calibrate the mass flow sensor, weigh grain harvested over a certain interval and then enter actual weight of grain harvested into the yield monitor. This interval might consist of one to several combine tank loads. Based on a defined approach, the yield monitor uses this information to fit a calibration curve or a series of factors, to the particular impact sensor, grain type and combine/sensor geometry. If any of these factors change, the system must be recalibrated. Changes in grain properties such as test weight and moisture content may require more frequent calibration.

Although the approach is similar from manufacturer to manufacturer, the quantity and nature of the internal calibration approach differs. While the factory or default calibration numbers provide a reasonable starting point, they are not a substitute for on-farm calibration. At the very least, one truckload of grain must be weighed. One manufacturer recommends weighing several individual combine tank loads of grain. This process is greatly simplified when a weigh wagon with digital readout is available for obtaining load weights.

(*iv*) **Grain Moisture Content:** A sample of grain from the moisture sensor or tank loading auger should be collected, analyzed and then compared with the moisture sensor reading from the instant in time when the grain passed over sensor to arrive at an accurate calibration offset. Use caution when adjusting moisture calibrations, particularly when considering the accuracy of the moisture device that will be used to determine the reference moisture content of the grain. Offsets vary with grain type, and each grain type requires calibration to determine the appropriate offset.

Operating a Combine Equipped with a Yield Monitor

The final appearance of a yield map depends on how the combine is operated. Frequent stopping or sudden changes in speed can cause erratic yield data due to the delay and smoothing phenomenon associated with combine separating system. The combine must be operated on a uniform swath width to ensure accurate yield data. You must enter the width of the header into the monitor manually to accurately calculate yield. Yield will be underestimated if fewer rows are harvested. Many yield monitors allow you to change the number of rows or the per cent of width harvested to correct yield for point rows or field edges. For best results, keep the mass flow rate of combine constant. This represents a constant flow of material moving through the combine. You can set the yield monitor display to show the instantaneous mass-flow rate (typically close to the calibration rate). This rate can be maintained near a constant rate by adjusting the travel speed to compensate for the amount of material entering the combine.

Limitations of Yield Monitors

The combine operation is dynamic and the flow rate of material processed can vary depending on entering and exiting the crop. These varying flow rates can influence the results of the yield monitor data. Since the yield monitor measures the rate at which clean grain is entering the grain tank, time delays between the time grain enters the combine header and the time it passes through clean grain elevator can be significant. Combines also smooth abrupt changes in yield hence, the yield monitor measures delayed averages of yield. The phenomena of time delays and smoothing are most obvious when a combine enters or leaves the crop at the ends of field. The combine, in the example above, has a delay of 15 seconds to process the entering crop and would travel 110 feet and harvest almost 0.04 of an acre before an accurate or stable yield is displayed on the yield monitor.

Most yield mapping software compensates for equipment delays caused by the combine and corrects the yield data. The resulting yield map will not be perfect, but it will be very adequate for observing the magnitude and location of yield variability. Yield data combined with mapping software and positional data are capable of producing a colorful map showing variations in grain yield and moisture. If these maps are to be of any real value, the data generated from them must be incorporated into the decision making, analysis, and overall planning process of the farm operation

STATUS OF PRECISION FARMING IN INDIA

The first thing that comes to mind is that the precision farming system is not for developing countries, especially India, where the farmers are poor, farming is mostly subsistent and the land holding size is small. But, this is far from the truth, as this approach has a large potential for improving the agricultural production in developing world. Imagine this situation where a farmer goes to his field with a Global Positioning System guided tractor. The GPS senses the exact location of tractor within the field. It sends signals to the computer fixed on to the tractor, which has a Geographical Information System, storing the soil nutrient requirement map in it.

GIS, in consultation with a decision support system would decide what the exact requirement of fertilizers for that location is. It then commands a variable rate fertilizer applicator, which is again attached with the tractor, to apply the exact dosage at the precise location of farm. But this is what precision farming means to large growers in the highly developed parts of the globe. To make it clear, precision farming is the system of matching of resource application and agronomic practices with soil attributes and crop requirements as they vary across a field. Though widely adopted in developed countries, the adoption of precision farming in India is yet to take a firm ground primarily due to its unique pattern of land holdings, poor infrastructure, lack of farmers inclination to take risk, socio-economic and demographic conditions.

Tata Kisan Kendra– The concept of precision farming being implemented by the TKKs has the potential to catapult rural India from the bullock-cart age into the new era of satellites and IT. TCL's extension services, brought to farmers through the TKKs, use remote-sensing technology to analyze soil and inform about crop health, pest attacks and coverage of various crops predicting the final output. This helps farmers adapt quickly to changing conditions. The result healthier crops, higher yields and enhanced incomes for farmers.

Government Organization– Precision Agriculture models are not complete, unless the parameters related to empowerment of the farmers especially small and marginal farmers are integrated. Now it is the turn of good news to the Indian farming community. Some of the research institutes such as ISRO, M.S. Swaminathan Research Foundation, Chennai, Indian Agricultural Research Institute, New Delhi, and Project Directorate of Cropping Systems Research, Modipuram, had started working in this direction and soon it will help the Indian farmers harvest the fruits of frontier technologies without compromising on the quality of land. According to the Exim Bank officials, though the research and development on PF is currently at a nascent stage in the country, the efforts being put on by the four research institutes were expected to turn the green revolution into an evergreen revolution. In this context, ISRO has also initiated Gramsat project in Orissa. In the line of JDCP, the Gramsat project aims at empowering the people especially the poor and marginalized by awareness building and access to information and services. Towards this, a network of one-way video and two-way audio forecasting the yield of mono and multiple crops is being done at NRSA. Acreage estimates and crop inventory is being done during *Kharif* and *Rabi* seasons for rice, which are the major crops grown in our India. Other crops like banana, chillies, cotton, maize, sugarcane and tobacco are also being inventoried. Satellite data can also delineate different crops that are grown in the same area and an inventory of each of the crops can be done.

ECONOMICS FEASIBILITY OF PRECISION FARMING IN INDIAN AGRICULTURE

Unlike some new technologies, there is no clear answer as to whether or not PA is economical in Indian agriculture.

- On the one hand there are depletions of ecological foundations of the agro-ecosystems, as reflected in terms of increasing land degradation, depletion of water resources and rising trends of floods, drought and crop pests and diseases. On the other hand, there is imperative socio-economic need to have enhanced productivity per unit of land, water and time.

- At present, 3 ha of rain fed area produce cereal grain equivalent to that produced in 1 ha. of irrigated. Out of 142 million ha net sown area, 92 million ha are under rain-fed agriculture in the county.

- From equity point of view, even the record agricultural production of more than 200 MT is unable to address food security issue. A close to 60 MT food grains in the storehouses of Food Corporation of India is beyond the affordability and access to the poor and marginalized, in many pockets of the country.

- Globally, there are challenges arising from the globalization, especially, the impact of WTO regime on small and marginalized farmers.

- Some other unforeseen challenges could be anticipated like global warming scenario and its possible impact on diverse agro ecosystems in terms of alterations in traditional crop belts, micro level perturbations in hydrologic cycle and more uncertain crop-weather interactions.

While some studies have reported positive returns to variable rate technology (VRT), others have reported costs higher than returns or no significant difference in returns.

- Precision Agriculture is a system, not a single piece of equipment or technology. A GPS by itself has little value to farmer. However, when combined with a yield monitor or a VRT, it may have value.

- Returns may be positive if costs can be spread over many applications. Specialized equipment which has limited uses, has greater risks associated with it than equipment that has many uses. A multi-use tractor likely pay for itself sooner than a new, single use machine.

- Precision Agriculture may not return on low valued commodities as it does on high valued specialty crops *i.e.* high revenue for grape than the wheat and paddy.

GPS controlled tractor guidance systems may affect when and how tractors are operated. Precision farming is useful in many situations in developing countries. Rice, wheat, sugar beet, onion, potato and cotton among the field crops and apple, grape, tea, coffee and oil palm among horticultural crops are perhaps most relevant. Some have a very high value per acre, making excellent cases for site-specific management. For all these crops, yield mapping is the first step to determine the precise locations of highest and lowest yield areas of the field. Researchers at Kyoto University recently developed a two-row rice harvester for determining yields on a micro plot basis (Iida *et al.,* 1998).

Subsidies on inputs outputs and mechanisms that prevent the price system from rationing limited resources are also common. The latter include state guaranteed crop prices, tariffs, import quotas, export subsidies. Inputs such as water and fossil fuels are usually sold at prices that are well below the real resource cost of their use, which consist not only production costs but also include scarcity value and costs of pollution. In such cases, the formulation of policies that reflect the real scarcity value of natural resources and penalize pollution and policies such as green payments for farmers adopting techniques that would lower environmental costs can promote the adoption of precision farming technologies (Branden *et al.*, 1994).

PROSPECTS AND SCOPE FOR INDIAN CONTEXT

Precision farming, though in many cases a proven technology, is still mostly restricted to developed (American and European) countries. Except for a few (Wang, 2001), there is not much literature to show the scope of its implementation in India. We feel that one of the major problems is small field size. In India more than 57.8 per cent of operational holdings have size less than 1 ha. However, in the states of Punjab, Rajastan, Haryana and Gujarat, more than 20 per cent of agricultural lands have operational holding size of more than 4 ha. These are individual field sizes. However, when we consider contiguous field with same crop (mostly under similar management practices) the field (rather simulated field) sizes are large. Using aerial data, it was found that in Patiala district of Punjab, more than 50 per cent of contiguous field sizes are larger than 15 ha. These contiguous fields can be considered a single field for the purpose of implementation of precision farming.

There is a scope of implementing precision farming for major food-grain crops such as rice, wheat, especially in the states of Punjab and Haryana. However, many horticultural crops in India, which are high profit making crops, offer wide scope for precision farming.

MISCONCEPTIONS ABOUT PRECISION AGRICULTURE

There are several mistaken preconceptions about precision agriculture.

(*i*) Precision agriculture is a cropping rather than an agricultural concept

This is due to cropping systems, in particular broad-acre cropping, being the face and driving force of precision agriculture technology. However precision farming concepts are applicable to all agricultural sectors from animals to fisheries to forestry. In fact it might be argued that precision farming concepts are more advanced in the dairy industry where the "site" becomes an individual animal, which is recorded, traced and fed individually to optimize production. These industries are just as concerned with improved productivity and quality decreased environmental impact and better risk management as the cropping industry, however, precision farming concepts have yet to be applied on the same scale in these areas. *For example* a grazer's use of advance warning meteorological data and market predictions to estimate fodder reserves and plan livestock numbers is a form of precision farming.

(*ii*) Precision agriculture in cropping equals yield mapping

Yield mapping is a crucial step and the wealth of information farmers are able to obtain from a yield map makes them very valuable. However they are only a stepping stone in a precision farming management system. The bigger agronomic hurdle lies in retrieving the information in yield map and using it to improve the production system. The advance of PA adoption in this country may soon be bottlenecked at this point, due to the lack of decision support systems to help agronomists and farmers understand their yield maps. Yield maps may not tell the whole story either with other data sources *e.g.* crop quality and soil maps, economic indicators or weather predictions, proving further information necessary for correct agronomic interpretations.

(*iii*) Precision agriculture equals sustainable agriculture

Precision agriculture is a tool to make agriculture more sustainable however it is not the total answer. Precision farming aims at maximum production efficiency with minimum environmental impact. Currently it is the potential for improved productivity and profitability that is driving precision farming rather than the more serious issue of long-term sustainability. Precision farming will not fix problems such as erosion and salinity by itself although it will help to reduce the risk of occurrence of these problems. Sustainable practices still need to be used in conjunction with precision farming.

OBSTACLES

There are many obstacles for the adoption of precision farming in developing countries in general and India in particular. Some are common to those in other regions but the others are specific to Indian conditions are (*i*) Culture and perceptions of the users, (*ii*) Small farm size, (*iii*) Lack of success stories, (*iv*) Heterogeneity of cropping systems and market imperfections, (*v*) Land ownership, infrastructure and institutional constraints, (*vi*) Lack of local technical expertise, (*vii*) Knowledge and technical gaps, (*viii*) Data availability and (*ix*) quality and costs.

CONCLUSION

Precision farming in many developing countries including India is in its infancy but there are numerous opportunities for adoption. We believe that progressive Indian farmers with guidance from the public and private sectors and agricultural associations, will adopt it in a limited scale as the technology shows potential for raising yields and economic returns on fields with significant variability and for minimizing environmental degradation. Although it is recognized that agriculture is a major polluter of the environment in many developing countries farmer's will not adopt precision farming unless it brings in more or at least similar profit as compared to traditional practice. The support from governments and the private sector during the initial stages of adoption is, therefore vital. It must be remembered that not all elements of precision farming are relevant for each and every farm. Likewise, not all farms are suitable to implement precision farming. Some growers are likely to adopt it partially, adopting certain elements but not others. Precision farming cannot be convincing if only environmental benefits are emphasized. On the other hand, its adoption would be improved if it can be shown to reduce the risk. We must be cautious, however, is not overselling the technologies without providing adequate product support. The adoption of precision farming also depends on product reliability, the support provided by manufacturers and the ability to show the benefits. Effective coordination among the public and private sectors and growers is, therefore, essential for implementing new strategies to achieve fruitful success.

REFERENCES

Barnes EM, Moran MS, Pinter PJJ and Clark TR (1996). Multispectral remote sensing and site-specific agriculture: examples of current technology and future possibilities. Published in Proc. of 3rd *Int. Conf. on Precision Agriculture*, June 23-26 (1996) Minneapolis, Minnesota, ASA. pp.843-854.

Branden, Bagemen J and Agkiosobud CF, (1994). Incentive-based non-point source pollution abatement in a re-authored clean water Act. Water Resource. Japan Agricultural Software Association, 1996. *Agriculture-related Software Book* (In Japanese). Rakuyu Shobo, Tokyo.

Gerhards R, SoÈkefeld M, Immermann C, Krohmann P and Kuhbauch W (2000). Precision weed controlled more than just saving herbicides. *Zeitschrift fur Pflanzenkrankheiten und Pflanzenschutz Sonderheft* **17**, 179- 186.

Hanson LD, Robert PC and Bauer M. (1995). Mapping wild oats infestation using digital imaginary for site-specific management. In Proc. *Site- Specific Management for Agricultural System.*, March 1994, Minneapolis, MN, ASA-CSA-SSSA, Madison, WI,pp.495-503.

Khosla R and Shaver T. (2001). Zoning in on nitrogen needs. Colorado State University. *Agronomy Newsletter* **21** (1) 24-26.

Khosla R, Fleming K, Delagado JA, Shaver TM and Westfall DG (2002). Use of site specific management zones to improve nitrogen management for precision agriculture. *J. Soil water conserve.* **75** (6) 513-518.

Ladha JK, Fischer AK, Hossain M, Hobbs PR, Hardy B. editors (2000). Improving the productivity and sustainability of rice-wheat systems of the Indo-Gangetic plains: a systematic synthesis of NARS-IRRI partnership research. IRRI Discussion paper series no. 40. Makati City (Philippines): *International Rice Research Institute.* 31p.

Lida M, Umeda M, Kaho T, Lee, CK and Suguri M. (1998). Measurement of Annual Crops. *International Conference on Precision Agriculture*, St. Paul. MN. 19-22, July 1998, ASA, CSSA, and SSSA, Madison, WI.

Moran MS, Inoue Y and Barnes EM. (1997). Opportunities and limitations for image -based remote sensing in precision crop management. *Remote Sensing of Environment.* **61**: 319-346.

National Research Council. (1997). Soil and water quality: *An agenda for agriculture. National Academy Press*, Wasington. DC.

Pierce FJ and Nowak P. (1999). Aspects of precision agriculture. *Advances in Agronomy.* V. 67. pp. 1-85.

Ravi N and Jagadeesha CJ. (2002). Precision Agriculture, *Training course on Remote Sensing and GIS applications in Agriculture*, May 27th -7th June, 2002, RRSSC-Bangalore, pp:225-228.

Ray SS, Panigrahy P and Parihar JS. (2001). Role of Remote Sensing for precision farming – with special Reference to Indian Situation Scientific Note SAC/RESA/ARG/AMD/ SN/01/2001, *Space Applications Center* (ISRO), Ahmedabad, pp: 1-21.

Shanwad UK, Patil VC, Dasog GS, Mansur CP and Shashidhar KC. (2002). Global Positioning System (GPS) in Precision Agriculture *The Asian GPS Conference* 2002, (October 24 - 25, 2002, New Delhi, India).

Singh CD. (2010). Basics of Tools and methods applied in precision farming. Lecture delivered in winter school on 'Enhancing input application efficiency for seeds, fertilizer and chemical using precision farming machinery, Decision support system and Electronic controllers for precision agriculture in vertisols' w.e.f. January 01-21, 2010 at CIAE Bhopal.

Taylor JC, Thomas G and Wood GA. (1997). Diagnosing sources of within field variation with remote sensing. *In Precision Agriculture'* 97, Vol. II: Technology, IT and Management (Ed. J.V. Stafford). BIOS Scientific Publishers Ltd., Oxford, UK. pp. 705-712.

Trimble, (2005). Precision Agriculture. (available at www.trimble.com)

Wang D, Dowell FE, Lacey RE., 2001. Single wheat kernel colour classification using neural networks. *Transaction of ASAE.* **42** (1) 233-40.

Wilson and Scot, (1982). Concepts of variable rate technology with considerations for fertilizer application. *Journal of Production Agriculture*, **7**(2) 195-201.

■

3

CHAPTER

Engineering Interventions in Precision Farming

☞ S.Mukesh[1], ☞ S.S. Mann[2] and
☞ Shiv Kumar Lohan Lohan[3]

PRECISION FARMING

Precision farming is a term used to describe the management of variability within field boundaries *i.e.*, applying agronomic inputs in the right place, at the right time and in the right quantity to improve the economic efficiency and diminish the environmental impact of crop production (Blackmore, 2004). Input rates are based on the needs for optimum production within-field location. Since over application and under application of agrochemicals are both minimized, this strategy has the potential for maximizing profitability and minimizing environmental impacts. Thus, Precision farming or precision agriculture is a farming management concept based on observing and responding to intra-field variations. It relies on new technologies like satellite imagery and information technology. It is also aided by farmers' ability to locate their precise position in a field using satellite positioning system like GPS.

Merits of Precision Farming

- The concept of "doing the right thing in the right place at the right time" has a strong intuitive appeal which gives farmers the ability to use all operations and crop inputs more effectively.
- More effective use of inputs results in greater crop yield and quality, without polluting the environment.
- Precision agriculture can address both economic and environmental issues that surround production agriculture today.

Constraints of Precision Farming

- High cost has proven difficult to determine cost benefits of precision agriculture management. At present, many of technologies used are in their infancy and pricing of equipment and services is hard to pin down.

---

[1] Assistant Scientist, Department of Farm Machinery & Power Engineering, CCS Haryana Agricultural University, Hisar-125 004

[2] Assistant Scientist, Department of Farm Machinery & Power Engineering, CCS Haryana Agricultural University, Hisar-125 004

[3] Assistant Research Enginner, Deptt. of Farm Machinery and Power Engineering, PAU, Ludhiana-141 001

- Lack of technical expertise knowledge and technology. The success of precision agriculture depends largely on how well and how quickly the knowledge needed to guide the new technologies can be found.
- Not applicable, difficult and costly for small land holdings
- Heterogeneity of cropping systems and market imperfections.

Concept of Precision Agriculture

Precision Agriculture is based on development and use of new technologies including new computerized equipment and information management systems for more effective crop production and environmental protection. It targets inputs and management practices to variable field conditions such as soil/landscape characteristics, pest presence and microclimate. Unlike traditional crop management, which assumes uniform field conditions and recommends average input application rates, it is an information intensive approach. A more holistic agricultural approach, it uses information technology to bring data from multiple sources to bear on decisions associated with agricultural production, logistics, marketing, finance, and personnel. Thus, precision agriculture is the modern management manner equipped with digitalizes techniques, information technologies, intelligent techniques and VRT. Here agriculture includes cropping, forest, fruit, husbandry, fishery, and so on. It aims for sustainable agriculture development by optimize invest of materials, lowest consumption of natural resource, poor environment pollution and high quality and production.

The practical implementation of precision farming is dependent on technological developments to provide, manage, and utilize the vast quantities of data required to understand spatial variations in crop yields and in the factors that affect yields. Today, low cost powerful computers, real-time controllers, variable rate application hardware, accurate location systems, and advances in sensor technology have combined to provide the technology to make precision farming a reality. The cyclic nature of precision farming approach used in crop production is given in Fig. 3.1. An initial step in this process is spatial measurement of those factors that

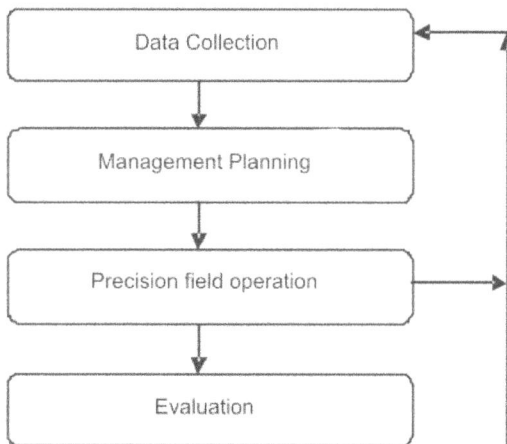

Fig. 3.1: *Basic precision farming approach*

limit or otherwise affect crop production. These variability data are then used to develop a management plan for the variable application of inputs such as fertilizers and herbicides. Inputs are applied in precision field operations. Finally, effectiveness of precision farming system is evaluated with respect to economics and environmental impacts. This evaluation becomes a part of data collection process for next cropping season. Multiple iterations through cycle allow for refinement of precision management plan in succeeding seasons.

Precision agriculture is a four-stage process using techniques to observe spatial variability and is discussed below.

GEOLOCATION OF DATA

Geolocating a field enables the farmer to overlay information gathered from analysis of soils and residual nitrogen and information on previous crops and soil resistivity. Geolocation is done in two ways.

- The field is delineated using an in-vehicle GPS receiver as the farmer drives a tractor around the field.
- The field is delineated on a basemap derived from aerial or satellite imagery. The base images must have right level of resolution and geometric quality to ensure that geolocation is sufficiently accurate.

CHARACTERIZING VARIABILITY

Intra and inter field variability may result from a number of factors. These include climatic conditions like hail, drought, rain, etc., soils (texture, depth and nitrogen levels), cropping practices (no-till farming), weeds and disease. Permanent indicators chiefly soil indicators provide farmers with information about main environmental constants. Point indicators allow them to track a crop's status, *i.e.* to see whether diseases are developing, if the crop is suffering from water stress, nitrogen stress, or lodging, whether it has been damaged by ice and so on. This information may come from weather stations and other sensors (soil electrical resistivity, detection with the naked eye, satellite imagery etc.). Soil resistivity measurements combined with soil analysis make it possible to precisely map agro-pedological conditions.

DECISION-SUPPORT SYSTEM

Using soil maps, farmers can pursue two strategies to adjust field inputs.

Predictive approach: based on analysis of static indicators (soil, resistivity, field history, etc.) during the crop cycle.

Control approach: information from static indicators is regularly updated during the crop cycle by

- Sampling: weighing biomass, measuring leaf chlorophyll content, weighing fruit, etc.

- Remote sensing: measuring parameters like temperature (air/soil), humidity (air/soil/leaf), wind or stem diameter is possible thanks to wireless sensor networks
- Proxy-detection: in-vehicle sensors measure leaf status this requires the farmer to drive around the entire field.
- Aerial or satellite remote sensing: multispectral imagery is acquired and processed to derive maps of crop biophysical parameters.

Decisions may be based on decision support models (crop simulation models and recommendation models) but in the final analysis it is up to the farmer to decide in terms of business value and impacts on the environment.

VARIABILITY REMOVAL

New information and communication technologies (NICT) make field level crop management more operational and easier to achieve for farmers. Application of crop management decisions calls for farming equipment that supports variable rate technology (VRT) *for example* varying seed density along with variable rate application (VRA) of nitrogen and phytosanitary products.

Precision agriculture uses special equipment on board the farmer's tractor

- Positioning system (*e.g.* GPS receivers that use satellite signals to precisely determine a position on the globe)
- Geographic information systems *i.e.,* software that makes sense of all the available data
- Variable-rate farming equipment (seeder, spreader).

TECHNOLOGY

Technologies include a vast array of tools of hardware, software and equipments.

1. Global Positioning System (GPS)

GPS provides continuous position information in real time, while in motion. Having precise location information at any time allows soil and crop measurements to be mapped. GPS receivers either carried to the field or mounted on implements allow users to return to specific locations to sample or treat those areas.

There are essentially three parts that make up GPS: the space segment, user segment, and control segment. The space segment is based on constellation of 24 active and 3 spare satellites orbiting the Earth. The control segment is a system of five monitoring stations located around the world with master control facility located at falcon air force base in Colorado. The user segment, which is the fastest growing segment, is made up of GPS receivers and user community. GPS receivers convert the satellites signals into position, velocity and time. This information is used for navigation, positioning, time dissemination and research. The GPS is owned and operated by the U.S. Department of Defense but is available for general use around the world. Briefly, here's how it works:

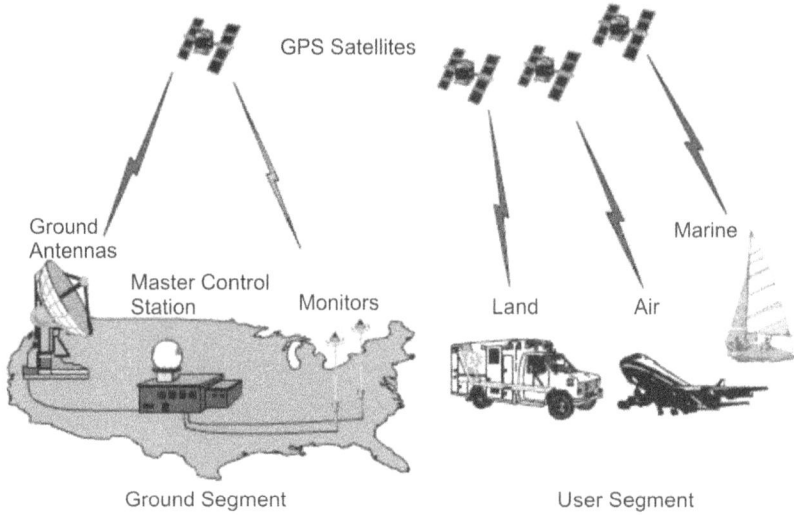

Fig. 3.2: *Space Segments*

- The 24 satellites that make up the GPS space segment are orbiting the earth about 12,000 miles above us. They are constantly moving, making two complete orbits in less than 24 hours. These satellites are travelling at speeds of roughly 7,000 miles an hour.

- GPS receivers take this information and use triangulation to calculate user's exact location. Essentially, GPS receiver compares the time a signal was transmitted by a satellite with time it was received. The time difference tells GPS receiver how far away the satellite is. Now, with distance measurements from few more satellites, the receiver can determine the user's position and display it on the unit's electronic map.

- A GPS receiver must be locked on to the signal of at least three satellites to calculate a 2D position (latitude and longitude) and track movement. With four or more satellites in view, receiver can determine the user's 3D position (latitude, longitude and altitude). Once user's position has been determined, GPS unit can calculate other information, such as speed, bearing, track, trip distance, distance to destination, sunrise and sunset time and more. GPS is widely available in the agricultural community and its potential is growing. Farm uses include mapping yields (GPS + combine yield monitor), variable rate planting (GPS + variable rate planter drive), variable rate lime and fertilizer application (GPS + variable rate spreader drive), variable rate pesticide application (GPS + variable rate applicator), field mapping for records and insurance purposes (GPS + mapping software) and parallel swathing (GPS + navigation tool).

DIFFERENTIAL GLOBAL POSITIONING SYSTEM (DGPS)

There are several things associated with the GPS system that can cause errors in GPS position information. The contributor of each source of error may vary depending on atmospheric and equipment conditions. With all these types of errors, it would appear that the system could not be all that accurate. However with the use of mathematics and modeling, some of errors can be eliminated or reduced. The receiver does the mathematics calculations and error corrections. Thus, error effect can be reduced depending on the capabilities of the receiver. In agricultural applications, the most common way to counteract GPS errors is by using Differential GPS or DGPS. In a DGPS system, GPS receiver is placed at an accurately known location as shown in Fig 3.3. This base station receiver will calculate the error between its actual location and the location computed from GPS signals. The error information is communicated to the rover receiver being used in the field, which is then able to correct the position information it computes from the GPS signals. Most differential corrections are provided by the U.S. Coast Guard, WAAS (the FAA's Wide Area Augmentation System), or by subscription to a private radio carrier or satellite link.

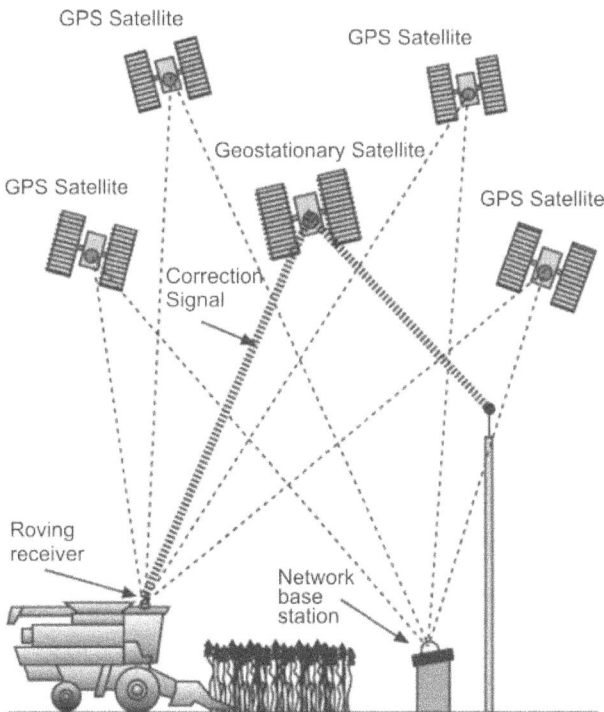

Fig. 3.3: *Differential GPS system*

DGPS is having a great impact on navigation in the agricultural industry it gives a producer the ability to know a specific location. Depending on the receiver used, this location can be found instantaneously. By knowing location, farmers can look at the field as a group of small zones and determine if the field is uniform or not (break the field up into smaller fields or grids). Computers and geographical information systems (GIS) enable producers to record location and other information. *For example* a yield monitor used with GPS allows a farmer to record yield for every location in the field. With this information practices that may improve efficiency and increase profitability can be considered.

GEOGRAPHIC INFORMATION SYSTEMS (GIS)

A geographic information system (GIS) integrates hardware, software, and data for capturing, managing, analyzing, and displaying all forms of geographically referenced information. GIS allows us to view, understand, question, interpret and visualize data in many ways that reveal relationships, patterns and trends in the form of maps, globes, reports and charts. A GIS helps to answer questions and solve problems by looking at the data in a way that is quickly understood and easily shared. GIS technology can be integrated into any enterprise information system framework.

Modern GIS technologies use digital information, for which various digitized data creation methods are used. The most common method of data creation is digitization, where a hard copy map or survey plan is transferred into a digital medium through the use of a computer aided design (CAD) programme and geo-referencing capabilities. With the wide availability of ortho-rectified imagery (both from satellite and aerial sources), heads up digitizing is becoming the main avenue through which geographic data is extracted. Heads-up digitizing involves the tracing of geographic data directly on top of the aerial imagery instead of by the traditional method of tracing the geographic form on a separate digitizing tablet (heads-down digitizing).

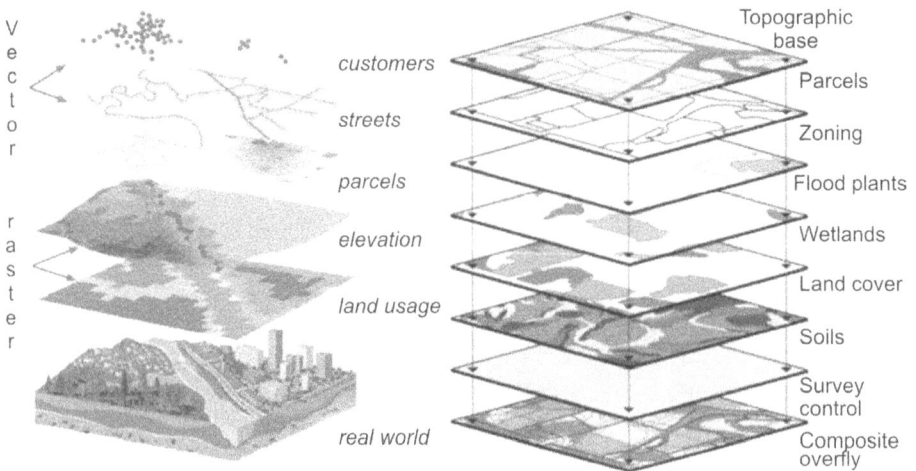

Fig. 3.4

Geographic information systems (GIS) are computer hardware and software that use feature attributes and location data to produce maps. An important function of an agricultural GIS is to store layers of information, such as yields, soil survey maps, remotely sensed data, crop scouting reports and soil nutrient levels.

REMOTE SENSING

It is the collection of data from a distance. Data sensors can simply be handheld devices, mounted on aircraft or satellite based. Remotely-sensed data provide a tool for evaluating crop health. Plant stress related to moisture, nutrients, compaction, crop diseases and other plant health concerns are often easily detected in overhead images. Remote sensing can reveal in season variability that affects crop yield and can be timely enough to make management decisions that improve profitability for the current crop. Generally, remote sensing refers to the activities of recording, observing, perceiving (sensing) objects or events at far away (remote) places. In remote sensing, the sensors are not in direct contact with the objects or events being observed. The information needs a physical carrier to travel from the objects/events to the sensors through an intervening medium. The electromagnetic radiation is normally used as an information carrier in remote sensing. The output of a remote sensing system is usually an image representing the scene being observed. A further step of image analysis and interpretation is required in order to extract useful information from the image. In a more restricted sense, remote sensing usually refers to technology of acquiring information about the earth's surface (land and ocean) and atmosphere using sensors onboard airborne (aircraft, balloons) or space borne (satellites, space shuttles) platforms.

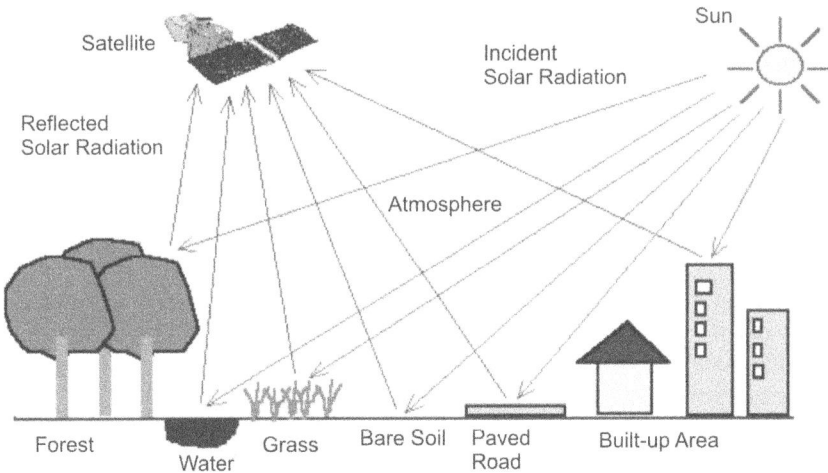

Fig. 3.5

Optical and Infrared Remote Sensing

In optical remote sensing, optical sensors detect solar radiation reflected or scattered from the earth, forming images resembling photographs taken by a camera high up in space. The wavelength region usually extends from the visible and near infrared (commonly abbreviated as VNIR) to the short wave infrared (SWIR). Different materials such as water, soil, vegetation, buildings and roads reflect visible and infrared light in different ways. They have different colours and brightness when seen under the sun. The interpretation of optical images require knowledge of the spectral reflectance signatures the various materials (natural or man-made) covering surface of the earth. There are also infrared sensors measuring the thermal infrared radiation emitted from the earth, from which the land or sea surface temperature can be derived.

Remote Sensing Images

Remote sensing images are normally in the form of digital images. In order to extract useful information from the images, image processing techniques may be employed to enhance the image to help visual interpretation and to correct or restore the image if the image has been subjected to geometric distortion, blurring or degradation by other factors. There are many image analysis techniques available and the methods used depend on requirements of specific problem concerned. In many cases, image segmentation and classification algorithms are used to delineate different areas in an image into thematic classes. The resulting product is a thematic map of study area. This thematic map can be combined with other databases of test area for further analysis and utilization.

Variable Rate Applicator

Grid soil samples are analyzed in the laboratory and an interpretation of crop input (nutrient and water) needs is made for each soil sample. Then the input application map is plotted by using entire set of soil samples. The input application map is loaded into a computer mounted on a variable rate input applicator. The computer uses the input application map and a GPS receiver to direct a product delivery controller that changes the amount or kind of input (fertilizer and water), according to the application map. The variable rate applicator has three components:

(i) Control computer, (ii) Locator and (iii) Actuator

Variable rate technology describes any technology which enables producers to vary the rate of crop inputs. VRT combines a variable rate control system with application equipment to apply inputs at a precise time or location to achieve site-specific application rates of inputs. A VRT system includes the use of a computer loaded with the appropriate field application software, controller, DGPS receiver, and control mechanism (*i.e.* hydraulic valve and motor) for the metering mechanism. A site-specific approach allows growers to apply products only where they are needed in field. Varying the application of inputs can reduce input and labor costs, maximize productivity and reduces the impact of over application that may have adverse effect on the environment. Examples of VR applications for agriculture include fertilizer, lime, seeding and pesticides.

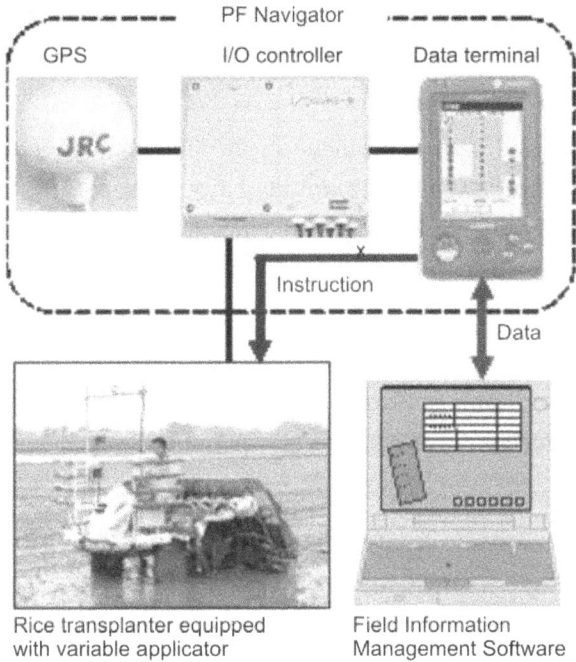

Fig. 3.6

Three different approaches exist to implement VRT: map based, sensor based, and manual. In the map based approach, a prescription map is generated based on soil analyses or other information and then used by VRT to control desired application rate within each zone. The sensor based approach is not discussed within this publication since it is new and under development by most companies. However, this type of system utilizes sensors to assess crop or field conditions to provide real time VRA of inputs. Manual control can also be used to vary application rates with operator responsible for changing rates on the controller during operation.

Uses of VRT

VRT continues to develop rapidly which coincides with advancements in electronic controls and communication. Application of VRT in crop production ~~can~~ includes

- Fertilizer (Macro and Micro-nutrients) and Lime
- Pesticides (herbicides, insecticides, and fungicides)
- Manure (Litter)
- Seeding
- Tillage (vary depth based on level of compaction)
- Irrigation

Benefits

Variable rate technology can provide several benefits

Economics

- Increased input efficiency - apply only what is needed
- Could reduce overall amount of inputs used
- Improved in-field equipment efficiency
- Improved crop yields through optimal use of inputs

Environmental

- Minimize over-application of inputs thereby reducing the risk of pesticide and fertilizer run-off or leaching into water sources
- Reduce application in environmentally sensitive areas

Constraints

Machinery can become more complex reducing reliability and increasing user frustration.

- VRA requires good equipment, management, calibration and proper maintenance
- VRA requires good knowledge of machinery
- Need to determine how to develop prescription maps
  - Assess field variability (*i.e.*, soil variability through intensive soil sampling) using either grids or management zones
  - Generate prescription maps
- Need to define overall goal for using VRT (*i.e.*, reduce costs, increase yields, improve environmental stewardship etc.)

COMBINE HARVESTERS WITH YIELD MONITORS AND MAPPING

Yield monitors continuously measure and record the flow of grain in the clean grain elevator of a combine. When linked with a GPS receiver, yield monitors can provide data necessary for yield maps. Yield measurements are essential for making sound management decisions. However, soil, landscape and other environmental factors should also be weighed when interpreting a yield map. Used properly, yield information provides important feedback in determining the effects of managed inputs such as fertilizer amendments, seed, pesticides and cultural practices including tillage and irrigation. Since yield measurements from a single year may be heavily influenced by weather, it is always advisable to examine yield data of several years including data from extreme weather years that helps in pinpointing whether observed yields are due to management or climate-induced.

Fig. 3.7

Advantages

- Efficient use of drying facility and increased harvest efficiency using accurate harvest information (yield and moisture content) for every field.
- Reliable operation report can be exchanged between contracted farmers.
- Easy and accurate fertilization design based on the information of the crop growth, soil fertility and various operations.

FUTURE SCOPE

- The application of information technology to crop management will significantly change methods of production and improve the efficiency of production system.
- Farm equipment manufacturers will introduce new and improved machines which can vary application rates of seeds, water, fertilizers, and pesticides in response to precision management plans.
- Yield monitors and other sensor systems will improve and increase in number.
- High resolution satellite images designed specifically for precision agriculture may become available in the near future.

- Many more producers will investigate the use of these machines and services, and will apply precision farming techniques to manage their crop production.

Promotion of precision farming at farm level

- Identify the niche areas for the promotion of crop specific precision farming.
- Creation of multidisciplinary teams involving agricultural scientists in various fields, engineers, manufacturers and economists to study the overall scope of precision agriculture.
- Promote the progressive farmers for precision farming technology who have sufficient risk bearing capacity.
- Encourage farmers to study of spatial and temporal variability of the input parameters using primary data at field level.
- Provide complete technical backup support to farmers to develop pilots or models, which can be replicated on a large scale.
- Pilot study should be conducted on farmer's field to show the results of precision agriculture implementation.
- Encourage the farmers to adopt water accounting protocols at farm level and to use of micro level irrigation systems and water saving techniques.
- Government legislation restraining farmers using indiscriminate farm inputs and thereby causing ecological and environmental imbalance would induce farmer to go for alternative approach.
- Creating awareness amongst farmers about consequences of applying imbalanced doses of farm inputs like irrigation, fertilizers, insecticides and pesticides.
- Policy support on procurement prices, efficient transfer of technology to the farmers, formulation of cooperative groups or self help groups since many of the precision agriculture tools are costly (GIS, GPS, RS, etc.).

REFERENCE

Blackmore Chris (2004) From one-off events to learning systems and committees of Practice in: (Pre) Proceedings of 6th, FSA European Symposium, 4-7 April, 2004. Vila Real, Portugal.

4

CHAPTER

Use of Spectral Sensors in Crop Production

☞ Manjeet Singh[1], ☞ Ankit Sharma[2]
☞ Bhupinder Singh[3], ☞ Vaibhav Suryawanshi[4] and
☞ Rajneesh Kumar[5]

INTRODUCTION

Innovations in electronics, computer science and sensor technologies strongly influence the use of engineering in agriculture. In particular sensor information may result in economical as well as ecological improvements. However, the interpretation of the corresponding sensor data for an agricultural point of view is high importance to find the right decision for plant or soil treatment, *for example*, sensors in the market are already used for fertilizer control (N-Sensor/Yara or Crop-Meter/Claas Agrosystems), measurement of maize plants during harvesting (Harvest Lab /John Deere or AutoScan/Krone) or driver assistance (such as Auto Fill/Claas or 3D-Scanner/New Holland).

In the 1980s, the potential for combining computers with image sensors provided opportunities for machine vision based guidance systems. During the mid-1980s, researches at Michigan State University and Texas A&M University were exploring machine vision guidance. Also during that decade, a programme for robotic harvesting of oranges was successfully performed at the University of Florida (Harrell *et al.*, 1990). In 1997, agricultural automation had become a major issue along with the advocacy of precision agriculture. The potential benefits of automated agricultural vehicles include increased productivity, increased application accuracy and enhanced operation safety. Additionally, the rapid advancement in electronics, computers and computing technologies has inspired renewed interest in the development of vehicle guidance systems. Various guidance technologies, including mechanical guidance, optical guidance, radio navigation and ultrasonic guidance, have been investigated (Reid *et al.*, 2000, Tillett 1991).

Having a look at field trials the implementation of sensors is of highest importance for interpreting plant breeding or modifications of agricultural processes in the field or machinery. Until now typically samples are evaluated by experts and analysed based on statistical methods (Thomas, 2006). However, this is very time consuming, generates high costs and may cause variations thereby resulting in

---

[1] Research Engineer, Department of Farm Machinery and Power Engineering, PAU Ludhiana.
[2] Assistant Professor (Agril. Engg.), Punjab Agricultural University, KVK, Mansa (Punjab)
[3] Research Associate, Department of Farm Machinery and Power Engineering, PAU Ludhiana.
[4] SMS, KVK, Mamurabad Farm, Jalgaon, Maharashtra.
[5] Sr. Research Fellow, Deptt. of Farm Machinery and Power Engineering, PAU, Ludiana.

uncertainties for the optimization of agricultural processes. As a consequence the implementation of sensors and corresponding system technology for data processing and storage has come into focus for field trials, as *for example* the development of measuring platforms for genetic studies as proposed by Montes *et al.,* (2007). Based on experiences in the development and application of sensors and electronic systems in agricultural environment, Ruckelshausen *et al.,* (1999) have developed tractor-mounted and autonomous plant phenotyping platforms which include the integration of high accuracy GPS information for individual plant detection and dynamic laboratory and field test setups for the evaluation of the sensor systems.

However, outdoor applications of sensors such as imaging devices are strongly influenced by complex and changing field and environmental conditions. In particular the robustness of sensor measurements may be limited by the influence of clouds, standing water, sun light, dust, humidity or vibrations. As a consequence the robustness of the sensor systems applied in dynamic field tests is of high importance.

Use of Near Infra Red (NIR) sensors

Crop fertiliser management can be fine tuned using NIR sensors to estimate the amount of biomass present and calculate future nitrogen requirements. NIR sensors measure the amount of light reflected back at them at specific wave lengths in order to make this estimate. NIR sensors can be integrated into systems to control fertiliser application at any point in the field.

Another use is crop scouting prior to fertiliser application to determine the optimum rate to be applied over the whole field. There are a number of NIR sensors available for crop monitoring but their outputs are not always equal. The various brands have slightly different properties such as the wave length measured and the size and shape of the area measured under the instrument.

Function of NIR sensors

NIR sensors have a light source and a sensor to measure the wavelengths of light reflected back.

For crops, the main wavelength bands of interest are; Chlorophyll band (550 – 590 nm), Visible band (670 – 690 nm) and NIR band (760 – 880 nm).

At wavelengths below the NIR band, a crop will absorb more light and reflect less as biomass increases. At wavelengths in the NIR band, the crop will absorb less light and reflect more as biomass increases. NIR sensors measure both below and in the NIR wavelength band, and the ratio between these measures is used to calculate indexes related to crop biomass. The most common indexes are; Normalised Difference Vegetation Index (NDVI): NDVI is a numerical indicator that uses the visible and near-infrared bands of the electromagnetic spectrum, and is adopted to analyze remote sensing measurements and assess whether the target being observed contains live green vegetation or not. The value of this index ranges from -1 to 1. The common range for green vegetation is 0.2 to 0.8.

Simple Ratio Index (SRI): SRI index is another old and well known Vegetation

Index. The SRI is the ratio of the highest reflectance. Absorption bands of chlorophyll makes it both easy to understand and effective over a wide range of conditions. The value of this index ranges from 0 to more than 30. The common range for green vegetation is 2 to 8. Green Difference Vegetation Index (GDVI), sometimes called the Chlorophyll Index (CI) is a measure of chlorophyll content in the plant.

These indexes are used to estimate properties such as shoot or tiller number, biomass and Green Area Index (GAI). If the full canopy closure has not been achieved, the reflectance from the soil needs to be taken into account for accurate estimates of crop biomass. The Soil Adjusted Vegetative Index SAVI will do this.

Other sensors available

Greenseeker

The greenseeker hand held data collection and mapping unit, is a crop research and consulting tool that provides useful data to determine normalized difference vegetative index (NDVI) and red to near infrared ratios (Fig. 4.1). This measures crop reflectance in the visible (656 nm) and NIR band (774 nm). It calculates various indexes, including NDVI, between 333 and 0.6 times per second. These data points can be used in conjunction with other agronomic references to index basic nutrient response, crop condition, yield potential, stress, pest and disease impact in a quantitative objective manner. The unit can be used to monitor changing field (crop, plant) conditions during the growing season or the effects of different levels of an input compared to a local standard.

The greenseeker hand held utilizes ntech's second-generation optical sensor. The unit's optical sensor captures the light reflectance of plants. The microprocessor circuit board analyzes the plant's reflected light. The data that is collected with the sensor can be downloaded to a personal computer in a text format that can be accessed by Microsoft excel.

Fig. 4.1: *A view of green seeker operating in the field*

Crop Circle

There are two Crop Circle models available. The ACS210 measures reflectance in two channels. One version measures reflectance in the chlorophyll (590 nm) and near infrared (NIR) (880 nm) bands. The other version measures the visible (690 nm) and NIR (880 nm) bands. The ACS470 is a three channel instrument (Fig. 4.2) which has a range of interference filters. Outputs include the direct reading from each channel and a range of vegetative indexes.

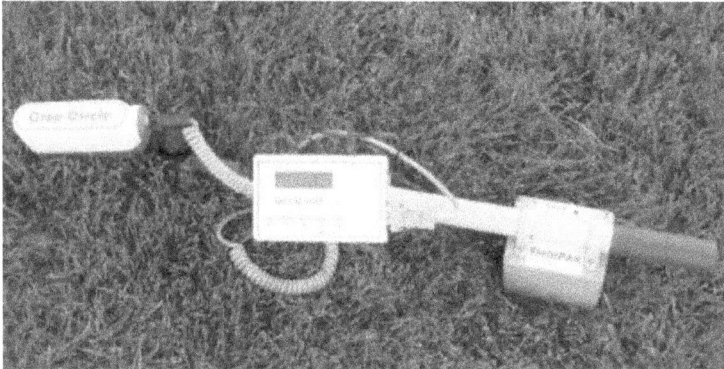

Fig. 4.2: *A stationary view of Crop Circle instrument*

N-Sensor

YARA N-Sensor YARA GmbH consists of two diode spectrometers, fiber optics and microprocessor in a hard shell, built on the roof of the vehicle. A spectrometer collects reflectivity at wavelengths from 620 to 1000 nm with four points, which are around the vehicle (Fig. 4.3). The area typically measured by scanner is approximately 50-100 m². Correction radiation occurs through the second spectrometer recognizing area sky at the same wavelength. Although the crop is scanned at a zenithal angle of oblique projection (average 640), the influence of solar azimuth better avoided by using special visual geometry. A fifth sensor positioned skywards measures the intensity of light allowing the sensor system to compensate for different light conditions while operating. The whole process of determining the crop's nitrogen requirement and application of the correct fertilizer rate happens instantaneously with no time delay. This enables 'real time agronomy' to be possible. The YARA N-Sensor system can easily be connected to a Differential Global Positioning System (DGPS) signal to allow location, sensor and application information to be plotted enabling the production of 'biomass' and nitrogen application maps for the field.

Before N Sensor is used a simple calibration procedure is required for each field. This process involves scanning a small area of crop to measure the average sensor reading. An optimum nitrogen rate is then calculated, for that area, using either farm practice, N Tester or N Plan. This optimum nitrogen rate is then keyed into the computer in the tractor cab and assigned to the average sensor reading. As the tractor passes over the field, the YARA N-Sensor will vary the rate around this average optimum, according to the crop's requirements. If required, the user can restrict the range of application rates by setting maximum and minimum levels. N-Sensor can measure the light reflectance of any crop but the system requires software to translate that reflectance information into an application rate. Currently, software has been developed for winter wheat, winter barley, spring wheat and spring barley, winter oilseed rape and potatoes. Ongoing research by YARA is focusing on use in terms of timings and other agronomic practices.

Fig. 4.3: *A view of N-Sensor operating in the field*

Cropspec

The Cropspec (Fig. 4.4) is mounted on top of the tractor and can read over a 3m width. Greenseeker and Crop Circle have a much smaller sampling area and it is usually recommended that two sensors are used to cover 24 m. On boom systems with liquid application (*e.g.* sprayer), at least four sensors are often used on a 24 m boom, this depends on the number of sections on the boom.

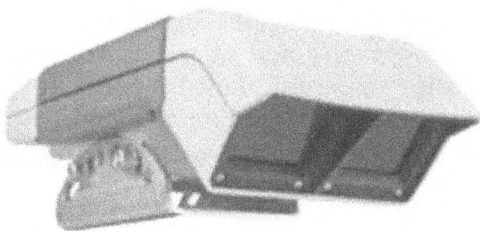

Fig. 4.4: *A view of Cropspec instrument*

FieldSpec Hand Held Spectroradiometer

The highly portable FieldSpec Hand Held instrument (Fig. 4.5) functions on a 512-channel, photo-diode array spectroradiometer covering the 325-1075 nm region, and offers spectral resolution of 3.5 nm at 700 nm. Most notably is its increased scanning speed. The Hand Held Pro scans three times faster than the FieldSpec. HandHeld operated by RS3 (pronounced "RS cubed"), the most powerful data acquisition and analysis software available for remote sensing, the Hand Held Pro is ideal for applications ranging from remote sensing, precision agriculture and forestry to industrial light measurement, and mining.

The Hand Held Pro measures 22 x 15 x 18 cm and runs on a standard lithium ion rechargeable battery pack. ASD's FieldSpec UV/ VIS/NIR spectroradiometers are laboratory-accurate instruments that allow users to make direct material identifications in the field rather than having to collect samples for later laboratory analysis. The portability and easy setup of these instruments makes it easier to move from site to site and collect more data in a shorter period of time.

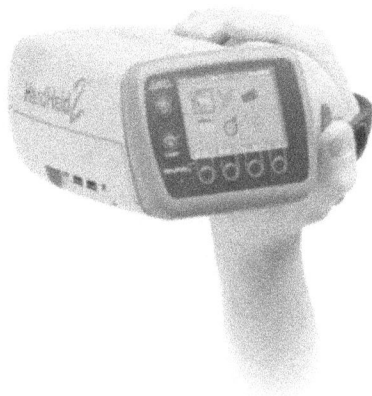

Fig. 4.5: *A view of Hand Held spectroradiometers*

FieldSpec Portable Spectroradiometer

The FieldSpec portable spectroradiometer (Fig. 4.6) offers the modular Goetz spectrometer engine with a spectral range from 350 nm to 2500 nm and is ideal for numerous remote sensing and research applications including : Airborne measurement, atmospheric research, climate effects, crops and soils research, forestry, ecology, and plant physiology research, general field spectrometry, geology and mineral analysis, ground truthing, hyperspectral and multispectral imagery and analysis, snow and ice research, light energy measurements, oceanography and inland water bodies research, plant breeding research, spectroradiometry and radiometric Calibration.

The ability to accurately perform reflectance and radiometric measurements of vegetation and soil in the field is critical to all of these

Fig. 4.6: *A view of portable spectroradiometer*

applications. The FieldSpec line of spectroradiometers offers a wide range of configuration options for both contact measurements (such as leaves or in a soil profile pit) and stand-off measurements (such as those needed to measure canopy reflectance). The FieldSpec uses a flexible fiber optic cable with several different accessories, giving researchers many options for acquiring critical data. Bringing a level of device portability, the FieldSpec also helps you work in some of the most remote regions of the planet.

VIS-NIR-EC-Force Probe

The VIS-NIR-EC-Force sensors provide unmatched soil profile information. The VIS-NIR spectrometers collect optical measurements through the sapphire window on the side of the probe as it moves into the soil profile. At the bottom of the probe is a cone-tip which contacts with soil for collecting dipole EC data. A load cell measures the insertion force required to push the probe into the soil. The heavy duty probe is constructed of 1" (2.54 cm) diameter probe rod.

When it's time to collect soil cores to calibrate sensor data to various soil properties, the advanced core sampling technology makes the job easy. The changeover from 4 sensor probe to core sampler takes less than a minute. The rack-and-pinion hydraulic side-shift provides lateral motion and the extend cylinder moves the probe forward or backward all controlled by handy, accessible lever controls, making it easy to collect cores adjacent to sensor probe locations. Hydraulic rotation, anchoring and hammer options are available.

The veris P4000 probe (Fig. 4.7) utilizes tools and technology made famous by Veris Technologies' sister company Geoprobe®, the worldwide leader in direct push systems for deep soil sampling. Standard coring equipment is a 0-35" (89 cm) sampler with 2" OD (5 cm). A complete line of cutting shoes, liners and other options are available.

Fig. 4.7: *A view probe type of soil senor in the field*

These new technologies provide rapid and efficient calibration to various soil properties, including carbon. The ability to investigate the soil profile with sensors significantly increases the amount and quality of soil information and minimizes lab costs.

VIS-NIR - Spectrophotometer Shank Type

At the heart of the VIS-NIR system is the optical shoe of the soil engaging part of the shank. On the bottom of the shoe is a nitrite-hardened wear plate containing a sapphire window. Inside the intricately machined housing is a tungsten halogen bulb that illuminates the soil through the window and an optic that directs reflected light into a fiber optic cable for transmission to the spectrometer. Internal shutters automatically actuate every 15 minutes to collect dark and reference spectral a critical step in assuring data quality.

The VIS-NIR shoe has a parallel linkage design so it is follows ground contours precisely and a toggle-trip design offers protection against rocks and other field obstacles. Covering disks bring loose soil over the slot, leaving the field smooth and level. The implement can be configured as a 3-point tractor mounted unit, or equipped with a front tongue in a pull-type mode. Soil electrical conductivity (EC) coulter-electrodes are optional, allowing EC data to be recorded in the same file as the spectral data.

The control and data acquisition software included with the system make VIS-NIR data collection as easy as possible. The user-friendly programme directs the operator through system calibration, data acquisition, initial data processing, and sampling. It alerts the operator to any malfunctions, preventing inadvertent collection of erroneous data and protecting your investment by shutting the system down if component humidity or temperature exceeds safe thresholds. Because soil samples are critical for calibrating spectral data, the software compresses and clusters the data using principal components analysis and establishes sampling locations that should optimize calibrations. A map of the field and sample sites appear and allow the operator to navigate to the sites.

Multispectral Imaging Sensor

Modern vision and imaging applications rely on interpretation of information acquired by an image sensor. Typically the sensor is designed to emulate human vision, resulting in a colour or monochrome image of the field of view as seen by the eye. This is accomplished by sensing light at wavelengths in the visible spectrum (400-700 nm). However, additional information can be gained by creating an image based on the light that is outside the sensitivity of the human eye. The information available can be maximized by combining information found in multiple spectral bands. The photonic spectrum includes energy at wavelengths ranging from the ultraviolet through the visible, near infrared, far infrared, and finally X-rays. The color image from a Charge Coupled Device (CCD) array is acquired by sensing the wavelengths corresponding to red, green and blue light. CCD sensors are capable of detecting light beyond the visible wavelengths out to 1100 nm. The wavelengths from 700 nm to 1100 nm are known as the near infrared (NIR) and are not visible to the eye. In standard colour video cameras the infrared light is usually blocked from the CCD sensor because it interferes with the quality of the visible image. Multispectral cameras give you access to the full power of the CCD's capabilities by providing one imaging array that performs color imaging and two more that sense the invisible light from 700-1100 nm. The wavelengths detected by each array can be further limited by adding narrowband optical filters in the imaging path. Combining the information from all three sensors provides image data in five spectral bands *i.e.* red, green, blue and the two infrared bands. The MS4100 high- resolution 3-CCD camera (Fig. 4.8) provides the ultimate digital imaging quality. Colour separating optics work in concert with large format progressive scan CCD sensors to maximize resolution, dynamic range and field of view. The MS4100 is available in three spectral configurations *i.e.* RGB for high quality colour imaging, CIR and RGB/CIR for multispectral applications. The HDTV one- inch sensor format provides the large pixel and sensing area needed to deliver wide coverage and high dynamic range. Advanced features such as exposure control and white balance maximize usability.

Fig. 4.8: *A view of multispectral camera*

Fig. 4.9: *Multispectral camera install on the tractor*

NIR sensors and fertilizer inputs

The normalized difference vegetative index (NDVI) can be used to estimate crop biomass. The amount and variability of crop biomass can then be used to fine tune fertilizer inputs. There are a number of approaches that can be used. NDVI measurements can be conducted as part of crop scouting (before fertilizer application) or in real-time (mounted on the fertilizer spreader). The former approach allows time to analyze why parts of the crop may have low NDVI, it may be completely unrelated to nutrient supply (*e.g.* water logging, disease or pest pressure). NDVI readings from reference strips of fully fertilized crop can be compared to those found across the crop. If variable rate fertilizer application is available, areas of low NDVI could then receive more fertilizer than areas of high NDVI. If variable rate technology was not available an optimal application rate for the entire paddock could be calculated. Another approach would be to compare changes in NDVI over time. A crop might be assessed before fertilizer application then assessed again a few weeks later to determine the amount of change in NDVI. Areas of low or high change can be checked and may need less or more fertilizer next time.

Keystone

The new technologies provide rapid and efficient calibration to various soil properties, including carbon. The ability to investigate the soil profile with sensors significantly increases the amount and quality of soil information and minimizes lab costs.

- NIR sensors use light reflectance from a range of wavelengths to calculate a number of crop indexes. The indexes correlate with parameters such as crop biomass, nitrogen content, shoot or tiller number and green leaf area.

- There are a number of sensors suitable for cropping. The most common are Green seeker, Crop circle, YARA N Sensor and Crop spec. However they all operate at different wavelengths and scales, so their outputs cannot be directly compared.

- Sensors can be used to estimate crop biomass and the amount and variability of biomass can then be used to fine tune inputs such as nitrogen fertilizer. This can be done by using the sensor either for crop scouting or on application equipment for real-time optimization of inputs.

- Multispectral imaging expands the camera's capability to include the power to image features that cannot be seen with the eye. By selectively combining both visible and infrared images, the available information from a field of view can be maximized.

CONCLUSIONS

Sensors are needed in site specific crop production for the controls of those properties of soil and plants which are spatially variable, essential for economy, environment and which cannot be recognized by the farmer during field work immediately. Experimental work to determine the best use of sensors in crop production is still in its infancy. Sensors for plant recognition and plant evaluation are based upon spectral evaluation in the VIS, NIR range and image processing. Machine vision gives promising results, but is expensive and sensitive in field work still. Multi spectral with low spatial resolution could be cost effective alternative in the development of sensors for the evaluation of crops. The use of crop sensors coupled with GPS technology allows inputs such as fertilizer to be optimized either at a paddock or within paddock scale. As the cost of the technology decreases and the cost of fertiliser increases, the value of using of precision farming tools such as NIR sensors will only increase.

REFERENCES

Harrell RC, Adsit PD, Pool TA and Hoffman R. **(1990)**. The Florida Robotic Grove Lab. *ASAE Transactions*, **33:** 391-399.

Montes JM, Melchinger AE, Reif JC., **(2007).** Novel throughput phenotyping platforms in plant genetic studies. Trends in Plant Science **12:** 433-436.

Reid JF, Zhang Q, Noguchi N and Dickson M. **(2000)**. Agricultural automatic guidance research in North America. *Computers and Electronics in Agriculture.* **25:** 155-167.

Ruckelshausen A, Dzinaj T, Gelze F, Kleine Hörstkamp S, Linz A, Marquering J. **(1999).** Microcontroller-based multisensor system for online crop/weed detection. Proceedings 1999 Brighton Conference, **2:** 601-606.

Thomas E. **(2006).** Feldversuchswesen. Verlag Eugen Ulmer Stuttgart, Germany, p. 387.

Tillett ND. (1991). Automatic guidance sensors for agricultural field machines: a review. *Journal of Agricultural Engineering Research.* **50**(33): 167-187.

■

PRECISION FARMING

Precision farming is a noval method of farming which tailors all inputs like water, fertilizer and pesticides etc. in a measured for to match verifying growth stages of each crop on the field. In other words it means optimal use of inputs in a location specific, field specific and crop specific approach to match verifying growth stages of each crop on the field in systematic manner by right method on right time at right place. In addition, it also takes adequate care of technology up-gradation and marketing support. Precision Farming adopts a location specific, field specific and crop specific approach. The objective is optimization of use of inputs to facilitate optimal output resulting in saving of valuable resources which are shrinking in modern era. Precision farming is one of the best alternate to satisfy one of the most important global goals of food security which is not just supply and demand, but it also considers food availability and safety, poverty alleviation and food affordability, global food trading and quality standards, and the environmental threshold and capacity; of ever increasing population with least use of resources.

ROLE OF PLANT GENETIC RESOURCES TO MEET CHALLENGES

Plant genetic resources are the biological basis of food and other essential uses for an ever increasing human population, and directly or indirectly, support the livelihoods of every individual on Earth. The plant genetic resources are part of the biodiversity that nurture people and encompass the diversity of genetic materials in both traditional varieties and modern cultivars, as well as wild relatives of crop plants and other natural plant species used as food. They are also the main source of raw materials for modern medicines such as antibiotics, heart drugs and pain killers, and also a myriad of other raw materials to sustain a variety of industries. Today, we are in a race between growing population and food production.

NEED OF MANAGEMENT OF PGR

Today, we are in a race between growing population and food production and and the pressure is on to improve agricultural production by developing food crops

---

[1] Assistant Professor,JNKVV, Regional Agricultural Research Station, Waraseoni, Balaghat (MP)
[2] Sr. Scientist Agronomy PDFSR (ICAR), Modipuram, Meerut (UP)
[3] Principal Scientist, RARS, Waraseoni, Balaghat (MP)

that can adapt to environmental changes and meet the growing food demands of a constantly increasing population. The challenge of feeding a world population growing by up to 160 people every minute remains daunting. It is forecast that, by 2050, world population will increase from the current level of ≈ 6 billion to more than 8 billion people. Feeding this population will require an astonishing increase in food production.

In fact, it has been estimated that the world will need to produce as much food during the next 50 years as was produced since the beginning of agriculture 10,000 years ago. Our staggering requirement for food must be viewed in the context of statistics that indicate that the area available for food production has remained constant since 1960. Despite some new land being brought into cultivation, soil erosion and urbanization have offset these gains. Financial support for agricultural research has decreased for the last several years and is expected to continue its slow decline. The International Food Policy Research Institute has predicted that, by the year 2020, maize and wheat are each expected to have an annual global demand of ≈ 775 million tons each and will be of critical consequence in the race between crop production and population growth.

Alarm bells are ringing in various quarters about India's vulnerability in sustaining food grain production to feed its ever growing population (about 15 million new faces are added every year). The country was transformed from a food deficient nation to a food sufficient nation. The nation appears to have achieved "food security". For many years, India has been comfortable in its ability to produce food and feed people. It has had surplus food grains too. Since 2003, India had to reverse the funnel and import more and more food grains every year, something it had not done in decades. India's position is precarious in food production because of erratic rainfall patterns. India has not exceeded 230 mt (2007-08) in the last one decade By 2021, however, India will need to produce 276 mt of food grains to feed its people. By 2050, the country will need to double its food grain production. If India fails to enhance production, leading to a huge gap between supply and demand, there could be social upheavals and rampant hunger and malnutrition. This should alert us all to the possibility of India slipping down the "food security" ladder.

Over the past few decades, Green Revolution has expanded food production all over the region. In the short term, the technology of chemical fertilizers and high-yielding varieties has been very successful, and has increased crop yields wherever it has been adopted. However today, several decades later, these yield increases from intensive high-input monoculture are becoming smaller, or even showing a reverse trend. A major cause is the pressure on the environment from modern intensive agriculture. Every country is suffering from similar problem of inappropriate use of resources, including the felling of forests, over use of slope lands and agricultural use of marginal lands leading to environmental deterioration. There is wide spread degradation of the agricultural resource base, with a serious decline in the quality of many agricultural soils, waterways and forests. There has been a decline, not only in the quality of resources used for agriculture, but also in their quantity, as water and land are diverted for industrial use.

NARROWING OF GENE POOL

It is estimated that the earth is home of approximately 240,000 species of plants including 30,000 edible plant species of which only about 5,000 have been carefully studied. This rich diversity that has already captured human attention now contains a large array of useful plant genes. Agriculture today is characterized by a sharp reduction in the diversity of cultivated plants. Though, as it is today, mankind, by and large, is dependent on 25 to 30 food plants, *viz.*, wheat, rice, maize, barley, oats, sorghum, millets, soybean, beans, pea, chickpea, peanut, banana, citrus, tomato, sugarcane, cassava, potato, sweet potato, yams and five oilseed types and a few beverages and only three species *viz.* rice, wheat and maize provide 60 % food for the mankind. In addition to the inter-specific reduction of crop diversity in agriculture, plant breeding contributes to diminution of the intra specific diversity, through development of adapted breeding populations, selection of the 'best' genotypes, development of genetically homogeneous cultivars and promotion of few widely adapted varieties. Further, the widespread adoption of high yielding crop varieties which has the potential to stimulate genetic erosion of landraces often used in their pedigrees and negatively affected socioeconomic and environmental. The lack of inter and intra specific genetic variability among cultivated crops can lead to Epidemics of pests and diseases; examples are the *Phytophthora infestans* infestation of potato in Western Europe in 1845-1846, the *Bipolaris maydis* disaster in T-cytoplasm maize in the USA in 1970 and the *Fusarium graminearum* epidemic in wheat and barley in the western USA (1994–1996) lack of adaptation to increasing abiotic stresses like drought or high ozone concentrations; lack of genetic variation for specific quality traits, *e.g.* starch quality in maize fatty acid composition or male sterility in oilseed reaching performance plateaux may be another risk.

EVERY GENE COUNTS

Traditionally, as plants evolved naturally in their fields, farmers made seasonal selections of which seeds to save and plant the next year, based on what worked best in their local environments. In the early nineteenth century, scientific advancement brought the ability to crossbreed more predictably. Today, modern biotechnology goes even further by providing plant breeders avenues to bring useful genes not only from other varieties but from other species into the mix. This means every crop variety has a potential use that extends far beyond a local farmer's field. The genes which are not useful today may be useful for tomorrow for the welfare of human beings.

THE WORLD DEVELOPMENTS FOR ACCESS OF PGR

Plant Genetic Resources for Food and Agriculture (PGRFA) in the 1960's and 1970's was one of "common heritage for humankind". Germplasm then generally was available without restrictions internationally, at a time when the impacts of the Green Revolution were being felt in high yielding varieties of wheat and rice. This period saw the initiation of the Consultative Group on International Agricultural Research (CGIAR) in 1971, the subsequent development of the International

Agricultural Research Centers (IARCs, now numbering 16), including the International Board of Plant Genetic Resources (IBPGR) in 1974, now the International Plant Genetic Resources Institute (IPGRI). The world then focused on building germplasm collections and genebanks, and the development of new high yielding varieties. Much of this work was stimulated by an increasing awareness of the narrow genetic base of advanced agriculture and consequent potential susceptibility to crop failures. This system made tremendous strides in germplasm collection and understanding of distribution and systematics of crops and their wild relatives (Hawkes *et.al.*, 2000). The passage of the Convention for the International Union for the Protection of New Varieties of Plants (UPOV) in 1961 legalized Plant Variety Protection (PVP) and introduced the concept of "breeders' rights. Subsequently, a Multilateral Trade Agreement made in 1995 under the Uruguay Round of negotiations included the Agreement on Trade-Related Aspects of Intellectual Property Rights (TRIPS), administered by the World Trade Organization (WTO). Under the obligations required by TRIPs, some countries have created *sui generis* forms of intellectual property rights for farmer varieties and related traditional knowledge. There are similar demands within the context of the meetings of World Intellectual Property Organization's (WIPO) Intergovernmental Committee on Intellectual Property and Genetic Resources. The Convention on Biological Diversity (CBD) recognises the sovereign rights all countries have over their natural resources and their authority to determine access to genetic resources within their borders for which they are the countries of origin. The International Treaty on Plant Genetic Resources for Food and Agriculture which came force in 2004, are the conservation and sustainable use of Plant Genetic Resources for Food and Agriculture and the fair and equitable sharing of the benefits arising out of their use, in harmony with the Convention on Biological Diversity, for sustainable agriculture and food security. This legally-binding treaty is called the *International Undertaking on Plant Genetic Resources* which is international agreement with the overall goal of supporting global food security, allows governments, farmers, research institutes and agro-industries to work together by pooling their genetic resources and sharing the benefits from their use in modern plant breeding and biotechnology - thus protecting and enhancing our food crops while giving fair recognition and benefits to local farmers who have nurtured these crops through the millennia. Although, this Treaty, only covered 35 food crops (including the *Brassica* complex which itself includes several crops from Cabbages to Rocket) and 29 forages crops species put into a common pool. These represent a small proportion of the 105 food crops of importance to food security, the many others that have nutritional significance and some 18,000 forages of value to food and agriculture. Onion, Garlic, Groundnut / Peanut, Oil Palm, Soybean, Tomato, Sugarcane and Minor Millets are among important crops missing from the list. The treaty facilitates access to those crops, makes them available free of charge for certain uses to researchers who agree to share any future commercial benefits from their use in modern plant breeding or biotechnology. This recognition and this benefit sharing are designed to ensure equity and encourage farmers to continue conserving and using the diversity in their fields. Therefore, the access to a range of genetic diversity is critical to the success of breeding programmes.

STATUS OF PLANT GENETIC RESOURCES

No country is self-sufficient in plant genetic resources; all depend on genetic diversity in crops from other countries and regions. International cooperation and open exchange of genetic resources are therefore essential for food security. The global effort to assemble, document, conserve and utilize these resources is enormous and the genetic diversity in the collections is critical to the world's fight against hunger. A total of 6.1 million accessions, including major crops, minor or neglected crop species, as well as trees and wild plants conserved in registered 1308 genebanks world-wide. World collections of major crops assembled by the IARCs and other major genebanks in the world. In India, NBPGR, with the network of its 10 regional stations located in diverse agro-climatic zones of the country, the 57 national active germplasm sites (NAGS), is the nodal organization in the country for acquisition, collection and management of both indigenous and exotic accessions for sustainable growth of agriculture. In addition to germplasm maintained by the NBPGR, several crop specific institutes and coordinated research projects, SAUs also hold germplasm.

GENETIC DIVERSITY IN INDIA

Genetic diversity is defined as variations in the genetic composition of individual within or among species. India being one of the 12-mega biodiversity countries possesses 11.9% of world flora. The Indian sub-continent is immensely rich in plant genetic resources of both crop species and wild relatives. PGR is fundamental to crop improvement programme and future nutritional security. The importance of plant genetic resources has increased significantly in the recent years with changing global scenario in material ownership and legal regimes with respect to access to PGR under the international agreements. India has a rich and varied heritage of biodiversity possesses 11.9% of world flora. Besides, North-East, Western Ghats, North–West Himalayas, Eastern Ghats, Andaman and Nicobar islands, two most important hot spots out of 25 listed, are present in India. In addition, India has 26 recognized endemic centres that are home to nearly one third of all the flowering plants described. India is homeland of 167 cultivated species and 329 wild relatives of crop plants and ranks seventh in terms of contributions to world agriculture.

ENDANGERED PLANT SPECIES

About 427 endangered plants species have been listed by the Botanical Survey of India in its publications on the floristic synthesis enumerated for Red Data book. This contributes to about 20 per cent of India's' total floristic wealth of higher plants. The human activities are eroding the prevalent biological resources and greatly reducing the biodiversity of the planet. Estimating the precise rate of loss or even the current status of the species, is stupendous task because no monitoring system, whatsoever, systematic, is likely to match the huge diversity in its existing form. Thus, much of the basic information remains lacking, especially on the species rich tropics wild flora and a large fraction of its wild fauna are threatened with, many of the verge of extinction. India has lost at least 50 per cent of its forest, polluted over 70 per cent of its water bodies, built or cultivated on much of its grassland and degraded many coastal areas in the past few decades. Further, to bio-prospecting this habitat, ill effects of destruction, hunting, over exploration and a host of other activities have together taken a heavy toll.

The nearly simultaneous emergence of the CBD and the Global Agreement on Trade and Tariffs (GATT) hinted at the demise of common heritage by stipulating national ownership of biological resources and pushing countries to adopt intellectual property for plant materials.

In the light of the fast developments at world flora for legal ownership of Plant Genetic Resources, It has become more important to acquire, collect, evaluate, safely conserve, document and catalogue our Plant Genetic Resources to meet future challenges and food security.

PLANT GENETIC RESOURCES

The sum totals of hereditary material *i.e.* all the alleles of various genes, present in a crop species and its wild relatives are referred to as germplasm. This is also known as genetic resources or gene pool or genetic stock. Genetic resources can be defined as all materials that are available for improvement of a cultivated plant species. In classical plant breeding, genetic resources may also be considered as those materials that, without selection for adaptation to the target environment, do not have any immediate use for the breeders (Hallauer and Miranda, 1981). Important features of plant genetic resources are given below.Genetic pool represents the entire genetic variability or diversity available in a crop species.

Germplasm consists of land races, modern cultivars, obsolete cultivars, breeding stocks, mutants, wild forms and wild species of cultivated crops.

Germplasm includes both cultivated and wild species and relatives of crop plants.

Germplasm is collected from centres of diversity, gene banks, gene sanctuaries, farmer's fields, markers and seed companies.

Germplams is the basic material for launching a crop improvement programme.

Germplasm may be indigenous (collected within country) or exotic (collected from foreign countries) assigned IC and EC numbers respectively.

According to the extended gene pool concept, genetic resources may be divided into primary gene pool, secondary gene pool, tertiary gene pool and isolated genes (Bretting and Widerlerhner, 1995).

Primary Gene Pool

The primary gene pool consists of the crop species itself and other species that can be easily crossed with it.

Secondary Gene Pool

The secondary gene pool is composed of related species that are more difficult to cross with the target crop *i.e.* where crossing is less successful (low percentage of viable kernels) and where crossing progenies are partially sterile.

Tertiary Gene Pool

The tertiary gene pool consists of species which can only be used by employing special techniques like embryo rescue or protoplast fusion.

The fourth class of genetic resources, isolated genes, may derive from related or unrelated plant species, from animals or micro-organisms. The importance of the different classes of genetic resources for crop improvement depends on the target crop species. In maize, *for example*, genetic variation in the primary gene pool is so large that the secondary or tertiary gene pools are rarely used. In rapeseed, on the other hand, genetic variation in the primary gene pool is small and breeders have to transfer important traits from *Brassica* species of the secondary and tertiary gene pool into the cultivated species (Maxted *et al.*, 2002).

COLLECTION OF GERMPLASM

The process of acquiring germplasm accessions from various sources is known as Collection of Germplasm. This can be done by (1) By arranging exploration expeditions (2) Procurement from other institutes and agencies, companies etc. in exchange and as gifts.

(1) Exploration and Collection

Explorations are the trips for the collection of various forms of crop plants and their wild related species. Explorations are the primary source of all kind of germplasm collections. Although existing potential of wild germplasm in India are tremendous, yet large areas are unexplored and untapped. A consortia approach would be desirable to augment, evaluate and sustainable use the wild plant species for improvement of cultivated crops. Germplasm enhancement is thus critical for sustainable agriculture in future and hence it would be necessary to meet these challenges successfully.

Priorities of collection

FAO assigned Code 'E' for threat based priority, and I, II, III for other lower levels of priorities whereas, need based priorities depend on the need of concern county.

Area of collection

Centre of origins, Primary centre of diversity, secondary centre of diversity, micro-centre of diversity, peripheral region of distribution of species and area where species has been introduced recently.

Sampling site

A primary survey should be made before proceeding for actual collection. Generally site for cultivated material should be scattered in the entire area while wild material should be collected in clusters. Although location of sampling sites depends on Soil pattern, changes in agricultural practices, changes in ecological, social and agricultural conditions in area.

Sampling procedure

The collection of plant materials from the sampling sites may be (1) random (2) selective with objective to detain the maximum amount of variability with the minimum number and size of samples. The distinctiveness of the materials during sampling is only considered in selective sampling, not in random sampling. Is the most suited for conservation efforts and is expected to capture the whole range of variability present in a species while only morphological variations are collected in selective sampling which does not represent the whole spectrum of variability. Hence, both are used as per the need and objective collection programme.

Sampling size

From each sampling site collection of 50-100 plants and 50 seeds from each plant has been suggested to capture whole range of variability.

Field records

Field records must be maintained during collection for their future references. Passport data include serial number, taxonomic name, collection site, date of collection and donor institute or place etc. Although expeditions of plant explorations are very tedious, time consuming and expensive, sometime poses hardship and threat to life by wild and poisonous animal. But these are the basic and incessant source of genetic diversity of all kinds of accessions collected.

(2) Procurement from Other Agencies

No country is self-sufficient in plant genetic resources all depend on genetic diversity in crops from other countries and regions. International cooperation and open exchange of genetic resources are therefore essential for food security. Indian agriculture has been enriched by introduction of valuable exotic germplasm. Hence, germplasm materials can be procured from other institutes and agencies, companies etc. in exchange or as gifts. Exchange can be done by importing or exporting this material. Under the import, plant introductions are major activity and played a glorified role in bringing Green revolution in India due introduction of semi dwarf, high input responsive introductions in wheat and rice. Likewise, hybrid era was also brought due to introduction of cytoplasmic-nuclear male sterility and fertility restoration genes. Further, soybean and sunflower became the major oilseed crops in our country. Some Plant Introduction may directly be useful in new environment for commercial cultivation without any alteration are known as primary Introductions and others need some changes in their genetic make up by breeding methods are known as secondary Plant Introductions. Number of crop species including pear, apple, walnut, guava, pineapple, papaya, cashew nut, litchi, cherries, grapes, mustard, mung, sesame, sorghum, red gram, millets, cotton, maize, groundnut, chillies, potato, sweet potato, cabbage, cauliflower, tea etc. had been introduced from different parts of globe by invaders, settlers, traders, travelers, explorers, naturalists and pilgrims.

Exchange and quarantine

Unregulated germplasm exchange activity may result in inadvertent introduction of insect-pests, diseases and weeds into country. There are examples enormous crop losses caused by the introduced pests as well weeds. The introduced exotic weed *Lantana camara* from Central America, *Parthenium hysterophorus* from Central and South America and *Phalaris minor* from Maxico have become a big threat to our crop production and environment. Likewise, *Aschochyta* introduced from Middle East caused large economic losses in grain yield of gram.

To prevent the entry of these diseases-pests as well as weeds there are legislative measures for the quarantine. As per plant, Fruit and Seed Order1989, the NBPGR, New Delhi is working as nodal agency for the quarantine processing of plant material imported for cultivation and research purposes and has ultimate responsibility for plant introduction activities in India. The quarantine offices are located at all seaports and international airports which thoroughly examine these imported materials. The material must bear a phytosanitary certificate from sender declaring the material is free from any disease, pest and weed otherwise quarantine department may reject or destroy the same.

The emergence of transgenic crops has raised serious concerns among the public regarding their effects on safety, health and environment as these issues have been discussed at various national and international fora. In India, as per regulatory mechanisms, NBPGR issues the import permit for introducing transgenic materials for research purpose and undertake quarantine processing of introduced transgenic materials. A National Containment Facility has been established at NBPGR under a collaborative project of ICAR and DBT. It has the objectives of the quarantine processing of imported transgenic, developing molecular probes for the detection of trans-genes (promoter and terminator sequences) and developing human resources in the field of bio-safety.

Germplasm registration

To counteract the challenges of global Intellectual Property Rights' regime, the ICAR has instituted a mechanism for the registration of experimentally developed germplasm of potential value. So far, 1614 proposals have been processed for the registration of such germplasm. Of these, 603 potentially valuable germplasm belonging to 115 crops have been registered.

Germplasm conservation

Conservation refers to protection of genetic diversity of crop plants from genetic erosion. There are two important methods of germpalsm conservation or preservation (i) *In-situ* conservation and (ii) *ex-situ* conservation. These are described below.

(i) *In - situ* Conservation

Conservation of germplasm under natural conditions is referred to as *in situ* conservation. This is achieved by protecting the area from – human interference, such an area is often called natural park, biosphere reserve or gene sanctuary.

NBPGR, New Delhi, established gene sanctuaries in Meghalaya for citrus, north Eastern regions for *Musa*, Citrus, *Oryza* and *Saccharum*. Gene sanctuaries offer the following advantage.

Merits

In this method of conservation, the wild species and the complete natural or seminatural ecosystems are preserved together.

In-situ (dynamic) conservation sought to recreate the products of the evolutionary process

A dynamic genebank (*In-situ*) strategy attempts to maintain interactions with the plant's natural environment, including pathogens, maintains gene flow (for outcrossing species), and maintains natural evolutionary forces, while a "static" strategy (as increases in greenhouses) does not.

This strategy serves many purposes, including conservation of a wide range of biodiversity, maintenance of ecosystems, and fosters societal goals such as recreation, environmental education and ecotourism etc.

Many problems associated with conservation of diversity and accumulation of deleterious alleles would be alleviated by conservation of genetic diversity *in-situ* (Maxted *et al.* 1997).

Demerits

Each protected area will cover only very small portion of total diversity of a crop species, hence several areas will have to be conserved for a single species.

The management of such areas also poses several problems.

There is a concern about loss of genetic diversity, especially in static genebanks, through genetic drift over serial increase cycles, genetic bottlenecks due to small sample sizes, and accumulation of deleterious mutations. This is a costly method of germplasm conservation.

(ii) *Ex - situ* Conservation

It refers to preservation of germplasm in gene banks. This is the most practical method of germplasm conservation. This method has following advantages.It is possible to preserve entire genetic diversity of a crop species at one place. Handling of germplasm is also easy.This is a cheap method of germplams conservation.Typically 25 to 100 seeds are required for regeneration in genebanks (Callow *et al.*, 1997).

Shortcomings

A critical issue affecting decisions of reliance on *ex situ* collections regards their longterm viability, maintenance of genetic diversity and accumulation of deleterious mutations in long-term storage.

Periodic regeneration is one of the major needs for *ex situ* collections. Furthermore, optimum regeneration requires detailed knowledge of the biology of the crop or crop relative regarding environmental requirements, pollination system, breeding system, and seed fecundity and longevity.

Their primary point was that genebanks conserved the products of evolution but ceased further evolutionary processes

This type of conservation can be achieved in the following 5 ways

(1) Seed banks

Germplam is stored as seeds of various genotypes. Seed conservation is quite easy, relatively safe and needs minimum space.Van Hintum *et al.*, (2000) has classified seeds, on the basis of their storability into two major groups.

(a) **Orthodox seeds:** Seeds which can be dried to low moisture content and stored at low temperature without losing their viability for long periods of time is known as orthodox seeds. Seeds of corn, wheat, rice, carrot, papaya, pepper, chickpea, cotton, sunflower etc. come under this group.

(b) **Recalcitrant:** Seeds which show very drastic loss in viability with a decrease in moisture content below 12 to 13% are known as recalcitrant seeds. This group includes the seeds of citrus, cocoa, coffee, rubber, oil palm, mango, jack fruit etc. Hence, it not possible to conserve these seeds in seed banks and therefore requires *In situ* conservation.

Seed Storage

Based on duration of storage, seed bank collections are further classified into three groups. (a) Base collections. (b) Active collections and (c) Working collection (d) Core Collections

(a) **Base or Principal or Whole Collections:** Seeds can be conserved under long term (50 to 100 years), at about -20°C with 5% moisture content in hermetically sealed containers. They are disturbed only for regeneration. These are used only when germplasm from other sources is not available for use.

(b) **Active Collection:** Seeds are stored at 0°C temperature and the seed moisture is between 5 and 8%. The storage is for medium duration, *i.e.*, 10-15 years. Germination test is carried out periodically to check the viability. These collections are used for evaluation, multiplication, and distribution of the accessions.

(c) **Working Collections:** Seeds are stored for 3-5 years at 5-10°C and the usually contain about 10% moisture. Such materials are regularly used in crop improvement programmes.

(d) **Core Collection:** At the outset, core collections were defined as a limited set of accessions representing, with a minimum of repetition, the genetic diversity of a crop species and its wild relatives. In the context of an individual genebank, a core collection consists of a limited number of the accessions of an existing collection. These are chosen to represent the genetic spectrum in the whole collection and should include as much of its genetic diversity as possible (Brown, 1995). Core collections can render the evaluation process more efficient because repetition of similar entries is avoided.

2. Plant Bank

(Field or plant bank) is an orchard or a field in which accessions of fruit trees or vegetatively propagated crops are grown and maintained.

Limitations: 1. Require large areas, 2. Expensive to establish and maintain such banks, and 3. Prone to damage from diseases, insect attacks and natural disasters.

3. Shoot Tip Banks

Germplasm is conserved as slow growth cultures of shoot-tips and node segments. Conservation of genetic stocks by meristem cultures has several advantages as given below.

Each genotype can be conserved indefinitely free from virus or other pathogens.

It is advantageous for vegetatively propagated crops like potato, sweet potato, cassava etc., because seed production in these crops is poor. Vegetatively propagated material can be saved from natural disasters or pathogen attack.

Long regeneration cycle can be envisaged from meristem cultures.

Regeneration of meristerms is extremely easy.

Plant species having recalcitrant seeds can be easily conserved by meristem cultures.

4. Cell and Organ Banks

Cryo-preservation (Greek word Kryo = frost) means 'preservation in frozen state'. Generally plant material is frozen and maintained at the temperature of liquid nitrogen (-196 °C). The storage at ultra low temperature retards or cease metabolic processes which increase longevity of material preserved such as the apical and lateral meristems, plant organs, seeds, cultured plant cells, protoplasts, calluses and somatic embryos. Cryo-preservation of plant cells, organs and tissues received widespread importance mainly due to emphasis given on *in vitro* manipulations with cultured cells.

Advantages:

The *in vitro* system is extremely suitable for storage of plant material as it requires small space to store.

It is easy to reproduce disease free plants from these conserved materials.

Material can be stored for indefinite period.

5. DNA Banks

In these banks, DNA segments from the genomes of germplasm accessions are maintained and conserved. First time successfully used an approach of introgression library in broadening the genetic base of tomato breeding material. Further introgression libraries in plants have been established in *Brassica oleracea* (Hawtin *et al.*, 1997).

GERMPLASM CHARACTERIZATION AND EVALUATION

Evaluation data refer to agronomic traits like grain yield, grain quality, lodging, and resistance to important pests and diseases as far as evaluated and is a continuous process whereas, Characterization data comprise scores for simple morphological traits like plant height, maturity date and thousand seed weight to descriptor states reflecting specific alleles for known genetic systems. Different people or institutions can be involved, such as genebanks, breeders, pathologists or physiologists searching for specific traits. IPGRI, Rome has developed model list of descriptors (= characters) for which germplasm accessions of various crops should be evaluated. The evaluation of germplasm is done in three different places *viz.* (1) in the field (2) in green house (3) in the laboratory depending upon concerned trait.

Need of Evaluation

To identify gene sources for resistance to biotic and abiotic stresses, earliness, dwarfness, productivity and quality characters.

To classify the germplasm into various groups.

To get a clear pictures about the significance of individual germplasm line.

Little is known about the extent of evaluation in individual genebanks, according to FAO (1996a) the percentage of evaluated accessions ranges from 5 to 100%. The use of PGR in crop improvement could be facilitated by systematic evaluation and documentation of the acquired data. Of particular importance are: information on valuable traits *e.g.* resistances and specific quality traits, reliable information on genotype x environment interactions and specific adaptation; information on general and specific combining abilities and affiliation to heterotic pools (if hybrid breeding is relevant) on-farm evaluation to gain information about farmer's perception, user-friendly information and documentation systems.

GERMPLASM CATALOGUING, DATA STORAGE AND RETRIEVAL

Gene bank accessions are described by passport and characterization data, and to a variable extent also by evaluation data. Additional notes can refer to seed viability, number and mode of regeneration or reproduction, and information about the distribution of the sample. Germplasm passport information exchange is facilitated by the internationally standardized list of multi-crop passport descriptors (MCPD), which have been developed jointly by IPGRI and FAO.

Each germplasm accession is given an accession number. This number is pre fixed in India, with either IC (Indigenous collection), EC (exotic collection) or IW (Indigenous wild). Information on the species and variety names, place of origin, adaptation and on its various feature or descriptors is also recorded in the germplasm maintenance records. Its compilation, storage and retrieval is now being done using special computer programmes. Catalogues of the germplasm collection for various crops are published by the gene banks. The amount of data recorded during evaluation is huge. Ideally, all data sets accompanying an accession are stored in a central database and are made available to the public.

Facts and information sources

World-wide, 1308 gene banks are registered in the WIEWS (World Information and Early Warning System on PGR) database and conserve a total of 6.1 million accessions, including major crops, minor or neglected crop species, as well as trees and wild plants. Of the 30 main crops, more than 3.6 million accessions are conserved *ex-situ* and little information exists about documentation and availability of materials that are maintained *in-situ*. The World Information and Early Warning System (WIEWS) on Plant Genetic Resources for Food and Agriculture (PGRFA) was established by FAO as a dynamic, world-wide mechanism to promote information exchange among member countries, by gathering and disseminating information on PGRFA, and as an instrument for the periodic assessment of the state of the world's PGRFA. The CGIAR (Consultative Group of International Agricultural Research) System-wide Information Network for Genetic Resources (SINGER) links the genetic resources information systems of the individual CGIAR centres around the world, allowing them to be accessed and searched collectively. SINGER contains key data of more than half a million individual accessions of crop, forage and agro-forestry genetic resources held in the Centre gene banks (http://www.singer.cgiar.org/). The International Plant Genetic Resources Institute (IPGRI) is the world's largest non-profit agricultural research and training organization devoted solely to the study and promotion of agricultural biodiversity (http://www.ipgri.cgiar.org).

Utilization of PGR

In order to enhance the utilization of PGR in crop improvement, the Global Plan of Action proposed following measures: expanded creation, characterization and evaluation of core collections increased genetic enhancement and base-broadening efforts development and commercialization of underutilized species development of new markets for local varieties and 'diversity-rich' products and concomitant efficient seed production and distribution comprehensive information systems for PGR promoting public awareness of the value of PGR for food and agriculture.

FUTURE AVENUE

(1) Role of Landraces in Diversification of Genetic Base

The role of landraces is often conceived as donor of specific useful genes for improvement of HYVs. Landraces grown in extreme areas *e.g.* semi-arid to arid regions in Asia and Africa, can represent important PGR in breeding for specific adaptation (Hawtin *et al.*, 1997). They can be donors for individual monogenic traits sources of new quantitative variation for specific adaptation to stress conditions and breeding population or crossing partner in the development of improved, locally adapted cultivars for the same or other marginal areas. Such vision is far too restrictive and diminutive, as compared with the great potential they retain for further progress of many agricultural areas, especially those exhibiting unpredictable, unfavourable conditions. Landraces or primitive cultivars, evolved under farmers' selection over millennia in the area, are composed by many different genotypes, each of which possess a slightly different expression of traits and are better adapted to slightly different conditions or combination of conditions than others.

(2) New Plant Resources (Under-utilized Crops)

New plant resources needed to be exploited in order to meet the growing needs of the human society which incidentally has depended only on a small fraction of crop plant. Accordingly, many of the under- utilized plants have a potential for improving Indian agriculture in diversified agro- ecological niches and have great potential for exploitation in view of the value of their economic products for use as food, fodder, medicine, energy and industrial purposes. Some of the exotic crops which have been found most promising for exploitation are guayule *(Parthenium argentatum)*, jojoba *(Simmondsia chinensis)* as sources for industrial rubber and oil respectively in arid situations, *Atriplex* spp, as forage resource for salt-affected soils and *(Leucaena leucocephala)* as a fodder tree source for semi-arid rainfed situations. Conversely, indigenous plant species such as rice bean *(Vigna umbellata)* and their introductions are being proved as promising new food sources.

(3) Role of Molecular Markers and Genome Research

The tools of genome research may finally unleash the genetic potential of our wild and cultivated germplasm resources for the benefit of the society. The utility of molecular markers and genome research in the context of using PGR for crop improvement include; Diversity studies to identify genetically similar or distinct accessions, and to determine individual degrees of heterozygosity and heterogeneity within populations of PGR. Genetic mapping to identify simply inherited markers in close proximity to genetic factors affecting quantitative traits (QTL), followed by marker-assisted selection (MAS) of desired genotypes in segregating populations. Exploitation of valuable QTL from PGR by advanced backcross QTL analysis to combine QTL analysis with the development of superior genotypes or by marker-assisted, controlled introgression of PGR into breeding materials through the development of introgression libraries. Association studies to mine directly the allelic diversity of PGR and to identify those alleles that are beneficial for important agronomic traits.

Initially, diversity studies were based on morphological and agronomical traits. The increasing availability of molecular marker systems opened up new possibilities for the diversity assessment of PGR intended to be used for crop improvement (Bretting and Widerlechner 1995, Karp *et al.*, 1997). For an efficient diversity assessment, molecular markers ideally need to be selectively neutral, highly polymorphic, co-dominant, well-dispersed throughout the genome and cost- and labour-efficient (Bretting and Widerlechner 1995 Van Treuen 2000). Genetic markers complying with these requirements are protein markers (*i.e.*, isoenzymes) and DNA markers such as restriction fragment length polymorphisms (RFLPs) and micro-satellites or simple sequence repeats (SSRs). Because the development of the latter two marker types requires prior knowledge of DNA sequences, a number of universal, dominant molecular marker types such as randomly amplified polymorphic DNA (RAPD) and amplified fragment length polymorphisms (AFLPs) have also been employed in PGR diversity studies. However, the latter are not suitable for such studies.

Seven pillars of food security wisdom

As observers of world food security strain to see whether another major crisis like that of 2007-08 looms on the horizon, CGIAR food policy experts are sending a strong message to developing and developed country governments as well as international organizations. Action must be taken now to prevent food price crises from becoming a recurring world nightmare in the decades to come.

Minimize food-fuel competition through policy reforms.

Strengthen social safety nets.

Make global trade more fair and transparent.

Create a global emergency physical grain reserve.

Promote and invest in agricultural growth, emphasizing improved small holder productivity.

Invest in realizing agriculture's potential for climate change adaptation and mitigation.

Set up an international working group to monitor the world food situation.

Most important with or without another major crisis, unrelenting food price inflation especially high food inflation in China and India and volatility will continue to cause great harm to the world's poor consumers.

Future thrust

The first phase of gene banking has been marvelously successful in collecting and preserving a broad range of diversity of cultivars and wild relatives of the major food crops, but still collection needs remain.

Over the long-term, continuing germplasm collections from centers of diversity would be a wise strategy to regenerate the natural diversity in *ex situ* collections, assuming that these collections persist.

Fiery need to explore land races, wild relatives, and under-utilized plants to offset future challenges like outbursts of epidemics.

There is serious backlog in needed regeneration cycles, hence, identified as one of the major needs for *ex situ* collection.

The low funding of gene banks, especially in developing nations and irreplaceable collections may be lost.

There is serious need for reduction of collections by identification of duplicates with the help of advance biotechnological tools.

Top priority must be assigned to effective evaluation of collected as well as material to be collected for better utilization.

Mapping of genetic diversity for more fruitful conservation. Optimum design of preserves would be determined in part by allowing sufficient population sizes to maintain gene flow and mitigate drift (Ellstrand and Elam, 1993).

Biotechnology can play an important role in genetic resources management as it follows the dissection of complex characters through the use of molecular markers.

India urgently needs to increase its outlay for agricultural R&D and simultaneously take appropriate measures to reduce population growth for secured future. Bio-safety issues related to transgenic must be tackled with proper policy.

REFERENCES

Bretting P K and Widerlechner MP. (1995). Genetic Markers and Horticultural Germplasm Management.*HortScience* **30:**1349-1356.

Callow J A, Ford-Lloyd BV and Newbury HJ. (1997). Overview. *In* Biotechnology and Plant Genetic Resources: Conservation and Use, edited by J. A. Callow, B. V. Ford-Lloyd and H. J.Newbury. Wallingford: CAB International.

Hawkes JG, Maxted N and Ford-Lloyd BV. (2000). The *Ex Situ* Conservation of Plant Genetic Resources. Dordrecht: Kluwer Academic Publishers.

Hawtin G, Iwanaga M and Hodgin T. (1997). Genetic Resources in Breeding for Adaptation. In: Tigerstedt PMA (ed.)18 B. I. G. Haussmann *et al.*, Adaptation in Plant Breeding. Amsterdam: Kluwer Academic Publishers, pp. 277–288.

Karp A, Edwards KJ, Bruford M, Funk S, Vosman B, Morgante M, Seberg O, Kremer A, Boursot P, Arctander P, Tautz D and Hewitt GM. (1997). Molecular Technologies for Biodiversity Evaluation: Opportunities and Challenges. *Nature Biotechnology* **15:** 625–628.

Karp A, Isaak PG and Ingram DS (1998) (eds). Molecular Tools for Screening Biodiversity. London: Chapman and Hall.

Maxted N, Ford-Lloyd BV and Hawkes JG (1997) (eds). Plant Genetic Conservation—The In Situ Approach. London: Chapman & Hall.

Maxted N, Guarino L, Myer L and Chiwona EA. (2002). Towards A Methodology for On-Farm Conservation of Plant Genetic Resources. *Genetic Resources and Crop Evolution* **49:** 31–46.

Van Hintum TJL, Brown AHD, Spillane C and Hodgkin T. (2000). Core Collections of Plant Genetic Resources. IPGRI Technical Bulletin **3:** 5–48.

Van Treuen (2000). Molecular Markers. Wageningen, The Netherlands: Centre for Genetic Resources (CGN), http://www.cgn.wageningen-ur.nl/pgr/research/molgen/.

■

6

CHAPTER

Scope for Application of Variable Rate Technology in India

☞ Sushil Sharma[1], ☞ S.S. Manhas[2] and
☞ R M Sharma[3]

INTRODUCTION : AN OVERVIEW

The agricultural production has become stagnant in the post-green revolution period and horizontal expansion of cultivable lands became limited due to burgeoning population and industrialization. The conventional agronomic practices follow a standard management option for a large area irrespective of the variability occurring within and among the field. Farmers have been applying various agricultural inputs like seeds, fertilizers, weedicides, pesticides and water based on recommendations emanating from research and field trials under specific agro-climatic conditions which have been extrapolated to a regional level. Since soil types, soil nutrient characteristics, weeds density, pests infestation and soil water vary not only between regions and between farms but also from plot to plot and within a field or plot, there is a need to take into account such variabilities to a particular crop. The goal is to obtain more efficient use of applied inputs to reduce any excess application that might cause environmental pollution and to improve economics. Agricultural technologies using the global positioning system and geographic information systems have started to change how farmers are managing crops. By utilizing these tools with an objective to manage existing field variability, variable-rate technology (VRT) has been evolved. Variable-rate technology combines a variable-rate (VR) control system with application equipment to apply inputs at a precise time or location to achieve site-specific application rates of inputs. VR mounted on equipment permits input application rates to be varied across fields in an attempt to site-specifically manage field variability. This type of strategy can reduce input usage and environmental impacts along with increasing efficiency and providing economic benefits. A VRT essentially comprises of a Locator (DGPS receiver), control computer and Actuator (VRA software and controller) which are integrated to make VRT work. Three different approaches exist to implement VRT: map-based, sensor-based and manual. The intra and inter field variability can be characterized by a number of factors like climatic conditions (hail, drought, rain etc.), soils (texture, depth, nitrogen levels), cropping practices (no-till farming), weeds

[1] Division of Agricultural Engineering, FoA, Main Campus, SKUAST-J, Chatha, Jammu-180009
[2] Asstt. Agronomist, Punjab Agricultural University Research Centre, Bhatinda (Punjab) INDIA
[3] Professor, Division of Fruit Science, SKUAST-J, Main Campus, Chatha-180009.Division of Agricultural Engineering, FoA, Main Campus, SKUAST-J, Chatha, Jammu-180009

and disease. These factors can be assessed by two types of indicators (Permanent indicators and point indicators). The permanent indicators are manly soil indicators which provide farmers with information about the main environmental constants. The point indicators helps them to track a crop's status *i.e.* to see whether crop is suffering from water stress or lodging, is developing diseases, if there is nitrogen stress, what is the weeds density. This information may come from weather stations and other sensors (soil electrical resistivity, detection with the naked eye, satellite imagery etc.). Soil resistivity measurements combined with soil analysis make it possible to precisely map agro-pedological conditions. By catering to this variability one can improve the productivity or reduce the cost of production and diminish the chance of environmental degradation caused by excess use of inputs.

INDIAN SCENARIO

India is a vast agrarian country with total land area of 329 million ha and arable land area of 166 million ha. The population of India is growing at the rate of 1.7 % per year. The required food grain production by the year 2025 is 300 million ton that will necessitate further intensification of cropping system (Anonymous, 1996). The horizontal expansion of cultivable lands is going to be limited day by day due to increasing population and industrialization. Thus, the only option left is to increase the vertical yield, which is possible only by adopting the intensive cropping system. Food production in India has increased spectacularly from 69 million tones in 1965 to 242 million tones at present due to adoption of intensive cropping system and other improved technologies of crop production. Now there are reports of increasing problems of environmental degradation in agricultural fields, especially in intensively cultivated regions such as the Indo-Gangetic plains. There is a great concern now about decline in soil fertility, decline in water table, rising salinity, resistance to many pesticide increase pesticide residues in Environment (Soil, Water and Air) and degradation of irrigation water quality (Anonymous, 2006). Today, over-exploitation of ground water is a serious problem in many regions of India. The water table in the north western part of India, *for example*, is receding at an annual rate from 0.2 to 0.5 m. Soil salinity and water loggings are the other problems that have already spread to several parts of the India like IGP. In some part of India like Punjab, the mean fertilizer use in rice and wheat has become much higher than the recommended dose such uses can sometimes result in increased NO_3 concentration in groundwater. World agriculture contributes about 4 per cent of total global carbon dioxide emissions, the most important greenhouse gas. Other green house gases contributing to atmosphere due to intensive agriculture practice are methane and nitrous oxide (Anonymous, 2006).

The uniform rate technology followed in conventional agriculture aggravate the problem of low productivity and environmental degradation. The conventional agriculture practices follow a standard management option for a large area, irrespective of the variability occurring within and among the field. Farmers have been applying various agricultural inputs like seed fertilizers, weedicides, pesticides and water, based on recommendations emanating from research and field trials and under specific agro-climatic conditions which have been extrapolated to a regional level. Since soil types, nutrient status, weeds density, pests infestation and soil water vary not only between regions and between farms but also from plot to plot within a field, so there is a need to take into account such variability for an intended crop. The goal is to obtain the efficient use of applied inputs to reduce any excess application that might cause environmental pollution & to improve cost of production. The input use efficiency can be improved by minimizing losses of water nutrient and avoiding excess use of fossil fuel by following conservation tillage practice and optimal allocation of these inputs synchronizing with the demand of the crops and cropping systems, instead of piecemeal approach of improving single input use efficiency there is need to find out synergistic combination among different inputs and integrated use of different inputs in a holistic manner to improve the overall input use efficiency. Variable rate technologies needs to be developed for maximizing the use efficiency of different inputs under integrated input management for different crops and cropping systems under different agro-climatic regions and soil types through multi-disciplinary approach. The VRT is the application of technologies and principles to manage spatial and temporal variability associated with all aspects of agricultural production for the purpose of improving crop performance and reducing environmental degradation/pollution.

VARIABILITY AND FACTOR AFFECTING VARIABILITY

The earth's surface could have been described simply, if it were the same every where. The environment however, is not like that, there is almost endless variety (Wenster and Oliver, 1990). The spatial and temporal variability of soil fertility, weeds and diseases exist as intra and inter fields. This intra and inter-field variability may result from a number of factors. These include, (i) Climatic conditions – These includes hail, drought rain etc., (ii) Soils – These includes texture, structure, depth, nitrogen levels etc. (iii) Cropping practices like no-till farming.

Variable Rate Technology requires information on spatial and temporal variability data on the Soil (Soil Texture, Soil Structure, Physical Condition, Soil Moisture, Soil Nutrients etc.), Crop (Plant Population, Crop Tissue Nutrient Status, Crop Stress, Weed Patches, Insect or Fungal Infestation, Crop Yield etc) and Climate (Temperature, Humidity Rainfall, Solar Radiation, Wind Velocity etc.).

NEED FOR VARIABLE RATE TECHNOLOGY

Variable rate technology is for assessing and managing field variability by doing the right thing in the right place at the right time. It increasing the effectiveness and efficiency of input use by maximizing use of minimum land units resulting into higher productivity and reducing the environmental pollution.

COMPONENTS OF VARIABLE RATE TECHNOLOGY

Variable rate technology combines a variable rate (VR) control system with application equipment to apply inputs at a precise time or location to achieve site specific application rates of inputs. VRT essentially comprises of following components. (i) Remote Sensing, (ii) Geographic Information Systems, (iii) Differential

Global Positioning System, (iv) Computer, (v) Software, (vi) Controller, (vii) Hydraulic Motor and Valve, and (viii) Control or Metering Mechanism.

REMOTE SENSING

The data of the farms is acquired by remote sensing to find the soil, vegetation and other parameters that are amenable. Remote sensing techniques play an important role in variable rate technology by providing continuous acquired data of agricultural crops. Remote sensors image vegetation which is growing on different soil types with different water availability, substrate, impact of cultivation and relief. These differences influence the state of the plants and cause heterogeneous regions within single fields. Hence, the heterogeneous vegetation acts as an interface between soil and remote sensing information, because vegetation parameters describing the state of the plants can be deduced from remote sensing imagery. The analysis of the variability occurring within the field was carried out by measuring soil and plant parameters through conventional methods as well as through spectral techniques using ground truth spectro-radiometer (350-1800 nm) and satellite data.

GEOGRAPHIC INFORMATION SYSTEMS

The Geographic Information Systems contributes significantly to VRT by allowing presentation of spatial data in the form of a map. In addition, GIS forms an ideal platform for the storage and management of model input data and the presentation of model results which the process model provides. Formally, GIS is an organized collection of computer hardware, software, geographic data and personal designed to efficiently capture, store, update, manipulate, analyze and display all forms of geographically referenced information (Anonymous,1997) Environmental Systems Research Institute.

DIFFERENTIAL GLOBAL POSITIONING SYSTEM

Differential Global Positioning System ensures doing the right thing, in the right place and at right time. The accuracy which is the important factor in VRT, demands for DGPS. The Global Positioning System technology provides accurate positioning system necessary for field implementation of VRT. It provides position information of the locations in the field which is used by the control software to adjust rates based on prescription map. It provides the ability to spatially log the actual rates applied for the generation of as applied maps. A GPS receiver with differential correction (WAAS, Starfire, OmniStar, RTK etc.) is usually used (Fig. 6.1).

(Source: Fulton *et al.*, 2009)

Fig. 6.1: *View of Differential Global Positioning System*

COMPUTER

A laptop, handheld computer or other computer system is used (Fig. 6.2) whereas some equipment manufacturers produce their own systems such as John Deere. This component serves as two purposes the user interface and has the ability to run the application control software.

(Source: Fulton *et al.*, 009)

Fig. 6.2: *View of Computer with Controller and Software*

SOFTWARE

It reads the uploaded prescription map, analyses it and determine the desired application rate based on field position and logging. Further it communicates the desired application rates to the controller.

CONTROLLER

The desired application rates communicated by software is processed and controlled by controller. It can be a separate system from the software and control mechanism. It implements the set point rate communicated by the software and ensures that the control mechanism (motor or actuator) is putting out the appropriate rate. It also uses feedback from ground speed radar (GSR) or other speed sensor to compensate for speed variations while also using a speed or position feedback from the control mechanism to ensure it is turning at the appropriate speed or positioned correctly.

HYDRAULIC MOTOR AND VALVE

The metering unit can be controlled by a motor, linear actuator, or another control device. The hydraulic motor uses a hydraulic control valve to adjust the flow rate to the motor thereby controlling speed of the motor (Fig. 6.3).

(Source: Fulton *et al.*, 2009)

Fig.6.3. *Hydraulic Motor and Valve*

CONTROL OR METERING MECHANISM

The feed rate is directly controlled by metering mechanism. Examples of metering mechanisms would be an apron chain, seed disk, or liquid injection system (Fig. 6.4).

(Source: *Fulton et al., 2009*)

Fig. 6.4: *Control or Metering Mechanism*

The computer coordinates the field operation. It has a map of desired activity as a function of geographic location. It receives the equipments current location from the locator which has a GPS in it and decides what to do based upon the map in its memory or data storage. It then issues the command to the actuator which does the input application (Ravi and Jagadeesha, 2002)

ASSESSING VARIABILITY

Assessing variability is the critical first step in VRT since it is clear that one cannot manage what one does not know. The processes and properties that regulate crop performance and yield vary in space and time. Adequately quantifying the variability of these processes and properties and determining when and where different combinations are responsible for the spatial and temporal variation in crop yield is the challenge in VRT (Mullu and Schepers, 1997). A yield map defines the spatial distribution of crop yield but does not explain the observed variability. Imagery of crop growth and development over the growing season can uncover the cause-effect relationship that explains not only the yield variation within a field but also the magnitude of yield observed I a particular growing environment (Schepers *et al*, 1996). For VRT agriculture to be useful, variation must be known of sufficient magnitude, spatially structures (on random) and manageable. Maps form one basis for precision management: real-time management forms the other basis. Use of management maps is more common and these can be categorized as condition maps, prescription maps and performance maps. Condition maps are measured or predicted using a broad array of technologies and techniques for estimating the spatial distribution of one or more properties of processes. Prescription maps are derived from one or more condition maps and form the basis for VRT (Sawyer, 1994). Performance maps record either inputs (fertilizers, pesticides, seed, energy etc.) or outputs (crop, yield and quality) and include derivatives of performance maps, such as profit maps (output-inputs).

Condition maps are a critical component of VRT and can be generated in your major ways, (i) Surveys, (ii) Interpolation of a network of point samples, (iii) High resolution sensing and (iv) Modeling to estimate spatial patterns.

Surveys

Surveys are purposeful inventories of specific quantities and have been particularly useful in natural resource management. A first glace, the soil survey should be an important assets to the principles of VRT agriculture. Farming by soils was initially thought to be a reasonable basis for precision management (Carr *et al.,* 1991). However, existing soil surveys have proved of limited value in explaining spatial variability observed within fields. Mausbach *et. al.,* (1993) state that the soil survey and its interpretations are not site specific. The value of the soil survey to VRT agriculture could be improved by intensifying map scales to fine scale resolutions needed in detailed environmental modeling applications or site-specific management (Moore *et. al.,* 1993). Nonetheless, new data sources (*e.g.* digial orthophotos, airborne and satellite imagery and yield maps) and analysis techniques (*e.g.* terrain analysis and yield maps) make it possible to map soils at needed resolution (Bell *et. al.,* 1995).

Interpolation of Point Samples

Another technique for assessing spatial variability involves sampling process. A network of points in some spatial arrangement is sampled and then interpolated to produce a spatial estimation (usually a map) of the whole area using a range of statistical procedures. These spatial statistical techniques can also be repeated over time to estimate the temporal variability (Mc Bratey *et al.,* 1997 and Stein *et al.,* 1997). To a large extent, sampling depends on the nature of the entity of interest. Soil sampling for soil survey, *for example,* is used to determine how much of the land is of a particular type or what proportion possesses some soil attribute noticed by Webster and Oliver (1990). For pest management, interest may be in obtaining insect pest density maps either within a field or within a region and overtime (Fleischer *et al.,* 1997). The goal of network or spatial sampling for precision agriculture is to provide an accurate and affordable map of the occurrence of a specific parameter to be managed. What this parameter is will depend on the nature of the cropping system and its biophysical context. Three important issues need to be addressed regarding spatial sampling for assessing variability; Sample unit, sample design arrangement and intensity and map accuracy.

High Resolution Sensing

Yield maps do not indicate the causes of the yield magnitude or its variability. However, it is the courses of variability that needs to be quantified if farmers are to adjust their management practices to specific conditions within a field at appropriate times during the growing season. High resolution remote sensing of the growing crop will reveal stresses that impact the crop during the growing season (Scheper *et al.,* 1996). Additionally, it is not physically or economically possible to accurately map certain soil properties, crop condition, or pest status without the use of high resolution sensing. The lower cost and ease of measure of high resolution sensors will be critical to the future success of VRT agriculture.

Modeling

Modeling is proposed as a important tool in precision agriculture to stimulate spatial and temporal variation I soil properties (Verhagen and Bouma, 1997), Pests (Kropff *et al.*, 1997), crop yield (Barnett *et al.*, 1997) and environmental performance of cropping systems (Verhagen *et al.*, 1995). Models have been developed and calibrated for specific purposes but have not been used extensively in spatial prediction. A major problem of models is the availability of inputs needed to run them.

Variable Rate Technology Implementation

The Variable Rate Technology can be implemented by three different approaches based on requirement and availability.

1. Map based
2. Sensor based
3. Manual

Map Based

The map based approach to VRA (variable rate application) is popular since soil testing analyses and other information can be used to easily develop the prescription maps. A prescription map is generated based on soil analyses or other information and then used by the VRT to control the desired application rate. Prescription maps can also be developed using additional information such as yield maps, historical management, soil texture or type, remote sensing (aerial or satellite), soil electrical conductivity and terrain features (elevation, slope etc.) The soil sample is usually named 'Farm Name - Field Name – Zone Name.' When soil analysis data is returned from the lab, the same naming convention is used, therefore allowing the GIS to match the data to the appropriate zone (Fig. 6.5). These maps can be generated using agricultural GIS software packages and can be either grid or zone-bases.

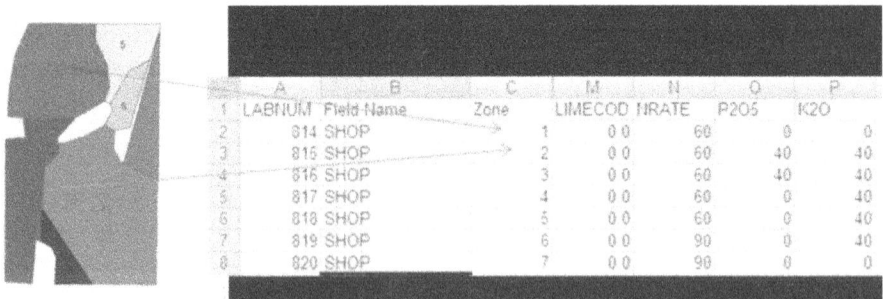

(Source: Norwood and Fulton, 2009)

Fig. 6.5: *Example of a prescription map with 7 defined zones. The spreadsheet on the right contains the soil test data with arrows indicating how zones are linked to the soil test results.*

Zone-Based

When zone maps are created, each zone within the field is automatically assigned a number (Fig 6.6). Zones can be used for numerous applications, such as: variable rate fertilizer (broadcast and sidedress) plant growth regulators, cotton defoliants, seeding rates.

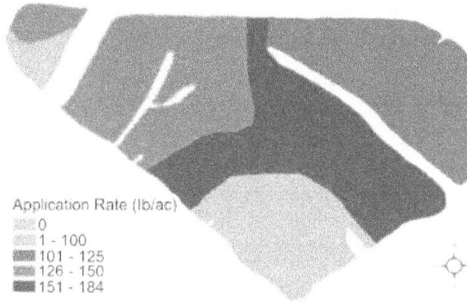

Application Rate (lb/ac)
0
1 - 100
101 - 125
126 - 150
151 - 184

(Source: Fulton, 2009)

Fig. 6.6: *An example of a zone-based prescription map.*

Grid-Based

In case of grid sampling, the boundary is typically divided into one ha grids using a desktop GIS or a PDA (handheld computer) loaded with an agricultural GIS software package (Fig. 6.7). Grids are usually arbitrary and do not account for any past or current knowledge of the field. Each grid is assigned a number, sometimes referred to as 'Feature ID.' Feature ID is comparable to the 'zone' column in case of zone based maps and is how the GIS package links the soil test results back to the geographical location.

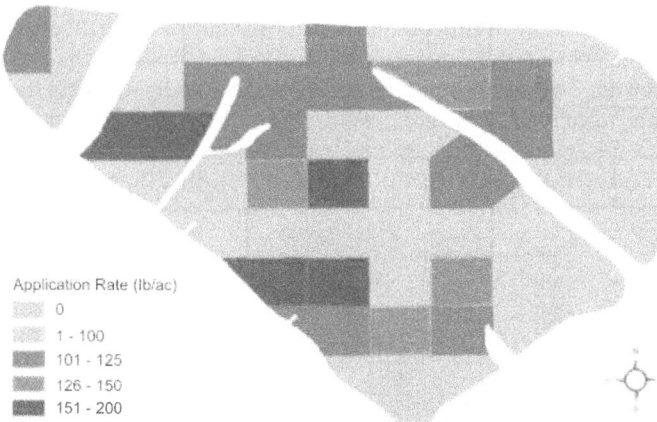

Application Rate (lb/ac)
0
1 - 100
101 - 125
126 - 150
151 - 200

(Source: Fulton, 2009)

Fig. 6.7: *An example of a grid-based prescription map*

Sensor-Based

The sensor-based systems are easier to use than most map-based systems besides this they can apply inputs such as nitrogen to better meet real time crop needs. The sensor based approach attempts to apply inputs more efficiently for meeting crop requirement at the time of application. This type of system utilizes sensors to assess crop or field conditions to provide real time variable rate application (VRA) of inputs. Sensors are devices that transmit an impulse in response to a physical stimulus such as heat, light, magnetism, motion, pressure and sound. Sensors can be contact or remote, ground based or space based and direct or indirect. Sensors have been developed to measure machinery, soil, plants, pests, atmospheric properties and water by sensing motion, sound, pressure, strain, heat, light and magnetism and relating these to properties such as reflectance, resistance, absorbance, capacitance and conductance. Several commercially available sensor-based, variable rate systems exist for efficiently managing inputs to maximize yields or returns. Sensors are critical to success in the development of a VRT for three important reasons.

(a) Sensors have fixed costs.

(b) Sensors can sample at very small scales of space and time.

(c) Sensors facilitate repeated measures.

Manual

Manual control can also be used to vary the application rates with the operator responsible for changing the rates on the controller during operation.

Comparison of Map with Sensor–Based Application

Variable rate application using precision agriculture technologies has generally followed one of two paths. One based entirely on map–based information and the other based on real–time sensors. The map–based approach allows use of historical information, while sensors allow us to assess in–season conditions. Map–based information is typically gathered with yield monitors, soil testing, soil maps or with data from sensors. The primary differences between map and sensor–based strategies are data analysis and interpretation. With map–based variable rate application, the practitioner must collect and analyze data for input to an expected crop response algorithm and then transfer the prescription map to a variable rate applicator. The prescription may appear as zones as shown in Fig 6.8 or in a grid format with smoother transitions. A GPS receiver locates the applicator's position on the map and the rate is then adjusted based on the prescription map as the applicator moves across the field.

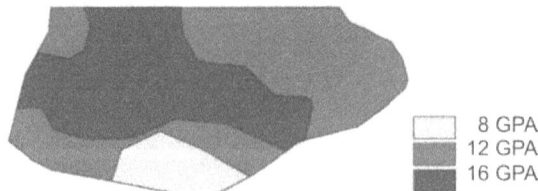

8 GPA
12 GPA
16 GPA

(Source: Taylor and Fulton, 2012)

Fig. 6.8: *Prescription map for a variable rate application using three distinct zones.*

The sensor–based approach to precision agriculture uses sensors to measure crop or soil properties in real–time as the applicator moves across the field. Data from the sensor is collected, processed and interpreted by an on–board computer which then sends a signal to a rate controller. One of the advantages of this approach is automating the data analysis and interpretation step versus the map–based strategy. A predetermined algorithm is used to convert the sensor information directly to an application rate. This algorithm is typically constant at a field scale and often at the regional scale. However, one challenge associated with this approach is that the prescribed rate is constantly changing (Fig. 6.9) as the applicator moves across the field requiring the rate controller to respond quickly.

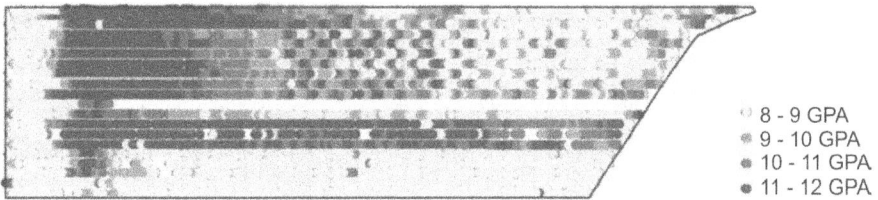

8 - 9 GPA
9 - 10 GPA
10 - 11 GPA
11 - 12 GPA

(Source: Taylor and Fulton, 2012)

Fig. 6.9: *Prescription map for sensor–based variable rate application*

CALIBRATING EQUIPMENT WITH VARIABLE-RATE TECHNOLOGY

Calibration of spreaders, planters, and sprayers is needed for uniform application, it is even more critical to calibrate VRT controlled equipment. The goal of calibration is to minimize application errors so target rates can be achieved with a certain level of confidence. The following details suggestions on how one might calibrate equipment with VRT.

Safety

First and foremost, be safe when calibrating equipment due to the nature of the materials being handled. Carefully read all warning and caution labels found on the containers of agricultural chemicals and products being used. For sprayers, it is recommended to only use water in the system unless the spray mixture output rate varies more than 5% from the output rate of water. For planters and spreaders, wear the necessary safety attire as recommended by product manufacturers.

Ground Speed Radar

VRT usually relies on either ground speed radar or a global positioning system (GPS) receiver as input for actual ground speed. If a GSR is used, yearly calibration of this sensor is important so the variable-rate controller will properly adjust application rates with ground speed variations. An improperly calibrated GSR will lead to application errors. Read either equipment or GSR literature for the proper calibration procedure. Typically, it is suggested to use a 400-feet run on level ground during GSR calibration. A second pass can be used to double-check the calibration.

(a)   (b)

(Source: Fulton *et al.*, 2009)

Fig. 6.10: *Illustration of Ground Speed Radar (a) and a course laid out for calibration (b)*

Expected Application Range

Prior to calibration of equipment with VRT, one must determine the range of planned application rates for the product(s) to be used during VR application. This range will be used to ensure that the software and hardware setup will properly operate over the expected range of rates once an acceptable setup is determined during calibration. *For example*, you might decide to vary your corn seeding rates from 24K to 36K seeds/ac. knowing this information, select the median rate (30K) to start calibrating the planter equipped with VRT.

Pre-Calibration Checks

Make sure all hardware and software are in proper operating condition. Replace any worn hardware, especially those controlling the metering and distribution of material. Pre –calibration checks are different for different applicators:

Granular / Dry Applicators

For spinner spreaders, this includes the divider, spinner-discs, and especially the spinner fins. For pneumatic applicators, check the metering mechanism, fan, tubes, and deflectors. Worn tubes can have an impact on material transportation to the deflectors while worn deflectors can impact distribution.

Sprayers

Select a nozzle that is capable of handling the expected application rates, spray pattern, ground speed, and pressure.

Planters

Check all drives and individual metering element(s) on each row unit. Ensure that the tank/bin is at least half full. If rates are going to be varied over a wide range, especially for lime application, check the distribution pattern at the median rate followed by a check at the minimum and maximum application rates. This procedure will ensure proper distribution over the range of desired application rates. If a distribution issue exists, the rate variation should be limited to a smaller range.

Calibration

Dry Product Density

If the intentions are to use variable-rate blended products, then make sure the products have similar density and particle size. It results into symmetric patterns over the expected application range (Fig. 6.11b). It is not advisable to blend products if these differences exist since pattern uniformity will be greatly affected over the range of application rates (Fig. 6.11a). This issue is more important for spinner spreaders than pneumatic applicators. Perform individual distribution tests for different granular products especially if they have different density and particle sizes. *For example*, potash will require a different calibration than lime.

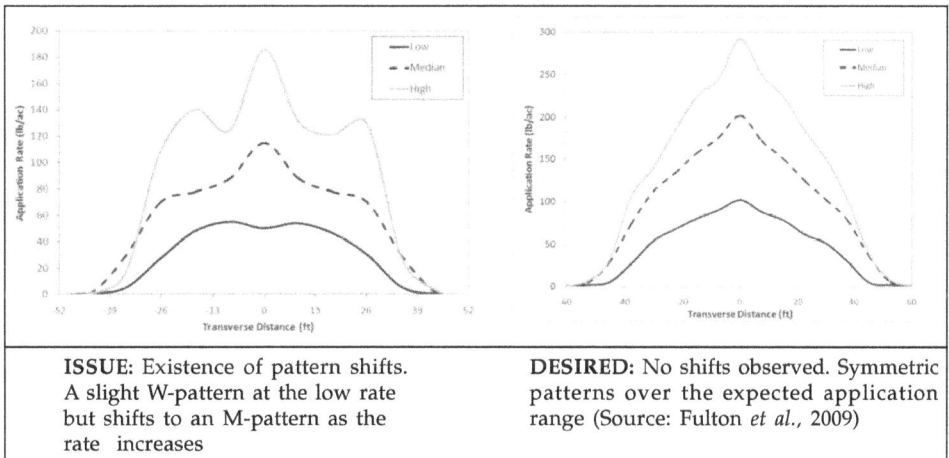

ISSUE: Existence of pattern shifts. A slight W-pattern at the low rate but shifts to an M-pattern as the rate increases	DESIRED: No shifts observed. Symmetric patterns over the expected application range (Source: Fulton *et al.*, 2009)

Fig. 6.11: *Distribution pattern results for a spinner spreader; a) undesirable pattern shifts and b) desired symmetric patterns.*

Sprayers

Take any pressure readings at the boom. Ensure nozzles are within 5% of the manufacturers operating specifications. If the spray mixture is significantly altered by the addition of adjuvants, it is recommended to compare the output mixed rate to water and ensure that the values are within 5 per cent. If 5 per cent exceeded, calibration must be conducted utilizing the spray mixture.

Planters

Ensure proper seed drop and that the planter settings/configuration can cover the range of expected seeding rates (Fig. 6.12). If the application range is too broad for the setup, then application at the upper and lower rates may deviate too much from the desired application rates and therefore generate application errors. For example, at planting, you plan to vary the seeding rates between 0 and 50,000 seeds/acre. The error in the 1-10,000 and 40,000-50,000 densities may be very high. Therefore, it is beneficial to narrow the range as much as possible based on the predetermined application rates. Along with a narrowed range, results can also be improved by conducting calibration at the median rate, or the most utilized rate. Once the VRT component is calibrated, the upper and lower rates of the narrowed range should be checked to ensure that calibration is between 5 per cent and 10 per cent.

(Source: Fulton *et al.*, 2009)

Fig. 6.12: *Planter calibration ensuring proper seed drop*

1. Application of Variable Rate Technology

Using sensor systems for variable rate application (VRA) is becoming more popular for cotton production (Taylor and Fulton, 2006). Three critical inputs for cotton production namely plant growth regulators, defoliants/boll openers and nitrogen can be very well monitored with the variable rate application technology.

Equipment

The equipment will consist of sensors and control interface, a display and control module, and an application rate controller (Fig. 6.13). Sensors can be used to measure soil or crop properties that are used to make in–season variable rate application to cotton.

Sensors and Vegetative Indices

Commercially available sensors operate above the crop and measure reflectance of different colours (wavelengths). The most common colours used for crop vegetation indices are red, near–infrared (NIR), green, and amber. Healthy, vigorous plants absorb red light and reflect near infrared light. The reflectance of these colours is used to calculate indices such as the Normalized Difference Vegetative Index (NDVI) or a simple ratio (RED/NIR). NDVI is determined from red and near infrared reflectance and is probably the most popular vegetative index. Other colours such as green or amber can be used in place of red when calculating NDVI. Though many vegetative indices can be used for VRA, only NDVI will be discussed for the remainder of this publication.

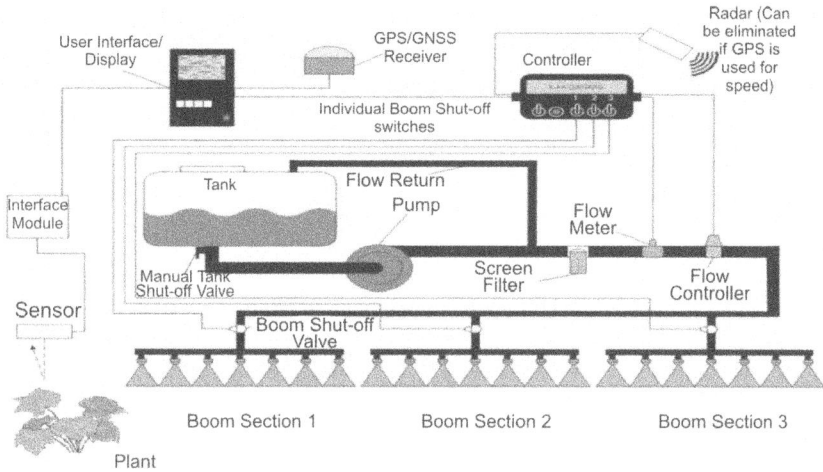

(Source: Taylor and Fulton, 2012)

Fig. 6.13: *Schematic of a sensor–based, variable rate application system for liquid products.*

Research has shown that reflective indices such as NDVI measured at the correct growth stage can be highly correlated with cotton yield. While NDVI values can range from 0 to 1, values below 0.3 or above 0.9 are of little value in crop production. When NDVI is below 0.3, there is generally not much green in the field of view (*i.e.,* more stubble, crop residue, or soil). Conversely, when values are greater than 0.9, everything in the field of view is green.

Prescriptions

Developing a sensor–based prescription happens in two steps. First, we have to determine the relationship between the plant property of interest and what the sensor measures. We must do the first step because commercially available sensors may not measure the plant property that is typically used to determine an application rate (*for example* plant height or per cent open bolls). The data shown in Fig. 6.14 is an example of developing the relationship between the sensor reading (NDVI) and per cent open bolls through research data. While the relationship is not perfect the trend is obvious and could likely be applied over a wide range of conditions with some acceptable error. The second step is to determine the application rate as a function of the sensor reading. For this we would convert the application rate for a given per cent open bolls to an application rate for the corresponding NDVI.

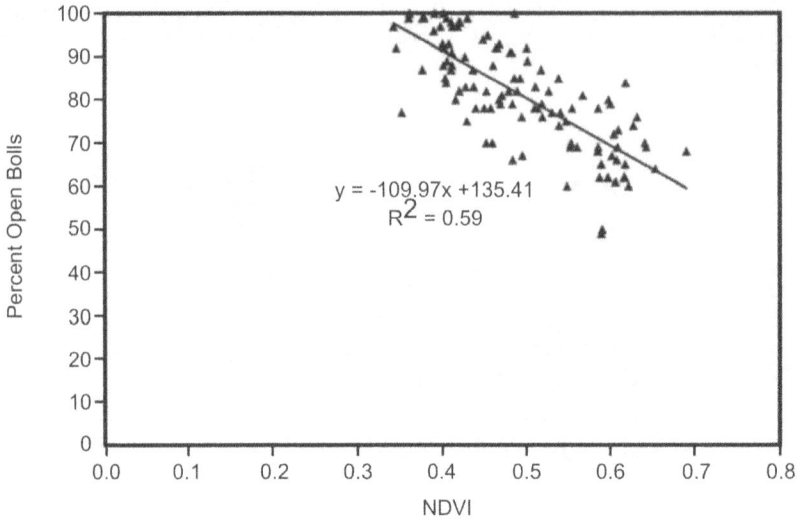

(Source: Taylor and Fulton, 2012)

Fig. 6.14: *Per cent open bolls as a function of NDVI taken from a long term fertility study in Altus, Okla. Data is combined from two sampling dates.*

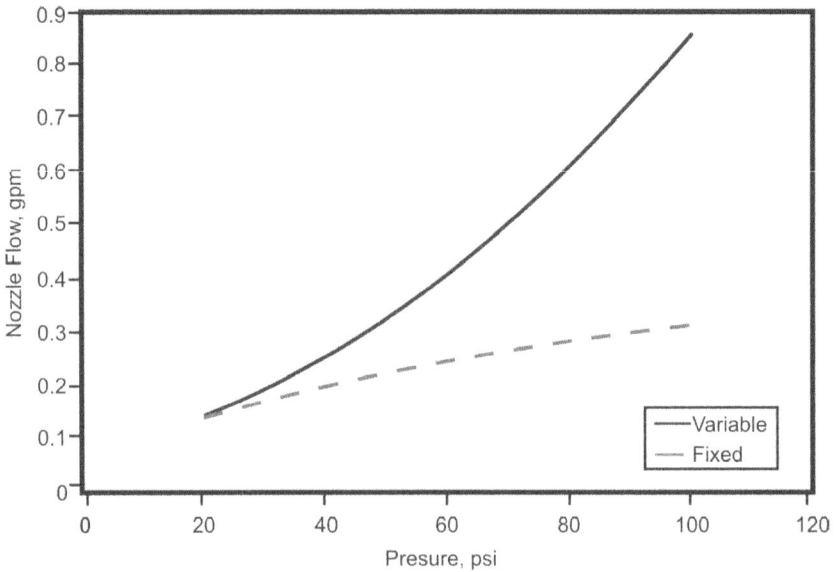

(Source: Taylor and Fulton, 2012)

Fig. 6.15: *Expected nozzle flow for fixed and variable orifice nozzles as a function of pressure. Note that from 20 to 100 psi, flow through the fixed orifice nozzle barely doubles, while flow through the variable orifice nozzle increases by almost a factor of 6.*

From a farmer or consultant perspective, the two steps that are outlined for developing prescriptions could be combined if the prescription is developed at the field level. You could measure NDVI at a location within a field and evaluate the crop to determine the application rate for the product at that location. Repeating this at various locations would yield a direct relationship between NDVI and the application rate. Another factor when determining prescriptions is whether upper or lower limits are desired. In other words do you want to set the maximum or minimum application rates so the system is restricted between these? *For example,* you may not feel comfortable applying less than 50 pounds per acre of nitrogen, so you set the lower limit at this rate. Then regardless of the prescription and sensor reading, the controller will not go below this rate. The same approach can be taken for the upper limit. These limits are established to ensure every part of the field receives some product, but that no part of the field receives excessive application of product.

There are three primary cotton inputs that can be variably applied with a sensor–based system: plant growth regulators, harvest aids, and nitrogen.

Plant Growth Regulators

The prescription for plant growth regulators (PGRs) is likely based on the relationship between NDVI and plant height or the height to node ratio. The NDVI typically increases with plant height. Therefore if the prescription is based on plant height, NDVI become a natural substitute. A PGR prescription based on NDVI is presented in Figure 6.16. Note the upper (10 gallons per acre [gpa]) and lower (5 gpa) limits on application rate in PGR prescription. Regardless of how low the NDVI may be, the prescribed rate will not go below 5 gpa. Likewise the prescribed rate will never exceed 10 gpa.

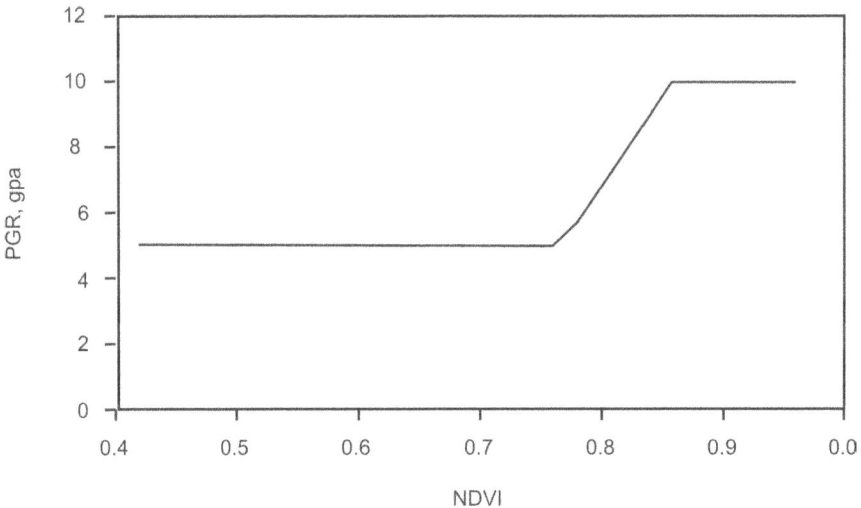

(Source: Taylor and Fulton, 2012)

Fig. 6.16: *An example variable rate prescription for a PGR as a function of NDVI*

One challenge associated with variable PGR application is that plant height and NDVI are related early in the season. However, as the canopy begins to close the sensor measured NDVI may reach a plateau and stay the same while plant height continues to increase. When this happens, it may be challenging to variably apply PGRs with a sensor system. Basically the entire field would receive the maximum prescribed rate applied uniformly.

Harvest Aids

The prescription for harvest aids such as defoliants and boll openers is likely based on the relationship between NDVI and per cent open bolls or nodes above cracked boll. The NDVI typically decreases as the per cent open bolls increases. Though the sensor is not actually measuring open bolls, NDVI is measuring something in the plant, like the natural desiccation, that is associated with open bolls. Therefore if the prescription is based on open bolls, NDVI become a natural substitute (Fig. 6.17). This prescription shows a linear increase in harvest aid application rate as NDVI increases. As with the **PGR** prescription in Figure 6.16 upper and lower limits were used to insure all areas receive some harvest aid, but no area gets more than the maximum rate established. In the case of harvest aids, NDVI can be an indicator of two things: more biomass or greenness in the plants and thus more leaves. In either case, a higher NDVI would indicate a greater need for harvest aids.

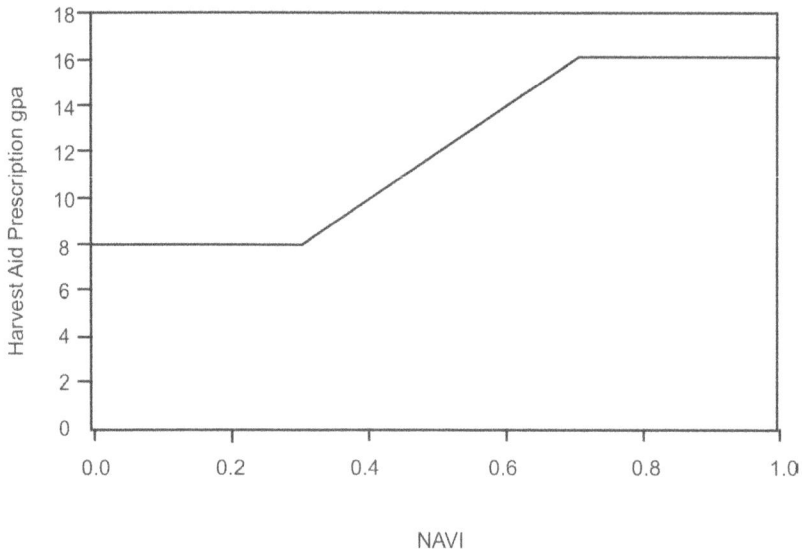

NAVI

Fig.6.17: *A harvest aid prescription based on NDVI. The lower limit is set at 8 GPA whereas the upper limit is set at 16 GPA.*

Nitrogen

Developing a prescription for sensor–based variable rate nitrogen application is more complex than PGRs or harvest aids. However, several universities have developed different approaches to sensor–based variable rate nitrogen on cotton. These are usually for side–dress application from first square to early flower. Some of these use a nitrogen rich reference strip that was developed for cereal grain production. The reference strip is an area where sufficient nitrogen is applied to insure that it is not limiting plant growth. This strip is then used to determine the environmental contribution of nitrogen or the maximum yield if nitrogen is not limiting growth prior to field application. Two variable rate prescriptions for nitrogen are shown in Fig. 6.18. This data are presented only to illustrate the conceptual differences between different approaches. The OK method is based on yield potential and a nitrogen rich strip. The MO method is based solely on the nitrogen reference strip. The MO method applies a high N rate on the lower NDVI areas of the field and no N on the higher NDVI areas. The OK method is more complex, but this doesn't affect the end–user because it is programmed in the on–board computer. The user needs to know the NDVI of the reference strip, the NDVI of the adjacent area, growing degree days, and the maximum yield potential. The limits of each prescription (maximum and minimum N rate and NDVI thresholds) are set based on regional experience or user preference. *For example*, cotton with an NDVI less than 0.3 will likely not respond to extra nitrogen.

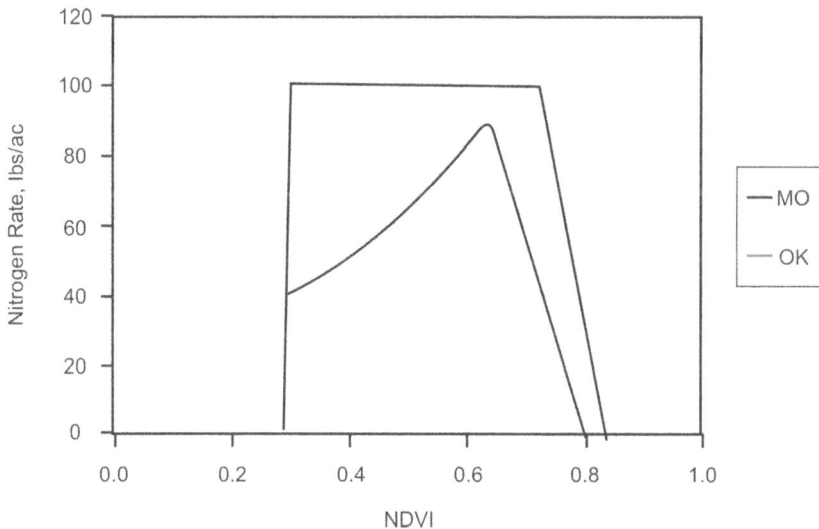

Fig.6.18: *Nitrogen rate derived from two nitrogen prescriptions based on: Missouri (MO) and Oklahoma (OK) methods. Note that in both cases, the prescription requires more information than NDVI alone.*

STEPS TO BE TAKEN FOR IMPLEMENTING VRT IN INDIA

In the present existent situation, the potential of VRT in India is limited by the lack of appropriate measurement and analysis techniques for agronomically important factors (Anonymous, National Research Council, 1997). High accuracy sensing and data management tools must be developed and validated to support both research and production. The limitation in data quality/availability has become a major obstacle in the demonstration and adoption of the variable rate technologies. The adoption of variable rate technologies needs combined efforts on behalf of scientists, farmers and the government. The following methodology could be adopted in order to operationalise precision farming in the country.

(*i*)   Creation of multidisciplinary teams involving agricultural scientists in various fields, engineers, manufacturers and economists to study the overall scope of variable rate technologies.

(*ii*) Formation of farmer's co-operatives since many of the variable rate technologies tools are costly (GIS, GPS, RS, etc.).

(*iii*) Government legislation restraining farmers using indiscriminate farm inputs and thereby causing ecological/environmental imbalance would induce the farmer to go for alternative approach.

(*iv*) Pilot study should be conducted on farmer's field to show the results of variable rate technologies implementation.

Creating awareness amongst farmers about consequences of applying imbalanced doses of farm inputs like irrigation, fertilizers, insecticides and pesticides.

LIMITATION OF ITS IMPLEMENTATION IN INDIA

VRT has bee mostly confined to developed countries. Reasons of limitations of its implementation I developing countries like India are:

(*i*)   Small land holding

(*ii*) Heterogeneity of cropping systems and market imperfections.

(*iii*) Lack of technical expertise knowledge and technology.

(*iv*) India spends only 0.3 per cent of its agricultural gross domestic product in research and development.

(*v*)   High cost of technology

(*vi*) No custom hiring operators.

Scope of VRT in India

In India, major problem is the small field size. More than 58 per cent of operational holdings in the country have size less than 1 ha. Only in the states of Punjab, Rajasthan, Haryana and Gujarat more than 20 per cent of agricultural lands have operational holding size of more than four hectare. There is a scope of implementing VRT for crops like rice and wheat especially in the states of Punjab and Haryana. Commercial as well as horticultural crops also show a wider scope for precision agriculture in the cooperative farms.

REFERENCES

Anonymous, (1996). The Production Yearbook. FAO, Rome, Italy.

Anonymous, (1997). National Research Council.

Anonymous, (1997). Understanding GIS: The ARC/INFO Method." Environmental Systems Research Institute (ESRI), ESRI/Wiley. Redlands. CA/New York.

Anonymous. (2006). Handbook of Agriculture, India Council of Agricultural Research

New Delhi, Fifth Edition, Chapter 3. Agriculture and Environment pp: 55-71.

Barnett V, Landau S, Colls JJ, Craigon J, Mitchell RAC and Payne RW, (1997). Predicting wheat yields: The search for valid and precise models. In "Precision Agriculture: Spatial and Temporal Variability of Environmental Quality" (J. Y. Lake. G. R. Bock, and J. A. Goode. Eds.), pp. 79-92. Wiley. New York.

Bell JC, Butler CA, Thompson JA. (1995). -Soil-terrain modeling for site-specific management. In "Proceedings of the Second International Conference on Sit Specific Management for Agricultural Systems, Bloomington/Minneapolis, MN, 27-30 March 1994" (P. C Robert, R. H. Rust. and W. E. Larson, Eds.), ASA Miscellaneous Publication. pp. 209-227 ASA. CSSA. and SSSA. Madison, WI.

Carr PM, Carlson GR, Jacobsen JS, Nielsen GA and Skogley EO, (1991). Farming soils, not fields: A strategy for increasing fertilizer profitability. J. Prod. Agric. 4, 57-61.

Fleischer SJ, Weisz R, Smilowitz Z and Midgarden D, (1997). Spatial variation in insect popu-lations and site-specific integrated pest management. In "The State of Site-Specific Management for Agriculture" (E J. Pierce and E. J. Sadler. Eds.), ASA Miscellaneous Publication, pp. 101-130. ASA. CSSA. and SSSA. Madison, WI.

Fulton J, Christian B, Ted T and Winstead T. (2009). Calibrating Equipment with Variable-Rate Technology: Precision Agriculture Series, Timely Information, Agriculture, Natural Resources & Forestry, Alabama Cooperative Extension System

Kropff MJ, Wallinga J and Lotz LAP. (1997). Modeling for precision weed management. In "Precision .Agriculture: Spatial and Temporal Variability of Environmental Quality" (J V. Lake, G. R. Bock and J. A. Goode, Eds.), pp. 182-200. Wiley, New York.

Mausbach MJ, Lytle DJ and Spivey LD. (1993). Application of soil survey information to soil specific farming. In "Proceedings of Soil Specific Crop Management: A Workshop on Research and Development Issues, 14-16 April 1992, Minneapolis, MN" Eds. (Robert PC, Rust RH and Larson WE), PP' 57-68. ASA, CSSA, and SSSA, Madison, WI.

McBratney AB, Whelan BM and Shatar TM. (1997). Variability and uncertainty in spatial, temporal, and spatiotemporal crop-field and related data. In "Precision Agriculture: Spatial and Tem-poral Variability of Environmental Quality" (Lake JV, Bock GR and Goode JA, Eds.), pp 141-160. Wiley, New York. -

Moore LD, Gessler PE, Neilsen GA and Peterson GA., 1993. Soils attribute prediction using terrain analysis. Soil Sci. Soc. Am. J. 57,443-452.

Mulla DJ and Schepers JS., (1997). Key processes and properties for site-specific management. In "The State of Site-Specific Management for Agriculture" (F. J. Pierce and E.J. Sadler, Eds.), ASA Miscellaneous Publication. pp. 1-18. ASA, CSSA, and SSSA, Madison, WI.

Taylor R and Fulton J, (2012). Sensor–Based Variable Rate Application for Cotton. http:// www.agmachinery.okstate.edu/PrecisionAgTech/CottonApplicationOctober2010 Share.pdf

Ravi N and Jagadeesha CJ., (2002). Precision Agriculture, Training course on Remote Sensing and GIS Applications in Agriculture, May 27th –7th June, 2002, RRSSC- Bangalore, pp: 225-228.

Sawyer JE. (1994). Concepts of variable rate technology with consideration for fertilizer application, J. Prod. Agric. 7, 195-201.

Schepers JS, Blackmer TM, Shah T and Christensen N. (1996). Remote sensing tools for site-specific management. In "Proceedings of the Third International Conference on Precision Agriculture, Minneapolis, MN, 23-26 June 1993 315-319. ASA, CSSA, and SSSA, Madison, WI.

Stein A, Hoosbeek MR and Sterk G., 1997. Space-time statistics for decision support to smart farming. In "Precision Agriculture: Spatial and Temporal Variability of Environmental Quality). (J. V. Lake, G. R. Bock., and J. A. Goode, Eds. p.p. 120-130. Wiley. New York.

Verhagen J and Bouma J., (1997). Modeling soil variability. In "The State of Site-Specific Management for Agriculture (F. J. Pierce and E. J. Sadler, Eds.), ASA Miscellaneous Publication, pp. 55 -67. ASA, CSSA, and SSSA. Madison, WI.

Verhagen A, Booltink HW and Bouma J., (l995a). Site- specific management: Balancing production and environmental requirements at farm level. Agric. Systems 49, 369-384.

Wenster R and Oliver MA. (1990). "Statistical Methods in Soil and Land Resource Survey:' Oxford Univ. Press, New York.

Norwood SH and Fulton J., (2009). Introduction to Prescription Maps for Variable-Rate Application Alabama Cooperative Extension System Alabama A&M And Auburn Uinversities, And Tuskegee University, County Governing Bodies And Usda Cooperating

Fulton J, Brodbeck C Winstead A, Tyson T and Norwood S., (2009). Calibrating Equipment with Variable-Rate Technology.

■

7

CHAPTER

Remote Sensing and Image Processing

☞ Purshotam Singh[1], ☞ Parmeet Singh[2],
☞ Narinder Panotra[3], ☞ B.A. Lone[4],
☞ Sameera Qayoom[5], Gurdeep Singh[6] and K.N.Singh[7]

INTRODUCTION

Remote sensing (RS), also called earth observation, refers to obtaining information about objects or areas at the Earth's surface without being in direct contact with the object or area. Humans accomplish this task with aid of eyes or by the sense of smell or hearing; so, remote sensing is day-today business for people. Reading the newspaper, watching cars driving in front of you are all remote sensing activities. Most sensing devices record information about an object by measuring an object's transmission of electromagnetic energy from reflecting and radiating surfaces. Remote sensing techniques allow taking images of the earth surface in various wavelength region of the electromagnetic spectrum (EMS). One of the major characteristics of a remotely sensed image is the wavelength region it represents in the EMS. Some of the images represent reflected solar radiation in the visible and the near infrared regions of the electromagnetic spectrum, others are the measurements of the energy emitted by the earth surface itself *i.e.* in the thermal infrared wavelength region. The energy measured in the microwave region is the measure of relative return from the earth's surface, where the energy is transmitted from the vehicle itself. This is known as *active remote sensing*, since the energy source is provided by the remote sensing platform. Whereas the systems where the remote sensing measurements depend upon the external energy source, such as sun are referred to as *passive remote sensing* systems.

Remote sensing can be defined as any process whereby information is gathered about an object, area or phenomenon without being in contact with it. Our eyes are

[1] Associate Professor, Division of Agronomy , Sher-e-Kashmir University of Agricultural Sciences & Technology of Kashmir, Srinagar-191 121

[2] Assistant Professor, Sher-e-Kashmir University of Agricultural Sciences & Technology of Kashmir, Srinagar-191 121

[3] Assistant Professor, Directorate of Extension Education , Sher-e-Kashmir University of Agricultural Sciences & Technology of Kashmir, Srinagar-191 121

[4&5] Assistant Professor, Division of Agronomy , Sher-e-Kashmir University of Agricultural Sciences & Technology of Kashmir, Srinagar-191 121

[6] Research Assosiate, Division of Agronomy , Sher-e-Kashmir University of Agricultural Sciences & Technology of Kashmir, Srinagar-191 121

[7] Professor and Head , Division of Agronomy , Sher-e-Kashmir University of Agricultural Sciences & Technology of Kashmir, Srinagar-191 121

an excellent example of a remote sensing device. We are able to gather information about our surroundings by gauging the amount and nature of the reflectance of visible light energy from some external source (such as the sun or a light bulb) as it reflects off objects in our field of view. Contrast this with a thermometer, which must be in contact with the phenomenon it measures, and thus is not a remote sensing device. Given this rather general definition, the term *remote sensing* has come to be associated more specifically with the gauging of interactions between earth surface materials and electromagnetic energy. However, any such attempt at a more specific definition becomes difficult, since it is not always the natural environment that is sensed (*e.g.*, art conservation applications), the energy type is not always electromagnetic (*e.g.*, sonar) and some procedures gauge natural energy emissions (*e.g.*, thermal infrared) rather than interactions with energy from an independent source.

STAGES IN REMOTE SENSING

- Emission of electromagnetic radiation, or **EMR** (sun/self- emission)
- Transmission of energy from the source to the surface of the earth, as well as absorption and scattering
- Interaction of **EMR** with the earth's surface: reflection and emission
- Transmission of energy from the surface to the remote sensor
- Sensor data output

REMOTE SENSING TECHNOLOGY

Remote sensing is the technique of deriving information about objects on the surface of the earth without physically coming into contact with them. This process involves making observations using sensors (cameras, scanners, radiometer, radar etc.) mounted on platforms (aircraft and satellites), which are at a considerable height from the earth surface and recording the observations on a suitable medium (images on photographic films and videotapes or digital data on magnetic tapes). Based on remote sensing a variety of data pertaining to identification of plant community, biomass estimation, shoreline changes, delineation of coastal landforms and tidal boundary, qualitative estimation of suspended sediment concentration, chlorophyll mapping, bathymetry of shallow waters, etc. can be collected and all these data can help in effective ecosystem management. Remote sensing imagery has many applications in mapping land-use and cover, agriculture, soils mapping, forestry, city planning, archaeological investigations, military observation, and geomorphological surveying, land cover changes, deforestation, vegetation dynamics, water quality dynamics, urban growth, etc.

Remote-sensing provides spatial coverage through the measurement of reflected, emitted and backscattered radiation, across a wide range of wavebands, from the earth's surface and surrounding atmosphere. Remotely sensed data may be obtained from the land surface across a wide range of wavebands, from the ultraviolet, visible, near-infrared, short-wave infrared, mid-infrared, thermal infrared, and microwave regions of the electromagnetic spectrum. These bands are located in "atmospheric windows" because there is a signal from the surface, total absorption (or scattering) of the light owing to atmospheric constituents does not occur. Each waveband provides different information about the atmosphere and land surface.

Reflective remote-sensing

The reflective portion of the electromagnetic spectrum (EMS) ranges nominally from 0.4 to 3.75 µm. Many remote-sensing instruments have channels situated in the red and near-infrared (NIR) wavelengths of the spectrum. These two reflective bands are often combined to produce vegetation indexes. The most common linear combinations are the simple ratio (NIR/red) and normalized difference vegetation index (NDVI); NDVI is derived from (NIR − red)/NIR + red). The effect of soil colour and quantization on overall reflectance can be distinguished using remotely sensed measurements such as Lambertian surface (one that has no angular dependency), with no shade present and two soils, one dark (5 % albedo) and the other bright (30 % albedo), with green grass (10 % albedo) in the red portion of the EMS. Most surfaces in the optical (reflective and thermal) portions of the EMS are strongly anisotropic. The geometry of the Sun, target and sensor, and the size, shape and spacing of elements (*for example*, trees over bare ground), control the amount of shadow contributing to the signal. This effect, termed the bidi-rectional reflectance distribution function (BRDF), is characteristic of vegetation structure in reflective remotely sensed images.

Thermal remote-sensing

Thermal remote-sensing is an observation of the status of the surface energy balance (SEB) at a specific time of day. The thermal portion of the EMS ranges nominally from 3.75 to 12.5 µm. The observed surface temperature is a function of the radiant energy emitted by the surface that is remotely sensed, be it land, ocean or cloud top. Models have been developed to allow surface temperature to be extracted from thermal remote-sensing data. During the daytime, the net radiation is usually dominated by incoming short-wave radiation from the Sun, and the amount that is reflected depends on the albedo of the surface. Remotely sensed thermal data are also recorded at night. At this time, the SEB is dominated by the release from the ground of heat that was absorbed during daylight hours, which is governed by how much heat was absorbed during the day and the rate at which it is released after sunset.

Microwave remote-sensing

The microwave portions of the EMS ranges nominally from 0.75 to 100 cm. Radio signals have wavelengths that are included in these bands. Radar is an active system that consists of sending a pulse of microwave energy and then recording the strength, and sometimes polarization, of the return pulses. The way the signal is returned provides information to determine characteristics of the landscape.

MODERN REMOTE SENSING TECHNOLOGY VERSUS CONVENTIONAL AERIAL PHOTOGRAPHY

The use of different and extended portions of the electromagnetic spectrum, development in sensor technology, different platforms for remote sensing (spacecraft, in addition to aircraft), emphasize on the use of spectral information as compared to spatial information, advancement in image processing and enhancement techniques, and automated image analysis in addition to manual interpretation are points for comparison of conventional aerial photography with modern remote sensing system.

During early half of twentieth century, aerial photos were used in military surveys and topographical mapping. Main advantage of aerial photos has been the high spatial resolution with fine details and therefore they are still used for mapping at large scale such as in route surveys, town planning, construction project surveying, cadastral mapping etc. Modern remote sensing system provide satellite images suitable for medium scale mapping used in natural resources surveys and monitoring such as forestry, geology, watershed management etc. However the future generation satellites are going to provide much high-resolution images for more versatile applications.

HISTORIC OVERVIEW

In 1859 Gaspard Tournachon took an oblique photograph of a small village near Paris from a balloon. With this picture the era of earth observation and remote sensing had started. His example was soon followed by other people all over the world. During the Civil War in the United States aerial photography from balloons played an important role to reveal the defence positions in Virginia. Likewise other scientific and technical developments this Civil War time in the United States speeded up the development of photography, lenses and applied airborne use of this technology. Table 7.1 shows a few important dates in the development of remote sensing. The next period of fast development took place in Europe and not in the United States. It was during World War I that aeroplanes were used on a large scale for photoreconnaissance. Aircraft proved to be more reliable and more stable platforms for earth observation than balloons. In the period between World War I and World War II a start was made with the civilian use of aerial photos. Application fields of airborne photos included at that time geology, forestry, agriculture and cartography. These developments lead to much improved cameras, films and interpretation equipment. The most important developments of aerial photography and photo interpretation took place during World War II. During this time span the development of other imaging systems such as near-infrared photography; thermal sensing and radar took place. Near-infrared photography and thermal-infrared proved very valuable to separate real vegetation from camouflage. The first successful airborne imaging radar was not used for civilian purposes but proved valuable for nighttime bombing. As such the system was called by the military 'plan position indicator' and was developed in Great Britain in 1941. After the wars in the 1950s remote sensing systems continued to evolve from the systems developed for the war effort. Colour infrared (CIR) photography was found to be of great use for the plant sciences. In 1956 Colwell conducted experiments on the use of CIR for the classification and recognition of vegetation types and the detection of diseased and damaged or stressed vegetation. It was also in the 1950s that significant progress in radar technology was achieved.

Table1: *Milestones in the History of Remote Sensing*

Year	Discovery/Development
1800	Discovery of Infrared by Sir W. Herschel
1839	Beginning of Practice of Photography
1847	Infrared Spectrum Shown by J.B.L. Fou cault
1859	Photography from Balloons
1873	Theory of Electromagnetic Spectrum by J.C. Maxwell
1909	Photography from Airplanes
1916	World War I: Aerial Reconnaissance
1935	Development of Radar in Germany
1940	WW II: Applications of Non-Visible Part of EMS
1950	Military Research and Development
1959	First Space Photograph of the Earth (Explorer-6)
1960	First TIROS Meteorological Satellite Launched
1962	Corona Satellite series (camera systems) initiated by the intelligence community.
1970	Skylab Remote Sensing Observations from Space
1972	ERIS - 1 Launched – First Landsat satellite.
1972	Rapid Advances in Digital Image Processing
1975	Publication of ASP's Manual of Remote Sensing
1982	Launch of Landsat -4 : New Generation of Landsat Sensors: TM
1986	French Commercial Earth Observation Satellite SPOT
1986	Development Hyperspectral Sensors
1990	Development of Global Remote sensing system; lidars.
First Commercial Developments in Remote Sensing	
1998	Towards Cheap One-Goal Satellite Missions
1999	Launch EOS : NASA Earth Observing Mission
1999	Launch of IKONOS, very high spatial resolution sensor system
2000	INSAT – 3B, the first satellite in the third generation INSAT – 3 series, launched by Ariane from Kourou Guyana
2001	The first development launch by GSLV-D1 with GSAT–1 on board from Sriharikota
2002	ISRO's Polar Satellite Launch Vehicle, PSLV–C4 successfully launched Kalpna–1 Satellite from Sriharikota.
2003	ISRO's Polar Satellite Launch Vehicle, PSLV–C5 successfully launched Resources at – 1 (IRS – P6)
2004	The first operational flight of (GSLV – F01) successfully launched EDUSAT
2005	Successful launch of INSAT–4A by Ariane, from Kourou French Guyana

Year	Discovery/Development
2008	PSLV – C–11 successfully launches Chandra Yaan–1
2009	PSLV–C-14 successfully launches seven satellites
2010	Successful launch of advanced communication satellite HYLAS (Highly Adaptable Satellite)
2011	PSLV–C18 successfully launches Mega-Tropiques, Jugnu, SRMSAT and Vesselsat–1
2012	PSLV–C19 successfully launches RISAT-1

FUNDAMENTAL CONSIDERATIONS

Energy Source

Sensors can be divided into two broad groups—*passive* and *active*. Passive sensors measure ambient levels of existing sources of energy, while active ones provide their own source of energy. The majority of remote sensing is done with passive sensors, for which the sun is the major energy source. The earliest example of this is photography. With airborne cameras we have long been able to measure and record the reflection of light off earth features. While aerial photography is still a major form of remote sensing, newer solid state technologies have extended capabilities for viewing in the visible and near-infrared wavelengths to include longer wavelength solar radiation as well. However, not all passive sensors use energy from the sun. Thermal infrared and passive microwave sensors both measure natural earth energy emissions. Thus the passive sensors are simply those that do not themselves supply the energy being detected. By contrast, active sensors provide their own source of energy. The most familiar form of this is flash photography. However, in environmental and mapping applications, the best example is RADAR. RADAR systems emit energy in the microwave region of the electromagnetic spectrum. The reflection of that energy by earth surface materials is then measured to produce an image of the area sensed.

Wavelength

Most remote sensing devices make use of electromagnetic energy. However, the electromagnetic spectrum is very broad and not all wavelengths are equally effective for remote sensing purposes. Furthermore, not all have significant interactions with earth surface materials of interest to us. The atmosphere itself causes significant absorption and/or scattering of the very shortest wavelengths. In addition, the glass lenses of many sensors also cause significant absorption of shorter wavelengths such as the ultraviolet (UV). As a result, the first significant *window* opens up in the visible wavelengths. Even here, the blue wavelengths undergo substantial attenuation by atmospheric scattering, and are thus often left out in remotely sensed images. However, the green, red and near-infrared (IR) wavelengths all provide good opportunities for gauging earth surface interactions without significant interference by the atmosphere. In addition, these regions provide important clues to the nature of many earth surface materials. Chlorophyll, for example, is a very strong absorber of red visible wavelengths, while the near-infrared wavelengths provide important clues to the structures of plant leaves. As a result, the bulk of remotely sensed images used in GIS-related applications are taken in these regions.

Extending into the middle and thermal infrared regions, a variety of good windows can be found. The longer of the middle infrared wavelengths have proven to be useful in a number of geological applications. The thermal regions have proven to be very useful for monitoring not only the obvious cases of the spatial distribution of heat from industrial activity, but a broad set of applications ranging from fire monitoring to animal distribution studies to soil moisture conditions. After the thermal IR, the next area of major significance in environmental remote sensing is in the microwave region. A number of important windows exist in this region and are of particular importance for the use of active radar imaging.The texture of earth surface materials causes significant interactions with several of the microwave wavelength regions. This can thus be used as a supplement to information gained in other wavelengths, and also offers the significant advantage of being usable at night (because as an active system it is independent of solar radiation) and in regions of persistent cloud cover.

Interaction Mechanisms

When electromagnetic energy strikes a material, three types of interaction can follow: reflection, absorption and/or transmission. Our main concern is with the reflected portion since it is usually this which is returned to the sensor system. Exactly how much is reflected will vary and will depend upon the nature of the material and where in the electromagnetic spectrum our measurement is being taken. As a result, if we look at the nature of this reflected component over a range of wavelengths, we can characterize the result as a *spectral response pattern*.

Spectral Response Patterns

A spectral response pattern is sometimes called a *signature*. It is a description of the degree to which energy is reflected in different regions of the spectrum. Most humans are very familiar with spectral response patterns since they are equivalent to the human concept of colour. The bright red reflectance pattern, *for example*, might be that produced by a piece of paper printed with a red ink. Here, the ink is designed to alter the white light that shines upon it and absorb the blue and green wavelengths. What is left, then, are the red wavelengths which reflect off the surface of the paper back to the sensing system (the eye). The high return of red wavelengths indicates a bright red, whereas the low return of green wavelengths in the second example suggests that it will appear quite dark. The eye is able to sense spectral response patterns because it is truly a multi-spectral sensor. Although the actual functioning of the eye is quite complex, it does in fact have three separate types of detectors that can usefully be thought of as responding to the red, green and blue wavelength regions. These are the *additive primary colours*, and the eye responds to mixtures of these three to yield a sensation of other hues.

In the early days of remote sensing, it was believed that each earth surface material would have a distinctive spectral response pattern that would allow it to be reliably detected by visual or digital means. However, as our common experience with colour would suggest, in reality this is often not the case. For example, two species of trees may have quite a different coloration at one time of the year and quite a similar one at another. Finding distinctive spectral response patterns is the key to most procedures for computer-assisted interpretation of remotely sensed imagery. This task is rarely trivial. Rather, the analyst must find the combination of spectral bands and the time of year at which distinctive patterns can be found for each of the information classes of interest.

The strong absorption by leaf pigments in the blue and red regions of the visible portion of the spectrum leads to the characteristic green appearance of healthy vegetation. However, while this *signature* is distinctively different from most non-vegetated surfaces, it is not very capable of distinguishing between species of vegetation—most will have a similar color of green at full maturation. In the near-infrared, however, we find a much higher return from vegetated surfaces because of scattering within the fleshy mesophyllic layer of the leaves. Plant pigments do not absorb energy in this region, and thus the scattering, combined with the multiplying effect of a full canopy of leaves, leads to high reflectance in this region of the spectrum. However, the extent of this reflectance will depend highly on the internal structure of leaves (*e.g.*, broadleaf versus needle). As a result, significant differences between species can often be detected in this region. Similarly, moving into the middle infrared region we see a significant dip in the spectral response pattern that is associated with leaf moisture. This is, again, an area where significant differences can arise between mature species. Applications looking for optimal differentiation between species, therefore, will typically involve both the near and middle infrared regions and will use imagery taken well into the development cycle.

Multispectral Remote Sensing

In the visual interpretation of remotely sensed images, a variety of image characteristics are brought into consideration: colour (or tone in the case of panchromatic images), texture, size, shape, pattern, context, and the like. However, with computer-assisted interpretation, it is most often simply color (*i.e.*, the spectral response pattern) that is used. It is for this reason that a strong emphasis is placed on the use of multispectral sensors (sensors that, like the eye, look at more than one place in the spectrum and thus are able to gauge spectral response patterns), and the number and specific placement of these spectral *bands*.

The LANDSAT satellite is a commercial system providing multi-spectral imagery in seven spectral bands at a 30 meter resolution. It can be shown through analytical techniques such as Principal Components Analysis, that in many environments, the bands that carry the greatest amount of information about the natural environment are the near-infrared and red wavelength bands. Water is strongly absorbed by infrared wavelengths and is thus highly distinctive in that region. In addition, plant species typically show their greatest differentiation here. The red area is also very important because it is the primary region in which chlorophyll absorbs energy for photosynthesis. Thus it is this band which can most readily distinguish between vegetated and non-vegetated surfaces.

Given this importance of the red and near-infrared bands, it is not surprising that sensor systems designed for earth resource monitoring will invariably include these in any particular multispectral system. Other bands will depend upon the range of applications envisioned. Many include the green visible band since it can be used, along with the other two, to produce a traditional false colour composite— a full colour image derived from the green, red, and infrared bands (as opposed to the blue, green, and red bands of natural colour images). This format became common with the advent of colour infrared photography, and is familiar to many specialists in the remote sensing field. In addition, the combination of these three bands works well in the interpretation of the cultural landscape as well as natural and vegetated surfaces. However, it is increasingly common to include other bands that are more specifically targeted to the differentiation of surface materials. *For example*, LANDSAT TM Band 5 is placed between two water absorption bands and has thus proven very useful in determining soil and leaf moisture differences. Similarly, LANDSAT TM Band 7 targets the detection of hydrothermal alteration zones in bare rock surfaces. By contrast, the AVHRR system on the NOAA series satellites includes several thermal channels for the sensing of cloud temperature characteristics.

Hyperspectral Remote Sensing

In addition to traditional multispectral imagery, some new and experimental systems such as AVIRIS and MODIS are capable of capturing *hyperspectral* data. These systems cover a similar wavelength range to multispectral systems, but in much narrower bands. This dramatically increases the number of bands (and thus precision) available for image classification (typically tens and even hundreds of very narrow bands). Moreover, hyperspectral signature libraries have been created in lab conditions and contain hundreds of signatures for different types of landcovers, including many minerals and other earth materials. Thus, it should be possible to match signatures to surface materials with great precision. However, environmental conditions and natural variations in materials make this difficult. In addition, classification procedures have not been developed for hyperspectral data to the degree they have been for multispectral imagery. As a consequence, multispectral imagery still represents the major tool of remote sensing today.

Sensor/Platform Systems

Given recent developments in sensors, a variety of platforms are now available for the capture of remotely sensed data. Here we review some of the major sensor/platform combinations that are typically available to the GIS user community.

Band 1, visible blue- 0.45-0.52 mm

Band 6, thermal infrared- 10.4-12.5 mm

Band 2, visible green- 0.52-0.60 mm

Band 3, visible red- 0.63-0.69 mm

Band 4, near-infrared- 0.76-0.90 mm

Band 7, middle-infrared- 2.08-2.35 mm

Band 5, middle-infrared- 1.55-1.75 mm

AERIAL PHOTOGRAPHY

Cameras and their use for aerial photography are the simplest and oldest of sensor used for remote sensing of the Earth's surface. Cameras mounted in light aircraft flying between 200 and 15,000 m capture a large quantity of detailed information. Aerial photos provide an instant visual inventory of a portion of the earth's surface and can be used to create detailed maps. Aerial photographs commonly are taken by commercial aerial photography firms which own and operate specially modified aircraft equipped with large format (23 cm x 23 cm) mapping quality cameras. Aerial photos can also be taken using small format cameras (35 mm and 70 mm), hand-held or mounted in unmodified light aircraft.

Camera and platform configurations can be grouped in terms of oblique and vertical. Oblique aerial photography is taken at an angle to the ground. The resulting images give a view as if the observer is looking out an airplane window. These images are easier to interpret than vertical photographs, but it is difficult to locate and measure features on them for mapping purposes.

Vertical aerial photography is taken with the camera pointed straight down. The resulting images depict ground features in plan form and are easily compared with maps. Vertical aerial photos are always highly desirable, but are particularly useful for resource surveys in areas where no maps are available. Aerial photos depict features such as field patterns and vegetation which are often omitted on maps. Comparison of old and new aerial photos can also capture changes within an area over time.

Vertical aerial photos contain subtle displacements due to relief, tip and tilt of the aircraft and lens distortion. Vertical images may be taken with overlap, typically about 60 per cent along the flight line and at least 20 per cent between lines.Overlapping images can be viewed with a stereoscope to create a three-dimensional view, called a *stereo model.*

Large Format Photography

Commercial aerial survey firms use light single or twin engine aircraft equipped with large-format mapping cameras. Large-format cameras, such as the Wild RC-10, use 23 cm x 23 cm film which is available in rolls. Eastman Kodak, Inc., among others, manufactures several varieties of sheet film specifically intended for use in aerial photography. Negative film is used where prints are the desired product, while positive film is used where transparencies are desired. Print film allows for detailed enlargements to be made, such as large wall-sized prints. In addition, print film is useful when multiple prints are to be distributed and used in the field.

Small Format Photography

Small-format cameras carried in chartered aircraft are an inexpensive alternative to large-format aerial photography. A 35mm or 70mm camera, light aircraft and pilot are required, along with some means to process the film. Because there are inexpensive commercial processing labs in most parts of the world, 35mm systems are especially convenient. Oblique photographs can be taken with a hand-held camera in any light aircraft; vertical photographs require some form of special mount, pointed through a belly port or extended out a door or window.

Small-format aerial photography has several drawbacks. Light unpressurized aircraft are typically limited to altitudes below 4000 m. As film size is small, sacrifices must be made in resolution or area covered per frame. Because of distortions in the camera system, small-format photography cannot be used if precise mapping is required. In addition, presentation-quality wall-size prints cannot be made from small negatives. Nonetheless, small-format photography can be very useful for reconnaissance surveys and can also be used as point samples.

Colour Photography

Normal colour photographs are produced from a composite of three film layers with intervening filters that act to isolate, in effect, red, green, and blue wavelengths separately to the different film layers. With colour infrared film, these wavelengths are shifted to the longer wavelengths to produce a composite that has isolated reflectances from the green, red and near-infrared wavelength regions. However, because the human eye cannot see infrared, a false colour composite is produced by making the green wavelengths appear blue, the red wavelengths appear green, and the infrared wavelengths appear red.

As an alternative to the use of colour film, it is also possible to group several cameras on a single aircraft mount, each with black and white film and a filter designed to isolate a specific wavelength range. The advantage of this arrangement is that the bands are independently accessible and can be photographically enhanced. If a colour composite is desired, it is possible to create it from the individual bands at a later time.

Clearly, photographs are not in a format that can immediately be used in digital analysis. It is possible to scan photographs with a scanner and thereby create multispectral datasets either by scanning individual band images, or by scanning a colour image and separating the bands. However, the geometry of aerial photographs (which have a central perspective projection and differential parallax) is such that they are difficult to use directly. More typically they require processing by special photogrammetric software to rectify the images and remove differential parallax effects.

Aerial Videography

Light, portable, inexpensive video cameras and recorders can be carried in chartered aircraft. In addition, a number of smaller aerial mapping companies offer videography as an output option. By using several cameras simultaneously, each with a filter designed to isolate a specific wavelength range, it is possible to isolate multispectral image bands that can be used individually, or in combination in the form of a colour composite. For use in digital analysis, special graphics hardware boards known as *frame grabbers* can be used to freeze any frame within a continuous video sequence and convert it to digital format, usually in one of the more popular exchange formats such as TIF or TARGA. Like small-format photography, aerial videography cannot be used for detailed mapping, but provides a useful overview for reconnaissance surveys, and can be used in conjunction with ground point sampling.

SATELLITE-BASED SCANNING SYSTEMS

Photography has proven to be an important input to visual interpretation and the production of analog maps. However, the development of satellite platforms, the associated need to telemeter imagery in digital form, and the desire for highly consistent digital imagery have given rise to the development of solid state scanners as a major format for the capture of remotely sensed data. The specific features of particular systems vary (including, in some cases, the removal of a true scanning mechanism). An idealized scanning system is presented that is highly representative of current systems in use. The basic logic of a scanning sensor is the use of a mechanism to sweep a small field of view (known as an *instantaneous field of view*—IFOV) in a west to east direction at the same time the satellite is moving in a north to south direction. Together this movement provides the means of composing a complete raster image of the environment. A simple scanning technique is to use a rotating mirror that can sweep the field of view in a consistent west to east fashion. The field of view is then intercepted with a prism that can spread the energy contained within the IFOV into its spectral components. Photoelectric detectors (of the same nature as those found in the exposure meters of commonly available photographic cameras) are then arranged in the path of this spectrum to provide electrical measurements of the amount of energy detected in various parts of the electromagnetic spectrum. As the scan moves from west to east, these detectors are polled to get a set of readings along the east-west scan. These form the columns along one row of a set of raster images—one for each detector. Movement of the satellite from north to south then positions the system to detect the next row, ultimately leading to the production of a set of raster images as a record of reflectance over a range of spectral bands. There are several satellite systems in operation today that collect imagery that is subsequently distributed to users. Several of the most common systems are described below. Each type of satellite data offers specific characteristics that make it more or less appropriate for a particular application.

In general, there are two characteristics that may help guide the choice of satellite data: *spatial resolution* and *spectral resolution*. The spatial resolution refers to the size of the area on the ground that is summarized by one data value in the imagery. This is the Instantaneous Field of View (IFOV). Spectral resolution refers to the number and width of the spectral bands that the satellite sensor detects. In addition, issues of cost and imagery availability must also be considered.

LANDSAT

The LANDSAT system of remote sensing satellites is currently operated by the EROS Data Center of the United States Geological Survey. This is a new arrangement following a period of commercial distribution under the Earth Observation Satellite Company (EOSAT) which was recently acquired by Space Imaging Corporation. As a result, the cost of imagery has dramatically dropped, to the benefit of all. Full or quarter scenes are available on a variety of distribution media, as well as photographic products of MSS and TM scenes in false colour and black and white. There have been seven LANDSAT satellites, the first of which was launched in 1972. The LANDSAT 6 satellite was lost on launch. However, as of this writing, LANDSAT 5 is still operational. LANDSAT 7 was launched in April, 1999.

LANDSAT carries two multispectral sensors. The first is the *Multi-Spectral Scanner* (MSS) which acquires imagery in four spectral bands: blue, green, red and near infrared. The second is the *Thematic Mapper* (TM) which collects seven bands: blue, green, red, near-infrared, two mid-infrared and one thermal infrared. The MSS has a spatial resolution of 80 meters, while that of the TM is 30 meters. Both sensors image a 185 km wide swath, passing over each day at 09:45 local time, and returning every 16 days. With LANDSAT 7, support for TM imagery is to be continued with the addition of a co-registered 15 m panchromatic band.

SPOT

The *Système Pour L'Observation de la Terre* (SPOT) was launched and has been operated by a French consortium since 1985. SPOT satellites carry two High Resolution Visible (HRV) pushbroom sensors7 which operate in multispectral or panchromatic mode. The multispectral images have 20 meter spatial resolution while the panchromatic images have 10 meter resolution. SPOT satellites 1-3 provide three multi-spectral bands: Green, Red and Infrared. SPOT 4, launched in 1998, provides the same three bands plus a short wave infrared band. The panchromatic band for SPOT 1-3 is 0.51-0.73_ while that of SPOT 4 is 0.61-0.68_.

All SPOT images cover a swath 60 kilometers wide. The SPOT sensor may be pointed to image along adjacent paths. This allows the instrument to acquire repeat imagery of any area 12 times during its 26 day orbital period. The pointing capability makes SPOT the only satellite system which can acquire useful stereo satellite imagery. SPOT Image Inc. sells a number of products, including digital images on a choice of magnetic media, as well as photographic products. Existing images may be purchased, or new acquisitions ordered. Customers can request the satellite to be pointed in a particular direction for new acquisitions.

IRS

The Indian Space Research Organization currently has 5 satellites in the IRS system, with at least 7 planned by 2004. These data are distributed by ANTRIX Corp. Ltd. (the commercial arm of the Indian Space Research Organization), and also by Space Imaging Corporation in the United States. The most sophisticated capabilities are offered by the IRS-1C and IRS-1D satellites that together provide continuing global coverage with the following sensors:

IRS-Pan: 5.8 m panchromatic

IRS-LISS38: 23.5 m multispectral in the following bands:

Green (0.52-0.59)

Red (0.62-0.68)

Near-Infrared (0.77-0.86)

Shortwave Infrared (1.55-1.7)

IRS-WiFS9: 180 m multispectral in the following bands:

    Red (0.62-0.68)

    Near-Infrared (0.77-0.86)

NOAA-AVHRR

The *Advanced Very High Resolution Radiometer* (AVHRR) is carried on board a series of satellites operated by the U.S. National Oceanic and Atmospheric Administration (NOAA). It acquires data along a 2400-km-wide swath each day. AVHRR collects five bands: red, near-infrared, and three thermal infrared. Spatial resolution of the sensor is 1.1 km and this data is termed Local Area Coverage (LAC). For studying very large areas, a resampled version with resolution of about 4 km is also available, and is termed Global Area Coverage (GAC). AVHRR may be "high" spatial resolution for meteorological applications, but the images portray only broad patterns and little detail for terrestrial studies. However, they do have a high temporal resolution, showing wide areas on a daily basis and are therefore a popular choice for monitoring large areas. AVHRR imagery is used by several organizations engaged in famine prediction and is an integral part of many early warning activities.

RADARSAT

RADARSAT is an earth observation satellite launched in November 1995 by the Canadian Space Agency. The data is distributed by RADARSAT International (RSI) of Richmond, British Columbia, Canada (or through Space Imaging in the US). Spatial resolution of the C-band SAR imagery ranges from 8 to 100 meters per pixel and the ground coverage repeat interval is 24 days.

The pushbroom sensor produces output like a scanner. However, there is no actual scanning motion. Rather, the sensor consists of a dense array of detectors— one for each raster cell in the scan line—that is moved across the scene like a pushbroom.

LISS = Linear Imaging and Self Scanning Sensor. Image format is approximately 140 km x 140 km.

Collection of stereo RADAR imagery. RADAR signals also penetrate cloud cover, thus accessing areas not available to other remote sensing systems. In contrast to other remotely sensed imagery, the returned RADAR signal is more affected by electrical and physical (primarily textural) characteristics in the target than by its reflection and spectral pattern, therefore requiring special interpretation and spatial georegistration techniques. Compared to other types of remotely sensed imagery, the use of RADAR data is still in its infancy, but has strong potential.

ERS

ERS-1 and ERS-2 (European Remote Sensing Satellite) were developed by the European Space Agency. These identical systems provide an interesting complement to the other commercial imagery products in that they offer a variety of Cb and RADAR imagery output formats. For GIS applications, the main output of interest is the side-looking airborne RADAR (SAR) output that provides 100 km wide swaths with a 30 meter resolution. This should prove to be of considerable interest in a variety of applications, including vegetation studies and mapping projects where cloud cover is a persistent problem.

JERS

The Japanese Earth Resource Satellite offers 18 m resolution L-band side-looking RADAR imagery. This is a substantially longer wavelength band than the typical C-band used in earth resources applications. L-band RADAR is capable of penetrating vegetation as well as unconsolidated sand and is primarily used in geologic, topographic and coastal mapping applications. JERS data is available in the United States from Space Imaging Corporation.

AVIRIS

AVIRIS is an experimental system developed by the Jet Propulsion Lab (JPL) that produces hyperspectral data. It captures data in 224 bands over the same wavelength range as LANDSAT.

MODIS

The MODIS sensor onboard the EOS AM-1 platform will provide a logical extension of the AVHRR by providing no fewer than 36 bands of medium-to-coarse resolution imagery with a high temporal repeat cycle (1-2 days). Bands 1 and 2 will provide 250 m resolution images in the red and near-infrared regions. Bands 3-7 will provide 500 m resolution multispectral images in the visible and infrared regions. Finally, bands 8-36 will provide hyperspectral coverage in the visible, reflected infrared, and thermal infrared regions, with a 1 km resolution.

DIGITAL IMAGE PROCESSING

Overview

Many of the techniques of digital image processing were developed in 1960s but the cost of processing in very high, however, with computing equipment of that era. That changed in the 1970s, when digital image processing proliferated as cheaper computers and dedicated hardware available. As a result of solid state multispectral scanners and other raster input devices, we now have available digital raster images of spectral reflectance data. The chief advantage of having these data in digital form is that they allow us to apply computer analysis techniques to the image data—a field of study called *Digital Image Processing*.

Digital Image Processing is largely concerned with four basic operations: *image restoration, image enhancement, image classification, image transformation. Image restoration* is concerned with the correction and calibration of images in order to achieve as faithful a representation of the earth surface as possible—a fundamental consideration for all applications.

Image enhancement is predominantly concerned with the modification of images to optimize their appearance to the visual system. Visual analysis is a key element, even in digital image processing, and the effects of these techniques can be dramatic.

Image classification refers to the computer-assisted interpretation of images—an operation that is vital to GIS. Finally, *image transformation* refers to the derivation of new imagery as a result of some mathematical treatment of the raw image bands. In order to undertake the operations listed in this section, it is necessary to have access to Image Processing software. IDRISI is one such system. While it is known primarily as a GIS software system, it also offers a full suite of image processing capabilities.

Image Restoration

Remotely sensed images of the environment are typically taken at a great distance from the earth's surface. As a result, there is a substantial atmospheric path that electromagnetic energy must pass through before it reaches the sensor. Depending upon the wavelengths involved and atmospheric conditions (such as particulate matter, moisture content and turbulence), the incoming energy may be substantially modified. The sensor itself may then modify the character of that data since it may combine a variety of mechanical, optical and electrical components that serve to modify or mask the measured radiant energy. In addition, during the time the image is being scanned, the satellite is following a path that is subject to minor variations at the same time that the earth is moving underneath. The geometry of the image is thus in constant flux. Finally, the signal needs to be telemetered back to earth, and subsequently received and processed to yield the final data we receive. Consequently, a variety of systematic and apparently random disturbances can combine to degrade the quality of the image we finally receive. Image restoration seeks to remove these degradation effects. Broadly, image restoration can be broken down into the two sub-areas of *radiometric restoration* and *geometric restoration*.

Radiometric Restoration

Radiometric restoration refers to the removal or diminishment of distortions in the degree of electromagnetic energy reg istered by each detector. A variety of agents can cause distortion in the values recorded for image cells. Some of the most common distortions for which correction procedures exist include: *uniformly elevated values,* due to atmospheric haze, which preferentially scatters short wavelength bands (particularly the blue wavelengths); *striping,* due to detectors going out of calibration; *random noise,* due to unpredictable and unsystematic performance of the sensor or transmission of the data; and *scan line drop out,* due to signal loss from specific detectors. It is also appropriate to include here procedures that are used to convert the raw, unitless relative reflectance values (known as digital numbers, or DN) of the original bands into true measures of reflective power (radiance).

Geometric Restoration

For mapping purposes, it is essential that any form of remotely sensed imagery be accurately registered to the proposed map base. With satellite imagery, the very high altitude of the sensing platform results in minimal image displacements due to relief. As a result, registration can usually be achieved through the use of a systematic *rubber sheet* transformation process10 that gently warps an image (through the use of polynomial equations) based on the known positions of a set of widely dispersed control points. This capability is provided in IDRISI through the module RESAMPLE. With aerial photographs, however, the process is more complex. Not only are there systematic distortions related to tilt and varying altitude, but variable topographic relief leads to very irregular distortions (differential parallax) that cannot be removed through a rubber sheet transformation procedure. In these instances, it is necessary to use photogrammetric rectification to remove these distortions and provide accurate map measurements11. Failing this, the central portions of high altitude photographs can be resampled with some success. RESAMPLE is a module of major importance, and it is essential that one learn to use it effectively. Doing so also requires a thorough understanding of reference systems and their associated parameters such as datums and projections.

Image Enhancement

Image enhancement is concerned with the modification of images to make them more suited to the capabilities of human vision. Regardless of the extent of digital intervention, visual analysis invariably plays a very strong role in all aspects of remote sensing. While the range of image enhancement techniques is broad, the following fundamental issues form the backbone of this area:

Contrast Stretch

Digital sensors have a wide range of output values to accommodate the strongly varying reflectance values that can be found in different environments. However, in any single environment, it is often the case that only a narrow range of values will occur over most areas. Grey level distributions thus tend to be very skewed. Contrast manipulation procedures are thus essential to most visual analyses.

Composite Generation

For visual analysis, color composites make fullest use of the capabilities of the human eye. Depending upon the graphics system in use, composite generation ranges from simply selecting the bands to use, to more involved procedures of band combination and associated contrast stretch. The IDRISI module COMPOSITE is used to construct three-band composite images.

Digital Filtering

One of the most intriguing capabilities of digital analysis is the ability to apply digital filters. Filters can be used to provide edge enhancement (sometimes called *crispening*), to remove image blur, and to isolate lineaments and directional trends, to mention just a few. The IDRISI module FILTER is used to apply standard filters and to construct and apply user-defined filters.

Image Classification

Image classification refers to the computer-assisted interpretation of remotely sensed images. The majority of image classification is based solely on the detection of the spectral signatures (*i.e.*, spectral response patterns) of land cover classes. The success with which this can be done will depend on two things: 1) the presence of distinctive signatures for the land cover classes of interest in the band set being used; and 2) the ability to reliably distinguish these signatures from other spectral response patterns that may be present.

There are two general approaches to image classification: *supervised* and *unsupervised*. They differ in how the classification is performed. In the case of supervised classification, the software system delineates specific landcover types based on statistical characterization data drawn from known examples in the image (known as training sites). With unsupervised classification, however, clustering software is used to uncover the commonly occurring landcover types, with the analyst providing interpretations of those cover types at a later stage.

Supervised Classification

The first step in supervised classification is to identify examples of the information classes (*i.e.*, land cover types) of interest in the image. These are called *training sites*. The software system is then used to develop a statistical characterization of the reflectances for each information class. This stage is often called *signature analysis* and may involve developing a characterization as simple as the mean or the range of reflectances on each band, or as complex as detailed analyses of the mean, variances and covariances over all bands.

Once a statistical characterization has been achieved for each information class, the image is then classified by examining the reflectances for each pixel and making a decision about which of the signatures it resembles most. There are several techniques for making these decisions, called *classifiers*. Most Image Processing software will offer several, based on varying decision rules. IDRISI offers a wide range of options falling into three groups depending upon the nature of the output desired and the nature of the input bands.

Hard Classifiers

The distinguishing characteristic of hard classifiers is that they all make a definitive decision about the landcover class to which any pixel belongs. IDRISI offers three supervised classifiers in this group: Parallelepiped (PIPED), Minimum Distance to Means (MINDIST), and Maximum Likelihood (MAXLIKE). They differ only in the manner in which they develop and use a statistical characterization of the training site data. Of the three, the Maximum Likelihood procedure is the most sophisticated, and is unquestionably the most widely used classifier in the classification of remotely sensed imagery.

Soft Classifiers

Contrary to hard classifiers, soft classifiers do not make a definitive decision about the land cover class to which each pixel belongs. Rather, they develop statements of the degree to which each pixel belongs to each of the land cover classes being considered. Thus, *for example*, a soft classifier might indicate that a pixel has a 0.72 probability of being forest, a 0.24 probability of being pasture, and a 0.04 probability of being bare ground. A hard classifier would resolve this uncertainty by concluding that the pixel was forest. However, a soft classifier makes this uncertainty explicitly available, for any of a variety of reasons. *For example*, the analyst might conclude that the uncertainty arises because the pixel contains more than one cover type and could use the probabilities as indications of the relative proportion of each. This is known as *sub-pixel* classification. Alternatively, the analyst may conclude that the uncertainty arises because of unrepresentative training site data and therefore may wish to combine these probabilities with other evidence before *hardening* the decision to a final conclusion. IDRISI offers three soft classifiers (BAYCLASS, BELCLASS and FUZCLASS) and three corresponding hardeners (MAXBAY, MAXBEL, and MAXFUZ). The difference between them relates to the logic by which uncertainty is specified—Bayesian, Dempster-Shafer, and Fuzzy Sets respectively. In addition, the system supplies a variety of additional tools specifically designed for the analysis of sub-pixel mixtures (*e.g.*, UNMIX, FUZSIG, MIXCALC and MAXSET).

Hyperspectral Classifiers

All of the classifiers mentioned above operate on multispectral imagery—images where several spectral bands have been captured simultaneously as independently accessible image components. Extending this logic to many bands produces what has come to be known as *hyperspectral* imagery. Although there is essentially no difference between hyperspectral and multispectral imagery (*i.e.*, they differ only in degree), the volume of data and high spectral resolution of hyperspectral images does lead to differences in the way that they are handled. IDRISI provides special facilities for creating hyperspectral signatures either from training sites or from libraries of spectral response patterns developed under lab conditions (HYPERSIG) and an automated hyperspectral signature extraction routine (HYPERAUTOSIG). These signatures can then be applied to any of several hyperspectral classifiers:

Spectral Angle Mapper (HYPERSAM), Minimum Distance to Means (HYPERMIN), Linear Spectral Unmixing (HYPERUNMIX), Orthogonal Subspace Projection (HYPEROSP), Absorption area analysis (HYPERABSORB) hyperspectral classifiers. An unsupervised classifier for hyperspectral imagery (HYPERUSP) is also available.

Unsupervised Classification

In contrast to supervised classification, where we tell the system about the character (*i.e.*, signature) of the information classes we are looking for, unsupervised classification requires no advance information about the classes of interest. Rather, it examines the data and breaks it into the most prevalent natural spectral groupings, or clusters, present in the data. The analyst then identifies these clusters as landcover classes through a combination of familiarity with the region and ground truth visits. The logic by which unsupervised classification works is known as *cluster analysis*, and is provided in IDRISI primarily by the CLUSTER module. CLUSTER performs classification of composite images (created with COMPOSITE) that combine the most useful information bands. It is important to recognize, however, that the clusters unsupervised classification produces are not information classes, but spectral classes (*i.e.*, they group together features (pixels) with similar reflectance patterns). It is thus usually the case that the analyst needs to reclassify spectral classes into information classes. *For example*, the system might identify classes for asphalt and cement which the analyst might later group together, creating an information class called pavement.

While attractive conceptually, unsupervised classification has traditionally been hampered by very slow algorithms. However, the clustering procedure provided in IDRISI is extraordinarily fast (unquestionably the fastest on the market) and can thus be used iteratively in conjunction with ground truth data to arrive at a very strong classification. With suitable ground truth and accuracy assessment procedures, this tool can provide a remarkably rapid means of producing quality land cover data on a continuing basis.

In addition to the above mentioned techniques, two modules bridge both supervised and unsupervised classifications. ISOCLUST uses a procedure known as *Self-Organizing Cluster Analysis* to classify up to 7 raw bands with the user specifying the number of clusters to process. The procedure uses the CLUSTER module to initiate a set of clusters that seed an iterative application of the MAXLIKE procedure, each stage using the results of the previous stage as the training sites for this supervised procedure. The result is an unsupervised classification that converges on a final set of stable members using a supervised approach (hence the notion of "self-organizing"). MAXSET is also, at its core, a supervised procedure. However, while the procedure starts with training sites that characterize individual classes, it results in a classification that includes not only these specific classes, but also significant (but unknown) mixtures that might exist. Thus the end result has much the character of that of an unsupervised approach.

Accuracy Assessment

A vital step in the classification process, whether supervised or unsupervised, is the assessment of the accuracy of the final images produced. This involves identifying a set of sample locations (such as with the SAMPLE module) that are visited in the field. The land cover found in the field is then compared to that which was mapped in the image for the same location. Statistical assessments of accuracy may then be derived for the entire study area, as well as for individual classes (using ERRMAT). In an iterative approach, the error matrix produced (sometimes referred to as a *confusion matrix*), may be used to identify particular cover types for which errors are in excess of that desired. The information in the matrix about which covers are being mistakenly included in a particular class (*errors of commission*) and those that are being mistakenly excluded (*errors of omission*) from that class can be used to refine the classification approach.

Image Transformation

Digital Image Processing offers a limitless range of possible transformations on remotely sensed data. Two are mentioned here specifically, because of their special significance in environmental monitoring applications.

Vegetation Indices

There are a variety of vegetation indices that have been developed to help in the monitoring of vegetation. Most are based on the very different interactions between vegetation and electromagnetic energy in the red and near-infrared wavelengths.

For example, the United Nations Food and Agricultural Organization (FAO) Africa Real Time Information System (ARTEMIS) and the USAID Famine Early Warning System (FEWS) programs both use continental scale NDVI images derived from the NOAA-AVHRR system to produce vegetation index images for the entire continent of Africa every ten days. While the NDVI measure has p roven to be useful in a variety of contexts, a large number of alternative indices have been proposed to deal with special environments, such as arid lands. IDRISI offers a wide variety of these indices (over 20 in the VEGINDEX and TASSCAP modules combined).

PRINCIPAL COMPONENTS ANALYSIS

Principal Components Analysis (PCA) is a linear transformation technique related to Factor Analysis. Given a set of image bands, PCA produces a new set of images, known as components, that are uncorrelated with one another and are ordered in terms of the amount of variance they explain from the original band set. PCA has traditionally been used in remote sensing as a means of data compaction. For a typical multispectral image band set, it is common to find that the first two or three components are able to explain virtually all of the original variability in reflectance values. Later components thus tend to be dominated by noise effects. By rejecting these later components, the volume of data is reduced with no appreciable loss of information.

Given that the later components are dominated by noise, it is also possible to use PCA as a noise removal technique. The output from the PCA module in IDRISI includes the coefficients of both the forward and backward transformations. By zeroing out the coefficients of the noise components in the reverse transformation, a new version of the original bands can be produced with these noise elements removed. Recently, PCA has also been shown to have special application in environmental monitoring. In cases where multispectral images are available for two dates, the bands from both images are submitted to a PCA as if they all came from the same image. In these cases, changes between the two dates tend to emerge in the later components.

More dramatically, if a time series of NDVI images (or a similar single-band index) is submitted to the analysis, a very detailed analysis of environmental changes and trends can be achieved. In this case, the first component will show the typical NDVI over the entire series while each successive component illustrates change events in an ordered sequence of importance. By examining these images, along with graphs of their correlation with the individual bands in the original series, important insights can be gained into the nature of changes and trends over the time series. The TSA (Time Series Analysis) module in IDRISI is a specially tailored version of PCA to facilitate this process.

OTHER TRANSFORMATIONS

IDRISI offers a variety of other transformations. These include colour space transformations (COLSPACE), texture calculations (TEXTURE), blackbody thermal transformations (THERMAL), and a wide variety of ad hoc transformations (such as image ratioing) that can be most effectively accomplished with the image calculator utility.

An archive dataset of monthly NDVI images for Africa is available on CD from Clark Labs. The Africa NDVI data CD contains monthly NDVI maximum value composite images (1982-1999), average and standard deviation of monthly NDVI images for each month over the same time period, monthly NDVI anomaly images, and ancillary data (DEM,land use and land cover, country boundaries and coast line) for Africa in IDRISI format. Contact Clark Labs for more information.

CONCLUSIONS

Remotely sensed data is important to a broad range of disciplines. This will continue to be the case and will likely grow with the greater availability of data promised by an increasing number of operational systems. The availability of this data, coupled with the computer software necessary to analyze it, provides opportunities for environmental scholars and planners, particularly in the areas of landuse mapping and change detection, that would have been unheard of only a few decades ago.

The inherent raster structure of remotely sensed data makes it readily compatible with raster GIS. Thus, while IDRISI provides a wide suite of image processing tools, they are completely integrated with the broader set of raster GIS tools the system provides.

■

8

Precision Irrigation Systems in Agriculture: A Perspective

☞ T. Ram[1], Dalamu[2], Ranveer Singh[3], K. N. Singh[4], D. Ram[5] and N.K. Jat[6]

CHAPTER

INTRODUCTION

Precision irrigation system involves the accurate and precise application of water to meet the specific requirements of individual plants or management units and minimize adverse environmental impact. The goal of the chapter is to define and describes precision irrigation; discuss the components of precision irrigation system, present case studies from production and research fields that illustrate opportunities and discuss critical research needs to fully implement precision irrigation. The chapter also provides an assessment of the role of current precision irrigation technologies in India along with various irrigation sensing approaches and their potential applications in agriculture.

Efficient water management plays an important role in irrigated agricultural systems (Kim and Evans, 2009). At present, agriculture consumes 70 per cent of the fresh water *i.e.*, 1,500 billion m^3 out of the 2,500 billion m^3 of water is being used each year (Goodwin and O'Connell, 2008). It is also estimated that 40 per cent of the fresh-water used for agriculture in developing countries is lost, either by evaporation, spills or absorption by the deeper layers of the soil, beyond the reach of plants' roots (Panchard *et al.*, 2007). Post green-revolution era in India agriculture is facing a technological fatigue for two reasons *viz.* (i) high rates of ground-water depletion and (ii) soil salinity due to excessive irrigation in some pockets. Thus efficient water management is a major concern in many cropping systems. Many planners as well as farmer associations are becoming conscious about water-audit and water utilization efficiency as the water resources are getting more and scarcer. Efforts of using micro-irrigation methods such as sprinkler and drip irrigation have been made in last three decades in many parts of the world. It has been reported that in

[1] Sr. Scientist, Agronormy, PDFSR (ICAR), Modipuram, Meerut (UP) - 250 110

[2] Crop Improvement Division, Central Potato Research Institute (CPRI), Shimla, 171 001 (Himachal Pradesh)

[3] Assistant Agricutlural Offiecr, Chirawa, Rajasthan

[4] Professor and Head , Division of Agronomy , Sher-e-Kashmir University of Agricultural Sciences & Technology of Kashmir, Srinagar-191 121

[5] Assistant Professor, Division of Soil Scienec , Sher-e-Kashmir University of Agricultural Sciences & Technology of Kashmir, Srinagar-191 121

[6] Scientist (Agronormy), PDFSR, Moodipuram, Meerut (UP) - 250 110

year 2005, 1.15 million ha land area was under micro-irrigation (drip and sprinkler) in India (Modak, 2009). There is no ideal irrigation method available which may be suitable for all weather conditions, soil structure and variety of crops cultures. In the semi-arid areas of developing countries, marginal and small farmers, who cannot afford to pay for powered irrigation, heavily depend on the rainfall for their crops. It is observed that farmers have to bear huge financial loss because of wrong prediction of weather and incorrect irrigation method. In light of a real need to improve the efficiency of irrigation systems and prevent the misuse of water, the focus is to develop a precision irrigation system which will enable farmers to optimize the use of water and only irrigate where and when need for as long as needed.

BENEFITS OF PRECISION IRRIGATION

Precision irrigation has the potential to increase both the water use and economic efficiencies by optimally matching irrigation inputs to yields in each area of field and either reducing the cost of inputs or increasing yield for the same inputs.

Precision irrigation system is a sustainable management of water resources which involves application of water to the crop at the right time, right amount, right place and right manner thereby helping to manage the field variability of water which in turn increasing the crop productivity and water use efficiency along with reduction in energy cost on irrigation. This concept utilizes a systems approach to achieve 'differential irrigation' treatment of field variation (spatial and temporal) as opposed to the 'uniform irrigation' treatment that underlies traditional management systems. Although significant progress has been made in irrigation technologies and in the implementation of irrigation management practices such as scientific irrigation scheduling but irrigation inefficiency remains the rule rather than the exception. Gains in water use efficiency can be achieved when water applications are precisely matched to the spatially distributed crop water demands, a central principle underlying precision irrigation (Sadler *et al.*, 2005). This distributed crop water demand is present in agricultural fields mainly because of variability in soil properties and topography (Evans *et al.*, 1996), but may also result from variable rainfall or crop variation associated with multiple crops planted in the same field or plants growing at different phenological stages induced by natural or manmade causes. Advances in precision irrigation on the farm have been limited primarily due to costs of site-specific control systems (Stone *et al.*, 2006) and because crop production functions needed to drive site-specific irrigation systems are largely unknown for the large diversity of soils (Sadler *et al.*, 2002). However, recent advances in a number of important technologies, particularly in chip technology, sensors and wireless radio frequency (RF), when integrated with information technologies and the internet, have made it possible to overcome these major limitations to precision irrigation (Pierce and Elliott, 2008).

CONCEPT OF PRECISION IRRIGATION

The concept of irrigation as an activity requiring some precision in implementation has been around since the introduction of irrigation scheduling and the first improvements in application system efficiencies. However, the specific term "precision irrigation" has only recently been introduced and has not been well defined. It has been variously used to describe variable rate irrigation applications controlled by a sensory input (Evans and Harting, 1999) or efficient application systems (Smith and Raine, 2000). However, neither of these uses adequately conveys that precision is required in both the accurate assessment of the crop water requirements and the precise application of the required volume at the required time. Similarly, the ability to spatially vary the water application within a management unit is not necessarily a requirement for precise irrigation as uniformity of application within a management unit may be preferred. Hence, it would seem more appropriate to define precision irrigation as the accurate and precise application of water to meet the specific requirements of individual plants or management units and minimize adverse environmental impact (Misra *et al.*, 2005 and Raine *et al.*, 2007). It also follows that an important characteristic of a precision irrigation system is that the timing, placement and volume of water applied should match plant water demand resulting in reduced non-transpiration volumetric losses (deep drainage and evaporation) and optimized crop production (yield quantity and quality) responses (Fig 8.1).

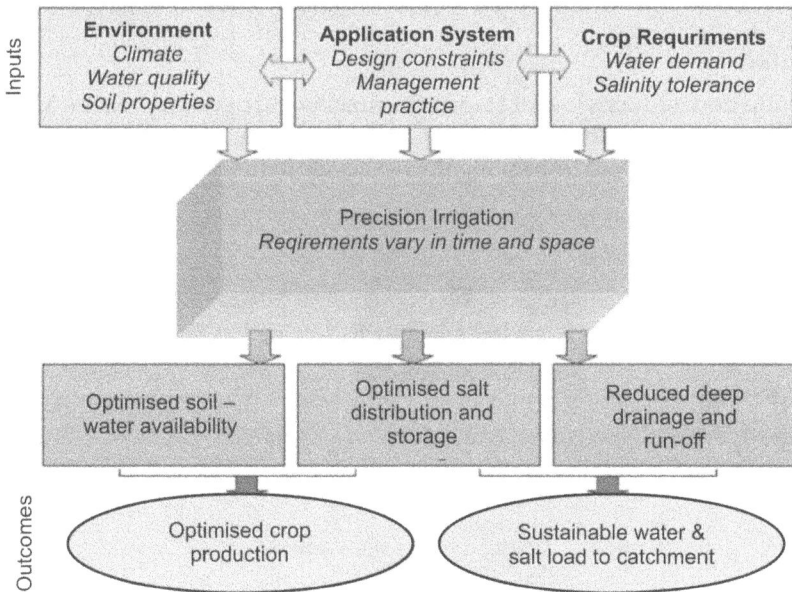

Fig. 8.1: *Inputs and outcomes associated with a precision irrigation system (Source: Raine et al., 2007)*

ELEMENTS OF PRECISION IRRIGATION SYSTEM

Precision irrigation is a management approach defined by the precision farming cycle. There are four essential elements in the process: data acquisition, interpretation, control and evaluation (Fig 8.2).

Data acquisition

A precision irrigation system requires ability to identify and quantify the variability *i.e.* spatial and temporal variability that exist in soil and crop conditions within a field and between fields. Existing technology is available to measure the various components of the soil-crop-atmosphere continuum many in real-time so as to provide precise and / or real time control of irrigation applications.

Interpretation

Data has to be collected, interpreted and analyzed at an appropriate scale and frequency. The inadequate development of decision support systems has been identified as a major bottle neck for the interpretation of real time data and adoption of precision agriculture (McBratney *et al.*, 2005).

Control

The ability to optimize the inputs and adjust irrigation management at appropriate temporal and spatial scales is an essential component of a precision irrigation system. Applying differential depths of water over a field will be dependent on the irrigation system. Automatic controllers with real time data should provide the most reliable and accurate means of controlling irrigation applications.

Evaluation

Evaluation is an important step in the precision irrigation process. Measuring the engineering, agronomic and economic performance of the irrigation system is essential for feedback and improving the next cycle in the system (Smith *et al.*, 2010).

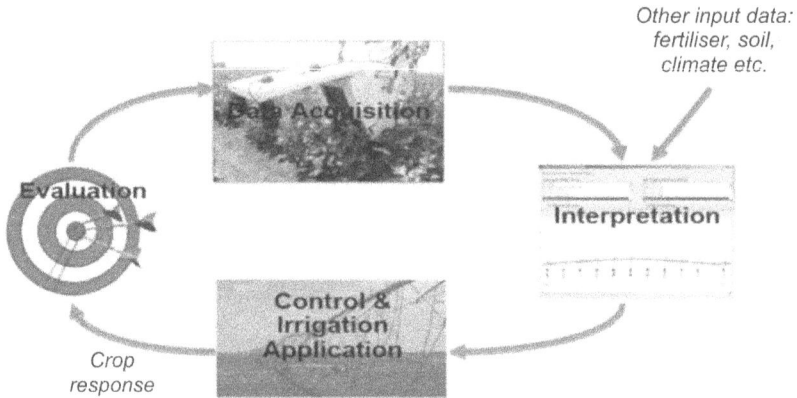

Fig. 8.2: *Elements of precision irrigation system*

TECHNOLOGY ASSOCIATED WITH PRECISION IRRIGATION

Integration of the various component technologies for precision irrigation stands out. Combining the crop and soil sensing with appropriate crop growth stimulation models to provide the seasonal decision making model is a necessary first step for all of the major crops. The advent of precision irrigation methods has played a major role in reducing the quantity of water required in agricultural and horticultural crops, but there is a need for new methods of automated and accurate irrigation scheduling and control. The early adopters found precision agriculture to be unprofitable and the instances in which it was implemented were few and far between. Further, the high initial investment in the form of electronic equipment for sensing and communication meant that only large farms could afford it. The technologies used are Remote Sensing (RS), Global Positioning System (GPS), Geographical Information System (GIS) and Wireless Sensor Network (WSN).

The technologies of GIS and GPS apart from being non-real-time, involved the use of expensive technologies like satellite sensing and also labor intensive. Over the last several years, the advancement in sensing and communication technologies has significantly brought down the cost of deployment and running of a feasible precision agriculture framework. However, a stand-alone sensor, due to its limited range, can only monitor a small portion of its environment but the use of several sensors working in a network seems particularly appropriate for precision agriculture. The technological development in made it possible to monitor and control various parameters in agriculture. Also recent advances in sensor and wireless radio frequency (RF) technologies and their convergence with the internet offer vast opportunities for application of sensor systems for agriculture. Emerging wireless technologies with low power needs and low data rate capabilities have been developed which perfectly suit precision agriculture (Wang *et al.*, 2006). The sensing and communication can now be done on a real-time basis leading to better response times. The wireless sensors are cheap enough for wide spread deployment and offer robust communication through redundant propagation paths (Akyildiz and Xudong, 2005). The have become the most suitable technology to monitor the agricultural environment.

WIRELESS SENSOR NETWORK (WSN)

Wireless sensor networks (WSN) is a network of small sensing devices known as sensor nodes or motes, arranged in a distributed manner, which collaborate with each other to gather, process and communicate over wireless channel about some physical phenomena. The sensor motes are typically low-cost, low-power, small devices equipped with limited sensing, data processing and wireless communication capabilities with power supply, which perfectly suites the precision agriculture/ precision irrigation (Stafford, 2000). A wireless sensor is a self-powered computing unit usually containing a processing unit, a trans-receiver and both analog and digital interfaces, to which a variety of sensing units such as temperature, humidity etc. can be adapted. The sensor nodes communicate with each other in order to exchange and process the information collected by their sensing units. If nodes communicate only directly with each other or with a base station, the network is single-hop. In some cases, nodes can use other wireless sensors as relays, in which case the network is said to be multihop. In a data-collection model, sensors communicate with one or several base stations connected to a database and an application server that stores the data and performs extra data-processing. The result is available typically via a web-based interface.Fig. 8.3 shows farm-scale WSN for precision irrigation scheduling in multiple research and commercial farming sites.

A number of WSN's with various topologies (*e.g.* star, mesh-network) have been developed and investigated by different researchers in the past decade (Ruiz-Garcia *et al.*, 2009), including WSN's for irrigation scheduling in cotton (Vellidis *et al.*, 2008), center-pivot irrigation (O'Shaughnessy and Evett, 2008) and linear-move irrigation systems (Kim *et al.*, 2008). The first reported greenhouse WSN was a bluetooth monitoring and control system developed by Liu and Ying (2003). Yoo *et al.* (2007) describes the deployment of a wireless environmental monitoring and control system in greenhouses. Wang *et al.*, (2008) also developed a specialized wireless sensor node to monitor temperature, relative humidity and light inside greenhouses. Lea-Cox *et al.*, (2007) reported on the early deployment of a WSN within a cut-flower greenhouse, where a number of soil moisture and environmental sensor nodes were deployed for real-time monitoring of crop production by the grower. With regards to large on-farm WSN deployments, Balendonck *et al.*, (2009) reported on the FLOW-AID project that has many of the same objectives that we are focused upon, *i.e.*, providing growers with a safe, efficient and cost-effective management system for irrigation scheduling. The FLOW-AID project is integrating innovative monitoring and control technologies within an appropriate decision support system (Balendonck *et al.*, 2007; Ferentinos *et al.*, 2003) that is accessible over the internet, to assist growers in long-term farm zoning and crop planning.

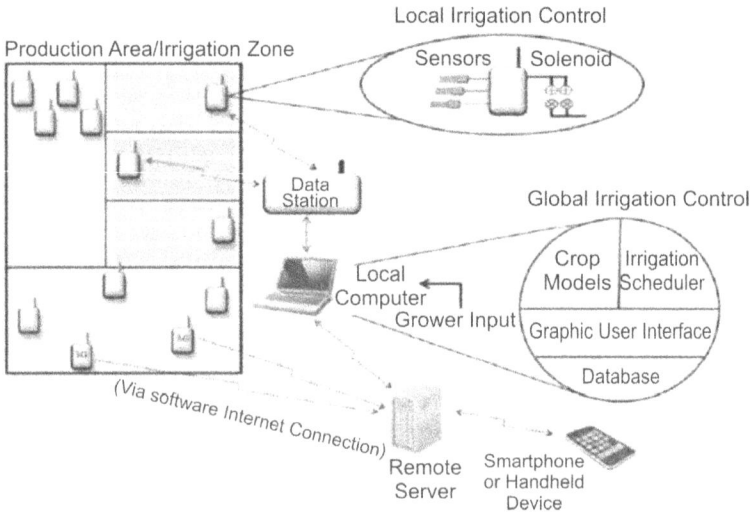

Fig. 8.3: *A schematic of a farm-scale wireless sensor network for precision irrigation scheduling (Source:* Balendock, 2009; Lea-Cox *et al.*, 2010)

It is especially focused on providing growers with regulated deficit irrigation and soil salinity management tools. To support shorter-term irrigation scheduling, a scheduling tool is being developed which allocates available water among several plots and schedules irrigation for each plot (Stanghellini *et al.*, 2007; Anastasiou *et al.*, 2008). To assist this advanced scheduling tool, a crop response model is being developed and used to predict crop stress (Balendonck *et al.*, 2009). Recently, described an application of a WSN for low-cost wireless controlled irrigation solution and real time monitoring of water content of soil. Data acquisition is performed by using solar powered wireless acquisition stations for the purpose of control of valves for irrigation. The designed system has 3 units namely: base station unit (BSU), valve unit (VU) and sensor unit (SU). The obtained irrigation system not only prevents the moisture stress of trees and salification, but also provides efficient use of fresh water resource. In addition, the developed irrigation method removes the need for workmanship for flood irrigation. The designed system was applied to an area of 8 deacres in a venue located in central Anatolia for controlling drip irrigation of dwarf cherry trees.

IRRIGATION SENSING APPROACHES

The main approaches to irrigation scheduling are based on the measurement of soil moisture (Bitelli, 2011), physiological measurements (Cifre *et al.*, 2005) or water balance calculations (Allen *et al.*, 1998). The conventional sensor-based approach has typically scheduled irrigation events on the basis of soil moisture status, whether using direct soil moisture measurements with capacitance or TDR-type sensors (Smith and Mullins, 2001), tensiometers (Smajstrla and Harrison, 1998) or soil-moisture water balance methods using daily ET estimates (Allen *et al.*, 1998). Some automated greenhouse systems have used load cells for the estimation of daily plant water use (Raviv *et al.*, 2000). However, these load cell systems have to be programmed to accurately correct for increasing total plant mass over the crop cycle, or to adjust to changes in wind and temperature changes, if deployed in outdoor environments. Nevertheless, if operated correctly, most of these systems enable much greater precision and improved water use efficiency over traditional time-based irrigation scheduling methods. Jones (2004) described the main sensor techniques that are currently used for irrigation scheduling or which have the potential for framing precision irrigation systems in the near future. The current irrigation sensing approaches are centered around using soil moisture sensing techniques, plant water sensing techniques or a combination of both techniques. Soil irrigation sensing approaches can either be based on direct measurement of soil moisture content (or water potential), or by using sensors to provide data for the water balance method, which accounts for inputs (rainfall, irrigation) and losses (ET, run-off and drainage) from the system. The emphasis on using soil moisture content for irrigation decisions has been based on the perception that water availability in the soil is what limits plant transpiration, and that irrigation scheduling should replace the water lost by plant water uptake and evaporation from the root zone (Jones, 2008).

PLANT BASED SENSING TECHNOLOGIES

There are a wide range of plant based sensing technologies available to identify the onset and severity of plant stress. These technologies can be broadly categorized into those requiring direct contact with the plant and those non-contact sensors that are proximally (*e.g.* hand-held or machine mounted) or remotely (*e.g.* airborne, satellite) mounted (White and Raine, 2008). The contact sensors provide detailed time-series data for individual plants, useful for understanding diurnal fluctuations. The proximal and remote sensors are more appropriate for collecting spatial data across field, farm or regional levels and hence, are more appropriate for assessing spatial variations in plant stress and application in a precision irrigation system.

Plant based sensors for irrigation typically measure plant responses that are related to moisture uptake (*e.g.* plant water status, sap flow), transpiration (*e.g.* canopy temperature, reflectance) or growth rate. Variations in these measures indicate crop stress which can be used to infer when to apply irrigation. However, plant based sensors do not provide any indication of the volume of irrigation water that is required to be applied. Hence, these techniques should be used in conjunction with either soil moisture measurements or simulation to confirm the irrigation requirements. It should also be noted that the level of crop stress observed is a complex function of soil, plant and atmospheric conditions. Hence the user needs to ensure that the crop stress observed is due to a root zone soil moisture deficit and not disease, pest or exceptional atmospheric conditions.

APPLICATIONS OF PLANT BASED SENSING TECHNOLOGIES

Applications of automated plant-based sensing are largely in the developmental stage, partly because it is usually necessary to supplement the plant-stress sensing by additional information (such as evaporative demand). In principle, with high-frequency on-demand irrigation systems one could envisage a real-time control system where water supply is directly controlled by a feedback controller operated by the stress sensor itself, so that no information on the required irrigation amount is needed. For such an approach, care will be necessary to take account of any lags in the plant physiological response used for the control signal. The use of expert systems, which integrate data from several sources, appears to have great potential for combining inputs from thermal or other crop response sensors and environmental data for a water budget calculation to derive a robust irrigation schedule. Among the various plant-based sensors that have been incorporated into irrigation control systems are stem diameter gauges sap-flow sensors (Schmidt and Exarchou, 2000) and acoustic emission sensors, though there has been most interest in the application of thermal sensors. *For example*, Kacira and colleagues (Kacira and Ling, 2001; Kacira *et al.,* 2002) have developed and tested on a small scale an automated irrigation controller based on thermal sensing of plant stress. Similar approaches have been applied in the field. *For example*, Evans *et al.* (2001) and Sadler *et al.* (2002) mounted an array of 26 infrared thermometers (IRTs) on a centre pivot irrigation system which they used to monitor irrigation efficiency, but had not developed the system to a stage where it could be used for fully automated control. Colaizzi *et al.* (2003) have tested another system that includes thermal sensing of canopy temperature on a large linear move irrigator (where the irrigator moves across the field). In another approach to the use of canopy temperature that makes use of the 'thermal kinetic window', Upchurch *et al.* (1990) and Mahan *et al.* (2000) have developed what they call a 'biologically identified optimal temperature interactive console' for the control of trickle and other irrigation systems based on canopy temperature measurements. In this direct control system, irrigation is applied as canopy temperature exceeds a crop-specific optimum. The development of thermal infrared imaging methods of irrigation control will be aided by the recent development of automated image analysis systems for extraction of the temperatures of leaf surfaces from thermal images, including shaded and sunlit leaves, soil, and other surfaces (Leinonen and Jones, 2004). Peters and Evett (2008) scheduled irrigations and controlled centre pivot and drip systems using a temperature-time threshold (TTT). The TTT method involves using infrared thermocouples to sense crop canopy temperatures continuously and remotely. If a threshold canopy temperature is exceeded for a predetermined threshold time, an irrigation is scheduled. Their work has shown the TTT method to be a viable alternative to traditional irrigation scheduling. Mounting an array of sensors on a centre pivot provides the means to manage spatially varied irrigations.

At present, use of remotes sensing to evaluate crop factors for use in irrigation scheduling is very common. In this case, the data have usually been processed to highlight differences in crop condition using a Normalised Difference Vegetation Index (NDVI). Various researchers have found relationships between NDVI and crop coefficients for a wide range of crops (Belmonte *et al.*, 2005; Johnson *et al.*, 2006; Trout and Johnson, 2007; Hornbuckle *et al.*, 2009). An alternative to the use of NDVI is the prediction of actual crop evaporation using remote sensing of the energy balance (Chavez *et al.*, 2009). Both approaches offer the means to obtain large scale, low cost, site specific crop evaporation data to assist in site specific irrigation management. The energy balance approach is still in the development stage whereas systems using NDVI are in use. Barnes *et al.* (2000) suggested that the synergy between remote sensing and crop simulation modeling was the future for management irrigations in precision agriculture. They demonstrated the capability by integration of the remotely sensed crop water stress index with the CERES-Wheat model to provide data on within-field variability in plant water requirements and yield response. Recently, McCarthy *et al.* (2007) demonstrated a prototype machine vision system for the purpose of determining real-time cotton plant irrigation requirement. The unit relies on plant geometrical parameters such as inter node length (*i.e.* the distance between successive branches on the main stem) as an indicator for water stress in cotton. However, with the aid of some not-yet automated procedures based on visual inspection, measurement of inter-node lengths to 3 per cent standard error has been demonstrated.

SOIL IRRIGATION SENSING APPROACHES

Soil irrigation sensing approaches can either be based on direct measurement of soil moisture content (or water potential), or by using sensors to provide data for the water balance method which accounts for inputs (rainfall, irrigation) and losses (ET, run-off and drainage) from the system. In these approaches, soil water content can be expressed either in terms of the energy status of the water in the soil (kPa) or as the amount of water in the substrate (per cent or m^3). Both methods have advantages and disadvantages. Soil/substrate matrix potential indicates how easily water is available to plants (Lea-Cox *et al.*, 2011), but it does not provide information on how much total water is present in the substrate. Conversely, volumetric water content indicates how much water is present in a substrate, but not if this water is extractable by plant roots. This is especially important for soilless substrates, since mixtures of different components means that substrates have very different water-holding capacities and moisture release curves (deBoodt and Verdonck, 1972). Sensors that estimate water content (*e.g.* capacitance and TDR-type sensors) tend to be more reliable than those sensors measuring water availability (tensiometers and psychrometers) (Jones, 2008; Murray *et al.*, 2004). A major disadvantage of almost all soil sensors, however, is their limited capability to measure soil moisture heterogeneity in the root zone, since they typically only sense a small volume around the sensor. Variation in soil water availability is well known, primarily as a function of variation in soil type, soil compaction and depth, among many sources of variation (*e.g.* organic matter content, porosity and rockiness). The use of large sensor arrays which may be necessary to get good representative readings of soil moisture tends to be limited by cost, but this could be overcome by sensor placement strategies. Soilless substrates are used by the nursery and greenhouse industry for a multitude of reasons, primarily to reduce the incidence of soil-borne pathogens, increase root growth, and reduce labor, shipping and overall costs to the producer (Majsztrik *et al.*, 2011). Over the years, many studies have shown large differences between soil and soilless substrates in the availability of water to root systems (Bunt 1961; deBoodt and Verdonck, 1972). Soilless substrates, which in most cases have larger particle sizes and porosity, tend to release more water at very low matrix potentials (Ψm=-1 to -40 kPa) which is 10 to 100 times lower than similar plant-available water tensions in soils (Lea-Cox *et al.*, 2011). Plant-available water (PAW) is the amount of water accessible to the plant, which is affected by the physical properties of the substrate, the geometry (height and width) of the container and the total volume of the container (Handreck and Black, 2002). Container root systems are usually confined within a short time after transplanting, and shoot: root ratios are usually larger than those of soil-grown plants, for similarly-aged plants. For all these reasons, maintaining the optimal water status of soilless substrates has been recognized as being critical for continued growth, not only because of limited water-holding capacity, but also because of the inadequacies of being able to accurately judge when plants require water (Karlovich and Fonteno, 1986). Although it is likely that mature plant root systems can extract substrate moisture at ψ_m less than -40 kPa, Leith and Burger (1989) and Kiehl *et al.*, (1992) found significant growth reductions at substrate Ψ_m as small as -16 kPa (0.16 Bar). This has major implications for choosing appropriate sensors for use in soilless substrates, as well as the measurement and automatic control of irrigation in these substrates.

Jones (2004) summarized the various types of soil moisture sensors available for irrigation scheduling (Table 8.1). The variety of soil moisture sensors (tensiometric, neutron, resistance, heat dissipation, psychrometric or dielectric) has continually evolved since then; the choices are now overwhelming, since each sensor may have specific strengths and weaknesses in a specific situation. Tensiometers have long been used to measure matric potential in soils (Smajstrla and Harrison, 1998) and in soilless substrates (Burger and Paul, 1987). Although tensiometers have proven to be valuable research tools, they have not been adopted widely in greenhouse and nursery production, mainly because of the problems with using them in highly porous soilless substrates. Tensiometers rely on direct contact between the porous ceramic tip and substrate moisture. If the substrate shrinks, or the tensiometer is disturbed, this contact may be disrupted. Air then enters and breaks the water column in the tensiometer, resulting in incorrect readings and maintenance issues (Zazueta *et al.*, 1994). A number of next generation soil moisture sensors have become available and provide precise data in a wide range of soilless substrates. These sensors determine the volumetric moisture content by measuring the apparent dielectric constant of the soil or substrate. These sensors are easy to use and provide highly reproducible data (Van Iersel *et al.*, 2011). The Decagon range of sensors are designed to be installed in soils or substrates for longer periods of time and all interface with Decagon's range of EM50 nodes, data station and data tracl software (http://decagon.com/products). Dielectric sensors generally require substrate-specific calibrations, because the dielectric properties of different soils and substrates differ, affecting sensor output. The conversion between water potential and volumetric water content (VWC) varies substantially with soil type. It is possible to inter-convert matric potential to volumetric water content (Lea-Cox *et al.*, 2011) for various sensors using substrate moisture release curves. However, such release curves are substrate-specific, and may change over time as the physical properties (*e.g.* pore size distribution) of the substrate changes (Van Iersel *et al.*, 2011) or root systems become more established. Fortunately, in most irrigation scheduling applications, the objective is simply to apply a volume of water that returns the soil moisture content to its original well-watered state. Changes in this total water-holding capacity (*i.e.* the maximum VWC reading) can easily be monitored for changes over time, *i.e.* after significant rainfall events or by periodically saturating the container with the embedded sensor.

More recently, hybrid 'tensiometer-like' sensors have been developed which use the principle of dielectric sensors (*e.g.* Equitensiometer, Delta T; MPS-1, Decagon Devices etc.) to determine the water potential of substrates (Van Iersel *et al.*, 2011). An advantage of such sensors is that they do not require substrate-specific calibrations, since they measure the water content of the ceramic material, not that of the surrounding soil or substrate. Unfortunately, the sensors that are currently available are not very sensitive in the matric potential range where soilless substrates hold most plant-available water (0 to -10 kPa). In addition, it is not clear whether these sensors respond quickly enough to capture the rapid changes that can occur in soilless substrates (Van Iersel *et al.*, 2011).

In a recent study, Hossain *et al.*, (2008) used EM38 to measure and map soil moisture content in the root zone of an irrigated cracking clay soil. They concluded that the apparent electrical conductivity (ECa) data produced by the commercially-available EM38 unit can be used with calibration to imply soil moisture in the root zone of agricultural crops. Padhi and Misra (2009) also suggested that the EM38 proved to be a useful technique for understanding paddock scale soil moisture variability in the root zone. Hedley and Yule (2009) also applied electromagnetic induction techniques for mapping soil water status and managing variable rate irrigation. The EM38 sensor was used to map soil variability and from this information management zones are identified.

The plant available water-holding capacity of each management zone was measured by taking soil samples between very wet and very dry. Aquaflex TDR soil moisture sensors were used to monitor soil moisture to 60cm soil depth, on an hourly basis in each management zone. This enabled the calibration of an EM map for plant available water-holding characteristics, so that a soil moisture map could be produced. A daily time step was then added to this soil moisture map so that as soils dried out, the zones which dried fastest, and therefore reached the trigger point for irrigation, were identified. The use of ground penetrating radar (GPR) for spatial mapping of soil moisture has received some attention in recent time.

MODELING AND IRRIGATION SENSING HYBRIDIZED APPROACHES

Previous studies with a variety of crop, ornamental and turf species have reported that the use of appropriate scheduling methods and precision irrigation technologies can save a significant amount of water, while maintaining or increasing yield and product quality (Bacci *et al.*, 2008; Blonquist *et al.*, 2006). Many of these empirical approaches have successfully incorporated environmental variables into various models, to further increase the precision of irrigation scheduling (Treder *et al.*, 1997). As an example of enhancing capability for precision water applications with knowledge of real-time plant water use, Van Iersel and his group conducted several studies (Burnett and Van Iersel 2008; Kim and Van Iersel, 2009). Van Iersel *et al.*, (2011) demonstrated that automated irrigation using soil moisture sensors allows for the very precise irrigation of greenhouse crops in soilless substrates. In addition, they maintained very low substrate moisture contents at very precise levels which advances our capability to use precision irrigation scheduling for regulated deficit irrigation (RDI) techniques (Jones, 2004), to increase fruit crop quality (Fereres *et al.*, 2003), and aid in precision nutrient (Lea-Cox *et al.*, 2001; Ristvey *et al.*, 2004) and disease management (Lea-Cox *et al.*, 2006). Most recently, Kim and Van Iersel (2011) have demonstrated that the measured daily evapotranspiration of petunia in the greenhouse can be accurately modeled with measurements of crop growth (days after planting, DAP), daily light integral (DLI), vapor pressure deficit (VPD) and air temperature. All these environmental fluxes obviously affect transpiration on a continuous basis. Ambient light affects plant water use due to its effects on evaporation and stomatal opening (Pieruschka *et al.*, 2010). Vapor pressure deficit is the driving force for transpiration and also affects stomatal regulation (Taiz and Zeiger, 2006) while temperature affects ET and plant metabolic activity (Van Iersel 2003). The importance of Kim and Van Iersel's empirical modeling approach is how they have demonstrated the sensitivity of plant water use to these four easily-measured variables. Thus, with a few inexpensive sensors (temperature, relative humidity and photosynthetic photon flux, PPF) and some simple software tools that can integrate these variables on short time-scales, it now appears possible to predict hourly plant water use for greenhouse crops with real precision. It should be noted however, that these models still require rigorous validation for production conditions. However, for these types of models to work in an external environment, it is likely that the complexity of our predictive water use models will have to increase, to incorporate additional variables. Water use by perennial woody crop species is much more complicated due to external environmental conditions (*for example*, how VPD and leaf temperature are affected by wind speed and boundary layer effects on canopies;

LAI effects on PPF interception). *For example*, Bowden *et al.* (2005) outlined an automated sensor-based irrigation system for nurseries that could calculate plant water consumption from species and genotype-specific plant physiological responses. The MAESTRA [Multi-Array Evaporation Stand Tree Radiation A] model (Wang and Jarvis, 1990) is a three-dimensional process-based model that computes transpiration, photosynthesis, and absorbed radiation within individual tree crowns at relatively short time (15-minute) intervals. The model is described fully by Bauerle, and Bowden (2011) and has been modified and previously validated to estimate deciduous tree transpiration (Bauerle *et al.*, 2002; Bowden and Bauerle 2008) and within-crown light interception (Bauerle *et al.*, 2004). The model applies physiological equations to sub-volumes of the tree crown and then sums and/or averages the values for entire canopies. Additionally, species-specific physiological values can be incorporated into model calculations, potentially yielding more accurate estimates of whole tree transpiration. The model holds potential advantages for nursery, forest, and orchard water use prediction in that structural parameters such as tree position, crown shape, and tree dimensions are specified. Bowden *et al.*, (2005) briefly illustrated how the model estimates of water use and plant water requirements are outputted from MAESTRA and used to make both irrigation decisions (command executed by a sensor node) and visualize model updates via a graphic user interface (Bauerle *et al.*, 2006). Within each 15-minute time step, the model adjusts transpiration based on interactions between environment, soil moisture and plant physiological response. The substrate moisture deficit calculation is described by Bauerle *et al.* (2002). An updated substrate moisture value is carried into the next time step for input into the substrate moisture deficit sub-routine. The calculated moisture deficit value is one of the input values required to calculate the amount of stomatal conductance regulation and hence, interacts with other equations to derive whole plant water use. Overall, this GUI (Bowden *et al.*, 2005; Bauerle *et al.*, 2006) provides a user friendly interface to a complex set of calculations. In this way, whole tree water use estimates can be rapidly visualized for either sensor node or human based irrigation decision management. Bauerle and his group are actively working to further refine the MAESTRA model by irrigation scheduling decision support system (Lea-Cox *et al.*, 2010).

ADVANTAGES OF PRECISION IRRIGATION

Precision irrigation has the potential to increase both the water use and economic efficiencies. It has been reported that precision irrigation (drip and sprinkler) can improve application efficiency of water to the tune of 80-90 per cent as against 40-45 per cent in surface irrigation method (Dukes and Scholberg, 2004). Results from case studies of variable rate irrigation showed water savings in individual years ranging from zero to 50 per cent. The potential economic benefit of precision irrigation lies in reducing the cost of inputs or increasing yield for the same inputs. Improved decision support systems and technology for real time monitoring and control would increase the utility of precision irrigation.

Water savings

The primary goal of precision irrigation is to apply an optimum amount of irrigation throughout the field. It is reported by many researchers as the most likely means of achieving significant water savings (Evans and Sadler, 2008). The site specific or variable rate irrigation is considered as a necessary or essential component of precision irrigation. Most researchers expect a reduction in water use on at least parts of fields, if not a reduction in the value aggregated over entire fields (Sadler *et al.*, 2005). It has been reported that variable rate irrigation could save 10 to 15 per cent of water used in conventional irrigation practice (Yule *et al.*, 2008). Hedley and Yule (2009) suggested water savings of around 25 per cent are possible through improvements in application efficiency obtained by spatially varied irrigation applications.

Yield and profit

The experimental studies carried out by King *et al.* (2006) for measuring the yield of potatoes under spatially varied irrigation applications reveals that yields were better in two consecutive years over uniform irrigation management. Similarly, Booker *et al.* (2006) analyzed yields and water use efficiency for spatially varied irrigation over four years for cotton. They concluded that cotton seems to be unpredictable to manage with spatially varied irrigation. These result are supported by the work of Bronson *et al.* (2006).

Table 8.1: Major irrigation sensing/scheduling approaches, along with their main advantages and disadvantages.

Irrigation sensing/scheduling approaches	Advantages	Disadvantages
I. Soil water measurement (a) Soil water potential (tensiometers, psychrometers, etc.) (b) Soil water content (gravimetric; capacitance/TDR; neutron probe)	Easy to practice; can be quite precise; at least water content measures indicate 'how much' water to apply; many commercial systems available; some sensors (especially capacitance and time domain sensors) readily automated	Soil heterogeneity requires many sensors (often expensive) or extensive monitoring programme (e.g. neutron probe); selecting position that is representative of the root-zone is difficult; sensors do not generally measure water status at root surface (which depends on evaporative demand).
II. Soil water balance calculations (Require estimate of evaporation and rainfall)	Easy to apply in principle; indicate 'how much' water to apply	Not as accurate as direct measurement; need accurate local estimates of precipitation/run-off; evapotranspiration

Irrigation sensing/scheduling approaches	Advantages	Disadvantages
		estimates require good estimates of crop coefficients (which depend on crop development, rooting depth, etc.); errors are cumulative, so regular recalibration needed.
III. Plant 'stress' sensing (Includes both water status measurement and plant response measurement)	Measures the plant stress response directly; integrates environmental effects; potentially very sensitive	In general, does not indicate 'how much' water to apply; calibration required to determine 'control thresholds'; still largely at research/development stage and little used yet for routine agronomy (except for thermal sensing in some situations).
(a) Tissue water status	It has often been argued that leaf water status is the most appropriate measure for many physiological processes (*e.g.* photosynthesis), but this argument is generally erroneous (as it ignores root-shoot signalling)	All measures are subject to homeostatic regulation (especially leaf water status), therefore not sensitive (isohydric plants); sensitive to environmental conditions which can lead to short-term fluctuations greater than treatment differences.
(i) Visible wilting	Easy to detect	Not precise; yield reduction often occurs before visible symptoms; hard to automate.
(ii) Pressure chamber (ψ)	Widely accepted reference technique; most useful if estimating stem water potential (SWP), using either bagged leaves or suckers	Slow and labour intensive (therefore expensive, especially for predawn measurements); unsuitable for automation.

Irrigation sensing/scheduling approaches	Advantages	Disadvantages
(iii) Psychrometer (ψ)	Valuable, thermodynami based measure of water status; can be automated	Requires sophisticated equipment and high level of technical skill, yet still unreliable in the long term.
(iv) Tissue water content (RWC, leaf thickness [γ- or β-ray thickness sensors], fruit or stem diameter)	Changes in tissue water content are easier to measure and automate than water potential measurements; RWC more directly related to physiological function than is total water potential in many cases; commercial micromorphometric sensors available	Instrumentation generally complex or expensive, so difficult to get adequate replication; water content measures (and diameter changes) subject to same problems as other water status measures; leaf thickness sensiti-vity limited by lateral shrinkage
(v) Pressure probe	Can measure the pressure component of water potential which is the driving force for xylem flow and much of cell function (*e.g.*, growth)	Only suitable for experimental or laboratory Systems
(vi) Xylem cavitation	Can be sensitive to increasing water stress	Cavitation frequency depends on stress prehistory; cavitation-water status curve shows hysteresis, with most cavitations occurring during drying, so cannot indicate successful rehydration.
(b) Physiological responses	Potentially more sensitive than measures of tissue (especially leaf) water status	Often require sophisticated or complex equipment; require calibration to determine 'control thresholds'.
(i) Stomatal conductance	Generally a very sensitive response, except in some anisohydric species	Large leaf-to-leaf variation requires much replication for reliable data.

Irrigation sensing/scheduling approaches	Advantages	Disadvantages
(a) Porometer	Accurate and bench-mark for research studies	Labour intensive so not suitable for commercial application; not readily automated (though some attempts have been made).
(b) Thermal sensing	Can be used remotely; capable of scaling up to large areas of crop (especially with imaging); imaging effectively averages many leaves; simple thermometers cheap and portable; well suited for monitoring purposes	Canopy temperature is affected by environmental conditions as well as by stomatal aperture, so needs calibration (*e.g.* using wet and dry reference surfaces).
(c) Sap-flow sensors	Sensitive	Only indirectly estimates changes in conductance, as flow is also very dependent on atmospheric conditions; requires complex instrumenta-tion and technical expertise; needs calibration for each tree and for definition of irrigation control threshold.
(ii) Growth rate	Probably the most sensitive indicator of water deficit stress	Instrumentation delicate and generally expensive.

PRECISION IRRIGATION TECHNOLOGIES IN INDIA

AgriSens irrigation system

AgriSens irrigation system was designed, developed and deployed at Indian Institute of Technology (IIT), Mumbai for its utility infield monitoring of grape crop performance (Das *et. al.*, 2010). In vineyard, there are different types of grapevines which requires different amount of soil moisture. Also, it is very difficult to manually control the irrigation required for particular type of grapevine. WSN based irrigation automation can tackle the problem and also help to save considerable amount of water. The moisture contents of the soil decide the actuator activation. If the threshold level of the soil moisture goes down below a certain level, the valve gets open. This threshold level has to be decided based on climate, topography and type of plant, etc, at the Agri-Information server (Desai *et al.*, 2008). The WSN System was designed to aid end users and researchers to analyze real time sensor data and assist in decision making and estimating crop water requirement. It was a web based application that could be accessed ubiquitously by the users thus providing a convenient and nimble tool. Since it was integrated with google map, it could provide location-based data. Moreover, this enabled the information to be displayed in a visually readable format.

Field Server (FS) system

Field Server (FS) is a WiFi (long-range communication) based self-organizing distributed sensing device with 24 channels to sense various weather, agricultural and environmental parameters. Sensors used in the present study include air-temperature, humidity, relative humidity, solar radiation, leaf wetness, soil moisture and CO_2 concentration. The sensory data collected at customized regular intervals gets transmitted and stored in the main/parent FS in the form of XML and with front end in HTML Java interfaces. The parent FS, which is equipped with Fit2PC, will store and transmits/shares all files through virtual private network (VPN) and Opera-Unite based cloud services to the GeoSense centralized server. The received FS raw data (XML) is appended to the sensory database at every one minute interval. To execute this function, the PHP based algorithm was developed to read new files. This new file is identified on the basis of file name *i.e.*, while storing the file in the database; modify the file name with system (Fit2PC) date and time in such a way that file can be identified and appended in the FS database. Also, algorithms were developed with PHP and Java languages to convert raw (analog to digital) to real (usable units) sensory data; and these data (raw as well as real) are stored in phpMyAdmin (SQL) database. A provision has been made to export the converted real data into open source consortium (OSC) data formats such as CodeGen, CSV, Excel, Word, Latex, Open Document Spreadsheet, Open Document Text, PDF, SQL, Texy Text, XML, YAML.

FIELD SERVER (FS) -BASED FLUX TOWER (FLTS) SYSTEM

Two FLTs were deployed in maize field to study the weather profiles and partitioning of energy into different fluxes (latent heat flux, sensible heat flux, ground heat flux). Each Flux Tower consists of three sensor modules with temperature, relative humidity and CO_2 concentration sensors at three different heights (1 m, 2 m and 3 m). Real time knowledge of weather profiles and energy fluxes allow farming community to calculate water requirement (ET), irrigation scheduling, pest and disease management, etc. (Karandikar Ketan *et al.*, 2011). Flux tower sensors were embedded with Field Server Engine (FSE) board (one of the components of FS) and are in parallel connection with FS with registered jack (RJ) 45 (RJ45) connectors. The associated FS collects and transmits sensory data to the designated server in the same manner as FS.

Field Twitter (FT) System

Field Twitter (FT) (Hirafuji *et al.,* 2011) comprised of Ardunio signal (transmitting to the internet clouds) through Fon and Algorithm process for Field Twitter data to the twitter environment.

Arduino: Arduino is an Open-Hardware electronics prototyping platform based on flexible, easy-to-use hardware and software. Ardunio is attached with an external handmade soil moisture sensor (probe) at a depth of 15 cm. This is the first attempt in the world in developing an open-hardware based cost effective sensing system and is particularly useful in developing countries where WSN is still a novice and costly technology (Fig. 8.4).

Signal transmission: In FT, the communication mode consumes more power than any other parts, as it has been customized into WiFi based communication system by using Fon (router) that helps in receiving the internet pockets (3G) from the Field Server. Subsequently, it Tweets/Transmits the attached sensory data either through gateway or in twitter (Hydbot01) environment. Anyone can follow the Field Twitter sensory data in Twitter social network in the name of "Hydbot01".

Software development: FT sensory data was stored in twitter database in the form of webpage, with XML syntax, which could be useful to maintain FT database. In FT web interface, the sensory data is available in raw format (analog to digital conversion – ADC). PHP-based algorithms have been developed for converting raw data into usable format (units) and store the data in GeoSense database.

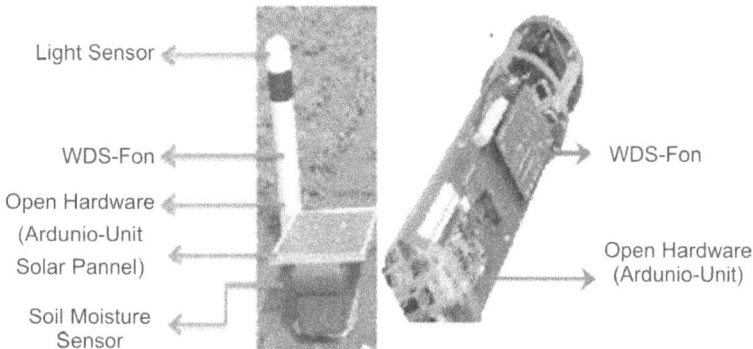

Fig. 8.4: Architecture of FieldTwitter (FT) system

In order to study the precision agriculture (crop yield modeling and irrigation aspects), experiments were carried out on maize (Dekalb Super 900 m cultivar), groundnut (TMV-2) and rice (MTU-1010) crops (Sudharsan *et al.,* 2012). Standard agriculture experimental design was laid out in the test bed, with different irrigation systems (rainfed, ridge & furrow, drip irrigation as per crop water requirement and life saving irrigation) in maize (*Karif*/monsoon agriculture season) and groundnut (*Rabi*/post-monsoon agriculture season) crops for precision irrigation and crop-yield modeling aspects. In addition, long-term rice experiments were also carried out with different dates of sowing and nitrogen levels to arrive at proper decision making to overcome climate change risks. The distributed wireless sensing systems (Field Server, Agrisens, Flux Towers and Field Twitter) were deployed under different crop experiments: Field Servers in rice, groundnut and maize fields for yield modeling. Agrisens were deployed in maize for nitrogen management and groundnut field for pest management, Flux Tower in maize crop field for crop energy balance studies and Field Twitter in the groundnut field for different date of sowing observations. Weather parameters from the weather station, which is close proximity to the experimental site, will augment the GeoSense researches and also help in validating the sensory data. Thus, these sensing systems are very useful to the rural farming community for ubiquitous decision making in precision agriculture aspects. In future, this system could be used in climate/environmental systems to understand the micro-climatic variations in real-time mode.

Challenges and Research Opportunities

There are several researchable issues to make precision irrigation, a viable and farmer-friendly technology. The major challenge include standardization of wireless sensing networks (WSN) protocols and communication frequencies, as they can be confusing for growers and researchers alike. Another challenge is working with large datasets. Thus, it is very essential to educate ourselves and users to the resolution required for optimum precision in each environment, keeping the ultimate use of the data in mind. The maintenance and calibration of sensors and equipment is an ongoing concern, particularly for growers who may be uncomfortable with the technology and equipment. Thus, opportunity for paid consultants to maintain and remotely monitor WSNs for optimum performance may solve the problem. Moreover, there is an urgent need to develop an online knowledge center, to provide assistance and guidance about various aspects of precision irrigation technologies, sensor use, strategies and best practices. Strong integration of better data analysis tools to handle large volumes of data from sensor networks and thorough user interface study on how growers actually use computer interfaces and to determine what features are needed are the major tasks to be done at priority. Predictive models for plant water use, environmental and disease management tools are rapidly being developed for growers, but there is a need to validate and verify these models for use in different environments. Incorporation of models into WSNs for decision-making appears to be relatively easy, but there are many details which have yet to be worked out. Although, there are many direct benefits of precision irrigation scheduling that can be accrued by the grower, such as saving on water, labor, electricity and fertilizer costs. However, there are many indirect (*e.g.* reduced disease incidence, fungicide costs) and societal benefits (reduced nutrient run-off, ground water consumption) that may have much larger benefits on long-term for all agricultural producers. Most importantly, there is a need to quantify the return on investment that a grower could expect to achieve, and to be able to scale those benefits for small and marginal farmers.

CONCLUSION

Extending the concept of precision irrigation to include spatially precise irrigation appears to have several potential opportunities for conservation of water and nutrients. The history of research confirms that precision irrigation is technologically feasible, if not yet economically advantageous. The considerations that might change this conclusion include increased awareness of need for water conservation because of drought, increased contention for short water supplies and possible future regulatory actions. The amount of water that could be conserved using precision irrigation remains a research topic. Though, the concept of precision irrigation or irrigation as a component of precision agricultural systems is still in its infancy, some more case studies similar to the one described for other crop-agriculture systems will go a long way in building faith in sensor based irrigation towards both saving precious water as well as soil-degradation due to excessive surface flood irrigation. It also remains to be seen through the field trials that precision irrigation can provide substantially greater benefits than traditional irrigation scheduling. The advances in wireless sensor networks, plant and soil irrigation sensing approaches have made some practical deployment possible for various agricultural operations on demonstration scale, which until a few years ago was considered extremely costly or labor intensive. Precision irrigation system with robust components such as, sensing agricultural parameters, identification of sensing location and data gathering, transferring data from crop field to control station for decision making and actuation and control decision based on sensed data will find application in future agriculture. Thus, the great potential of integrating the precision farming with WSN to interpolate over a large area for spatial decision making need to be tapped for making Indian agriculture well placed in future.

REFERENCES

Akyildiz IF and Xudong W. (2005). A Survey on Wireless Mesh Networks, IEEE Communication Magazine, Vol. **43**, pp. S23-S30.

Allen RG, Pereira LS , Raes D and Smith M. (1998). Crop Evapotranspiration: Guidelines for Computing Crop Water Requirements. FAO Irrigation and drainage paper no.56. FAO, Rome, Italy.

Anastasiou A, Savvas D, Pasgianos G, Stangellini C, Kempkes F and Sigrimis N. (2008). Decision Support for Optimised Irrigation Scheduling. *Acta Hort.* **807**:253–258.

Ardunio. (2011). "Open-source hardware and software" Available at *http://www.arduino.cc/* accessed on (02/01/2012).

Bacci L, Battista P and Rapi B. (2008). An Integrated Method for Irrigation Scheduling of Potted Plants. *Sci. Hortic.* **116**:89-97.

Balendonck J, Stanghellini C, and Hemming J. (2007). Farm Level Optimal Water Management: Assistant for Irrigation under Deficit (FLOW-AID). *In: Water Saving in Mediterranean Agriculture & Future Research Needs*, Bari, Italy, Lamaddalena N. & Bogliotti, C. (Eds.), CIHEAM, Bari, Italy. pp. 301-312.

Balendonck J, Stanghellini C, Hemming J , Kempkes FLK and van Tuijl, BAJ. (2009). Farm Level Optimal Water Management: Assistant for Irrigation under Deficit (FLOWAID). *Acta Hort* **807**: 247-252.

Barnes EM , Pinter PJ , Kimball BA , Hunsaker DJ , Wall GW and LaMorte RL. (2000). Precision irrigation management using modelling and remote sensing approaches. Proceedings of the 4th Decennial National Irrigation Symposium. Phoenix, AZ, American Society of Agricultural Engineers.

Bauerle WL and Bowden JD. (2011). Separating Foliar Physiology from Morphology Reveals the Relative Roles of Vertically Structured Transpiration Factors within Red Maple Crowns and Limitations of Larger Scale Models. *J. Exp. Bot.* **62**:4295-4307.

Bauerle WL, Bowden JD, McLeod MF & Toler JE. (2004). Modeling Intracrown and Intracanopy Interactions in Red Maple: Assessment of light transfer on carbon dioxide and water vapor exchange. *Tree Physiol.* **24**:589-597.

Bauerle WL , Post CJ , McLeod M.E , Dudley JB and Toler JE. (2002). Measurement and Modeling of the Transpiration of a Temperate Red Maple Container Nursery. *Agric. For Meteorol.* **114**:45-57.

Bauerle WL , Timlin DJ , Pachepsky Ya A and Anantharamu S. (2006). Adaptation of the Biological Simulation Model MAESTRA for Use in a Generic User Interface. **Agron. J.** **98**:220-228.

Belmonte, AC, Jochum, AM, Garcia, AC, Rodriguez, AM and Fuster, PL (2005) Irrigation management from space: Towards user friendly products. *Irrigation and Drainage Systems*, **19**: 337-353.

Bitelli M. (2011). Measuring Soil Water Content: A Review. *HortTechnology* 21: 293-300.

Blonquist, JJM, Jones, SB & Robinson, DA (2006). Precise Irrigation Scheduling for Turfgrass using a Subsurface Electromagnetic Soil Moisture Sensor. *Agricultural Water Management* 84:153-165.

Booker JD , Bordovsky J, Lascano RJ and Segarra E . (2006). Variable Rate Irrigation on Cotton Lint Yield and Fiber Quality. *Belt wide Cotton Conferences*, San Antonio, Texas.

Bowden JD, Bauerle WL , Lea-Cox JD and Kantor GF. (2005). Irrigation Scheduling: An Overview of the Potential to Integrate Modeling and Sensing Techniques in a Windows-based Environment. *Proc. Southern Nursery Assoc. Res. Conf.* **50**:577-579.

Bowden JD and Bauerle WL. (2008). Measuring and Modeling the Variation in Species-Specific Transpiration in Temperate Deciduous Hardwoods. *Tree Physiol.* **28**:1675-1683.

Bronson, KF, Booker, JD, Bordovsky, JP, Keeling, JW, Wheeler, TA, Boman, RK, Parajulee, MN, Segarra, E and Nichols RL. (2006). Site-specific Irrigation and Nitrogen Management for Cotton Production in the Southern High Plains. *Agronomy Journal*, **98**, 212-219.

Bunt AC. (1961). Some Physical Properties of Pot-Plant Composts and their Effect on Plant Growth *Plant and Soil* **13**:322-332.

Burger DW and Paul JL. (1987). Soil Moisture Measurements in Containers with Solid State, Electronic Tensiometers. *Hort Science* **22**:309-310.

Burnett SE and van Iersel MW. (2008). Morphology and Irrigation Efficiency of *Gaura Lindheimeri* Grown with Capacitance-Sensor Controlled Irrigation. *Hort Science* **43**:1555-1560.

Chavez JL , Gowda PH , Howell TA and Copeland KS. (2009). Radiometric surface temperature calibration effects on satellite based evapotranspiration estimation. *International journal of Remote sensing*, **30**(9): 2337-2354.

Cifre J, Bota J, Escalona, JM , Medrano H and Flexas J. (2005). Physiological Tools for Irrigation Scheduling in Grapevine (*Vitis vinifera* L.). *Agric. Ecosys. Environ.* **106**:159-170.

Colaizzi PD, Barnes EM, Clarke TR, Choi CY, Waller PM, Haberland J, Kostrzewski M. (2003). Water stress detection under high frequency sprinkler irrigation with water deficit index. *Journal of Irrigation and Drainage Engineering* **129**, 36–42.

Das Ipsita, Shah NG and Merchant SN. (2010). AgriSens: *Wireless Sensor Network in Precision Farming: A Case study*. LAP Lambert Academic Publishing, Germany, ISBN: **978**-3-8433-5525-4, Germany

deBoodt M and O Verdonck. (1972). The Physical Properties of Substrates in Horticulture. Acta Hort. **26**:37-44.

Desai UB, Merchant SN and Shah NG. (2008). Design and Development of Wireless SensorNetwork for Real Time Remote Monitoring. *Unpublished Project Report* submitted to Ministry of Information Technology, Govt. of India.

Dukes MD and Scholberg JM. (2004). Automated Subsurface Drip Irrigation Based on Soil Moisture. ASAE Paper No. 052188.

Evans DE, Sadler EJ, Camp CR, Millen JA. (2001). Spatial canopy temperature measurements using center pivot mounted IRTs. Proceedings of the 5th International Conference on Precision Agriculture, Bloomington, Minnesota, July, 2000, 1–11.

Evans RG and Sadler E. (2008). Methods and Technologies to Improve Efficiency of Water Use. *Water Resources Research*, **44:**

Evans RG and Harting GB. (1999). Precision irrigation with center pivot systems on potatoes. In: Proceedings of American Society of Civil Engineers 1999 International Water Resources Engineering Conference. R. Walton and R.E. Nece (eds.) August 8-11. Seattle, Washington. American Society of Civil Engineers, Reston,Virginia. CD-ROM, no pagination.

Evans RG, Han S, Kroeger MW and Schneider SM. (1996). Precision centre pivot irrigation for efficient use of water and nitrogen. Precision Agriculture, Proceedings of the 3rd International Conference, ASA/CSSA/SSSA, Minneapolis, Minnesota, June 23-26, pp. 75-84.

Ferentinos KP, Anastasiou A, Pasgianos GD , Arvanitis KG and Sigrimis N. (2003). A DSS as a tool to optimal water management in soilless cultures under saline conditions. *Acta Hort.* **609**:289-296.

Fereres E, Goldhamer DA and Parsons L R (2003). Irrigation Water Management of Horticultural Crops. *Hort Science* **38**:1036-1042.

Goodwin I and O'Connell MG (2008). The Future of Irrigated Production Horticulture - World and Australian Perspective. *Acta Horti*, **792:** 449–458.

Handreck K and Black N. (2002). Growing Media for Ornamental Plants and Turf (3rd Ed.). *Univ. New South Wales Press*, Sydney, Australia.

Hedley CB and Yule I J. (2009). A method for spatial prediction of daily soil water status for precise irrigation scheduling. *Agricultural Water Management*, **96**(12): 1737-1745.

Hirafuji H, Yochi H , Kura T , Matsumoto K , Fukatsu T , Tanaka K, Shibuya Y, Itoh A, Nesumi H , Hoshi N , Ninomiya S, Adinarayana J , Sudharsan D , Saito Y , Kobayashi K and Suzuki T. (2011). Creating –High-Performance/Low-cost Ambient Sensor Cloud System usingInternational Journal of Database Management Systems (IJDMS) Vol.4, No.1, February 2012 OpenFS (Open Field Server) for High-throughput Phenotyping, *Proceedings of the SICE Annual Conference, IEEE Catalog No. CFP11765-DVD*, pp 2090-2092.

Hornbuckle JW , Car NJ, Christen EW , Stein TM and Williamson B. (2009). IrriSatSMS Irrigation water management by satellite and SMS - A utilisation framework. CRC for

Irrigation Futures Technical Report No. 01/09, CSIRO Land and Water Science Report No. 04/09.

Hossain MB Lamb DW, Lockwood PV and Frazier P. (2008). Field determination of soil moisture in the root zone by multi-height EM38 measurements. 1st Global Workshop on High Resolution Digital Soil Sensing and Mapping, 5-8 Feb., Sydney (CDROM), Huguet J-G, Li SH.

G. (1992). Specific micromorphometric reactions of fruit trees to water stress and irrigation scheduling automation. *Journal of Horticultural Science* **67**, 631–40.

Hubbard SS, Redman JD and Annan AP (2003). Measuring soil water content with ground penetrating radar: A review. Vadose Zone Journal, **2**: 476-491.

Johnson L , Pierce L , Michaelis A , Scholasch T and Nemani R. (2006). Remote sensing and water balance modeling in california drip-irrigated vineyards. Proceedings, ASCE World Environmental & Water Resources Congress, Omaha NE, 21-25, May year 33.

Jones HG. (2004). Irrigation Scheduling: Advantages and Pitfalls of Plant-Based Methods. *J. Exp. Bot.* **55**:2427-2436.

Jones HG. (2008). Irrigation Scheduling-Comparison of Soil, Plant and Atmosphere Monitoring Approaches. *Acta Hort.* **792**: 391-403.

Kacira M Ling PP and Short T.H. (2002). Establishing crop water stress index (CWSI) threshold values for early, non-contact detection of plant water stress. Transactions of the American Society of Agricultural Engineers **45**: 775–780.

Kacira M and Ling P.P. (2001). Design and development of an automated non-contact sensing system for continuous monitoring of plant health and growth. Transactions of the American Society of Agricultural Engineers **44**: 989–996.

Karandikar Ketan, J Adinarayana, D Sudharsan, AK Tripathy, Abhishek Kodilkar, UB Desai, SN Merchant, K Tanaka, S Ninomiya, T Kiura, M Hirafuji, D Raji Reddy and G Sreenivas (2011). Energy Balance and Weather Profile Studies using Sensor Network based Flux Tower- A preliminary Approach" Conference on Geosptial Technologies and its Application (Geomatrix-11) 26-27 February 2011, IIT, Mumbai, India, pp: 97-107.

Karlovich PT and Fonteno WC. (1986). The Effect of Soil Moisture Tension and Volume Moisture on the Growth of *Chrysanthemum* in Three Container Media. *J. Amer. Soc. Hort. Sci.* **111**:191-195.

Kiehl PA , Liel JH & Buerger DW. (1992). Growth Response of Chrysanthemum to Various Container Medium Moisture Tensions Levels. *J. Amer. Soc. Hort. Sci.* **117**:224-229.

Kim Y and Evans RG. (2009). Software design for wireless sensor-based site-specific irrigation. *Comput. Electron. Agric.*, **66**: 159-165

Kim J and van Iersel MW. (2011). Abscisic Acid Drenches can Reduce Water Use and Extend Shelf Life of Salvia Splendens. *Sci. Hort.* **127**: 420–423.

Kim J and van Iersel MW. (2009). Daily Water Use of Abutilon and Lantana at Various Substrate Water Contents. Proc. Southern Nursery Assn. Res. Conf. **54**:12-16.

Kim Y, Evans RG and Iversen WM. (2008). Remote Sensing and Control of an Irrigation System using a Distributed Wireless Sensor Network. *IEEE Trans. Instrum. Meas.* **57**:1379-1387.

King BA and Kincaid DC. (1999). Variable flow sprinkler for site-specific water and nutrient management. ASAE paper No. 962074, St Joseph, MI.

King BA, Stark JC and Wall R W. (2006). Comparison of Site-Specific and Conventional Uniform Irrigation Management for Potatoes. *Applied Engineering in Agriculture,* **22(5)**, 677-688.

Lea-Cox J.D, Arguedas-Rodriguez FR , Amador P , Quesada G and Mendez CH. (2006). Management of the Water Status of a Gravel Substrate by Ech20 Probes to Reduce *Rhizopus* Incidence in the Container Production of *Kalanchoe blossfeldiana. Proc. Southern Nursery Assoc. Res. Conf.* **51**: 511-517.

Lea-Cox JD, Kantor GF , Anhalt J., Ristvey AG and Ross DS. (2007). A Wireless Sensor Network for the Nursery and Greenhouse Industry. *Proc. Southern Nursery Assoc. Res. Conf.* **52**: 454-458.

Lea-Cox JD , Arguedas-Rodriguez FR , Ristvey AG and Ross DS. (2011). Relating Real time Substrate Matric Potential Measurements to Plant Water Use, for Precision Irrigation. *Acta Hort.* 891:201-208.

Lea-Cox JD, Kantor GF , Bauerle WL , van Iersel MW , Campbell C , Bauerle TL , Ross DS, Ristvey AG , Parker D, King D , Bauer R , Cohan SM , Thomas P. Ruter JM , Chappell M , Lefsky M , Kampf S and L Bissey. (2010). A Specialty Crops Research Project: Using Wireless Sensor Networks and Crop Modeling for Precision Irrigation and Nutrient Management in Nursery, Greenhouse and Green Roof Systems. *Proc. Southern Nursery Assoc. Res. Conf.* **55**: 211-215.

Lea-Cox JD , Ross DS and Teffeau KM. (2001). A Water and Nutrient Management Planning Process for Container Nursery and Greenhouse Production Systems In Maryland. *J. Environ. Hort.* 19:230-236.

Leinonen I and Jones HG. (2004). Combining thermal and visible imagery for estimating canopy temperature and identifying plant stress. *Journal of Experimental Botany* **55**: 1423–1431.

Leith JH and Burger DW. (1989). Growth of Chrysanthemum using an irrigation system Controlled by Soil Moisture Tension. *J. Amer. Soc. Hort. Sci.* **114**: 387-397.

Liu G and Ying Y. (2003). Application of Bluetooth Technology in Greenhouse Environment, Monitor and Control. *J. Zhejiang Univ., Agric Life Sci.* 29:329-334.

Mahan JR, Burke JJ, Upchurch DR and Wanjura DF. (2000). Irrigation scheduling using biologically-based optimal temperature and continuous monitoring of canopy temperature. *Acta Horti.* 537: 375–381.

Majsztrik JC , AG Ristvey and JD Lea-Cox. (2011). Water and Nutrient Management in the Production of Container-Grown Ornamentals. *Hort. Rev.* 38:253-296.

McBratney AB , Whelan B , Ancev T and Bouma J. (2005). Future directions of precision agriculture. *Precision Agriculture,* **6**: 7-23.

McCarthy CL , Hancock NH and Raine SR. (2007). Field measurement of plant geometry using machine vision. Fifth International Workshop on Functional Structural Plant Models (SFPM07), Napier, New Zealand, November, pp 19.1-19.3.

Misras R , Raine SR , Pezzaniti D , Charlesworth P and Hancock NH. (2005). A Scoping Study on Measuring and Monitoring Tools and Technology for Precision Irrigation. CRC for Irrigation Futures, Irrigation Matter Series No. 01/05.

Modak CS. (2009). Unpublished Report from Land and Water Management Institute, Aurangabad, Maharashtra, India.

Murray JD , Lea-Cox JD and Ross DS. (2004). Time Domain Reflectometry Accurately Monitors and Controls Irrigation Water Applications *Acta Hort.* 633:75-82.

O'Shaughnessy SA and Evett SR (2008). Integration of Wireless Sensor Networks into Moving Irrigation Systems for Automatic Irrigation Scheduling. Amer. Soc. Agric. Biol. Eng. 29 June - 2 July, 2008. Providence, RI. Paper # 083452, p. 21.

Padhi J and Misra RK. (2009). Monitoring spatial variation of soil moisture in crop fields with EM38. Irrigation and Drainage Conference 2009, Irrigation Australia Ltd, Swan Hill, Vic, 18-21 Oct.

Panchard J, Rao S, Prabhakar TV, Hubaux JP and Jamadagni H. (2007). Common Sense Net: A Wireless Sensor Network for Resource-Poor Agriculture in the Semi-arid Areas of Developing Countries. Information Technologies and International Development, **4**(1): 51–67.

Peters R T and Evett SR. (2008). Automation of a center pivot using the temperature time-threshold method of irrigation scheduling. *J. Irrig. Drain. Engr.*, **134**: 286-291.

Pierce FJ and Elliott TV. (2008). Regional and on-farm wireless sensor networks for agricultural systems in Eastern Washington. Computers and Electronics in Agriculture, **61**: 32–43.

Pieruschka R. Huber G and Berry J.A. (2010). Control of Transpiration by Radiation. Proc. Natl. Acad. Sci. **107**:13372-13377.

Raine SR , Meyer WS , Rassam DW, Hutson JL and Cook FJ. (2007). Soil-water and solute movement under precision irrigation: knowledge gaps for managing sustainable root zones. *Irrigation Science*, **26**(1): 91-100.

Raviv M , Lieth JH and Wallach R. (2000). Effect of Root-Zone Physical Properties of Coir and UC Mix on Performance of Cut Rose (cv. Kardinal). *Acta Hort.* **554**:231-238.

Ristvey AG , Lea-Cox JD and Ross DS. (2004). Nutrient uptake, partitioning and leaching losses from container-Nursery Production Systems. *Acta Hort.* **630**:321-328.

Ruiz-Garcia L, Lunadei L, Barreiro P and Robla JI. (2009). A Review of Wireless Sensor Technologies and Applications in Agriculture and Food Industry: State of the Art and Current Trends. *Sensors* 9: 4728-4750.

Sadler EJ, Camp CR, Evans DE, Millen JA. (2002). Corn canopy temperatures measured with a moving infrared thermometer array. Transactions of the American Society of Agricultural Engineers **45**: 581–591.

Stone KC , Sadler EJ , Millen JA , Evans DE and Camp CR. (2006). Water flow from a precision irrigation system. *Applied Engineering in Agriculture*, **22**(1), 73–78.

Sadler EJ, Camp CR, Evans DE, Millen JA. (2002). Corn canopy temperatures measured with a moving infrared thermometer array.Transactions of the American Society of Agricultural Engineers **45**: 581–591.

Sadler EJ , Evans RG , Stone KC and Camp CR. (2005). Opportunities for conservation with precision irrigation. *Journal of Soil and Water Conservation*, **60**(6): 371-379.

Schmidt U and Exarchou E. (2000). Controlling of irrigation systems of greenhouse plants by using measured transpiration sum. *Acta Horti.* **537**: 487–494.

Smajstrla AG and Harrison DS. (1998). Tensiometers for Soil Moisture Measurement and Irrigation Scheduling. Univ. Fl. IFAS Ext. Cir. No. 487. p. 8

Smith KA and Mullins C. (2001). Soil and Environmental Analysis: Physical Methods. 2nd Ed. Marcel Decker, New York.

Smith RJ and SR Raine. (2000). A prescriptive future for precision and spatially varied irrigation. Nat. Conf. Irrigation Association of Australia. Melbourne.

Smith RJ , Baillie JN, McCarthy AC , Raine SR and Baillie CP. (2010). Review of Precision Irrigation Technologies and their Application. National Centre for Engineering in Agriculture Publication 1003017/1, USQ, Toowoomba.

Stafford JV. (2000). Implementing Precision Agriculture in the 21st century. *Journal of Agricultural Engineering Research*, **76:** pp 267-275.

Stanghellini C , Pardossi A and Sigrimis N. (2007). What Limits the Application of Wastewater and/or Closed Cycle in Horticulture? *Acta Hort.* **747**:323-330.

Sudharsan D , Adinarayana J , Ninomiya S , Hirafuji M. and T. Kiura T. (2012). Dynamic real time distributed sensor network based database management system using XML, JAVA and PHP technologies. *International Journal of Database Management Systems* 4: 1 DOI: 10.5121/ijdms.2012.4102 9.

Taiz L and Zeiger E. (2006). Plant Physiology. *Sinauer Associates, Inc.* Sunderland, MA.

Treder J. Matysiak B. Nowak J and Treder W. (1997). Evapotranspiration and Potted Plants Water Requirements as Affected by Environmental Factors. *Acta Hort.* **449**:235-240.

Trout TJ and Johnson LF. (2007). Estimating crop water use from remotely sensed NDVI, crop models, and reference ET. Proc USCID Fourth International Conference on Irrigation and Drainage, pp. 275-285.

Upchurch DR,Wanjara DF, Mahan JR. (1990). Automating trickle irrigation using continuous canopy temperature measurements. *Acta Horti.* **278**: 299–308.

Van Iersel MW. (2003). Short-Term Temperature Change Affects the Carbon Exchange Characteristics and Growth of Four Bedding Plant Species. *J. Amer. Soc. Hort. Sci.* **128**:100-106.

Van Iersel MW, Dove S and Burnett SE. (2011). The Use of Soil Moisture Probes for Improved Uniformity and Irrigation Control in Greenhouses. *Acta Hort.* **893**:1049-1056.

Vellidis G , Tucker M , Perry C , Wen C and Bednarz C. (2008). A Real-Time Wireless Smart Sensor Array for Scheduling Irrigation. *Comput. Electron. Agric.* **61**:44-50.

Wang C , Zhao CJ, Qiao XJ , Zhang X and Zhang YH. (2008). The Design of Wireless Sensor Networks Node for Measuring the Greenhouse's Environment Parameters. *Computing. Technol. Agric.* **259**:1037-1046

Wang N, Zhang N and Wang M. (2006). Wireless Sensors in Agriculture and Food Industry-Recent Development and Future Perspective: Review. *Computers and Electronics in Agriculture*, **50**: 1–14.

Wang YP and Jarvis PG. (1990). Description and Validation of an Array Model-MAESTRO. *Agricultural Forest Meteorology.* **51**:257–280.

White S.C. and Raine S.R. (2008). A Grower Guide to Plant Based Sensing for Irrigation Scheduling. National Centre for Engineering in Agriculture, NCEA Publication 1001574/6.

Yoo S, Kim J , Kim T , Ahn S , Sung J and Kim D. (2007). A2S: Automated Agriculture System Based on WSN. In: ISCE 2007. IEEE International Symposium on Consumer Electronics. Irving, TX, USA.

Yule IJ , Hedley CB and Bradbury S. (2008). Variable-rate irrigation. 12th Annual Symposium on Precision Agriculture Research and Application in Australia, Sydney.

Zazueta FS , Yeager T, Valiente JI and Brealey JA. (1994). A Modified Tensiometer for Irrigation Control in Potted Ornamental Production. *Proc. Soil Crop Sci. Soc. Fla.* **53**:36-39.

9

CHAPTER

Laser Guided Land Leveling and Grading for Precision Farming

☞ Shiv Kumar Lohan[1], ☞ Harminder Singh Sidhu[2] and ☞ Manpreet Singh[3]

L and leveling is necessary for taking up good agronomic, soil and crop management practices. Land leveling saves irrigation water, facilitates field operations and increases yield and quality of the produce. Leveled land also helps in mechanization of various field operations. Land leveling may be carried out to produce a level land without any slope, such as followed in case of rice or a slope may be given in one direction or two perpendicular directions.

This chapter explains the different ways to level fields and the agricultural and financial benefits of good land leveling as part of land preparation. The unevenness in land level (sometimes called surface topography) within a field has a major effect on crop management and crop yields. Unevenness in land level results in uneven water coverage. Uneven water coverage means that more water is needed to wet up the soil for land preparation and plant establishment reducing the effective time available to complete these tasks. Unevenness in land level results in uneven crop stands, increased weeds and uneven maturing crops. All of these factors result in reduced yields and reduced grain quality. Effective land leveling will improve crop establishment and care, reduce the amount of effort required to manage the crop and will increase both grain quality and yields.

WHY LASER LAND LEVELING?

Laser leveling is a laser guided precision leveling technique used for achieving fine leveling with the desired grade on the agricultural field. Laser leveling uses a laser transmitter unit that constantly emits 360^0 rotating beam parallel to the required field plane. This beam is received by a laser receiver fitted on a mast on the scraper unit. The signal received is converted in to cut and fill level adjustments and the corresponding changes in scraper level carried out automatically by a two way hydraulic control valve. Laser leveling maintains the grade by automatically performing the cutting and filling operations. The field is cultivated and planked before using the laser land leveler. A grid survey is performed using grade rod to identify highs and lows in the field and mean grade is found. A grid spacing of

[1&3] Deptt. of Farm Machinery & Power Engineering PAU, Ludhiana - 141 001
[2] Borlaug Institute for South Asia, Punjab Agricultural University Campus, Ludhiana- 141001

10m x 10m is maintained for accurate land survey, however this spacing can be varied depending upon the size of the field. For practical purpose and with experience, grid surveys can be done by pacing off the distances. A map is then drawn to indicate high areas which require soil to be cut and the lows which require soil to be added (Sidhu *et al*, 2007).

LASER LAND LEVELING AND ITS IMPORTANCE IN PRECISION FARMING

Precision agriculture involves optimum application rate and placement of inputs so as to maximise crop yields, reduce inputs and subsequently input cost and also reduce operational costs. Precision land leveling through operation of laser guided land leveler ensures accurate land leveling and grading, ensuring the following:

- Uniform application of water in different parts of the field.
- Prevents accumulation of water in depressions.
- Uniform sowing depth leading to uniform emergence pattern.
- Higher crop yields due to the above factors.
- Laser guided land leveler gives a maximum average deviation of 1-2 cm as compared to 5-10 cm in conventional leveling.

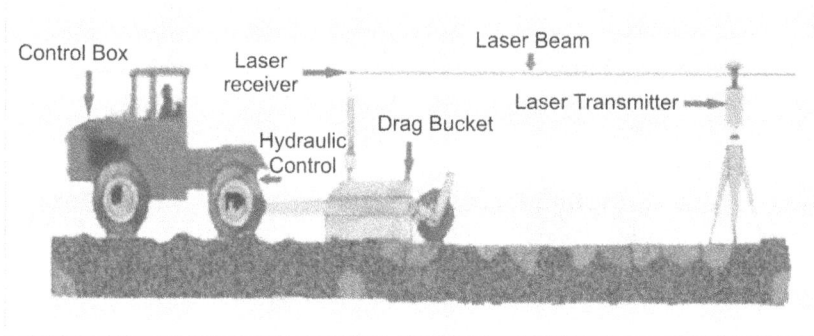

Fig. 9.1: *Principle of laser land leveling*

Benefits of Land Leveling

- Increase in water application efficiency up to 50 per cent.
- Reduces the amount of water required for land preparation(20 – 30 per cent), hence saving in energy (diesel/electricity)
- Better crop stand due to even application of fertilizers and other inputs *i.e.* improvement in crop yield by 10 to 15 per cent.
- Less area under bunds/ridges *i.e.*, increase in 8 to 10 per cent area under the crop.
- Reduced labour requirement for irrigation.

- Totally automatic (less load on operator)
- Effective land leveling reduces the work in crop establishment and crop management, and increases the yield and quality.
- Level land improves water coverage that improves crop establishment
- Reduces weed problems
- Improves uniformity of crop maturity
- Decreases the time to complete tasks
- More level and smooth soil surface
- Reduction in time and water required to irrigate the field
- More uniform distribution of water in the field
- More uniform moisture environment for crops
- More uniform germination and fast growth of crops
- Reduction in seeds rates, fertilizer, chemicals and fuel used in cultural co-operation
- Improved field used traffic ability (for subsequent operations)

INCREASING WEED CONTROL AND CROP YIELD

Research has shown a large increase in rice yield due to proper field leveling. A large part of this increase is due to improved weed control. It improves water coverage from better land leveling and reduces weeds by up to 40 per cent. This reduction in weeds results in less time for weeding. A reduction from 21 to 5 labour-days per hectare is achieved (Jat *et al.*, 2006).

Seeding Practices

Leveling reduces the time taken for planting, transplanting and direct seeding. Land leveling provides greater opportunity to use direct seeding. The possible reduction in labour by changing from transplanting to direct seeding is approximately 30 person-days per hectare.

Efficiency of Water Use

Rice farmers using animals or two-wheel tractors rely on water to accumulate in the field before starting land preparation. The average difference in height between the highest and lowest portions of rice fields in Asia is 160 mm. This means that in an unleveled field an extra 80 mm to 100 mm of water must be stored in the field to give complete water coverage. This is nearly an extra 10 per cent of the total water requirement to grow the crop. The economics of land leveling and the initial cost of land leveling for using contractors and machinery is high. The costs of leveling vary according to the topography, the shape of the field and the equipment used.

FINANCIAL BENEFITS OF LAND LEVELING

Although the initial cost of land leveling is an extra expense, a cash flow over the years shows that financial benefits do result from land leveling. There are many financial benefits can be gained through land leveling. What it does not include are the benefits of being able to direct seed, plow the field on time, harvest evenly ripened crop and shed flood waters more rapidly. While poor farmers may have problems financing a contractor to level land, it is quite possible for all farmers to level part of their land each year using animals and harrows during the normal plowing cycle.

Components of laser Leveler

The laser control system requires a laser transmitter, a laser receiver, an electrical control panel and a twin solenoid hydraulic control valve.

The Laser Transmitter

The laser transmitter transmits a laser beam which is intercepted by the laser receiver mounted on the leveling bucket. The control panel mounted on the tractor interprets the signal from the receiver and opens or closes the hydraulic control valve which will raise or lower the bucket.

Laser Emitter

The laser emitter unit sends continuous self-leveled laser beam signal with 360° laser reference up to a command radius of 300-400 m (depending upon it's range) for auto guidance of the receiving unit. The laser emitter is mounted on a tripod stand placed just outside the field to be laser leveled and high enough to have unobstructed laser beam travel. Different working components and controls on the laser emitter unit include laser emission indicator, low battery assembly and manual mode indicator for setting of desired grade.

Fig. 9.2: *Laser Emitter*

Laser Beam Receiver

The laser receiver mounted on the scraper, is an omni-directional (360°) receiver that detects the position of the laser reference plane and transmits it to the control box. Further, this control box directs the double actuating hydraulic valve for desired upward and downward movement of scraper blade to obtain the leveled field. The grade position LED's indicate the position of the machine's blade relative to the plane of the laser light from the laser emitter. These lamps function in the same way as the grade position lamps on the control box mounted on tractor except they flash instead of lighting solidity.

Fig. 9.3: *Laser Beam Receiver*

Control Box

The control box is to be mounted on the tractor so that the operator can easily access the switches and view the indicator lamps. The control box has the main control unit for actuating the double acting hydraulic valves. The control box receives and processes signals from the laser receiver mounted on the bucket. It displays these signals to indicate the drag bucket's position relative to the finished grade. The control box is set to automatic position, it provides electrical output for driving the hydraulic valve to operate scraper automatically.

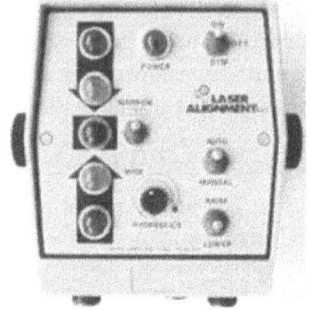

Fig. 9.4: *Control Box*

Hydraulic valve assembly

The valve assembly regulates the flow of tractor hydraulic oil to the hydraulic cylinder to raise and lower the scraper blade. The oil supplied by the tractor's hydraulic pump is normally delivered at 2000-3000 psi pressure. As the hydraulic pump is a positive displacement pump and always pumping more oil than required, a pressure relief valve has also been provided in the system to return the excess oil to the tractor reservoir. The solenoid control valve controls the flow of oil to the hydraulic ram which raises and lowers the bucket. The

Fig. 9.5: *Hydraulic valve assembly*

desired rate at which the bucket could be raised and lowered is dependent on the operating speed and amount of oil supplied to the delivery line.

Laser eye

Laser eye is to be mounted on the grade survey rod for obtaining the level of the field. It contains a laser receiving panel and when the laser emitted by the laser emitter panel falls in the center of this eye, a continuous beep sound indicates the level of that specific point with respect to the laser emitter.

The grade of that point is the read from grade rod.

Fig. 9.6: *Field Survey before laser leveling*

Source of Power

A four-wheel tractor is required to drag the leveling bucket. The size of the tractor can vary from 30-500 hp depending on the time restraints and field sizes. In Asia tractors ranging in size from 30-100 hp have been successfully used with laser-controlled systems. It is preferable to have a four wheel drive tractor than two wheel drive and the higher the horsepower the faster will be the operation. Power shift transmissions in the tractor are preferred to manual shift transmissions.

Ploughing

The fields will require plowing before and after land leveling. Depending on the amount of soil that must be cut it may also be necessary to plow during the leveling operation. Disc, moldboard or tine plows can be used.

Drag Bucket

The leveling bucket can be either three-point linkage mounted or pulled by the tractor's drawbar. Pull type systems are preferred as it is easier to connect the tractor's hydraulic system to an external hydraulic ram than connect to the internal control system used by the three-point-linkage system. Bucket dimensions and capacity will vary according to the available power source and field conditions. A 60 hp tractor will pull a 2 m wide x 1 m deep bucket in most soil types. The design specifications for the bucket should match the available power from the tractor.

Design of laser guided land leveler

Design Principles

- The mapping of the field for reduced levels should be accurate, for this the maximum error permissible should be 2 mm at 100 m distance.
- The laser system should be selected based on the requirements-leveling without any slope as required for rice fields, leveling in the plane of desired slope in one direction and leveling in the plane of desired slope in two perpendicular directions as required for drainage operations.
- The laser guided land leveler is used for final leveling and grading after rough leveling has been carried out. Even then the deviation from average level is sometimes 10 cm or more. The laser receiver should be able to receive signals when deviation is 15 cm.
- Range of the laser transmitter should be at least 300 m.
- The control box of the laser leveler should be able to operate on DC power obtained from tractor battery of 12 V.
- The hydraulic system should be designed so that the once desired signal is received from the laser system, the drag scraper should move down or up to desired level in shortest possible time. In other words, the response time of the leveler should be as less as possible.
- The size of the drag scraper should be matching to the power of the tractor.

Selection of laser system specifications based on design principles

- Select the type of laser transmitter based on requirements- level, single slope or dual slope.

- Self leveling arrangement of transmitter.
- Range of laser transmitter 300 m or more.
- Minimum width of sensing portion of receiver should be 30 cm for receiver mounted on drag scraper.
- Minimum width of sensing portion of receiver used for mapping should be 20 cm or more.
- Control box should be operated by 12 V DC current obtained from tractor battery.

Hydraulic System Design

The hydraulic system supplies oil to raise and lower the levelling bucket. A pressure relief valve is needed in the system to return the excess oil to the tractor reservoir. If this relief valve is not large enough or it malfunctions, damage can be caused to the tractor's hydraulic pump. Components of external hydraulic system are, (i) Solenoid operated directional control valve, (ii) Pressure relief valve if tapping is taken other than auxiliary hydraulic outlet of tractor, (iii) Hydraulic cylinder and (iv) Hydraulic hose pipes, couplers etc.

- Solenoid operated directional control valve should be having capacity more than maximum flow rate of hydraulic system.
- For 50 to 60 hp tractors, flow rate is about 25-35 l/min.
- Selection of hydraulic cylinder is carried out with a view to minimise response time.
- Since the laser guided land leveler is designed for final precision leveling, the maximum depth of operation is considered as 15 cm.

Field Operation and Calibration

Land leveling makes possible the use of larger fields. Larger fields increase the farming area and improve operational efficiency. Increasing field sizes from 0.1 hectare to 0.5 hectare increases the farming area by between 5 per cent and 7 per cent. This increase in farming area gives to farmer option to reshape the farming area that can reduce operating time by 10 per cent to 15 per cent. The field should be preferably plowed from the center of the field outwards. Plow when the soil is moist, because if it is plowed dry a significant increase in tractor power is required and large clod sizes may result. Cut up or remove surface residues to aid soil flow from the bucket (Chandiramani, 2007)

Following are the guidelines for setting up the laser emitter, laser receiver, laser eye and control box on tractor.

- The laser emitter and laser eye should be fully charged or equipped with replicable batteries with sufficient battery back-up before taken to field.
- The electrical connections of control box on tractor and double actuating valve should be made properly.

- Choose a location in the field for the laser emitter (to be fixed on tripod) where obstruction, such as trees and buildings, passages etc. do not block the plane of laser light. The laser receiver on scraper should be able to sense the plane of laser light all times.

- As far as possible, set up the laser emitter and receiver at a height above the tractor's canopy or any cab or roll over protection attachment to avoid any blocking to the plane of laser beam as the machine/equipment moves around the field.

- Fix the laser eye on the graded rod for the field level survey (as mentioned below) which is essential for estimating the quantum of work and find the level plane to maintain as per needs.

- Sufficient number of iron or wooden pegs should be arranged for marking the different points in field during survey.

Land survey

A topographic survey should be conducted to record the high and low spots in the field. The mean height of the field can be calculated by taking the sum of all the readings and dividing by the number of readings taken. Then, using a field diagram and the mean height of the field, determine how to move soil effectively from the high to low areas.

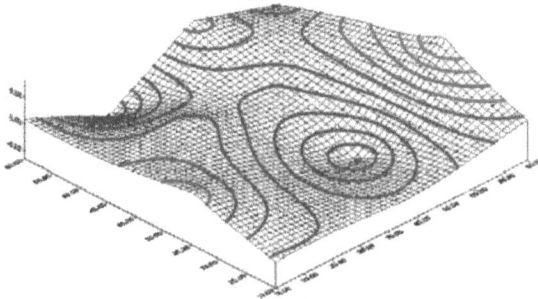

Fig.9.7: *Contour map and digital elevation map in Gurusar Kaonke village (Punjab).*

Field evaluation

The evaluation of levelling a field involves the following steps:

- The laser controlled bucket should be positioned at a point that represents the mean height of the field.

- The cutting blade should be set slightly above ground level (1-2 cm)

- The tractor should then be driven in a circular direction from the high areas to the lower areas in the field.

- To maximize working efficiency, as soon as the bucket is near filled with soil the operator should turn and drive towards the lower area. Similarly, when the bucket is near empty the tractor should be turned and driven back to the higher areas.

- When the whole field has been covered, the tractor and bucket should then do a final levelling pass in long runs from the high end of the field to the lower end.
- Re survey to make sure that the desired level of precision has been attained.
- The fields should not require further major leveling works for at least 8 years.

Limitations

- High cost of the equipment/laser instrument.
- Need for skilled operator to set/adjust laser settings and operate the tractor.
- More efficient for regularly sized and shaped field.

Unevenness of the soil surface has a major impact on the germination, stand and yield of crops through nutrient water interaction and salt and soil moisture distribution pattern.

Land leveling is a precursor to good agronomic, soil and crop management practices. Resource conserving technologies perform better on well-leveled and laid-out fields. Farmers has recognize this and therefore, devote considerable attention and resources in leveling their fields properly. However, traditional methods of leveling land are not only more cumbersome and time consuming but more expensive as well. Very often most rice farmers level their fields under ponded water conditions. The others dry level their fields and check level by ponding water. Thus, in the process of a having good leveling in fields, a considerable amount of water is wasted. It is a common knowledge that most of the farmers apply irrigation water until all the parcels are fully wetted and covered with a thin sheet of water. Studies have indicated that (20-25 per cent) amount of irrigation water is lost during its application at the farm due to poor farm designing and unevenness of the fields. This problem is more pronounced in the case of rice fields. Unevenness of fields leads to inefficient use of irrigation water and also delays tillage and crop establishment options. Fields that are not level have uneven crop stands, increased weed burdens and uneven maturing of crops. All these factors tend to contribute to reduced yield and grain quality which reduce the potential farm gate income. Effective land leveling is meant to optimize water use efficiency, improve crop establishment, reduce the irrigation time and effort required to manage crop.

Laser land leveling seeks to explain the benefits of land leveling in fields, particularly rice fields, and help develop skills of farmers and operators in using laser technology to achieve a level field surface. It is also intended to enable the users to identify and understand the working of the various components of a laser-controlled land leveling system, undertake a topographic survey using a laser system, set up and use a laser-controlled leveling system and troubleshoot a laser-controlled leveling system. It is hoped that the users (farmers and service providers) will find this information useful in adopting resource conserving technology as a precursor to several other improved agronomic, soil and crop management practices.

PROCEDURE FOR TESTING OF LASER GUIDED LAND LEVELER

The variables need to be measured for testing of laser guided land leveler are, accuracy of leveling, subjective methods, contour maps, surface charts, measurement of variations in reduced levels and calculating standard deviation of reduced levels, leveling index, forward speed, field capacity, capacity of the leveler, Fuel consumption and cost of operation.

EVALUATION METHODOLOGY

- Reduced levels of grid points (10X10 m) are taken prior to and after leveling.
- In case of leveling the slope is 00 in both directions.
- In case of grading, the laser transmitter is adjusted for slope in both directions. The laser beams are then transmitted in plane of the grade set. Reduced levels readings are taken before and after grading with respect to this plane.
- While grading, observations on forward speed, field capacity and fuel consumption are taken.
- From readings of reduced levels, mean and standard deviation of reduced levels are calculated.
- The numerical difference between mean reduced level and actual reduced levels are calculated. The average value gives the leveling index.

Break even Usage of Laser Land Leveler

The initial cost of the Laser leveler is quite high so this type of service should be available on custom hiring to the farmers. PAU has already been advocating the use of costly farm machinery through cooperative societies, custom hiring or contracting to make these services available to all categories of farmers to reduce its operating cost because it improves the annual use and thus reduce the cost of farming. Fig. 9.8 presents the break even usage of laser land leveler.

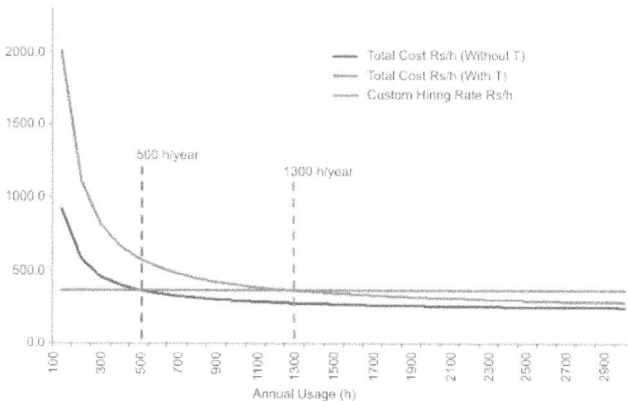

Fig. 9.8: *Break even Usage of Laser Land Leveler*

GROWTH OF LASER LEVELING IN PUNJAB

The work on laser land leveling was started in rice crop initially from 2005-06 from 8 units of machine at PAU and various farmer's field in farmer's participatory research mode. Few machinery modifications has also been done like hitch system of scraper bucket was modified from it's V shape to Y shape which improved it's turning radius by 27 per cent for it's easy manure-ability in the small fields and hydraulic control system of the scraper bucket was simplified for easy, simplified and leak proof operation and machine working. The main intention was to authenticate the replicated trial's results (in terms of water saving and better yields etc.) with the farmer's managed fields. Then more efforts were made for the popularization of this technology by organizing trainings for farmers, contractors and Govt officials, demonstrations, field days and workshops for all the stake holders. Machinery manufacturers were also enrolled in to the potential future growth of the technology to maintain the price & quality simultaneously. In Punjab the area covered under laser leveling has been increased tremendously in last few years (Fig. 9.9).

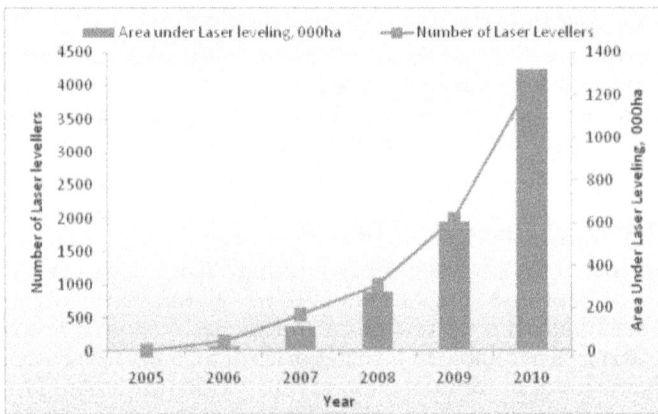

Fig. 9.9: *Growth of Laser Land Leveling area in Punjab.*

(*Source*: Sidhu *et al.*, 2007& 2012)

REFERENCES

Chandiramani M, Kosina P, Jones. (2007).Cereal Knowledge Bank. (knowledgebank. cimmyt.org). IRRI-CIMMYT alliance.

Jat ML, Chandna P, Gupta R, Sharma SK and Gill MA. (2006). Laser Land Leveling: A Precursor Technology for Resource Conservation. Rice-Wheat Consortium Technical Bulletin Series 7. New Delhi: Rice-Wheat Consortium for the Indo-Gangetic Plains.

Sidhu HS, Mahal JS, Dhaliwal IS, Bector V, Manpreet S, Sharda A, Singh T. (2007). Laser land levelling: A boon for sustaining Punjab agriculture. Farm Machinery Bulletin 2007/01.Department of Farm Power & Machinery, PAU, Ludhiana.

Sidhu HS. (2012). Laser Land Leveling – A Technology for Resource Conservation. ppt. BISA project. Punjab.

10

CHAPTER

Effective Water Resources Management Using Efficient Irrigation Scheduling

☞ Rohitashw Kumar[1],
☞ Sheikh Modasir[2] and ☞ T. Ram[3]

INTRODUCTION

World consumed about 80% of the water in agricultural sector (Hutson, 2000). A quantitative study of water transport in the soil–root system is an important exercise for determination of crop water requirement, optimum irrigation scheduling and sound water management policies (Wang and Smith, 2004). The boundary between soil and the plant root system is a major hydrologic interface across which over 50 per cent of evapotranspiration take place. The sustainable irrigation practices requires better understanding of the biophysical processes of root- water uptake in soil and transpiration from plant canopies (Steve *et al.*, 2006). Efficient water management of crops requires accurate irrigation scheduling which, in turn, requires the accurate measurement of crop water requirement. Irrigation is applied to replenish depleted moisture for optimum plant growth. Mathematical models are useful tools to estimate the water requirement of crop, which is essential information required to design or choose best water management practices (Kumar *et al.*, 2012). Soil moisture uptake by plant roots is a dynamic process influenced by the plant characteristics, atmospheric demand, soil moisture availability and its hydraulic properties. Run-off from over-irrigation may affect water quality parameters such as pH, total suspended solids and dissolved oxygen other negative impacts associated with over irrigation include: wastage of water, energy and reduced crop yields.

SOIL - WATER BALANCE

A soil water balance model incorporated into a data acquisition system is a powerful tool for scheduling and optimizing irrigation. Advancements in computer microprocessors, memory and software development tools have improved data acquisition methods and made data acquisition system integration more reliable and cost effective. The soil water balance models incorporate inputs of soil moisture, water application and evapotranspiration. Irrigation scheduling is a technique for timely and accurate application of water to the crop (Jensen, 1980). Irrigation scheduling is

[1] Sr. Engineer, Division of Agricultural Engineering, Sher-e-Kashmir University of Agricultural Sciences & Technology of Kashmir, Srinagar-191121 (J&K)

[2] Division of Agricultural Engineering, Sher-e-Kashmir University of Agricultural Sciences & Technology of Kashmir, Srinagar-191121 (J&K)

[3] Sr. Scientist Agronomy, PDFSR (ICAR), Modipuram, Meerut (UP)

important for planning and decision-making activity which affect the farm management practices. Irrigation scheduling has been described as the primary tool to improve water use efficiency, increase crop yields, increase the availability of water resources and provoke a positive effect on the quality of soil and groundwater (FAO, 1996).

PAN EVAPORATION FOR IRRIGATION

The technique of using pan evaporation for irrigation scheduling has been extensively tested by many researchers worldwide (Khalil, 1996, Ashraf *et al.*, 2002 and Khalil *et al.*, 2006) and it was proven that it can save up to 20 per cent of the applied irrigation water. It is a useful indicator for quantifying the impact of irrigation scheduling decisions with regard to water management (FAO, 2003). The climate change urgently needs to be assessed at the farm level, so that the poor and vulnerable farmers dependent on agriculture can be appropriately targeted in research and development activities on poverty alleviation (Jones and Thornton, 2003). Therefore, water is essential element for plant growth (Figure 10.1 & 10.2).

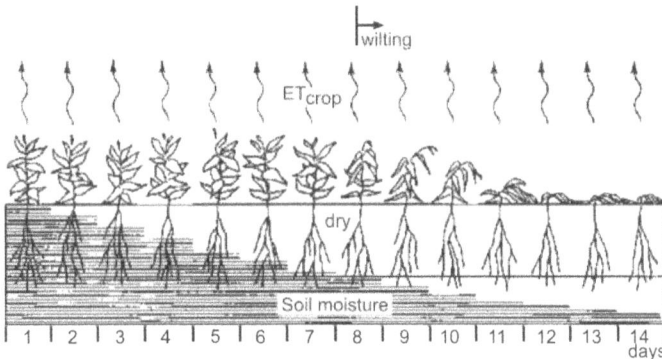

Fig. 10.1: *Effect on plant in absence of rainfall and irrigation*

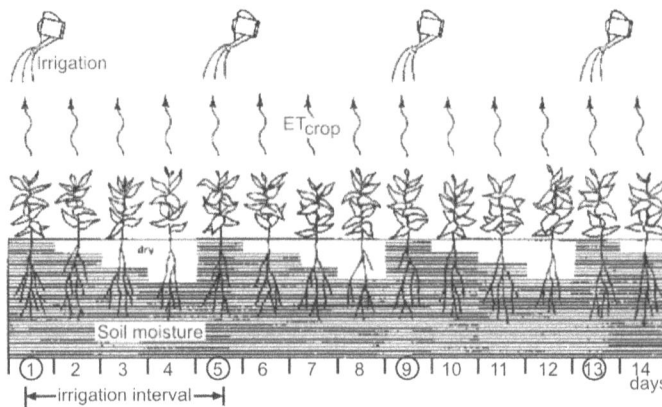

Fig. 10.2: *Irrigation water is applied regularly so that the plants do not suffer from water shortage*

Soil texture

The soil moisture budgets, first describe the components and the hydrological conditions of soil. In general, inorganic soil is composed of mixes of sands, silts and clays. Sands, silts and clays differ not only by particle size distribution, but also in the atomic arrangement and charge distribution at the molecular level (McBride 1994). Soil geomorphology is the process by which sands and silts chemically and physically transform into clays as the soil ages (Birkeland, 1999). Sands, silts, clays and organics represent the solid particle composition of soil while air and water fill the pore spaces between the solid particles. When soil is completely saturated with water, the porosity will be equal to the volumetric soil moisture content (Warrick, 2003). The amount of organics in soil will also affect the bulk density and the porosity. The hydrologic properties of soil play an important role in a crops ability to transpire water through their root systems. Knowledge of volumetric soil moisture content (θ, m^3 m^{-3}) is important input into the soil water balance model. Permanent wilting point (θ_{pw}) is the soil moisture level at which plants can no longer take water from the soil. Plant transpiration and direct evaporation will decrease the moisture level in soil to a point below θ_{pw} and, in some cases, down to near dryness.

Field Capacity (θ_{fc})

Field capacity (θ_{fc}) is defined as the threshold point at which the soil pore water will be influenced by gravity. Above field capacity, the gravitational force will overcome the capillary forces suspending the moisture in the pores of the soil allowing for down movement of water in the soil column. Field capacity and permanent wilting point are heavily influenced by soil textural classes, particularly clay content (Rowell, 1994). The available water capacity (θ_{ac}) of soil is the water that is available to a plant. It represents the range of soil moisture values that lie above the permanent wilting point and below the field capacity.

$$\Theta_{pw} < \theta_{ac} < \theta_{fc}$$

The values for permanent wilting point and field capacity for common soil textural classes (Rowell, 1995) are shown in Table 10.1.

Table 10.1: Permanent wilting point and field capacity for common soil textural classes

Soil texture	Field capacity	Permanent wilting point
Sand	0.12	0.04
Loamy Sand	0.14	0.06
Sandy Loam	0.23	0.1
Loam	0.26	0.12
Silt Loam	0.3	0.15
Silt	0.32	0.165
Sandy Clay Loam	0.33	0.175
Silty Clay Loam	0.34	0.19
Silty Clay	0.36	0.21
Clay	0.36	0.21

Water application

Soil moisture data provides information about moisture status in the root zone. The measured application of water is the amount of water applied to the crops using different irrigation methods. An important factor for quantifying the water budget is the evapotranspiration rate (ET). Evapotranspiration is the total volume of water which is transpired out of the soil by the plant plus the amount of water lost to evaporation (Allen *et al.*, 1998). The factors that affect the ET rate include wind, temperature, relative humidity and solar radiation. Based on the Penman Monteith model for ET estimations, ET is not measured directly for an individual crop, but rather it is determined from a standard reference grass and then adjusted for different crops and plants with a crop coefficient (Allen, 1998). The evapotranspiration for a reference grass is referred to as the potential evapotranspiration (ET_0). Potential evapotranspiration values will vary regionally and seasonally and are available in the literature. It is important to irrigate at a rate that is less than the infiltration rate of the soil. The infiltration rates of soils based on soil textural classes (Brouwer, 1988) are given in Table 10.2.

Table 10.2: Infiltration rates of soils based on soil textural classes

Soil Texture	Infiltration Rate (inches/hour)
Sand	1.5 or more
Sandy Loam	1 to 1.5
Loam	0.5 to 1
Clay Loam	0.25 to 0.5
Clay	0.05 to 0.25

The infiltration of water into soil will vary with texture, but it will also depend on soil moisture, vegetation, bulk density and soil geomorphologies factors.

Data acquisition

Data acquisition systems are the most effective tools for identifying and reaching soil moisture and water application targets for irrigation optimization. Soil moisture probes at different depths in the soil column are referred to as a soil profile. There should be at least one soil profile for every irrigation regime. Irrigation regimes are determined by crop type, crop age, soil type, slope, and irrigation methods .The irrigator can then view the real time data and make decisions about when to irrigate based on the soil moisture targets and the rate of water consumption by the crop from the evapotranspiration rate (ET).

Water management

Water management is an important aspect for sustainable crop production. Both its shortage and excess affects the growth and development of the plants, yields and quality of produce. Efficient irrigation systems and water management practices can help in farm profitability in an era of limited, higher-cost water supplies. Efficient water management may also reduce the impact of irrigated production on offsite water quantity and quality. However, measures to increase water use efficiency may not be sufficient to achieve environmental goals in absence of other adjustments within the irrigated sector. Water savings through improved management of irrigation supplies are considered essential to meet the future water needs. Irrigation is the most significant use of water, accounting for over 95 per cent of freshwater withdrawals. However, expanding water demands for municipal, industrial, recreational, and environmental purposes increasingly compete for available water supplies.

Irrigation scheduling

Different irrigation methods whether furrow, border strip, sprinkler or drip, all are very important and required large investment. The keys to effective, efficient irrigations are knowing when, how much, and how to irrigate. Modern irrigation scheduling techniques help to know when and how much to irrigate. Irrigation scheduling is a generic term applied to any technique/practice that is intended to aid the farmer in determining when and how much to irrigate. There are a number of ways in which the different techniques could be categorized. A common way is to group them as either "water budget" or "graphical/sensor-based" methods. Irrigation scheduling involves the application of irrigation water based on a systematic monitoring of crop soil-moisture requirements. Sophisticated scheduling methods based on sensors, microprocessors, and computer aided decision tools may be used to determine the optimal timing and depth of irrigation to meet changing crop needs over the production season. Various methods are available to assess crop water needs. Crop water requirements can be indirectly estimated through climate variables. Local weather-station data including temperature, humidity, wind speed and solar radiation are applied in formulas to calculate crop water requirements for a wide range of crops. Soil moisture available for plant growth may also be measured directly through periodic soil testing. Soil probes are used to obtain soil samples at various depths for "feel and visual" evaluation. More sophisticated devices such as tensiometers, neutron probes, and various electrical conductivity devices can be used to accurately quantify the amount of water removed from the soil profile. The main points emerged in irrigation scheduling are:

- Irrigation scheduling is the decision of 'when' and 'how' much water to apply to a field.
- Its purpose is to maximize irrigation efficiencies by applying the exact amount of water needed to replenish the soil moisture to the desired level.
- Irrigation scheduling saves water and energy.
- All irrigation scheduling procedures consist of monitoring indicators that determine the need for irrigation.

The purpose of irrigation scheduling is to determine the exact amount of water to apply to the field and the exact timing for application.

Irrigation criteria Vs. Irrigation scheduling

Irrigation criteria are the indicators used to determine the need for irrigation. The most common irrigation criteria are soil moisture content and soil moisture tension. The final decision depends on the irrigation criterion, strategy and goal. To illustrate irrigation scheduling, consider a farmer whose goal is to maximize yield. Accurate water application prevents over or under irrigation. Over irrigation wastes water, energy and labor; leaches expensive nutrients below the root zone, out of reach of plants and reduces soil aeration, and thus crop yields. Under irrigation stresses the plant and causes yield reduction.

The total amount of available water in the effective root zone, when the effective root zone is at field capacity is called available water holding capacity (AWHC),

Thus, AWHC = field capacity - permanent wilting point

Management allowed depletion (MAD)

The amount of water that is allowed to be used by the plant between irrigations is referred to as management allowed depletion. Water going out of the effective root zone is primarily deep percolation (from excess irrigation or infiltrated rain) and crop evapotranspiration (crop water use).Water manage in root zone by identify water going into and out of the system, measure amount of water and control these amounts.

Water budget irrigation scheduling

All water going into and out of the effective root zone can be measured in terms of a depth of water. Thus, if the amounts of the different types of water going into and out of the root zone can be identified, simple addition and subtraction can indicate the net change in root zone water content. All of the water going into and out of the effective root zone can be measured in terms of a depth of water. The water budget equation is given as

Water end = Water start + IRR + Rain - ET_c – Deep percolation + Flux net

Identifying the values and solving this equation is the basis for water budget irrigation scheduling. Water budget irrigation scheduling chooses starting point soil moisture in the effective root zone. Then, the equation is solved on a daily basis, considering the amounts of water moving into and out of the root zone for that day.

Using the water budget equation to schedule (predict) irrigation

The water budget equation can be solved on a real-time basis. The basics of soil-water-plant relationships show that the individual crop water use could be estimated if a reference evapotranspiration and a crop coefficient are available.

$ET_c = ET_o * K_c$

Where

ET_c = evapotranspiration of the crop

ET_o = reference evapotranspiration

K_c = the crop coefficient that relates the crop's water use to the reference ET_o

Advantages of irrigation scheduling

The irrigation scheduling offers several advantages, which are given as under:

- It enables the farmer to schedule water rotation among the various fields to minimize crop water stress and maximize yields.

- It reduces the farmer's cost of water and labor through less irrigation, thereby making maximum use of soil moisture storage.
- It lowers the fertilizer costs by holding surface runoff and deep percolation (leaching) to a minimum.
- It increases the net return by increasing crop yields and crop quality.
- It minimizes water-logging problems by reducing the drainage requirements.
- It assists in controlling root zone salinity problems through controlled leaching.
- It results in additional returns by using the "saved" water to irrigate non-cash crops that otherwise would not be irrigated during water-short periods.

Methods of irrigation scheduling

Irrigation scheduling used for water management strategies to prevent over-application of water while minimizing yield loss due to water shortage or drought stress. Irrigation scheduling will ensure that water is applied to the crop when needed and in the required quantity. Effective scheduling requires knowledge of the following factors:

- Soil properties,
- Soil-water relationships and status,
- Type of crop and its sensitivity to drought stress,
- Stage of crop development,
- Status of crop stress,
- Potential yield reduction if the crop remains in a stressed condition,
- Availability of a water supply and
- Climatic factors such as rainfall and temperature.

There are three ways to determine when irrigation is needed

- Measuring soil water,
- Estimating soil water using an accounting approach and
- Measuring crop stress.

Soil water can be measured using a range of devices including tensiometers, which measure soil water suction, electrical resistance blocks (also called gypsum blocks or moisture blocks), which measure electrical resistance that is related to soil water by a calibration; neutron probes, which directly measure soil water. Once the decision to irrigate has been made, it is important to determine the amount of water to apply. Irrigation needs are a function of the soil water depletion volume in the effective root zone, the rate at which the crop uses water and climatic factors. Accurate measurements of the amount of water applied are essential to maximizing irrigation efficiency. The quantity of water applied can be measured by such devices as a totalizing flow meter that is installed in the delivery pipe.

Soil moisture measuring techniques

Soil water can be measured by the methods that determine the soil water content or the soil water potential. Soil water content is the amount of water per unit volume of soil or weight of dry soil. Soil water potential is the force necessary to remove the next increment of water from the soil. Different measurement methods have particular strengths and weaknesses.

For example, the gravimetric method is very accurate, but it is very slow and many samples are needed for each field and site specific interpretations are necessary. Therefore, water content in soil determined by various methods given as (i) Feel method, (ii) Gravimetric (iii) Neutron probe and (iv) Time domain reflectometry (TDR).

Soil water potential or soil water tension is determined by following methods (i) Tensiometers, (ii) Gypsum blocks and (iii) Granular matrix sensors (GMS)

Soil Moisture Budget

All irrigation scheduling methods consist of an irrigation criterion that triggers irrigation and an irrigation strategy that determines how much water to apply. Irrigation scheduling methods differ by the irrigation criterion. A common and widely used irrigation criterion is the soil moisture status. Table 10.3 compares different methods of irrigation scheduling by monitoring soil moisture content or tension. The methods given below described how to measure or estimate the irrigation criterion.

Table 10.3: Different methods of irrigation scheduling

Method	Measured parameter	Equipment needed	Irrigation criterion	Advantages	Disadvantages
Hand feel and appearance of soil	Soil moisture content by feel	Hand probe	Soil moisture content	Easy to use; simple; can improve accuracy with experience	Low accuracies; field work involved to take samples.
Gravimetric soil moisture sample	Soil moisture content by taking samples	Auger, caps, oven.	Soil moisture content	High accuracy	Labour intensive including field work; time gap between sampling and results
Tensiometers	Soil moisture tension	Tensiometers including vacuum gauge	Soil moisture tension	Good accuracy; instantaneous reading of soil moisture tension	Labour to read; needs maintenance; breaks at tensions above 0.7 atm

(Contd..)

(Contd..)

Method	Measured parameter	Equipment needed	Irrigation criterion	Advantages	Disadvantages
Electrical resistance blocks	Electric resistance of soil moisture	Resistance blocks AC bridge (meter)	Soil moisture tension	Instantaneous reading; works over larger range of tensions; can be used for remote reading	Affected by soil salinity; not sensitive at low tensions; needs some maintenance and field reading.
Water budget approach	Climatic parameters: temperature, radiation, wind, humidity and rainfall; depending on model used to predict ET.	Weather station or available weather information.	Estimation of moisture content	No field work required; flexible; can forecast irrigation needs in the future; with same equipment can schedule many fields.	Needs calibration and periodic adjustments, since it is only an estimate calculations; cumbersome without computer.
Modified atmometer	Reference ET.	Atmometer gauge.	Estimate of moisture content.	Easy to use, direct reading of reference ET.	Needs calibration; it is only an estimation.

Criteria for scheduling irrigation

The purpose of irrigation scheduling is to determine the exact amount of water to apply to the field and the exact timing for application. The amount of water applied is determined by using a criterion to determine irrigation need and a strategy to prescribe how much water to apply in any situation. Several approaches for scheduling irrigation have been used by scientist and farmers given as under.

Soil moisture depletion approach

The available soil moisture in the root zone is a good criterion for scheduling irrigation. When the soil moisture in root zone is decline to a particular level which is different for different crops, it is also replenished by irrigation. For practical purpose, irrigation should be started when about 50 per cent of the available moisture in the soil root zone is depleted. The available water is the soil moisture, which lies between field capacity and wilting point. The relative availability of soil moisture is not the same from field capacity to wilting point stage and since the crop suffers before the soil moisture reaches wilting point, it is necessary to locate the optimum point within the available range of soil moisture, when irrigation must be scheduled to maintain crop yield at high level. Soil moisture deficit represents the difference in the moisture content at field capacity and that before irrigation. This is measured by taking into consideration the percentage, availability, tension and resistance.

Plant basis or plant indices

As the plant is the user of water, it can be taken as a guide for scheduling irrigation. The deficit of water will be reflected by plants itself such as dropping, curling or rolling of leaves and change in foliage colour as indication for irrigation scheduling. However, these symptoms indicate the need for water. They do not permit quantitative estimation of moisture deficit. Growth indicators such as cell elongation rates, plant water content and leaf water potential, plant temperature leaf diffusion resistance etc. are also used for deciding when to irrigate.

Climatological approach

Evapotranspiration mainly depends upon the climate. The amount of water lost by evapotranspiration is estimated from climatological data and when ET reaches a particular level, irrigation is scheduled. The amount of irrigation given is either equal to ET or fraction of ET. Different methods in climatological approach are IW/CPE ratio method and pan evaporimeter method. In IW/CPE approach, a known amount of irrigation water is applied when cumulative pan evaporation (CPE) reaches a predetermined level. The amount of water given at each irrigation ranges from 4 to 6 cm. The most common being 5 cm irrigation, scheduling irrigation is at an IW/CPE ratio of 1.0 with 5 cm. Generally, irrigation is given at 0.75 to 0.8 ratios with 5 cm of irrigation water.

Critical growth approach

In each crop, there are some growth stages at which moisture stress leads to irrevocable yield loss. These stages are known as critical periods or moisture sensitive periods. If irrigation water is available in sufficient quantities, irrigation is scheduled whenever soil moisture is depleted to critical moisture level say 25 or 50 per cent of available soil moisture. Under limited water supply conditions, irrigation is scheduled at moisture sensitive stages and irrigation is skipped at non-sensitive stages. In cereals, panicle initiation, flowering and pod development are the most important moisture sensitive stages.

Practices for efficient irrigation water transport

Irrigation water transportation systems that move water from the source of supply to the irrigation system should be designed and managed in such a manner that minimizes evaporation, seepage, and flow-through water losses from canals and ditches. Delivery and timing need to be flexible enough to meet varying plant water needs throughout the growing season. Transporting irrigation water from the source of supply to the field irrigation system can be a significant source of water loss and cause of degradation of both surface water and ground water.

Practices for drainage water management

Drainage water from an irrigation system should be managed to reduce deep percolation, move tail water to the reuse system, reduce erosion, and help control adverse impacts on surface water and groundwater. A total drainage system should be an integral part of the planning and design of an efficient irrigation system. This may not be necessary for those soils that have sufficient natural drainage abilities.

CONCLUSIONS

The water requirements of crops are dependent on evapotranspiration (ET), soil chemistry and the crop's maximum allowable depletion (MAD). Direct measurements of root zone soil moisture, water application along with evapotranspiration (ET) values and soil textures, can be used in a soil water balance model that can significantly optimize irrigation efficiency. Advancements in computer microprocessors, memory and software development tools has improved data acquisition methods and made data acquisition system integration more reliable and more cost effective. The information for proper soil water management for both commercial and small farmers who are growing different crop under supplementary irrigation and full irrigation scheduling can be used to increase water productivity of crops. The aim of this chapter is to advocate the use of water budget irrigation scheduling technology in crops as a water management tool in irrigated agriculture to increase water use efficiency (WUE) and reduce input costs. Precision irrigation will optimize irrigation by minimizing the waste of water and energy, while maximizing crop yields. The most effective method for determining the water demands of crops is based on the real time monitoring of soil moisture, and direct water application used in conjunction with the information about soil hydrological properties and evapotranspiration. The water management tools use information about evapotranspiration, soil and the crop to set specific irrigation targets. These irrigation targets will help the irrigator to optimize the amount of water used for irrigation. Optimization of irrigation water will increase crop yields while conserving water resources.

REFERENCES

Allen RG, Pereira LS, Raes D and Smith M. (1998). Crop evapotranspiration Guidelines for computing crop water requirements - FAO Irrigation and drainage paper 56. Rome. Accessed /2004.

Ashraf MY, Sarwar G, Ashraf M, Afaf R, Sattar A (2002). Salinity induced changes in α-amylase activity during germination and early cotton seedling growth. *Biol Plantarum,* **45:** 589–591.

Blrkeland, PW. (1999). Soils and Geomorphology. Paperback, ISBN-10:0195078861.

Brouwer, F. (1988). Determination of broad scale land use changes by climate and soils. International Institute for applied Systems analysis, A-2361 Laxenburg, Austria.

FAO (1990). Guidelines for profile description. 3rd Edition. Rome.

FAO, FAOSTAT agriculture data, 2003. http://apps.fao.org/. Irrigation and Drainage Paper No. 24", FAO. Rome, Italy. pp. 156.

Hutson, SS. (2000). U.S. Geological Survey, Estimated Use of Water in the United States, USGS Circular 1268.

Jensen, ME. (1980). Water consumption by agricultural plants. Chapter 1. In: T.T. Kozlowski (Ed): Water Deficits and Plant Growth Vol. II (pp 1–22). Academic Press, New York.

Jones, PG., Thornton, P K. (2003). The potential impacts of climate change on maize production in Africa and Latin America in 2055. *Global Environmental Change*, **13:** 51–59.

Khalil, H K. (1996). Nonlinear Systems (2nd ed.). Englewood CliMs, NJ: Prentice-Hall.

Khalil, MAK, Shearer, MJ (2006). Decreasing of emissions methane from rice agriculture. In: Solvia CR, Takahaski, J, kreuzer M (eds). Greenhouse gases and animal agriculture: an update International congress séries, 1293, Elsevier, The Netherlands, pp. 33–41.

Kumar R, Jat MK, Shankar V. (2012). Methods to estimate reference crop evapotranspiration-A review. *Journal of Water Science and Technology* (IWA publishing, UK), **66**(3): 525–535.

Mc Bride MB. (1994). Environmental Chemistry of Soils. Oxford University Press.

Rowell DL. (1994). Soil Science Methods and Applications. John Wiley & Son.

Steve, RG, Kirkham, MB and Brent, EC. (2006). Root uptake and transpiration: From measurements and models to sustainable irrigation. *J. Agri. Water Manag.*, **86:** 165-176.

Wang, E and Smith, CJ. (2004). Modeling the growth and water uptake function of plant root systems: A review. *Aust. J. Agric. Res.*, **55**(5): 501–523.

Warrick, AW. (2003). Soil Water Dynamics. Robens Centre for Public and Environmental Health University of Surrey Guildford, *The Environmentalist* 02/2004; **24**(1): 59–60. DOI: 10.1023/B:ENVR.0000046450.62059.62.

■

11

CHAPTER

Site Specific Varietal Improvement Strategies for Rice Breeding in Temperate Regions

☞ Gulzar Singh Sanghera[1],
☞ Subhash Chander Kashyap[2] and ☞ F.A. Aga[3]

INTRODUCTION : AN OVERVIEW

Modern intensive cropping in irrigated agro-ecosystem has brought a substantial increase in the production of food grains. However, with increasing population, limited resources, uncertain climatic conditions, food security needs becomes more daunting for agricultural scientists. The technologies are required which can increase the grain yield of the crops and raise the income of farmers without deteriorating the quality of environment. Many factors of green revolution like irrigation, input responsive genotypes, fertilizers and other agrochemicals etc. have almost exhausted in most part of country. To achieve, 109 million (156 m t in 2030 as per ICAR vision 2030) tons target by 2025, we need to concentrate our research efforts on long term as well as short term strategies, where the later ones places more emphasis on effective transfer of the available technology to the farmers. We must understand the importance of technological innovations. Therefore, increasing productivity of different crops in a sustained manner, thus bringing about 'evergreen' revolution. To retain and make use of traditional wisdom, farmers and agriculturists (Rice breeders) must optimally utilize technological advancements in agriculture for site specific varietal improvement programme. The objectives and priorities of genetic improvement of crop varieties should be defined for each region and cropping system by using the outputs from a preliminary characterization of the systems conducted in different sites. To harness sustainable production and productivity of different crops, adaptation to environmental conditions and crop management will be regarded as the most important objectives to ensure future adoption of new varieties. Increase in yield is always a priority for farmers. Moreover, grain characteristics must satisfy local requirements for household consumption and for sale of harvest surpluses. Finally, resistance to major insect pests and diseases is often a priority for site specific varietal improvement, especially in areas where small farmers cannot access or have difficulties in accessing credit to buy pesticides. When seeking a specific adaptation the direct incorporation of new germplasm, whether landrace or exotic, into an existing improved population could induce a setback in genetic progress in productivity, plant height, earliness or grain quality, thus moving away from the objectives. However, the possibility of incorporating

[1 & 2] Mountain Research Centre for Field Crops, SKUAST-K, Khudwani, Anantnag – 192 102
[3] Associate Professor, Division of Agronomy, SKUAST-K, Srinagar – 190 025

well-adapted local varieties into these populations was considered in terms of contributing adaptation to specific environmental conditions, stable resistance to diseases or a specific grain quality for household consumption. An integrated pre-breeding procedure leading to the production of novel genetic combinations (designer genotypes) and participatory breeding jointly with farming families for the development of location-specific varieties would help to combine genetic efficiency and diversity in a mutually reinforcing manner. This will help to avoid the danger inherent in spreading single genotypes over large areas and also sustainable agriculture needs for sustenance of site specific varieties. Therefore, in present chapter attempt has been made to compile various conventional and non-conventional breeding approaches that are required for development of location specific crop varieties along with some case studies.

RICE PRODUCTION

Rice has been the staple food for seventeen countries in Asia and the pacific, eight countries in Africa, seven countries in Latin America and Caribbean and one in Near East. Rice provides two thirds of calories intake of more than three billion people in Asia and one-third calorie intake of nearly 1.5 billion people in Africa and Latin America (FAO, 1995), when all developing countries are considered together rice provides 27 per cent dietary energy supply and 20% dietary protein. Globally 150 m ha are planted to rice with an average production of 400 mt/year. In India, it occupies about 24 per cent of total gross cropped area thereby contributing 43% of total food grain production and 46 per cent of total cereal production. India produces about 97 mt of rice over an area of about 43 m ha with a productivity level of 2.25 t/ ha which is much below the world average of 3.9 t/ha (Anonymous, 2010). China is next to India in rice area (28.7 m ha) with an average productivity of 6.2 t/ha. In the State of Jammu & Kashmir rice, being the principal food crop, is cultivated across diverse agro-ecological situations extending from sub-tropical region (>1000 m asl) of Jammu to temperate cold high altitude areas (1800-2500 m amsl) of Kashmir (Ahmad *et al*, 1999). During 2009, 0.56 million tonnes of rice were grown from 0.26 m ha with an average yield of about 0.22 t/ha (Anonymous, 2009).

In Kashmir the production environment is typically temperate with cold winters and mild summers characterized by short growing season, long days, cool nights and occurrence of wide differences between maximum and minimum temperatures (Sanghera *et al*, 2011). Compared with average yields of temperate regions of world, the rice productivity is disappointingly very low (0.26 t/ha) in Kashmir (Sanghera and Wani, 2008). The rice area in Kashmir can be broadly categorized into two distinct zones with entirely different agro-climatic conditions. The higher belts, comprising areas between 1800-2500 m amsl, constitute 10-12% of total rice area. The growing season is restricted to 125-135 days. The lower belt or valley basin enjoys relatively favourable growing conditions of 140-145 days and lies between 1500-1800 m amsl. The varietal requirement of two zones is different with *Japonica* rices being more predominantly grown in upper belts and *Indica* in the valley basin (Sanghera and Wani, 2008).

OBJECTIVES AND GENETIC IMPROVEMENT OF RICE VARIETIES

The objectives and priorities of genetic improvement of crop varieties should be defined for each region and cropping system, using the outputs from a preliminary characterization of the systems conducted in different sites. To harness sustainable production and productivity of different crops, adaptation to environmental conditions and crop management will be regarded as the most important objectives to ensure future adoption of new varieties. Increase in yield is always a priority for farmers. Moreover, grain characteristics must satisfy local requirements for household consumption and for sale of harvest surpluses. Improvements in rice farming will help not only farmers, farm workers and their families but also the millions of poor people who are totally dependent on rice. An increase in rice production is strongly needed and it should come primarily from higher yields from existing rice fields in the irrigated and favorable rainfed lowland ecosystems, as opportunities to increase rice planting area are limited. It is even more difficult to achieve higher rice yields in less favorable agricultural environments. The good news is that a wealth of valuable knowledge on rice farming exists from the research efforts. However, if technologies developed by rice research are not effectively delivered to farmers their livelihoods are unlikely to improve. The challenge is to move this research-based rice farming knowledge from the research centers to the doorsteps of rice farmers with special reference to rice grown under temperate conditions.

DIVERSITY IN RICES

Cultivation and selection by farmers for centuries under varied growing conditions has resulted in a myriad of rice varieties. About 1, 20,000 distinct rice varieties exist in the world, which are grown in more than 100 countries. From its subtropical origins rice is now cultivated between 55^0 N in China and $36\ ^0$ S in Chile.

Rice varieties differ from each other in growth duration, photoperiod sensitivity, grain size, shape and colour and endosperm properties. Varieties also differ in the level of tolerance/ resistance to abiotic and biotic stresses (Khush, 1997). On the basis of ecological differentiation, two major eco-geographic races exist *viz. Indica* and *Japonica*. They differ in morphological and serological characters as well as intervarietal fertility (Glaszmann, 1987). The traditional *Indica* rice varieties, widely grown throughout the tropics and subtropics, are tall and heavy tillering with long, narrow, light green leaves. The *Japonica* varieties have narrow dark green leaves, medium height tillers and short to intermediate plant height. It is usually grown in cooler subtropics and temperate climates. *Javanica* rice belongs to the Japonica race of *Oryza sativa* and thus is morphologically similar to *Japonicas*. However, *Javanica* have wider and more pubescent (hairy) leaves, in addition, the grain frequently has hair-like awns. Rice belongs to Poaceae or grass family. The genus *Oryza* to which cultivated rice belongs has twenty-one wild and two cultivated species. Nine of the wild species are tetraploid and remaining are diploid.

ASSESSING GENETIC DIVERSITY USING MARKERS

A marker must be polymorphic, that is, it must exist in different forms so that chromosome carrying the mutant gene can be distinguished from the chromosome with normal gene by the form of marker it also carries. This polymorphisms in markers can be detected at three levels *i.e.* phenotype (morphological), differences in proteins (biochemical) and differences in the nucleotide sequences of DNA (molecular). These markers differ in the level at which they detect genetic variation, extent of polymorphisms, degree of environmental stability, number of loci, molecular basis of polymorphism, practicality and amenability to statistical estimation of population genetic parameters.

MORPHOLOGICAL MARKERS

Chang (1976) and Takahashi (1984) had recognized three eco-species in *Oryza sativa i.e., Japonica, Javanica* and *Indica*. Matsuo (1952) and Oka (1953) did the genetic differentiation among these three eco species by morphological and phenological analysis. Kinoshita (1986) constructed the first genetic linkage map of rice with morphological markers.

Morphological markers have been routinely used to analyze genetic diversity but it is associated with several disadvantages *i.e.,* limited number of morphological characters available for analysis and these characters are influenced by environmental factors also. As a consequence other different plant genotypes can not be unequivocally distinguished. The genetic basis of most morphological variation is generally unknown and hence these markers; could not be proved desirable for genetic differentiation.

BIOCHEMICAL MARKERS

Biochemical markers offer greater diversity as compared to the morphological markers. Glaszmann (1987) examined 1688 rice cultivars from different Asian countries for allelic frequencies at 15 isozymes loci and analyzed the data by a multivariate analysis. The results showed that 95 per cent of the cultivars fall into six groups, the remaining 5 per cent being scattered over intermediate positions. Group-II corresponded to the *Indica* and group VI to the *Japonica*. The aromatic rices of Indian subcontinent fall into Group-V. Isozymes markers have been applied to assess the genetic variation but these markers represent only a minor portion of the genome and hence undesirable for practical breeding (Second, 1982). The limitation with isozymes is, that these are unavailable in the germplasm. Generally only 30-40 loci can be mapped in a genome as in the case of maize and tomato. Another potential short coming of isozyme markers is that they are also considered to be phenotypic markers since the assay is for gene product and not the gene itself. As a result, this can cause problems when expression of particular isozymes is developmentally regulated or dependent on external stimuli for expression.

MOLECULAR MARKERS

In the last two decades, a rapid progress has been made towards the development and application of molecular marker technology in plant genome analysis. Molecular markers are considered best for the analysis of genetic diversity and cultivar identification since they are indifferent to development stage or environment. Molecular marker analysis has been widely used for varietal identification and know about the cases of adulteration and even the level of adulteration (Bligh *et al.,* 1999). However, to obtain fingerprints of closely related varieties or to have some characteristic differences, it is essential to obtain a very high level of polymorphism. So, it has become necessary to establish common basis for assessing the relative effectiveness of the various markers systems currently available for identification and discrimination between interspecific groups of rice.

WHY SITE SPECIFIC VARIETIES

The rice production constraints in Kashmir as noted above are essentially those encountered elsewhere in temperate regions of the world. Apart from low temperature, water shortages as a result of aberrant weather conditions, susceptibility to biotic stresses particularly blast and socio-economic, technical, management, institutional, technology transfer and adoption/ linkage problems limit the rice production (Sanghera and Wani, 2008). The HYV's bred for situations with assured nustrient and water supply could not largely replace the traditional land races having tolerance to local adverse growing conditions in the region. Even promising HYV's for favourable conditions could not be grown as seed of the desired varieties are in short supply and are not available on time. High humidity prevailing in the region during most part of the year causes quality deterioration of the seed (Bandey and Sanghera, 2002). Diverse and variable rice-growing ecologies prevail even in a small geographical area of a village Panchayat or a Block level. Upland, favourable shallow rainfed lowland, unfavourable deep water and flood prone area, swampy land and hilly area co-exist together in a block requiring completely different problem solving approaches (Sanghera and Zarger, 2002). Heavy, erratic and torrential rain causes recurrent floods in many parts of the region. Inherent poor response of the local land races to high dose of fertilizer, its poor recovery due to various type of losses restrain the farmers from investing more on fertilizer. The major problems which hinder the rice production in Kashmir are essentially those of temperate regions associated with condition of low temperature (cold), high altitude, short growing season and insect pest and diseases (Sanghera and Zarger, 2002). The low temperature is the greatest concern of rice growers in the region. Rice crop is exposed to injury by cold from early seedling stage to the ripening stage. The problem is more severe in higher belts. The effects of cold are manifested as poor and delayed germination, stunted growth, seedling discoloration, tip burning and reduced tillering during vegetative phase. At the reproductive stage stress effects are considered more severe which result in delayed heading, poor panicle exertion, prolonged flowering period due to irregular heading spikelet degeneration, impaired seed fertility and abnormal grain formation (Sanghera *et al.,* 2011, Kaneda and Beachell, 1974). The critical low temperature for inducing sterility ranges between 15-17°C in highly cold tolerant varieties and 17-19°C for cold sensitive varieties at the meiotic stage of crop growth (Sanghera *et al.,* 2011). Although farmers have tried to deploy varieties possesses tolerance to low temperature and recommended cultural practices such as planting date and water depth during panicle development rice crop still suffers from high grain sterility due to variable climate. The dependence on yield for achieving the level of production needed to met the household food requirements makes it imperative to develop varieties having high yield potential with tolerance to low temperate and other stresses.

Rice crops are found in various growth stages favouring perpetuation of different insects and pathogens. Among biotic stresses, diseases and weeds are of greater significance, whereas insect pests have a minor influence on rice production in Kashmir. Blast, brown spot and sheath blight are important rice diseases in the Valley. However, blast is the most devastating and of wide spread occurrence when favourable climatic conditions prevail for the pathogen and, if left unchecked, could inflict heavy yield losses. This frequently causes severe incidences of diseases like blast, sheath blight and sheath rot (Bhat *et al.*, 2001). Continuous cultivation of just one variety with high seed rate and nitrogen application coupled with high relative humidity (>90%) and moderate temperature (22-26°C) promote disease development. Management practices aiming at controlling collateral hosts (*Echinochola causgalli, Cypress* sp. etc.) and a combination of low population density and nitrogen fertilization are advocated to check the disease. Seed treatment and economic spray schedules devised to minimize the impact of disease are not without economic, health and environmental implications. Further, insecticides and fungicides to control these are not popular and are in short supply in the region.

Therefore, use of resistant varieties offers an easy, economic and most eco-friendly approach for managing the rice blast (Barman *et al.*, 2004). Nevertheless, because of the high variability of the pathogen genetic resistance generally breaks down shortly after new varieties are released. This envisages the germplasm enhancement through a systematic and well planned resistance breeding programme to cope with the problem. Using international blast differentials (Dular, Raminad Str.3, Caloro, Kanto-52, Zenith, NP 125, Usen and Sha-tiao-tsao) two races of blast ID-I and IC 17 have been identified in the valley (Anwar *et al.*, 2011). Identification of resistant / tolerant donors against blast and their utilization in varietal improvement continues to be the integral part of rice breeding programme. It is thus imperative to find ways and means to lift the present yield level, optimize the use of various inputs such as water and fertilizer, in order to make the rice production efficient, cost effective and suitable for resource poor farmers, sustainable and environment friendly.

DEVELOPMENT STRATEGIES FOR IMPROVED PRODUCTION

Improving crop management will have major impacts on production but there are still several countries that do not have adequate varieties or access to improved germplasm. In addition, there are countries that have access to elite germplasm but lack the means to evaluate genetic material. Furthermore, few countries have established technology transfer programmes. Most of the small regions and province are deficient in all of the above. These have limited human and financial resources for rice research and development and it would be cost-ineffective for them to work independently. The solution is collective action in which states join forces and share the costs of rice research and development. It is apparent that immediate production gains can be obtained by bridging the yield gap in irrigated rice. The yield gap is significant in all countries and is mainly a function of inadequate crop management. Bridging the yield gap requires an effective technology transfer programme. Variety improvement is also important, especially in countries that have weak national programmes and limited access to improved germplasm. However, the possibility of incorporating well adapted local varieties into these populations was considered in terms of contributing, *for example*, to adaptation to specific environmental conditions, stable resistance to diseases or a specific grain quality for household consumption. These choices were based on the results of the diagnosis and the preliminary *in situ* (PVS) trials of diverse lines and varieties. The incorporation of local varieties as contributors of local adaptation genes into the exotic populations may be carried out in two ways, according to the type of population and the time needed to develop new varieties. It should be emphasized that when seeking a specific adaptation or insect pest resistance, the direct incorporation of new germplasm, whether landrace or exotic, into an existing improved population could induce a setback in genetic progress in productivity, plant height, earliness or grain quality, thus moving away from the objectives as defined by (Chaves, 1997). Hence, new materials should first be incorporated through intermediate populations and then, after several crosses between these materials and the improved population, introduced into the principal population (Gallais, 1990). Morais *et al.,* (2000) presented the alternative of crossing each new parent with the population and evaluating the individual crosses, before mixing the seeds of all individual combinations. The yield potential of modern rice varieties has reached the plateau. There are also signs of yield decline in rice as found at various experimental farms and some research locations. Earlier, rice yields reaching 8-9 t/ha at farm but now it has not exceeded 6-7 t/ha in dry season and 5-6 t/ha in the wet season (Casman and Pingali, 1993). Similarly findings have been reported from other rice growing areas of the world therefore, the challenge before rice scientists is how to sustain long term rice production. Fortunately, a yield gap between potential rice yields and actual on-farm yields still exists and improved site specific varieties will make an important contribution to closing this gap.

ELEMENTS OF A BREEDING STRATEGY

Genetic improvement is a basic component of national policies for raising agricultural production in many countries. Although breeding programmes run at international centers of the CGIAR (Consultative Group on International Agricultural Research) system have made an important contribution to major food crops, national breeding programmes (public or private) maintain a fundamental role for crop improvement in their target region (usually applicable to the whole country), particularly as the exploitation of specific adaptation and yield stability characteristics are concerned, owing to their better knowledge of and easier access to, local germplasm and cropping environments. These programmes aim to make correct decisions on a number of issues which comprise a breeding strategy (Simmonds, 1991). Decisions may concern, in particular

- Adaptation strategies, yield stability and other (*e.g.* crop quality) targets
- Genetic resources forming the genetic base (indigenous or exotic, from traditional or improved varieties)

- Techniques for the recombination and introgression of useful genetic variation
- Variety type (*e.g.* single-cross hybrid, double-cross hybrid, improved population or synthetic variety, with regard to outbred species)
- Breeding plan and selection procedures (selection environments, indirect selection criteria, presence and extent of participatory breeding, experimental designs etc.)

The definition of a strategy with respect to G x E interactions may require decisions on most of these elements, namely, adaptation strategy and stability targets, genetic resources, variety type, breeding plan and selection procedures. Initial decisions may change with time as a consequence of new opportunities offered by scientific progress, experimental evidence, available funding, food security policies, changes in national seed systems, international cooperation etc. but they should remain consistent with the breeding objective. *For example*, the inconsistency between targeting also unfavourable areas and adopting genetic resources and selection procedures producing material specifically adapted to favourable environments has contributed to the partial failure of a number of breeding programmes carried out in the 'Green Revolution' context (Simmonds, 1991 and Ceccarelli, 1994).

RAISING THE YIELD CEILING

According to several estimates there is a need to produce 30% more rice by 2030 to satisfy the growing demand without negatively affecting the sustainability of rice production (Brown, 1996). It is projected that the population will increase 93% during this period in the developing world, where most countries depend on rice as a staple food. At present, the rice production in Jammu and Kashmir is just 0.59 million tonnes per year as against the total requirement of 1.24 million tonnes. This is not sufficient to feed 0.85 million people (Anonymous, 2006).

A huge task is now proposed to rice breeders: to develop varieties with higher yield potential so that more rice can be produced from suitable lands. On the other hand, rice cultivars will have to tolerate several abiotic stresses like low or high temperature, submergence, drought and salinity. Further, studies were performed aiming to explore the natural diversity regarding stress tolerance (Ali *et al.*, 2006). This may be another strategy allowing the adaptation of rice farming to environments formerly not suitable to its use. The fundamental concern of rice breeders is to produce varieties with superior yields. Grain yield is the result of several components that are determined at various stages in the growth of rice (Yoshida, 1981), while changes to any components can affect yield. Therefore, breeder is concerned with special agronomic, morphological and physiological characteristics that make a plant suitable for a particular production method or growing environment.

Rice production was increased remarkably in the 1960s with the introduction of the semi-dwarfing gene, into traditional *indica* and *japonica* cultivars resulting in the "green revolution" in rice and probably the most well known triumph of agricultural research (Khush, 1999a). This recessive gene, *sd1* or OsGA20ox2, was detected in the Taiwanese native semi-dwarf variety, Dee-geo-woo-gen. The modification of plant architecture through the introduction of semi-dwarfism resulted in lodging resistance and increased harvest index and two to three times higher yield potential than that of varieties available prior to green revolution. The development and release of the variety IR8 derived from Dee-geo-woo-gen had twice the yield potential of the traditional, tall varieties (Khush, 2005). This and other new high-yielding varieties also presented multiple resistances to diseases and insects and consequently higher yield stability (Khush, 1999b).

In Kashmir major increases in rice production have occurred during the last 25 years, due to adoption of high yielding semi-dwarf varieties and improved management practices (Sanghera and Zarger, 2002). The salient characteristics of recommended rice varieties in Kashmir bred at SKUAST (K)-Rice Research and Regional Station, Khudwani during 1950 to 2005 is given in Table 11.1. However, the rate of increase of rice production is lower (1.5 per cent per year) than the rate of increase of population (1.8 per cent per year). If this trend is not reversed, severe food shortage will occur in 1st quarter of century as it has been estimated that demand of rice will exceed production by next 25-30 years (Anonymous, 2006). To further increase yield potential of rice under temperate Kashmir conditions, prospects of new strategies like hybrid rice, ideotype breeding, cold tolerance and some other approaches that are still not well explored are described below?

Table11.1: Characteristics of recommended rice varieties in Kashmir, 1950 to till date

Cultivar/ Variety	Year of release	Maturity (Days)	Yield Potential (t/ ha)	Grain type	Reaction to blast	Reaction to lodging
Low altitude varieties						
China 1039	1953	142-145	5.0-5.5	MB	MR	S
China 1007	1956	147-152	5.0-6.0	MB	R	R
China 988	-do-	147-152	5.0-6.0	MB	MR	R
China 972	-do-	147-152	5.0-5.5	MB	MR	R
Low altitude varieties						
K-60	1962	---	5.0-5.5	MB	MR	R
K-65	1966	-----	5.0-5.5	MB	MR	R
SKAU-5	1978	140-142	5.8-6.2	MB	MR	MR
SKAU-23	1996	140-142	6.0-6.5	MB	MR	MR
SKAU-27	1996	140-142	6.0-6.5	MB	MR	MR
Shalimar Rice-1	2005	140-150	6.5-7.0	MB	R	R
High altitude varieties						
Shenei	1967	---	3.0-3.5	B	MR	MR
China 971	1967	142- 150	3.0-3.5	B	MR	MR
K-78-13	1974	140-150	3.8-4.0	B	MR	MR
K-332	1982	130-140	4.0-4.5	B	MR	R
K-429	2000	135-140	4.2-4.7	B	MR	R

B= Bold, MB= Medium bold, R= Resistant and MR= Moderately resistant (Source: Sanghera and Zarger, 2002)

NEW PLANT TYPE FOR HIGHER YIELD POTENTIAL

To increase yield potential of rice beyond what was achieved by the high-yielding semi-dwarfs, breeders conceptualized a new plant type in the late 1980s. Modern semi- dwarf varieties derived from IR8 produce a large number of unproductive tillers and excess leaf area, causing mutual shading and reducing sink size and canopy photosynthesis (Dingkuhn *et al.*, 1991 and Virk *et al.*, 2004). Plant physiologists have suggested that physical environment in the tropics is not a limiting factor for increasing rice yields. Maximum yield potential was estimated to be 9.5 t ha^{-1} during wet season and 15.9 t ha^{-1} during dry season (Yoshida, 1981).

It is well known that yield is a function of total dry matter or biomass and the harvest index. Therefore, to further increase the yield potential of rice, either increase the total biomass production or harvest index or both. The IRRI Scientists conceptualized a plant type to increase the biomass to about 23 t ha^{-1} and harvest index to 0.55. Such a plant should produce a grain yield of about 12.5 t ha^{-1} or an increase of 25 % over the yield of existing high yielding varieties (Khush, 1995). The harvest index can be increased by increasing the production of energy stores in the grain or by increasing the sink size. The sink size can be increased by (i) Large number of spikelets per panicle or ear, (ii) Increased spikelet filling (High spikelet fertility), (iii) Slow leaf senescence, (iv) Maintenance of healthy root system and (v) Increased lodging resistance.

Biomass can be increased by both genetic manipulations as well as by better management practices. Varietal characteristics for increasing the biomass include (i) Establishment of desirable leaf canopy structure, (ii) Rapid leaf area development, (iii) Rapid nutrient uptake and (iv) Increased lodging resistance.

Breeding efforts to develop a "New Plant-Type" (NPT) started in 1990 with the objective of developing an improved germplasm with low tillering capacity, few unproductive tillers, 200–250 grains per panicle, plant height of 90–100 cm, thick and sturdy stems, dark green, thick and erect leaves, vigorous root system, 100–130-d growth duration and increased harvest index from 0.3 to 0.5 (Khush, 1995). This work started with tropical *Japonica* germplasm identified as donors of these traits from Indonesia, Malaysia, Thailand, Myanmar, Laos, Vietnam, and the Philippines (Virk *et al.*, 2004). These NPT donors had the desired ideotype however most were susceptible to tropical diseases and insects, characteristics that they inherited to their progeny. An additional difficulty was that farmers and consumers in tropical rice-growing countries prefer varieties with long and slender grains and intermediate to high amylose content (*Indica* subspecies) (Virk *et al.*, 2004). Consequently, new modern high-yielding *Indica* varieties/ elite lines were included in the hybridization program and this 2nd generation of NPT lines had improved yield and higher resistance to diseases and insects. A major breakthrough in the NPT project has been the inclusion of *Japonica* germplasm in a breeding program for tropical regions. Crosses between the two sub species had so far very limited success but there are notable exceptions, such as Tongil rice in Korea and Mahsuri in the rainfed lowland tropics (Virk *et al.*, 2004). Typically, traits from *Indica* germplasm are being incorporated into tropical *Japonica* breeding programs due to narrow genetic bases in *Japonica* cultivars and the higher yield potential of *Indica*. From the ideotype breeding, it became evident that the introgression of *Japonica* genetic background into *Indica* gene pool can lead to improvements in secondary traits such as lodging resistance and nutritional quality. Similarly to what has been done using tropical *Japonica*, should we also start using temperate *Japonica* to "mingle" for NPT breeding? This strategy could be exploited to improve rice breeding in several temperate areas world wide (Negrao *et al.*, 2008).

The main aim to work on new plant type (super rice) was to improve physiological aspect of modern rice for increasing yield potential. The potential yield of this would reach 15 t/ha, compared with the present potential yield of 10t / ha. It is expected that new plant type (NPT) lines with all the desirable traits will help to feed 300 million more rice consumers (Khush, 1999b), when planted widely in tropics and subtropics regions. However, these lines could not be adopted for growing in the temperate areas as they lack cold tolerance, one of the most important adaptability traits of temperate rices. Thus, NPT lines could be useful parents for increasing the yield potential of temperate rice. In this regard collaboration has been established with the rice improvement programme of International Rice Research Institute, Philippines. Preliminary studies on the adoption and stability of different NPT lines from international Network for Genetic Evaluation of Rice (INGER) nurseries *viz* International Rice Cold Tolerant Nursery (IRCTN), International Irrigated Rice Observational Nursery (IIRON) and International Rice Temperate Observational Nursery (IRTON) have been reflected under Kashmir conditions (Sanghera *et al.*, 2011) which can be used to breed high yielding NPT varieties adapted to local conditions of Kashmir valley.

HYBRID BREEDING FOR INCREASED YIELD POTENTIAL

Hybrid rice offers yet another opportunity to boost the yield of rice. Hybrid rice has a yield advantage of 15-20 per cent over the conventional high yielding varieties (Virmani *et al.*, 1993). Commercial exploitation of hybrid rice has been demonstrated in China, which enabled it to reduce its rice area from about 34.4 mha in 1978 to about 31.98 mha in 1988 and at the same time increased its rice production from 136.9 Mt to 169. 1 Mt during the same period (Ma & Yuan, 2003 and Tran & Nguyen, 1998). China is the leading producer of hybrid rice in the world (Swaminathan, 2006).

The success story of hybrid rice in China aroused spurts of interests at the IRRI, Philippines and in many national rice research programmes to intensify research on hybrid rice. The trials conducted at IRRI, Philippines and in several national programmes *viz*. India, Vietnam and Malaysia, the rice hybrids out yielded the best check variety (Virmani, 1996). Many heterotic rice hybrids have been released for commercial cultivation in countries other than China (Takita 2003 and Mishra, 2003). Though a few high yielding rice varieties have been released in the past, rice productivity in the valley has reached a plateau and chances of further yield enhancement are scanty due to low genetic variability in hill cultures (Sanghera and Zarger, 2002). However, hybrid rice offers an opportunity to boost the yield of rice under valley conditions as hybrid rice has a yield advantage of 15-20 per cent over the conventional high yielding varieties (Virmani *et al.*, 1996). A good number of hybrids have been released in India (Mishra, 2003) but the initial evaluation conducted on these hybrids and their parental lines introduced from Directorate of Rice Research, Hyderabad has shown that hybrids developed from tropical and sub tropical areas as such were not suitable for cultivation under temperate condition of Kashmir (Sanghera *et al.*, 2003). The positive standard heterosis in grain yield reported in rice hybrids is attributed to the increased dry matter production due to increased leaf area index, higher crop growth rate and harvest index, high spikelet number and increased 1000 grain weight (Peng *et al.*, 2003). Rice hybrids showing positive heterosis for adverse temperature, soil conditions and water regime will be of immense importance in developing rice hybrids for stress environments. The heterosis in the hybrids between modern high yielding *indica* varieties and the new plant type *japonicas* is even high approaching (Virmani *et al.*, 1996). Such hybrids may yield as much as 15 t/ha. Thus, the combined approaches of varietal modification and hybrid vigour exploitation should help us to increase the yield potential of irrigated rice by 50 % (Virmani and Kumar, 2004). Recently, Sanghera *et al.*, (2010a) had developed SKAU 7A and SKAU 11A temperate (CMS) lines that can be utilized for development of hybrids for temperate conditions. Further, some putative restorers and maintainers to these cyto-sterile lines have been identified (Sanghera *et al.*, 2010b) so the following research strategies can be explored to further increase rice production in temperate areas like Kashmir.

COMBINATION OF IDEOTYPE AND HETEROSIS

Since Johan and Jeroen (2003) proposed the ideotype concept and several models have been proposed for super high-yielding rice: low-tillering and large panicle model by Khush (1996) the bushy type and rapid-growing model by Huang (2003) the ideal plant type and huge rice model by Yang *et al.*, (1996) and the heavy panicle model by Zhou *et al.*, (1997). The new plant type (NPT) being developed by IRRI might raise current yield by 20–25%. These models, yet to be proven in practice, provide the leading concepts for super high-yielding programs since they are based on certain theories and practical experiences. Based on the characteristics of in hybrid rice breeding, Yuan (1997) proposed a morphological model of super high-yielding rice in terms of (i) plant height, (ii) the uppermost three leaves, (iii) plant type, (iv) panicle weight and number, (v) leaf area index and ratio of leaf area to grains, and (vi) harvest index of above 0.55.

USE OF INTER-SUB-SPECIFIC HETEROSIS

As the heterosis of intersubspecific hybrids is much stronger than that of intervarietal hybrids, its use is one of the most feasible approaches for realizing super high yield. To exploit intersubspecific super high-yielding hybrid rice, the development of various lines with wide compatibility, especially lines with a broad spectrum of compatibility is important (Wan and Ikehashi, 1996). The differentiation of the rice sub species *Indica* and *Japonica* is ancient and there is a sterility barrier associated with the hybrids between these two sub species. However, the discovery of wide compatibility gene has great potential to overcome *Indica-Japonica* hybrid sterility (Ikehashi and Araki, 1984). In addition, the fine location of the hybrid sterility locus *S-5* responsible for hybrid sterility in crosses between *japonica* and *indica* subspecies will be very useful for marker-assisted selection (MAS) of wide compatibility varieties (Zhang *et al.*, 2007). Development of *Indica/Japonica* hybrid varieties have shown immense yield potential compared with intra-subspecific rice hybrids due to the genome diversity (Gao *et al.*, 2005). It appears that this man made "gene pool blend" is enhancing the genetic variability deployed throughout the decades, in the majority of rice lands. In the *Japonica* sub species the advances in hybrid rice is problematic because there are few effective restorer lines and the yield advantage is not as large as in *Indica* inbred rice which may be due to the narrow genetic base of *Japonica* gene pool and unstable sterility of CMS lines (Tao *et al.*, 2002). Therefore, in order to promote *Japonica* hybrid rice development, it is essential to identify suitable *Japonica* restorer lines. The genetic analysis of fertility restoration genes (*Rf*) in *Indica* group is difficult because of variations in number and mapping positions of the various *Rf* genes involved. Probably *Indica* varieties may contain additional unidentified genes that influence pollen fertility (Tada, 2007 and Najeeb *et al.*, 2011). By incorporating WC genes into restorer lines and male sterile lines of *Indica, Japonica*, or intermediate type, each with different growth duration, various super high yielding hybrids will be developed for different ecological environments.

UTILIZATION OF FAVOURABLE GENES FROM WILD SPECIES

Wide hybridization is an important research strategy used for broadening the rice gene pool. It is often difficult to improve rice cultivars further by using the *O. sativa* genetic resources. Rice yields have been stagnant for 15–20 years in the rice producing regions where farmers have successfully adopted green-revolution technologies (Tilman *et al.*, 2002). The genetic diversity present in the wild relatives was crucial for the green revolution. An *sd1* gene variant that does not induce short-stature was preserved in the wild ancestor as hidden variation. Farmers selected this allele together with genes interacting with it, in order to obtain a higher yield, an identical variant has been maintained in several landraces (Nagano *et al.*, 2005). Additionally, several wild species having the AA genome (*O. sativa* L. f. *spontanea, O. perennis* and *O. glumaepatula*) have been an important source of CMS, a major resource to breed commercial rice hybrids (Brar and Khush, 1997). Nevertheless, several constraints such as low cross ability, increased sterility, and limited recombination between chromosomes of wild and cultivated rice have inhibited the transfer of useful genes (Brar and Khush 2002). Xiao *et al.*, (1996 and 1998) analysed BC_2 test cross families from the interspecific cross (*O. sativa* x *O. rufipogon*) and found that *O. rufipogon* alleles at marker loci RM5 on chromosome 1 and RG256 on chromosome 2 were associated with an 18 and 7 per cent increase in grain yield per plant. A total of 68 significant QTL were identified and of these, 35 (51 per cent) had beneficial alleles derived from the phenotypically inferior *O. rufipogon* parent. Moncada *et al.*, (2001) used advanced backcross QTL analysis on *O. sativa* x *O. rufipogon* derivatives and found that certain regions of rice genome harbour genes which are useful in a range of environments. Molecular markers can be used to identify QTL from wild species responsible for transgressive segregation. Tagging of specific genes/QTLs introgressed from wild species allows the use of MAS (Rahman *et al.*, 2007). In the past years, several studies have reported the identification and tagging of valuable genes and QTLs. These include genes for yield enhancement (Xiao *et al.*, 1996), resistance to bacterial blight including *Xa21* (Khush *et al.*, 1990), *Xa23* and *Xa27(t)* (Song *et al.*, 1995 and Zhang *et al.*, 1998), resistance to BPH (Ishii *et al.*, 1994) and resistance to blast through the gene *Pi-9t* (Amantebordeos *et al.*, 1992).

Hybrids between cultivated rice and AA genome wild species can be produced through normal procedures. Hybrids between rice and distantly related wild species on the other hand, are usually difficult to produce, low crossability and abortion of hybrid embryos are common features in such crosses. Hybrids have been produced through embryo rescue between elite breeding lines or varieties and several accessions of wild species representing BBCC, CC, CCDD, EE, FF, GG, HHJJ and HHKK genomes. A number of useful genes for resistance to brown planthopper (BPH), white backed plant hopper (WBPH), bacterial blight (BB), blast and tungro disease have been transferred from wild species to rice (Jena *et al.*, 1991 and Brar and Khush, 2002).

With the available tools offered by genomics, explore intensively the reservoir of wild relatives and incorporate useful genes into commercial *Japonica* varieties grown in temperate regions. These alien genes are often very different from the ones available in the cultivated rice and therefore, can expand the gene pool of rice which is of particular interest in the *Japonica* subspecies.

GENOTYPE X ENVIRONMENT INTERACTIONS IN BREEDING STABLE VARIETIES

The present superior-yielding varieties exhibit variable performance because of a high proportion of G x E interaction. There is a need to identify and release stably yielding varieties even on a specific-area basis, instead of relatively less stable varieties on a wide-area basis. There are strong genotypic differences among varieties for this interaction as well as methods for selecting varieties that are more stable across environments. Prior to releasing varieties, it is possible to select varieties with a stable performance even in unfavorable environments or management regimes.

DEFINING ADAPTATION STRATEGIES

Setting adaptation strategies for breeding programmes and defining recommendation domains for cultivars are distinct objectives. As such, they may require partly different analytical approaches and provide different results with regard to the definition of sub regions. The same data set may be analysed with both objectives in mind. However, the adaptation strategy objective focuses on the responses of a set of genotypes to obtain indications and generate predictions relative to future breeding material that may be produced from the genetic base of which the tested genotypes are assumed to be a representative sample. Assessing the value of a specific adaptation strategy, implying a distinct selection programme for each sub region rather than a unique selection programme for the whole target region is of obvious interest to globally-oriented breeding programmes of large seed companies or international research centres, where the target region may include more than one country and very different environments. In this case, each sub region may include several countries. Specific adaptation, however, may also prove a valuable target for national breeding programmes, for which the yield gain derived from exploitation of interaction effects within the country can also help face the increasing competition exerted on local seed markets by international seed companies. For public institutions, the breeding of diversified, specifically adapted germplasm can be a major element of a research policy enforcing sustainable agriculture (Bramel-Cox *et al.*, 1991 and Ceccarelli, 1994) by,

- Maximizing the potential of different areas by fitting cultivars to an environment instead of altering the environment (possibly with costly or environment-unfriendly inputs, such as pesticides, fertilizers and irrigation) to fit widely adapted cultivars, and

- Safe guarding crop biodiversity by increasing the number of varieties under cultivation, with positive implications for the stability of production at national level.

Furthermore, specific breeding may facilitate the technological adaptation of varieties by fixing characteristics of specific interest to sub-regions (for small-grain cereals, short straw for intensive cereal farming and long straw for extensive cereal-livestock systems; for cereal or food legume crops, different grain quality characteristics etc.).

STABLE PERFORMING VARIETIES

Agricultural techniques, following an assessment of their economic performance in regional yield trials, can be recommended: either widely over the target region or specifically for one sub region (Perrin, 1976 and Shaner *et al.*, 1982). For specific recommendation, the recommendation domain of a given technique may be defined on the basis of geography alone, or also on the basis of farming practices (*e.g.* irrigated or rainfed cropping) or socio-economic constraints. In all cases, the information obtained from previous testing is exploited for predicting yield responses in coming years and, most frequently, in new locations. Superior yielding varieties are available (Chaudhary, 1996), which can take farmer's yield to 8.0 tonnes/ha if grown properly. But their performance is variable due to higher proportion of Genotype × Environment (G × E) interaction. G × E interaction is a variety dependent trait (Kang, 1990, Gauch, 1992 and Chaudhary, 1996). While the genetic reasons of stability in the performance may be difficult to understand, resistance to biotic and abiotic stresses, and insensitivity to crop management practices are the major reasons. There is a need to identify and release stable yielding varieties even on a specific area basis, as against relatively less stable but on a wide area basis. There are strong genotypic differences among varieties for this interaction, providing opportunities for selecting varieties which are more stable across environments and methods are available to estimate these (Kang, 1990 and Gauch, 1992). Thus, two varieties with similar yield may have different degrees of stability. During the final selection process, before release, it is possible to select varieties which are more stable and thus giving stable performance even in poorer environments or management regimes.

FARMER'S PARTICIPATORY VARIETAL SELECTION

Peculiarity of Kashmir province lies in its peculiar rice-growing ecology even in a small geographical area. The taste and requirements are specific as they prefer medium bold rice varieties with high amylose content. Therefore, it is wise to involve farmers in selecting the appropriate varieties from a few selected good performing ones so as to suit their local need. The first 5-10 varieties from co-ordinated trials can be multiplied and a mini-kit consisting of 2 kg seed of each variety along with local checks for various duration groups can be distributed at block levels and farmers' group can be invited and associated to select the best ones. This will allow the selection of location, specific best varieties and also their spread with minimal effort. In addition, it will provide feedback to the researchers about the problems and requirements of farmer. The present day's technologies are not percolating to the farmers, as their problems are not properly understood by the researchers. The farmer's participatory varietal selection approach opens the frontier to the researcher in properly accessing the difficulties of the farmers and their requirements. Thus the research gaps whenever exist can easily be filled up.

CULTIVAR RECOMMENDATIONS

For definition of an adaptation strategy and yield stability targets, the contribution of any of the tested genotypes (considered as a sample of the relevant genetic base) is equally important. Emphasis is therefore placed on the estimation of genotypic and genotype-environmental components of variance, and location similarity is assessed on the basis of adaptation patterns for all genotypes. Conversely, the most important information for variety recommendation concerns the response of, and comparison between, high-yielding genotypes. There are four possible conclusions, implying a general recommendation for the target region or a specific recommendation for distinct sub-regions and, in both cases, the inclusion or exclusion of yield stability in the assessment of genotype merit.

CONCLUSIONS

A quantum jump in rice yield was possible using semi dwarf cultivars that responded to increased use of fertilizers, pesticides, herbicides and a most of other chemicals, along with water. In India, rice production grew at 1.11 per cent from 1994 to 2001 (Venkataramani, 2002). The biological pathways for raising the ceiling to yield included both an increase in total biomass and higher harvest index. It is essential that the capacity of the plant to produce higher biomass per day is enhanced, because the scope for yield improvement through the harvest index pathway has been practically exhausted. But to raise the yield ceiling by breaking the yield barrier, new approaches need to be implemented vigorously. These could be feasible by using the concepts of hybrid rice and the new plant type (Yang *et al.*, 2007). In addition to the morphological and physiological attributes, tolerance or resistance to a wide range of biotic and abiotic stresses will also be necessary. Several donors from tropical germplasm are available in cultivars with good genetic backgrounds (Negrao *et al.*, 2008). The greatest source of rice improvement research is the availability of a wide range of germplasm in the International Rice Gene Bank at IRRI as well as in the *ex situ* gene banks of several countries like India, China, and Japan. With the advent of molecular breeding technologies, it is also becoming possible to transfer genes from wild *Oryza* species. Stabilization of already achieved yield levels by developing host plant resistance through the integration of conventional and molecular breeding approaches; raising yield frontiers further in irrigated areas through exploitation of hybrid vigour and restructuring of plant types and special efforts to improve the production potential of ecologically handicapped rice areas should continue to receive major research thrust to achieve the envisaged enhanced rice production.

REFERENCES

Ahmad N, Sanghera GS, Zarger MA and Rather MA (1999). Status of rice production in World, India and Kashmir. Paper presented in training on "Rice Production Technology" at Directorate of Extension Education, SKUAST (K) - Shalimar. pp 1-15.

Ali AJ, Xu JL, Ismail AM. (2006). Hidden diversity for abiotic and biotic stress tolerances in the primary gene pool of rice revealed by a large backcross breeding program. *Field Crops Res.* 97:66–76.

Amantebordeos A, Sitch LA and Nelson R. (1992). Transfer of bacterial-blight and blast resistance from the tetraploid wild rice, *Oryza minuta* to cultivated Rice, *Oryza sativa*. *Theor. Appl. Gen.* **84**:345–54.

Anonymous. (2006). A bridged account of area, production and productivity since independence. Government of Jammu and Kashmir, Directorate of Agriculture, Kashmir. pp. 5-7.

Anonymous. (2009). Statistical Digest. Directorate of Economics and Statistics, Government of Jammu and Kashmir.

Anonymous. (2010). Annual Progress Report, Crop Imrovement, Directorate of Rice Research, Rajendranagar, Hyderbad, AP.

Anwar A, Teli MA, Bhat GN, Parray GA, Wani SA. (2011). Characterization of physiological races of *Pyricularia grisea* in temperate agro-ecosystem of Kashmir, India. *Indian Phytopathology* **64**:52-56.

Bandey AH and Sanghera GS. (2002). Seed quality assurance during production, processing and storage. Presented in a short course on "Recent Advances in Rice Production Technology in the Hills" held on 1-10 August, 2002, at Division of Plant Breeding and Genetics, SKUAST (K)- Shalimar. pp 32-41.

Barman SR, Gowda M, Venu RC and Chattoo B B. (2004). Identification of a major blast resistance gene in rice cultivar "Tetep". *Plant Breeding* **123**: 300-302.

Bhat GN, Anwar A, Sanghera GS and Rather MA. (2001). Occurrence of fungal diseases and their severity on rice under Kashmir conditions. National Seminar on "Agriculture and Environment" at SKUAST, Shalimar, March 28-29, 2001. pp. 57.

Bligh HF, Blackshell NW, Edwards KJ and McClung AM. (1999). Using Amplified Fragment Length Polymorphisms and Simple Sequence Length Polymorphisms to identify cultivars of brown and white milled *Rice*. *Crop Sci.* **39**:1715-1721.

Bramel-Cox, P1, Barker T, Zavala-Garcia F and Eastin JD. (1991). Selection and testing environments for improved performance under reduced-input conditions. pp. 29-56, In: Plant Breeding and Sustainable Agriculture, Considerations for objectives and methods (D.A. Sleeper, T.C. Barker and P.J. Bramel-Cox, eds.). CSSA Special Publication No. 18.

Brar DS and Khush GS. (1997). Alien introgression in rice. *Plant Mol. Biol.* **35**:35–47.

Brar DS and Khush GS. (2002). Transferring genes from wild relatives into rice In: Kang MS (ed) quantitative genetics, genomics and plant breeding CAB International Wallingford (UK), 197–217.

Brown LR. (1996). Tough Choices: facing the challenge of food scarcity. New York, USA.

Casman K and Pingali P. (1993). Extrapolating trends from long-term experiments to farmers fields: the case of irrigated rice systems in Asia, Paper presented at the Working Conference on "Measuring Sustainability using Long Term Experiments", Rothamsted Experiment Station.

Ceccarelli S. (1994). Specific adaptation and breeding for marginal conditions. *Euphytica* **77**: 205-219.

Chang TT. (1976). Rice. *In: Evaluation of Crop Plants* (Simmonds, N.W., Ed.), Longman, London and New York. 98-104.

Chaudhary RC. (1996). Internationalization of elite germplasm for farmers: Collaborative mechanisms to enhance evaluation of rice genetic resources. In: New Approaches for Improved use of Plant Genetic Resources; Fukuyi, Japan; pp. 26.

Chaves LJ. (1997). Criterios para escoger progenitores para un programa de selección recurrente. In E.P. Guimaraes, ed. Selección recurrente en arroz, pp. 13-24. Cali, Colombia, CIAT.

Dingkuhn M, Penning de Vries FW and Datta SK. (1991). Concepts for a new plant type for direct seeded flooded tropical rice In: Institute IRR (ed) Direct-seeded flooded rice in the tropics Los Ban os (Philippines), pp. 17–38.

FAO. (1995). World Rice Information, issue No.1. FAO, Rome, Italy.

Gallais A. (1990). Théorie de la sélection en amélioration des plantes. Collection des sciences agronomiques. Paris, Masson. pp. 588.

Gao LZ, Zhang CH, Chang LP. (2005). Microsatellite diversity within *Oryza sativa* with emphasis on *indica- japonica* divergence. *Genet. Res.* **85**:1–14.

Glaszmann JC. (1987). Isozymes and classification of Asian rice varieties. *Theor Appl Gen* **74**:21–30.

Huang Y. (2003). Construction and advancement of rice ecological breeding system. *World Sci. Tech. Res.* **4**:1-8. (In Chinese.)

Ikehashi H and Araki H. (1984). Varietal screening of compatibility types revealed in F$_1$ fertility of distant crosses in rice. *Jap. J. Breed.* **34**:304–13.

Ishii T, Brar DS, Multani DS and Khush GS. (1994). Molecular tagging of genes for brown plan thopper resistance and earliness introgressed from *Oryza australiensis* into cultivated rice, *O. sativa. Genome* **37**: 217-21.

Jena KK, Multani DS and Khush GS. (1991). Monosomic alien addition lines of *Oryza australiensis* and alien gene transfer. *Rice Gen.* **2**:728-29.

Johan DP, Jeroen RV. (2003). Breeding by design. *Trends Plant Sci.* **8**:330-334.

Kaneda C and Beachell H M. (1974). Response of indica/ japonica rice hybrids to low temperatures. *SABRAO J.* **6**: 17–32.

Kang MS. (1990). Genotype-by-Environment Interaction in Plant Breeding. Louisiana State University, Agricultural Centre, Baton Rouge, Louisiana, USA.

Khush GS. (1995). Breaking the yield frontier of rice. *Geo Journal* **35** (3): 324-32.

Khush GS. (1999a). New plant type of rice for increasing the genetic yield potential. In: Rice Breeding and Genetics, Research priorities and challenges, Nanda, J.S.(ed.) pp. 99-108.

Khush GS. (1999b). Green revolution: preparing for the 21st century. *Genome* **42**: 646–55.

Khush GS (2005). What it will take to feed 5.0 billion rice consumers in 2030. *Plant Mol Biol.* **59**:1–6.

Khush GS, Bacalangco E, Ogawa T. (1990). A new gene for resistance to bacterial blight from *O. longistaminata. Rice Genet. Newsl.* **7**:121–23.

Khush GS. (1996). Prospects of and approaches to increasing the genetic yield potential of rice. In: Evenson RE et al, editors. Rice research in Asia: progress and priorities. Wallingford (UK): CAB International and IRRI. pp 59-71.

Khush GS. (1997). Origin, dispersal, cultivation and variation of rice. *Plant Mol. Biol.* **35**: 25 - 34.

Kinoshita T. (1986). Standardization of gene symbols and linkage maps in rice. *In: Rice Genet.* Proc. Int. Rice Gen., Symp. May 1965. International Rice Research Institute, Los Banos, Philippines. 215-228.

Matsuo T. (1952). Genecological studies on cultivated rice. *Bull. Nat. Inst. Agric Sci. Jpn.* **3**. 101-111.

Mishra B. (2003). Rice research in India- Major achievements and future thrusts. Presented in training on "Advances in hybrid rice technology" Directorate of Rice Research, Hyderabad, 10-30, Sept. 2003, pp1-15.

Moncada P, Martinez CP, Borrero J, Chatel M, Gauch HJ, Guimaraes E, Tohme J and McCouch SR. (2001). Quantitative trait loci for yield and yield components in an *Oryza sativa x Oryza rufipogon* BC2F2 population evaluated in an upland environment. *Theor Appl Genet* **102**: 41-52.

Morais OP, Castro EM, SantAna EP and Neto FP. (2000). Evaluación y selección de los progenitores: Población CG2 de arroz de terras altas. *In:* E.P. Guimaraes, ed. *Avances en el mejoramiento poblacional en arroz,* pp. 210-220. San Antônio de Goiás, Brazil, Embrapa Arroz e Feijao.

Nagano H, Onishi K and Ogasawara M. (2005). Genealogy of the "Green Revolution" gene in rice. *Genes Genet. Syst.* **80**:351–56.

Najeeb S, Zargar MA, Rather AG, Hassan B, Sheikh FA, Ahanger MA, Razvi SM, Dar ZA and GS Sanghera. (2011). Hybrid sterility and role of wide compatibility variety in different genetic backgrounds of rice (*Oryza sativa* L.) under temperate conditions of Kashmir. *Applied Biological Research* **13(1)**:

Negrao S, Oliveira MM. Jena KK and Mackill D (2008). Integration of genomic tools to assist breeding in the *japonica* subspecies of rice. *Mol Breed* (In press).

Oka HI. (1953). Variation of various characters and character combinations among rice varieties. *Japan. J. Breed,* **3**: 33-43.

Peng S, Yang J, Laza RC, Sanico AL, Visperas RM, Son TT. (2003). Physiological bases of heterosis and crop management strategies for hybrid rice in the tropics. In: Virmani SS, Mao CX, Hardy B, editors. Hybrid rice for food security, poverty alleviation, and environmental protection. Proceedings of the 4th International Symposium on Hybrid Rice, Hanoi, Vietnam, 14-17 May, 2002. Los Banos (Philippines): International Rice Research Institute. pp 153-172.

Perrin RK. (1976). From agronomic data to farmer recommendations: An Economic Training manual. CIMMYT, Mexico City.

Rahman ML, Chu SH and Choi MS. (2007). Identification of QTLs for some agronomic traits in rice using an introgression line from *Oryza minuta*. *Mol Cells* **24**:16–26.

Sanghera GS Ahmad N, Zarger MA. and Rather MA (2003). Studies on the performance of some CMS lines under temperate conditions of Kashmir. *SKUAST J Res* **5**: 121-24.

Sanghera GS and Wani SH. (2008). Innovative approaches to enhance genetic potential of rice for higher productivity under temperate conditions of Kashmir. *The Journal of Plant Science and Research* **24**: 99-113.

Sanghera GS and Zarger MA. (2002). Breeding strategies for Improvement of Yield, Maturity, Cold and Drought in Rice. Presented in a Seminar on "Strategies to Combat Biotic and Abiotic Stresses in Rice" at Directorate of Research, SKUAST (K) - Shalimar, January 9, 2002.

Sanghera GS, Hussaini AM, Anwer A and Kashyap SC. (2011). Evaluation of some IRCTN rice genotypes for cold tolerance and leaf blast disease under temperate Kashmir conditions. *Journal of Hill Agriculture* **2**(1): 28-32

Sanghera GS, Husaini AM, Rather AG, Parray GA and Shikari AB. (2010a). SKAU 7A and SKAU 11A: New cold tolerant CMS lines from Kashmir, India. International Rice Research Notes **35**: 1-3.

Sanghera GS, Husaini AM, Parray GA, Rather AG, Shikari AB and Wani SA. (2010b). Generation of cold tolerant CMS lines of rice and identification of maintainers/restorers for hybrid rice development in Kashmir. *The Indian Journal of Crop Sciences* **5**(1-2): 143-146

Second G. (1982). Origin of the genetic diversity of cultivated rice (*Oryza* spp): study of polymorphism scored at 40 isozyme loci.*Jpn.J.Genet,* **57**:25- 57.

Shaner WW, Phillip PF and Schmehl WR. (1982). Farming Systems Research and Developnent : Guidelines for Developing Countries. Westview Press. Boulder, Colorado 1982.

Simmonds NW. (1991). Selection for local adaptation in a plant breeding programme. *Theor. Appl. Genet.* **82**:363-367.

Song WY, Wang GL and Chen LL. (1995). A receptor kinase like protein encoded by the rice disease resistance gene, *Xa21*. *Sci* **270**:1804–06.

Tada Y. (2007). Effects of *Rf-1, Rf-3* and *Rf-6(t)* genes on fertility restoration in rice (*Oryza sativa* L.) with WA- and BT-type cytoplasmic male sterility. *Breed Sci* **57**:223–29.

Takahashi N. (1984). Differentiation of ecotypes in Oryza sativa. In: *Biology of Rice* (Tsunoda S and Takanashi N Eds.). Japan Sci. Soc. Press. Tokyo/ Elsevier, Amsterdam. 31-67.

Takita T. (2003). Hybrid rice research and development in Japan. In: Virmani S S, Mao C X, Hardy B, editors. Hybrid rice for food security, poverty alleviation, and environmental protection. Proceedings of the 4th International Symposium on Hybrid Rice, Hanoi, Vietnam, 14-17 May 2002. Los Banos (Philippines): International Rice Research Institute. pp 337-340.

Tao D, Xu P and Li J. (2002). Inheritance of *japonica* upland rice restoration lines and their restoration gene mapping. *Rice Gen News* **19**:23–26.

Tilman D, Cassman KG and Matson PA. (2002). Agricultural sustainability and intensive production practices. *Nature* **418**:671–77.

Venkataramani G. (2002). Policies need to be farmer-friendly. pp. 5–7. In: Survey of Indian agriculture. The Hindu, Tamil Nadu, India.

Virk PS, Khush GS, Peng S. (2004). Breeding to enhance yield potential of rice at IRRI: the ideotype approach. *IRRN* **29**:5–9.

Virmani SS (1996). Hybrid rice. *Adv. Agron.* **57**: 377-462.

Virmani SS and Kumar I. (2004). Development and use of hybrid rice technology to increase rice productivity in the tropics. *IRRN* **29**:10-20.

Virmani SS, Prasad MN and Kumar I. (1993). Breaking yield barrier of rice through exploitation of heterosis. In: Muralidharan, K and Siddiq, E A (Eds.). New frontier in rice research, pp. 76-85.

Wan J, Ikehashi H. (1996). Two new loci for hybrid sterility in cultivated rice. Theor. Appl. Genet. 92:183-190.

Xiao JH, Grandillo S, Ahn SN and McCouch SR. (1996). Genes from wild rice improve yield. *Nature* **384**:223–24.

Xiao J, Li J, Grandillo S, Ahn SN, Yuan L, Tanksley SD and McCouch SR. (1998). Identification of trait-improving quantitative trait loci alleles from a wild rice relative. *Genetics* **150**: 899-909.

Yang W, Peng S, Laza RC, Visperas RM and Dionisio-Sese ML. (2007) Grain yield and yield attributes of new plant type and hybrid rice. *Crop Sci* **47**:1393–1400.

Yoshida S. (1981). Fundamentals of rice crop science. IRRI, P.O.Box 933, Manila Philippines.

Yuan L. (1997). Hybrid rice breeding for super high yield. Hybrid Rice 12(6):1-6.

Zhang Q, Lin SC and Zhao CL. (1998). Identification and tagging a new gene for resistance to bacterial blight (*Xanthomonas oryzae pv oryzae*) from *O. rufipogon*. *Rice Gen News* 15:138–42.

Zhang XJ, Chen YZ, Wei YP. (2007). Fine location of the S-5 locus responsible for wide compatibility in rice using SSR markers. *Cereal Res Com* 35:1–10.

Zhou K, Wang X, Li S, Li P, Li H, Huang G, Liu T, Shen M. (1997).The study on heavy panicle type of inter-subspecific hybrid rice (*Oryza sativa* L.). Sci. Agric. Sin. 30(5):91-93.

■

12

CHAPTER

Expanding Horizons of Precision Farming Driven Crop Protection

☞ **Sudheer Kumar[1] and ☞ Prem Lal Kashyap[2]**

INTRODUCTION TO PRECISION FARMING

Precision farming (PF) is the art and science of using advanced, innovative, cutting-edge, site-specific techniques and technologies to manage spatial and temporal variability in farm fields to enhance productivity, efficiency and profitability of agricultural production systems (Mondal *et al.*, 2011). The protection of crops against insect pests and plant pathogens is one of the biggest challenge in today's agriculture in order to achieve efficiency in production with respect to production costs and yield and to minimize the environmental impact. The most common plant protection method in a growing arable crop is use of pesticides as plant protection products. Excessive use of pesticides for insect pest and disease management, increases production costs and raises the risk of toxic residue levels in the crops. Disease control can be more efficient if the spatial and temporal distribution of disease incidence within fields could be identified and specified. Thus, methods for assessing the spatial distribution of disease in fields at early stages of disease development would greatly assist in the decision making process, specifically when and how much fungicide to apply and possibly where if the disease is concentrated in certain areas of the paddock. Moreover, disease forecast systems inevitably rely on disease measurement as an influential variable in the prediction models (Parker *et al.*, 1995).

The development and optimization of protocols for the precise and pre-symptomatic detection of diseases, and non-invasive evaluation of genotype-specific pathogen resistance enabling selection of more promising genotypes in breeding programmes is an important task. Preliminary measures that assist accurate site-specific procedures, especially selection of resistant cultivars to biotic stresses imposed by pathogens and insect pests (Lütticken, 2000) framed as integral component of modern precision crop protection (Auernhammer, 2001). In this context, appropriate genotypes may contribute to reducing variability in the crop and that due to diseases, both of which are important factors in precision agriculture (Zhang *et al.*, 2002). In breeding programs, resistance of new cultivars is tested over several years in field experiments that require high inputs of time and money (Schnabel *et al.*, 1998). In general, the severity of plant disease is assessed visually

[1 & 2] Scientists, National Bureau of Agriculturally Important Microorganisms, Mau, Uttar Pradesh-275101

with ordinal rating scales. This time consuming, qualitative evaluation can be subjective and does not indicate invisible damage to photosynthetic mechanism of the plant. Furthermore, the rapid advances of genetic engineering and tailored plant breeding, with their potential impact on precision agriculture (Auernhammer, 2001), underline the demand for rapid, objective and more precise methods of screening to quantify resistance to stress (Steiner *et al.*, 2008; Scholes and Rolfe, 2009).

Recent developments in ground-based optical sensor technology and remote sensing techniques may provide for an objective, time and cost-effective method of recording and mapping disease intensity for large crop areas (Qin and Zhang, 2005; Devadas *et al.*, 2009). These tools have the potential to enable direct detection and assessment of the foliar symptoms caused by diseases and pests in afflicted fields so that treatment can be applied in a site-specific way to the infested areas only (West *et al.*, 2003). The remotely sensed image provided the spatial data necessary for delineation and management of homogenous classes within a field (Taylor *et al.*, 2007). The principles of precision agriculture driven crop protection can be readily applied to monitor and control the spread of the disease through the creation of management zones. Furthermore, there is the potential to use these management zones to optimize pesticide application.

PRECISION FARMING DRIVEN CROP PROTECTION: A CONCEPT

Precision farming is a farming system concept which involves the development and adoption of knowledge based technical management systems with the main goal of enhancing productivity, efficiency and profitability. It is based on information derived from global positioning satellite systems and electronic monitoring, and processed through the geographical information system (GIS) (Gebbers and Adamchuk, 2010). This allows to take into account the heterogeneity of farmers' fields in space and time, and to adapt cultural practices to heterogeneity through variable rates in planting, chemical applications, irrigation, and though just-in-time application of treatments. Precision agriculture provides a means to monitor the food production chain and manage both the quantity and quality of agricultural produce. Over all, the aim of precision farming is to gather and analyse information about variability of soil and crop conditions in order to maximize efficiency of crop inputs within small area of farm field. Contrarily, precision crop protection (PCP) is the branch of PF that relies upon intensive sensing of environmental conditions and computer processing of the resulting data to inform decision-making and control insect pest and diseases. PCP technologies typically connect global positioning systems (GPS) with satellite imaging of fields to remotely sense crop pests and diseases and then automatically adjust levels of pesticide applications as the tractor moves around the field.

At present, PCP is a grand challenge within precision agriculture and offers great potential to minimize the costs and environmental impact of fungicide use. A site-specific crop management enriched with a high density of spatial and temporal information with regard to the status of any crop growth-relevant parameter is of paramount importance for effective control of any plant disease. Disease monitoring and decision-making process is the fundamental origin for a site-specific management of spatially and temporally variable diseased field sites (Steiner *et al.*, 2008). Currently, there are two different approaches which are being employed for site specific fungicide application. These are based on indirect decision-making by assessing canopy density or crop growth stage (Dammer *et al.,* 2008) and direct disease detection (West *et al.,* 2003). These ultramodern strategies in plant production and crop protection are closely related to innovative technologies. Near-range and remote sensing, like hyper and multispectral sensors or thermographs in precision pest management hold multiple opportunities to enhance the productivity of agricultural production systems and to reduce the environmental burden from pesticides. Real-time decision based on the information of the sensing system- 'spray or don't spray' can control cultural practices (Stafford, 2000). Due to high control costs and the environmental impact of fungicides, a site-specific application according to precision farming techniques *i.e.,* monitor and manage spatially-variable fields site-specifically (Stafford, 2000) is need of hour. Therefore, a precise, reproducible, and time-saving disease monitoring method is essential (Bock *et al.,* 2010).

ELEMENTS OF PRECISION CROP PROTECTION

Precision farming relies on three main elements- information, technology and decision support systems. Timely and accurate information is modern farmers' most valuable resources. This information should include data on crop characteristics, soil properties, fertility requirements, weather predictions, weed and pest populations, plant growth responses, harvest, yield, post-harvest processing and marketing projections. Precision farmers must find, analyse and use the available information at each step in the crop system. An enormous database is available on the internet. The data is both accessible and precision farmers must assess how new technologies can be pragmatic to their operations. *For example*, the personal computer (PC) can be used to effectively organize analyses and manage data. Record keeping is easy on a pie and information from past years can be easily accessed. GIS and other types of application software are readily available and most are easy to use. Another technology that precision farmers use is the GPS. GPS allows producers and agricultural consultants to locate specific field positions within a few feet of accuracy. As a result, numerous observations and measurements can be taken at a specific position. GIS can be used to create field maps based on GPS data to record and assess the impact of farm management decision. Data sensors used to monitor soil properties, crop stress, growth conditions, yield or post harvesting processing are neither available or under development. These sensors provide the precision farmer with instant (real-time) information that can be used to adjust or control operational inputs. Decision support combines traditional management skills with precision farming tools to help farmers make the best management choices or prescriptions for their crop production system.

TOOLS FOR PRECISION CROP PROTECTION

In order to gather and use information effectively, it is important for anyone considering precision farming to be familiar with the technological tools available. These tools include hardware, software and recommended practices.

Remote Sensing (RS)

Remote sensing is the science of making inferences about material objects from measurements, made at distance, without coming into physical contact with the objects under study (Liaghat and Balasundram, 2010). A remote sensing system consists of a sensor to collect radiation and a platform - an aircraft, balloon, rocket, satellite or even a ground-based sensor-supporting stand - on which a sensor can be mounted. Currently a number of aircraft and spacecraft imaging systems are operating using remote sensing sensors. Some of the current image systems from spacecraft platform include Indian Remote Sensing Satellites (IRS), French National Earth Observation Satellite (SPOT), IKONOS etc. at present, remote sensing holds great promise for precision agriculture because of its potential for monitoring spatial variability over time at high resolution (Moran *et al.*, 1997). Various workers (Hanson *et al.*, 1995) have shown the advantages of using remote sensing technology to obtain spatially and temporally variable information for precision farming driven crop protection. This tool provides a possible way to detect the incidence and severity of the disease rapidly. In particular, several successful studies have focused on detection and identification of infectious damage with spectral measurements and analysis as described in Table 12.1.

Table 12.1: Important applications of precision agriculture driven crop protection in sustainable agriculture

Field of application	Description	Reference
Disease detection and diagnosis	Detection of *Puccinia recondita* and *Blumeria graminis* in wheat canopies by reflectance measurement.	Lorenzen and Jensen (1989); Franke *et al.* (2005)
	Detecting damage caused by Penicillium digitatum in mandarins using hyperspectral computer vision system.	Gómez-Sanchis *et al.* (2008)
	Distinction between infected and non-infected wheat plots using normalized difference vegetation index (NDVI).	Jacobi and Kühbauch (2005)
	Assessment of the severity of bacterial leaf blight in rice using canopy hyperspectral reflectance.	Yang (2010)
	Detection of citrus canker in citrus plants using laser induced fluorescence spectroscopy.	Lins *et al.* (2009)
	Detection of mountain pine beetle infestation in British Columbia (Canada) using QuickBird imagery.	Coops *et al.* (2006)
	Detection of tulip breaking virus (TBV) in tulips using optical sensors.	Polder *et al.* (2010)
	Distinguishing nitrogen deficiency and fungal infection of winter wheat by laser-induced fluorescence.	Tartachnyk *et al.* (2006)

(Contd...)

Field of application	Description	Reference
Rapid screening and evaluation of crop germplasm	Estimation of resistance of potato cultivars to Phytophthora infestans using used an imaging chlorophyll fluorescence system	Schnabel *et al.* (1998)
	Assessment of different levels of resistance of sugar-beet cultivars towards *Cercospora beticola* using imaging chlorophyll fluorescence and thermal and video cameras.	Chaerle *et al.* (2007)
	Discrimination between susceptible and resistant barley leaves infected with powdery mildew using electron transport rate (ETR), non-photochemical quenching (NPQ) and effective PSII quantum yield.	Swarbrick *et al.* (2006)
	Discrimination between differences in the level of resistance of wheat cultivars to *Puccinia triticina* using pulse amplitude modulated (PAM) chlorophyll fluorescence imaging.	Bürling *et al.* (2010)
	Discriminating between rust-infected and rust-free leaves using ten spectral vegetation indices.	Devadas *et al.* (2009)
	Detection and identification of yellow rust in wheat using *in-situ* spectral reflectance measurements and airborne hyperspectral imaging.	Huang *et al.* (2007)
Disease mapping	Mapping of basal stem rot disease spectrum in oil palms in North Sumatra with QuickBird imagery.	Santoso *et al.* (2011)
	Spatial analyses of basal stem rot disease using geographical information system.	Azahar *et al.* (2008)
Real time applications of pesticides	Determination of the optimal date of a fungicide application against *Septoria leaf blotch* caused by Septoria tritici in wheat.	Nicolas (2004)
	Early pathogen detection under different water status and the assessment of spray application in vineyards through the use of thermal imagery.	Stoll *et al.* (2008)

Geographic Information System (GIS)

It is a computerized data storage and retrieval system, which can be used to manage and analyze spatial data relating crop productivity and agronomic factors. It can integrate all types of information and interface with other decision support tools. GIS can display analyzed information in maps that allow (i) better understanding of interactions among plant, pathogen and their interaction with environment and (ii) decision-making based on such spatial relationships. Many types of GIS software with varying functions and low prices are now available. Many farm information systems (FIS) are available, which use simple programmes to create a farm level database. One example of such FIS is LORIS (Local Resources Information System) which consists of several modules, enable the data import, generation of raster files by different gridding methods, the storage of faster information in a database, the generation of digital agro-resource maps and the creation of operational maps etc. (Schroder *et al.*, 1997). A comprehensive farm GIS contains base maps such as topography, soil type, N, P, K and other nutrient levels, soil moisture, pH, etc. Data on crop rotations, tillage, nutrient and pesticide applications, yields, etc. can also be stored. GIS is also useful to create fertility, weed and pest intensity maps, which can then be used for making maps that show recommended application rates of nutrients or pesticides.

Global Positioning System (GPS)

It is a satellite-based navigation system that can be used to locate positions anywhere on the earth (Aronoff, 1989). GPS provides continuous (24 hours/day), real-time, three- dimensional positioning, navigation and timing worldwide in any weather condition. GPS was originally intended for military applications, but in the 1980s, the government made the system available for civilian use. There are no subscription fees or setup charges to use GPS. Any person with a GPS receiver can access the system, and it can be used for any application that requires location coordinates.

Development of publicly available GPS has opened new doors in opportunities for spatial data meter of an actual site in the field. The GPS positional accuracy when used in single receiver mode (autonomous navigation) can be degraded by various error sources. The positional accuracy of the GPS can be of the order of 20 m. In order to achieve the required accuracies, especially needed for precision farming, GPS has to be operated in a differentially corrected positioning mode *i.e.* DGPS. In DGPS, the errors computed by a reference station, which is located in a known place, is transmitted to the mobile user and error correction is done to improve the accuracy. The most common use of GPS in agriculture is for yield mapping and variable rate fertilizer/pesticide applicator. It is important to find out the exact location in the field to assess spatial variability and site-specific application of the inputs. The horizontal positional accuracy of GPS can be of the order of 20m. GPS operating in differential mode are capable of providing location accuracy of 1m. The availability of GPS approaches to farming will allow all field-based variables to be tied together. This tool has proven to be the unifying connection among field variables such as weeds, crop yield, soil moisture, and remote sensing data.

Yield monitoring and mapping

Grain yield monitors continuously measure and record the flow of grain in the clean-grain elevator of a combine. When linked with a GPS receiver, yield monitors can provide data necessary for yield maps. Yield measurements are essential for making sound management decisions. However, soil, landscape and other environmental factors should also be weighed when interpreting a yield map. Used properly, yield information provides important feedback in determining the effects of managed inputs such as fertilizer, lime, seed, pesticides and cultural practices including tillage and irrigation. Yield measurements from a single year may be heavily influenced by weather. Examining yield information records from several years and including data from extreme weather years helps in determining if the observed yield level is due to management or is climate-induced.

Crop scouting

Crop scouting is being advocated as another process that can benefit from precision agricultural technologies. Using a GPS receiver on an all-terrain vehicle or in a backpack, a location can be associated with observations associated with insect pest and disease infestation, making it easier to return to the same location for treatment. These observations also can be helpful later when explaining variations in yield maps.

Information management and Decision support systems

Adoption of precision agriculture requires the joint development of management skills and pertinent information databases. Effectively using information requires a farmer to have a clear idea of business objectives and of the crucial information necessary to make decisions. Effective information management requires more than record-keeping analysis tools or a GIS. It requires an entrepreneurial attitude toward education and experimentation. In this context, decision support systems provide a framework for integrating database management systems, analytical models, and graphics, in order to improve decision-making process. These systems are designed to help growers to solve complex spatial problems and to make decision concerning to irrigation scheduling, fertilization, use of crop growth regulators and other chemicals. Spatial decision support systems have evolved in parallel with decision support systems. The land-cover map derived from the satellite data and data on the environment and pest population could be used to develop a GIS knowledge-base. GIS could then be integrated with a mathematical model for predicting the risk of insect pest and diseases throughout the region. This combination of the GIS with the model (Intelligent GIS) provides a decision-support tool that could be made available to pest managers over the Internet.

Decision support system would contain a variety of utilities to link the knowledge base on turf grass landscapes, pest populations, and the environment for the study area with the predictive model. Pest managers who log into the web site would use a point and click graphical user interface to query the knowledge base and predictive model, and the results will be delivered to them in the form of maps (*i.e.* contour maps) and graphs of the projected occurrence of insect pest and pathogen populations throughout the region. Pest managers then will be able to use the system to decide on a course of action that might involve further surveillance or preventative management. Similarly, crop simulation models are also helpful for consultants, researchers, and other farm advisors to determine the pattern of field management that optimizes production or profit. However, the effective use of these tools requires their evaluation in fields to be optimized, their integration with other information tools such as GIS, geostatistics, remote sensing, and optimization analysis. Crop simulation models like CERES (maize, wheat, rice, sorghum, barley and millet) CROPGRO (soybean, peanut, dry bean and tomato), SUBSTOR (potato), CROPSIM (cassava) and CANEGRO (Sugarcane) models has been developed by researchers from several countries. These models respond to weather, soil water holding and root growth characteristics, cultivar, water management, nitrogen management, and row spacing/plant population. Also decision support system like, DSSAT incorporates crop/soil/weather models, data input and management software, and analysis programmes for optimizing production or profit for homogenous fields. DSSAT also includes links to GIS and remote sensing information, which allows mapping of spatially variable inputs across a field and mapping of predicted outputs from the models, such as yield, nitrogen leaching, water use, pesticide use etc. The site specific yield potentials can be estimated determining spatial pattern crop and land information and using it in above simulation models.

Variable rate technology (URT)

It is the most advanced component of PF technologies, provides "on-the-fly" delivery of field inputs. A GPS receiver is mounted on a truck so that a field location can easily be recognized. An in-vehicle computer, which contains the input recommendation maps, controls the distribution valves to provide a suitable input mix by comparing to the positional information received from the GPS receiver. Current commercial VRT systems are either map-based or sensor-based (NRC, 1997). The map based VRT systems require a GPS/DGPS geo-referenced location and a command unit that stores a plan of desired application rates for each field location. The sensor-based VRT systems do not require a geo-referenced location but include a dynamic control unit, which specifies application through real time analysis of soil and/or crop sensor measurements for each field location. New VRT systems like manure applicator being developed at Purdue University may soon enable precise application of manure in cropping systems. There are two methods of VRT *i.e.* the first method, Map-based, includes the following steps like grid sampling a field, performing laboratory analyses of the soil samples, generating a site-specific map of the properties and finally using this map to control a variable-rate applicator. During the sampling and application steps, a positioning system, usually DGPS (Differential Global Positioning System) is used to identify the current location in the field. The second method, Sensor-based, utilizes real-time sensors and feedback control to measure the desired properties on-the-go, usually soil properties or crop characteristics, and immediately use this signal to control the variable rate applicator.

PRECISION CROP PROTECTION STRATEGIES AND THEIR APPLICATIONS

Modern tool-box for monitoring plant disease

Remote sensing technologies are one basic tool for precision agricultural practice which can provide an alternative to visual disease assessment (Nutter *et al.*, 2010). Many researchers have shown the prospective of remote sensing techniques in the field of plant disease detection by harnessing spectral sensor systems for detection of fungal diseases (Zhang *et al.*, 2003, Moshou *et al.*, 2004, Steddom *et al.*, 2005, Franke and Menz, 2007 and Huang *et al.*, 2007). To implement these sensors into precision plant protection technologies, they have to be robust, low-cost, and preferably real-time sensing (Zhang *et al.*, 2002).

The sensor evolution in remote sensing started from multispectral sensors to hyperspectral sensors and transforming to ultraspectral sensors (Meigs *et al.*, 2008). These devices provide a magnitude of information over the covered spectral range. But depending on the measured object and aim just few regions of spectral range are of interest. Narrow spectral bands of hyperspectral sensors with a spectral resolution up to 1 nm are highly correlated to each other, redundant information is being measured. Likewise, understanding of spectral characteristics of the object and of signal-object interaction is elementary for optimization of remote sensing sensors for disease detection.

Spectral reflectance measurements are used for non-destructive assessment of physiological status of vegetation (*e.g.* pigment content, leaf area), and to discriminate crop species or to detect the impact of stress such as plant diseases, drought stress or nutrition deficiencies (Blackburn, 2007). Leaf reflectance of sunlight in the visible (VIS 400–700 nm) and near infrared (NIR 700-1000 nm) are driven by multiple interactions, involving the scattering of light as a result of leaf surface and internal cellular structures, and radiant energy absorption induced by leaf chemistry. The function described by the ratio of the intensity of reflected light to the illuminated light for each wavelength forms the leaf/canopy spectral signature (Jones *et al.*, 2003). Consequently, biophysical and biochemical attributes of vegetation can be derived by reflectance spectra. Optical methods such as hyper-spectral imaging and non-imaging sensors have been proved to be a useful tool to detect changes in plant vitality (Hatfield *et al.*, 2008). The best results for identifying diseases were obtained in the visible and near-infrared range of spectrum.

Disease symptoms often result from physiological changes in plant metabolism brought about by the pathogen. Impact of plant diseases on the physiology and phenology of plants, however, varies with the host-pathogen interaction and may cause modifications in pigments (Pinter *et al.*, 2003), water content, functionality of tissue or the appearance of pathogen-specific structures. In fact, these individual impacts may alter the spectral pattern of the plant. Knowledge of the physiological effect of diseases on metabolism and structures of plants are beneficial for hyper-spectral discrimination of healthy and diseased leaf and canopy elements (Moran *et al.*, 1997). Using a quadratic discriminating model based on reflectance, Bravo *et al.*, (2003) classified yellow rust infestation on winter wheat with a reliability of 96 per cent. Steddom *et al.*, (2005) demonstrated that multispectral disease evaluation can be used to measure necrosis caused by CLS in sugar beets. Other researchers successfully used spectral data to detect *Phytophthora infestans* on tomato (Zhang *et al.*, 2002) or *Venturia inaequalis* on apple trees (Delalieux *et al.*, 2007), and *Dothistroma septospora* on pine trees (Coops *et al.*, 2003). Damages to crops caused by virus diseases (Naidu *et al.*, 2009) or insects (Board *et al.*, 2007,Carrol *et al.*, 2008, Xu *et al.*, 2007 and Yang *et al.*, 2007) could also be detected using spectral sensors. However, most of these studies used airborne data for discrimination between mature disease symptoms and healthy leaves at an advanced level of infection. The detection of a specific plant disease and discrimination between healthy and diseased plants were the main focus of several research groups. To bring this research forward into field, there are still some difficulties and open questions. First, from the technical side it is still open, which spatial and spectral resolution is required and following which sensor systems harbors the optimal specifications for disease detection (Steiner *et al.*, 2008). Second, an early detection, even before visible symptoms appear, was realized only by few working groups using different technical and analytical approaches (Rumpf *et al.*, 2010). Third, the assessment of the disease severity or quantification of diseases has to be implemented in further studies. Larsolle and Muhammed (2007) classified disease severity from hyperspectral reflectance in wheat and barley, compared to visual assessments using a nearest neighbor classifier with an accuracy of 86.5 per cent. Fourth, the sensor system should be able to differentiate between different kinds of stresses, especially different diseases. Most stress factors, such as diseases, nutrient deficiency or water stress induce symptoms with little distinguishing spectral characteristics (Stafford, 2000). Moshou *et al.*, (2006) discriminated between yellow rust infection and nitrogen deficiency and Qin *et al.*, (2009) using hyperspectral near range imaging differentiated citrus canker from different kinds of citrus diseases on grapefruit. Since most of the published studies have used non-imaging hyperspectroscopy, the application of hyperspectral imaging focusing on spectral information of disease symptoms is limited. Bravo *et al.*, (2003) used in-field spectral images for an early detection of yellow rust infected wheat. Later on, Luo *et al.*, (2008) identified and separated yellow rust from water and nutrient stress with the normalized difference vegetation index and physiological reflectance index (PRI), which are important vegetation spectral indices. Using the advantage of support vector machine technique, Wang *et al.*, (2007) classified winter wheat plants into different degrees of disease severity effectively. In addition, Huang *et al.*, (2007) discovered that PRI responded sensitively to the physiological changes in winter wheat plant caused by yellow rust. Results showed that severity of disease was estimated with considerable accuracy with ground measurements by both spectrometer and airborne hyperspectral images. The coefficients of determination (R^2) were 0.97 and 0.91, respectively. Balasundaram *et al.*, (2009) and Qin *et al.*, (2009) developed a hyperspectral imaging approach to detect canker lesions on citrus fruits. In other studies hyperspectral imaging has been successfully applied for quality assessment of pickling cucumbers, maize kernels, poultry carcasse or apples (Ariana *et al.*, 2006 and Nansen *et al.*, 2008). Nansen *et al.*, (2009) analyzed hyperspectral data cubes for the detection of insect induced stress in wheat plants, and Polder *et al.*, (2010) have combined different optical sensors for the detection of tulip breaking virus. By now, hyperspectral imaging is more widespread in the field of monitoring food security and quality. Though the use of reflectance measurements in plant pathology research started about two decades ago, this is still a new technology, not fully tested or adapted to the needs of plant disease detection and severity assessment (Bock *et al.*, 2010).

Indices approach can enhance the capability of multispectral remote sensing for disease discrimination at the field level (Zhang *et al.*, 2005). A 5-index image was used to identify late blight disease, caused by the fungal pathogen *Phytophthora infestans*, in tomato fields based on information from field-collected spectra and linear combinations of the spectral indices. This 5-cluster scheme successfully separated the diseased tomatoes from the healthy ones before economic damage happened. Recently, Zhang *et al.*, (2011) presented a novel approach by constructing a spectral knowledge base (SKB) of diseased winter wheat plants, which takes the airborne images as a medium and links the disease severity with band reflectance from environment and disaster reduction small satellite images (HJ-CCD) accordingly. Through a matching process with a SKB, we estimated the disease severity with a disease index (DI) and degrees of disease severity. The proposed approach was validated against both simulated data and field surveyed data. Estimates of DI (%) from simulated data were more accurate, with a coefficient of determination (R^2) of 0.9 and normalized root mean square error (NRMSE) of 0.2. The overall accuracy of classification reached 0.8 with a kappa coefficient of 0.7. Validation of the estimates against field measurements showed that there were some errors in the DI value with the NRMSE close to 0.5. The result of classification was more encouraging with an overall accuracy of 0.77 and a kappa coefficient of 0.58. For the matching process, mahalanobis distance performed better than the spectral angle in all analyses in this study. Although lot of experiments has been done till now, results proved that precision agriculture demands more resolution, accuracy and correlation among remotely sensed data and agronomic parameters. May be in next decade more sophisticated, cheap, accurate remote sensing technology along with genetically engineered 'smart crop' can make justice to demand of precision farming.

The high-resolution multispectral images also showed great potential in monitoring winter wheat diseases such as powdery mildew and leaf rust with an accuracy of 88.6 per cent (Franke and Menz, 2007). However, in general, hyperspectral and high-resolution satellite or airborne observations are relatively rare and costly. Therefore, for monitoring large areas, moderate resolution satellite images that have broader cover and shorter revisit time cycles are more likely to be used widely. Coops *et al.* (2006) demonstrated the use of QuickBird imagery to detect red attack damage caused by infestation of the mountain pine beetle in British Columbia and Canada.

Laser-induced fluorescence (LIF) of plants has been explored as a tool in vegetation studies for the past two decades (Buschmann and Lichtenthaler, 1998). The LIF may be a more accurate indicator of the physiological state of plants than other optical techniques because laser light is strongly monochromatic. Therefore, its specificity and selectivity may be able to detect the impacts of several environmental factors causing plant stress at several growth stages. For technical and safety reasons ultraviolet excitation (UV) has been preferred to visible excitation to monitor vegetation (Cerovic *et al.*, 1999). The UV excitation of green leaves induces two distinct types of fluorescence: a blue–green fluorescence (BGF) in the 400–600 nm range, which is due to several biological components (Buschmann and Lichtenthaler, 1998) and chlorophyll fluorescence (ChlF) in the red to near infrared region (650–800 nm) of the spectrum (Buschmann and Lichtenthaler, 1998). The relative intensities of these two fluorescent bands obtained using UV excitation are very sensitive to intrinsic leaf properties and environmental factors (Govindjee, 1995). On the contrary, visible excitation induces mainly the chlorophyll fluorescence in 650–800 nm region of spectrum. The most important aspect of LIF is that the technique is nondestructive and nonintrusive to the plant biochemistry, physiology and ecology. In addition, it is easy and fast to use for many purposes in both the laboratory and field (Govindjee, 1995). Studies using chlorophyll fluorescence emission have been applied successfully to detect mineral deficiencies, water and temperature stress and pathogens in plants (Buschmann and Lichtenthaler, 1998). Recently Lins *et al.*, (2009) used laser induced fluorescence spectroscopy tool to discriminate between mechanical stress and citrus canker stress caused by *Xanthomonas citri* sub sp. *citri*.

Real-time controlled site specific pesticide application

Site-specific pesticide application is one of the most challenging tasks of precision farming. Applying pesticides only on diseases infested areas protects the environment and reduces the pesticide costs. Also crop disturbance is reduced and thereby, yield is increased on non-treated areas. To control the site-specific application of pesticides, information about disease distribution is required. To obtain accurate information about plant diseases within a field, sensors have to be positioned as close as possible to the field's surface. Ideally, it should be mounted son the vehicle being used, at a minimal distance from the target. Information from sensor is either transformed into an application map and used for indirect control of sprayer (off-line method) or used directly to operate sprayer (on-line or real-time method). The real-time method requires an application system that permits the variation of pesticide type and concentration on-the-go. This is possible with direct injection systems, however, the spatial accuracy of application depends directly on system's response time. If the sensors are mounted directly on sprayer boom, distance between sensor and nozzles is limited. If we assume that it does not exceed 1 m and usual speed of the application vehicle is from 2 to 5 m s^{-1} (7–18 km h^{-1}), the time available for the whole process of signal sensing, processing and preparation of water pesticide mixture cannot exceed 0.5-0.2 seconds. The time taken for signal sensing and processing depends on the software and electronic hardware.

Real-time controlled pesticide application requires a fast reacting system, enabling continuous variation of pesticide type and concentration based on a control signal (Vondricka and Lammers, 2009). Direct nozzle injection system (NDIS) is the best way to inject pesticides into the carrier centrally in the hydraulic system of a field sprayer or in every boom section is to inject the pesticide at each nozzle. Direct nozzle injection has an advantage over boom section injection in terms of a reduced transport lag time. However, there are problems with the homogeneity of the mixture in direct nozzle injection (Zhu *et al.*, 1998). In boom injection systems, mixing is not a major concern because the pesticide has sufficient time to mix with carrier before being discharged through spray nozzles. With direct nozzle injection the time for mixing is significantly reduced (Rockwell and Ayers, 1996). Another important disadvantage of NDIS is increase of costs compared to boom injection due to the parts required to deliver the pesticide to each nozzle. However, this is only way that real-time controlled application can be made to work.

Using the high sensitivity to temporal and spatial variation in leaf temperature, ground based remote sensing may have important applications in disease forecast models or reducing the amount of pesticide applied and optimizing pesticide application efficiency. Stoll *et al.*, (2008) demonstrated the use of thermal imagery to determine the target area of spray application of single nozzle positions amongst the canopy. After spraying on a cloudless day with a maximum temperature of 28°C the canopy temperature reverted within 4 min to initial temperature. Spray coverage of the same canopy section of the grapevine row was monitored over a period of 6 min for each nozzle position. Analyses of the temperature differences for the sections of the canopy where spray coverage occurred were assessed immediately after sprayer had driven past the target area. Thus, the spatial temperature differences can be used to clearly identify the position where the spray hits or misses the target from a technical point of view, thermal imagery has the potential to assess the evenness of spray coverage within a canopy, hence optimizing pesticide application efficiency.

Rapid screening and evaluation of crop germplasm

The rapid advances of genetic engineering and tailored plant breeding, with their potential impact on precision agriculture, underline the demand for rapid objective and more precise methods of screening to quantify resistance to stress (Steiner *et al.*, 2008 and Scholes and Rolfe, 2009). Chlorophyll fluorescence devices are one of the powerful tools to evaluate plant resistance against plant pathogens (Hunsche *et al.*, 2010). Schnabel *et al.*, (1998) used an imaging chlorophyll fluorescence system to estimate the resistance to stress from infection by *Phytophthora infestans* on leaf disks of potato cultivars. It is rare for entire leaves to be affected at once by pathogens (Lichtenthaler and Miehe, 1997). Therefore, changes in plant physiology may be suitably recorded by imaging fluorescence systems that provide spatially resolved information. On this basis, the responses of susceptible and resistant barley leaves infected with powdery mildew have been assessed by electron transport rate (ETR), non photochemical quenching (NPQ) and effective PSII quantum yield Y(II) (Swarbrick *et al.*, 2006). However, this work focused on the observation of metabolic consequences and started with the occurrence of visual symptoms. Similarly, Chaerle *et al.* (2007) assessed the plant-fungus interaction of *Cercospora beticola* on attached leaves, leaf stripes and leaf disc assays of sugar-beet cultivars with different levels of resistance using imaging chlorophyll fluorescence and thermal and video cameras. Recently, Bürling (2011) adopted pulse amplitude modulated (PAM) chlorophyll fluorescence imaging technique for the early detection of leaf rust (*Puccinia triticina*) infection in susceptible and resistant wheat (*Triticum aestivum* L.) cultivars.

Mapping and assessment of disease spectrum

The synoptic impression provided by remote sensing technology is promising for monitoring and assessing the disease spectrum in farm. Remote sensing tool provides data with high spatial, spectral and temporal resolutions. Specifically, QuickBird satellite imagery, which provides panchromatic imagery at 0.6m resolution and multispectral imagery at 2.5 m resolution, can distinguish individual oil palms with fronds of 6–8 m (Santoso *et al.*, 2011). The red band of QuickBird imagery (630–690nm) can be used to delineate areas infected by basal stem rot by using an approximation based on non oil palm areas or dead oil palms. It provided a mapping accuracy of 96 per cent and was 84 per cent accurate in identifying dead palms. Shafri and Hamdan (2009) used various vegetation indices and red edge techniques obtained from airborne hyperspectral imagery to detect and map oil palm trees that were infected by basal stem rot in Peninsular Malaysia. Imagery taken at various times could also provide additional data that may enhance further understanding of dynamics of palm's health and nutrition. Other high resolution imagery sources that can record information over large areas at a reasonable cost are also useful (such as SPOT and Ikonos) (Nguyen *et al.*, 1995). A new remote sensing product, Rapid Eye, is now available with 5-m spatial resolution and a red-edge band. The red-edge band, which is the region of rapid change in reflectance of chlorophyll in the near infrared range, can improve the accuracy for mapping disease. In future, research based on Quick Bird satellite imagery should be useful for providing site-specific information to plantation managers to assess the damage caused by plant pathogens, to develop a strategic plan for infection control over the whole plantation, to identify areas that need to be controlled by showing the precise locations and treatment of individual plant. The author felt that multi-temporal analysis of the imagery could provide a further understanding of disease dynamics and pattern of spread over time in near future.

PRECISION CROP PROTECTION: INDIAN PERSPECTIVES

Integrating farmer knowledge, precision crop protection tools and crop simulation modelling to evaluate management options for poor performing patches in cropping fields can be an excellent option for country like India. So far, expanding horizon of precision driven crop protection in India is in its infancy but there are numerous opportunities for adoption. There is a hope that progressive Indian farmers with guidance from public and private sectors and agricultural associations, will adopt it in a limited scale for demonstrations as the technology shows potential for raising yields and economic returns on fields with significant variability, and for minimizing environmental degradation. The support from governments and the private sector during the initial stages of adoption is therefore vital. It must be remembered that not all elements of precision farming are relevant for each and every farm. For instance, introduction of variable rate applicators is not always necessary or the most appropriate level of spatial management in Indian farms. Likewise, not all farms are suitable to implement precision farming. Its adoption would be improved if it can be shown to reduce the risk against the natural calamities. Effective coordination among the public and private sectors and growers is therefore, essential for implementing new strategies to achieve fruitful success.

The study on precision agriculture has already been initiated in India, in many research institutes. Indian Space Research Organization (ISRO), Ahmedabad has started experiment in the Central Potato Research Station farm at Jallandhar, Punjab to study the role of remote sensing in mapping the variability with respect to space and time. M S Swaminathan Research Foundation, Chennai, in collaboration with NABARD, has adopted a village in Dindigul district of Tamil Nadu for variable rate input application. Indian Agricultural Research Institute has drawn up a plan to do precision farming experiments in the institutes' farm. Project Directorate for Farming Systems Research (PDFSR), Modipuram, Meerut (UP) in collaboration with Central Institute of Agricultural Engineering (CIAE), Bhopal also initiated variable rate input application in different cropping systems. In coming few years precision farming may help the Indian farmers to harvest the fruits of frontier technologies without compromising the quality of land.

FUTURE ROADMAP

Precision agriculture has created scope of transforming the traditional agriculture, through the way of proper resource utilization and management to an environmental friendly sustainable agriculture. For the future information driven crop protection as a combination of geospatial and agricultural data management will encourage the actual utilization of precision agriculture applications. Current research on precision agriculture for crop production focuses on the development of sensors for remote detection of crops and soil in real time. Relevant field parameters like soil properties, topography, water status, crop micro-climate, nutritional status, weeds, and pests and diseases as well as yield can be monitored and estimated. Integration of different remote sensing techniques and image analysis in combination with a global positioning system will be an essential step towards online application. In future, application of artificial neural networks, genetic algorithms, fuzzy logic, wavelet techniques, decision tree, smart microprocessors, genetically engineered plant, biosensors along with other future development in the field of precision agriculture will act as driving wheel of new millennium agriculture transformation. Translation of remote sensing data, GIS techniques and precision farming database information in to implementable schemes at field level and absorption of technology at the grass root level by the actual beneficiaries still remains a greater challenge. These technologies should infiltrate in to agricultural sector at micro level for greater and sustainable benefits. The low temporal resolution of current sensor systems with high spatial resolution is a restrictive factor for practical implementation. The launch of future observation systems with improved repetition rates can open up a wider field of application. Thus, near-range detection systems and remote sensing data could be complementary tools for precision agriculture applications. Remote sensing images can be valuable as a source of additional information for the generation of fungicide application maps, particularly at growth stages when fungal infections are to be expected.

At present, many of the technologies used are in their infancy, and pricing of equipment and services is hard to pin down. This can make our current economic statements about a particular technology dated. Precision agriculture can address both economic and environmental issues that surround production agriculture today. It is clear that many farmers are at a sufficient level of management that they can benefit from precision management. Questions remain about cost-effectiveness and most effective ways to use the technological tools we now have, but the concept of "doing the right thing in the right place at the right time" has a strong intuitive appeal. Ultimately, the success of precision agriculture driven crop protection depends largely on how well and how quickly the knowledge needed to guide the new technologies can be found.

CONCLUSION

This century is the century of biotechnology and information technology revolution. Future Precision Agriculture will be an offspring of these two technologies with a rich heritage of relatively old, satellite based technologies of last century. PA has created scope of transforming the traditional agriculture, through the way of proper resource utilization and management, to an environmental friendly sustainable agriculture. Application of artificial neural networks, genetic algorithms, fuzzy logic, wavelet techniques, decision tree, smart microprocessors, genetically engineered plant, biosensors along with other future development areas already discussed will make PA not only suitable for developed countries but also for developing countries, if applied properly, and can work as a tool to destroy the distance between developed world and the rest. A number of concepts and technologies that will make up tomorrow's precision agriculture are still emerging. Over last two decades PF was introduced, its objectives and capabilities have changed dramatically. Originally, it was mainly seen as a technology to manage heterogeneous fields. The challenges then were seen as developing technologies that would allow mapping spatial variability and adjusting inputs accordingly. Over time, PF has evolved into a general management concept to reduce decision uncertainty caused by uncontrolled variation, with widely ranging applications and scales of management. It became evident that managing temporal variation is as important as managing spatial variation. The major challenge to be tackled in the future is that of making the interpretation process more automatic, generic, and mechanistic rather than relying on empirical, location-specific remote sensing solutions for crop stress management.

REFERENCES

Ariana DP, Lu R and Guyer DE. (2006). Near-infrared hyperspectral reflectance imaging for the detection of bruises on pickling cucumbers. *Computers and Electronics in Agriculture* **53**: 60-70.

Aronoff S. (1989). Geographic information systems: a management perspective. *Ottawa: WDL Publications* 294 pp.

Auernhammer H. (2001). Precision agriculture-the environmental challenge. *Computers and Electronics in Agriculture* **30**: 31-43.

Azahar TM, Boursier P and Seman IA. (2008). Spatial analysis of basal stem rot disease using geographical information system. In: *Map Asia 2008, 18–20 August, 2008, Kuala Lumpur, Malaysia*.

Balasundaram D, Burks TF, Bulanon DM, Schubert T and Lee WS. (2009). Spectral reflectance characteristics of citrus canker and other peel conditions of grapefruit. *Post-harvest Biology and Technology* **51**: 220–226.

Blackburn GA. (2007). Hyperspectral remote sensing of plant pigments. *Journal of Experimental Botany* **58**: 844–867.

Board JE, Maka V, Price R, Knight D and Baur ME. (2007). Development of vegetation indices for identifying insect infestations in soybean. *Agronomy Journal* **99**: 650–656.

Bock CH, Poole GH, Parker PE and Gottwald TR. (2010). Plant disease severity estimated visually, by digital photography and image analysis, and by hyperspectral imaging. *Critical Reviews in Plant Science* **29**: 59–107.

Bravo C, Moshou D, West J, McCartney A and Ramon H. (2003). Early disease detection in wheat fields using spectral reflectance. *Biosystems Engineering* **84**:137–145.

Bürling K, Hunsche M and Noga G. (2010). Quantum yield of non-regulated energy dissipation in PSII (Y(NO)) for early detection of leaf rust (*Puccinia triticina*) infection in susceptible and resistant wheat (*Triticum aestivum* L.) cultivars. *Precision Agriculture* **11**:703–716.

Buschmann C and Lichtenthaler HK. (1998). Principles and characteristics of multi-colour fluorescence imaging of plants. *Journal of Plant Physiology* **152**: 297–314.

Carrol MW, Glaser JA, Hellmich RL, Hunt TE, Sappington TW, Calvin D, Copenhaver K and Fridgen J. (2008). Use of spectral vegetation indices derived from airborne hyperspectral imagery for detection of European corn borer infestation in Iowa corn plots. *Journal of Economic Entomology* **101**: 1614–1623.

Cerovic ZG, Samson G, Morales F, Tremblay N and Moya I. (1999). Ultraviolet-induced fluorescence for plant monitoring: Present state and prospects. *Agriculture and Environment* **19**: 543–578.

Chaerle L, Hagenbeek D, De Bruyne E and Van Der Straeten D. (2007). Chlorophyll fluorescence imaging for disease resistance screening of sugar beet. *Plant Cell Tissue and Organ Culture* **91**: 97–106.

Coops NC, Johnson M, Wulder MA and White JC. (2006). Assessment of QuickBird high spatial resolution imagery to detect red attack damage due to mountain pine beetle infestation. *Remote Sensing of Environment* **103**: 67-80.

Coops N, Stanford M, Old K, Dudzinski M, Culvenor D and Stone C. (2003). Assessment of Dothistroma needle blight of *Pinus radiata* using airborne hyperspectral imagery. *Phytopathology* **93**: 1524-1532.

Dammer KH, Wollny J and Giebel A. (2008). Estimation of the leaf area index in cereal crops for variable rate fungicide spraying. *European Journal of Agronomy* **28**: 351–360.

Delalieux S, van Aardt J, Keulemans W, Schrevens E and Coppin P. (2007). Detection of biotic stress (*Venturia inaequalis*) in apple trees using hyperspectral data: non-parametric statistical approaches and physiological implications. *European Journal of Agronomy* **27**: 130–143.

Devadas R, Lamb DW, Simpfendorfer S, and Backhouse D. (2009). Evaluating ten spectral vegetation indices for identifying rust infection in individual wheat leaves. *Precision Agriculture* **10**:459–470.

Franke J and Menz G. (2007). Multi-temporal wheat disease detection by multi-spectral remote sensing. *Precision Agriculture* **8**: 161–172.

Franke J, Menz G, Oerke E-C and Rascher U. (2005). Comparison of multi- and hyperspectral imaging data of leaf rust infected wheat plants. In M. Owe and G. D'Urso (Eds.), *Remote sensing for agriculture, ecosystems, and hydrology VII: Proceedings of the SPIE*, Vol. **5978**:1–11.

Gebbers R and Adamchuk VI. (2010). Precision Agriculture and Food Security. *Science* **327**: 828-831.

Gómez-Sanchis J, Gómez-Chova L, Aleixos N, Camps-Valls G, Montesinos-Herrero C and Molto' E. (2008). Hyperspectral system for early detection of rottenness caused by *Penicillium digitatum* in mandarins. *Journal of Food Engineering* **89**: 80–86

Govindjee (1995). Sixty-three years since Kautsky: Chlorophyll *a* fluorescence. *Australian Journal of Plant Physiology* **22**: 131–160.

Hanson LD, Robert PC and Bauer M. (1995). Mapping wild oats infestation using digital imagery for site specific management. In Proc. *Site-Specific Mgt. for Agric. Syst. 27-230, March, 1994, Minneapolis, MN, ASA-CSA-SSSA, Madison, WI,* pp.495-503.

Hatfield LJ, Gitelson AA, Schepers SJ and Walthall LC. (2008). Application of spectral remote sensing for agronomic decisions. *Agronomy Journal* **100**: 117–131.

Huang W, Lamb d W, Niu Z, Zhang EY, Liu EL and Wang EJ. (2007). Identification of yellow rust in wheat using *in-situ* spectral reflectance measurements and airborne hyperspectral imaging. *Precision Agriculture* **8**:187–197.

Huang S, Vleeshouwers V, Visser RGF and Jacobsen E. (2005). An accurate *in vitro* assay for high throughput disease testing of *Phytophthora infestans* in potato. *Plant Disease* **89**:1263-1267.

Hunsche M, Bürling K, Saied AS, Schmitz-Eiberger M, Sohail M and Gebauer J. (2010). Effects of NaCl on surface properties, chlorophyll fluorescence and light remission, and cellular compounds of *Grewia tenax* (Forssk.) Fiori and *Tamarindus indica* L. leaves. *Plant Growth Regulation* **61**: 253–263.

Jacobi J and Kühbauch W. (2005). Site-specific identification of fungal infection and nitrogen deficiency in wheat crop using remote sensing. In: *Proceedings of the 5th European Conference on Precision Agriculture, edited by J.V. Stafford* (Wageningen Acadamic Publishers, Netherlands), 73–80.

Jones HG, Archer N, Rotenburg E and Casa R. (2003). Radiation measurement for plant ecophysiology. *Journal of Experimental Botany* **54**: 879–889.

Kobayashi T, Kanda E, Kitada K, Ishiguro K and Torigoe Y. (2001). Detection of rice panicle blast with multispectral radiometer and the potential of using airborne multispectral scanners. *Phytopathology* **91**: 316–323.

Larsolle A and Muhammed HH. (2007). Measuring crop status using multivariate analysis of hyperspectral field reflectance with application to disease severity and plant density. *Precision Agriculture* **8**:37–47.

Liaghat S and Balasundram SK. (2010). A Review: The role of remote sensing in precision agriculture. *American Journal of Agricultural and Biological Sciences* **5**:50-55.

Lichtenthaler HK and Miehé J. (1997). Fluorescence imaging as a diagnostic tool for plant stress. *Trends in Plant Science* **2**: 316–320.

Lins EC, José Belasque Jr. and Marcassa ELG. (2009). Detection of citrus canker in citrus plants using laser induced fluorescence spectroscopy. *Precision Agriculture* **10**:319–330.

Lorenzen B and Jensen A. (1989). Changes in leaf spectral properties induced in barley by cereal powdery mildew. *Remote Sensing of Environment* **27**: 201–209.

Lütticken RE. (2000). Automation and standardisation of site specific soil sampling. *Precision Agriculture* **2**:179–188.

Luo JH, Huang WJ, Wei CL, Huang MY, Chen YH and Wang JH. (2008). Quantitative identification of stripe rust and common stress on winter wheat: Application of hyper-spectrum. *Journal of Natural Disasters* **17**: 115–118.

Mahlein AK, Steiner U, Dehne HW and Oerke EC. (2010). Spectral signatures of sugar beet leaves for the detection and differentiation of diseases. *Precision Agriculture* **11**: 413–431.

Stoll M, Schultz HR, Baecker G and Berkelmann-Loehnertz B. (2008). Early pathogen detection under different water status and the assessment of spray application in vineyards through the use of thermal imagery. *Precision Agriculture* **9**:407–417.

Meigs AD, Otten LJ and Cherezova TY. (2008). Ultraspectral imaging: A new contribution to global virtual presence. *IEEE Aerospace and Electronic Systems Magazine* **23**:11–17.

Mondal P, Basu M and Bhadoria PBS. (2011). Critical review of precision agriculture technologies and its scope of adoption in India. *American Journal of Experimental Agriculture* **1**: 49-68.

Moran MS, Inoue Y and Barnes EM. (1997). Opportunities and limitations for image based remote sensing in precision crop management. *Remote Sensing Environment* **61**:319-346.

Moshou D, Bravo C, Wahlen S, West J, McCartney A, Baerdemaeker J and Ramon H. (2006). Simultaneous identification of plant stresses and diseases in arable crops using proximal optical sensing and self-organising maps. *Precision Agriculture* **7**:149–164.

Moshou D, Bravo C, West J, Wahlen S, McCartney A and Ramon H. (2004). Automatic detection of yellow rust in wheat using reflectance measurements and neural networks. *Computers and Electronics in Agriculture* **44**:173–188.

Naidu RA, Perry EM, Pierce FJ and Mekuria T. (2009). The potential of spectral reflectance technique for the detection of grapevine leafroll-associated virus-3 in two red-berried wine grape cultivars. *Computers and Electronics in Agriculture* **66**:38-45.

Nansen C, Kolomiets M and Gao X. (2008). Considerations regarding the use of hyperspectral imaging data in classifications of food products, exemplified by analysis off maize kernels. *Journal of Agricultural and Food Chemistry* **56**: 2993–2938.

Nansen C, Tulio M, Swanson R and Weaver DK. (2009). Use of spatial structure analysis of hyperspectral data cubes for detection of insect-induced stress in wheat plants. *International Journal of Remote Sensing* **30**: 2447-2464.

Nguyen HV, Lukman F, Caliman JP and Flori A. (1995). SPOT image as a visual tool to assess sanitary, nutrient and general status of estate oil palm plantation. In: *A. Ibrahim & M. J. Ahmad (Eds.), Proceedings of the 1993 PORIM International Palm oil Congress (pp. 548–554). Kuala Lumpur: Palm Oil Research Institute Malaysia.*

Nicolas H. (2004). Using remote sensing to determine of the date of a fungicide application on winter wheat. *Crop Protection* **23**:853–863.

NRC. 1997. Precision Agriculture in the 21st Century Geospatial and information techniques in crop management. *National Academy Press, Washington DC. 149pp.*

Nutter F, van Rij N, Eggenberger SK and Holah N. (2010). Spatial and temporal dynamics of plant pathogens. *In: E.C. Oerke, R. Gerhards, G. Menz and R.A. Sikora, eds., Precision Crop Protection – the Challenge and Use of Heterogeneity. Springer, Dordrecht, Netherlands, pp. 27–50.*

Nutter FW and Littrell RH. (1996). Relationship between defoliation, canopy reflectance and pod yield in the peanut-late leafspot pathosytem. *Crop Protection* **15**:135–142.

Parker SR, Shaw MW and Royle DJ. (1995). The reliability of visual estimates of disease severity on cereal leaves. *Plant Pathology* **44**: 856–864.

Pinter PJ, Hatfield JL, Schepers JS, Barnes EM, Moran MS, Daugthry CST and Upchurch DR. (2003). Remote sensing for crop management. *Photogrammetric Engineering and Remote Sensing* **69**:647-664.

Polder G, van der Heijden GWAM, van Doorn J, Clevers JGPW, van der Schoor R, Baltissen AHMC. (2010). Detection of the tulip breaking virus (TBV) in tulips using optical sensors. *Precision Agriculture* 11:397-412.

Qin J, Burks TF, Ritenour MA and Bonn WG. (2009). Detection of citrus canker using hyperspectral reflectance imaging with spectral information divergence. *Journal of Food Engineering* 93: 183–191.

Qin Z and Zhang M. (2005). Detection of rice sheath blight for in-season disease management using multispectral remote sensing. *International Journal of Applied Earth Observation and Geoinformation* 7:115–128.

Rockwell AD and Ayers PD. (1996). A variable rate, direct nozzle injection field sprayer. *Applied Engineering in Agriculture* 12: 531–538.

Rumpf T, Mahlein AK, Steiner U, Oerke EC, Dehne HW and Plümer L. (2010). Early detection and classification of plant diseases with Support Vector Machines based on hyperspectral reflectance. *Computers and Electronics in Agriculture* 74: 91–99.

Santoso H, Gunawan T, Jatmiko RH, Darmosarkoro W and Minasny B. (2011). Mapping and identifying basal stem rot disease in oil palms in North Sumatra with QuickBird imagery. *Precision Agriculture* 12:233–248.

Schnabel G, Strittmatter G and Noga G. (1998). Changes in photosynthetic electron transport in potatoes cultivars with different field resistance after infection with *Phytophthora infestans*. *Journal of Phytopathology* 146: 205–210.

Scholes JD and Rolfe SA. (2009). Chlorophyll fluorescence imaging as tool for understanding the impact of fungal diseases on plant performance: A phenomics perspective. *Functional Plant Biology* 36: 880–892.

Schroder D, Haneklaus S and Schung E. (1997). Information management in precision agriculture with LORIS. *In Precision Agriculture'97, Vol.II: Technology, IT and Management (Ed. J.V. Stafford). BIOS Scientific Publishers Ltd., Oxford, UK. pp.821-826.*

Shafri HZM and Hamdan N. (2009). Hyperspectral imagery for mapping disease infection in oil palm plantation using vegetation indices and red edge techniques. *American Journal of Applied Sciences* 6:1031–1035.

Sharp EL, Perry CR, Scharen AL, Boatwright G, Sands DC and Lautenschlager LF. (1985). Monitoring cereal rust development with a special radiometer. *Phytopathology* 75:936–939.

Stafford JV. (2000)s. Implementing Precision Agriculture in the 21st Century. *Journal of Agricultural Engineering and Research* 76: 267-275.

Steddom K, Bredehoeft MW, Khan M and Rush CM. (2005). Comparison of visual and multispectral radiometric disease evaluations of *cercospora leaf spot* of sugar beet. *Plant Disease* 89:153–158.

Steiner, U, Bürling K and Oerke EC. (2008). Sensorik fu¨reinen pra¨zisierten Pflanzenschutz. *Gesunde Pflanzen* 60:131–141.

Stoll M and Jones HG. (2007). Thermal imaging as a viable tool for monitoring plant stress. *Journal International des Sciences de la Vigne et du Vin* 41:77-84.

Stoll M, Schultz HR and Loehnertz BB. (2008). Exploring the sensitivity of thermal imaging for *Plasmopara viticola* pathogen detection in grapevines under different water status. *Functional Plant Biology* 35: 281–288.

Swarbrick PJ, Schulze-Lefert Pand Scholes JD. (2006). Metabolic consequences of susceptibility and resistance (race-specific and broad-spectrum) in barley leaves challenged with powdery mildew. *Plant Cell and Environment* 29:1061–1076.

Tartachnyk II, Rademacher EI and Kühbauch EW. (2006). Distinguishing nitrogen deficiency and fungal infection of winter wheat by laser-induced fluorescence. *Precision Agriculture* 7:281–293

Taylor JA, McBratney AB and Whelan BM. (2007). Establishing management classes for broadacre agricultural production. *Agronomy Journal* 99:1366–1376.

Vondricka J and Schulze LP. (2009). Real-time controlled direct injection system for precision farming. *Precision Agriculture* 10:421–430

Wang HG, Ma ZH, Wang T, Cai CJ, An H and Zhang LD. (2007). Application of hyperspectral data to the classification and identification of severity of wheat stripe rust. *Spectroscopy and Spectral Analysis* 27: 1811–1814.

West JS, Bravo C, Oberti R, Lemaire D, Moshou D and McCartney HA. (2003). The potential of optical canopy measurement for targeted control of field crop diseases. *Annual review of Phytopathology* 4: 593–614.

Xu HR, Xing YB, Fu XP and Zhu SP. (2007). Near-infrared spectroscopy in detecting leaf miner damage on tomato leaf. *Biosystems Engeneering* 96: 447–454.

Yang CM. (2010). Assessment of the severity of bacterial leaf blight in rice using canopy hyperspectral reflectance. *Precision Agriculture* 11:61–81.

Yang CM, Cheng CH and Chen RK. (2007). Changes in spectral characteristics of rice canopy infested with brown plant hopper and leaf folder. *Crop Science* 47:329–335.

Zhang J, Huang W, Li J, Yang G, Luo J, Gu X, Wang J. (2011). Development, evaluation and application of a spectral knowledge base to detect yellow rust in winter wheat. *Precision Agriculture* 12. DOI: 10.1007/s11119-010-9214-1.

Zhang M, Qin Z, Liu X and Ustin SL. (2003). Detection of stress in tomatoes induced by late blight disease in California, USA, using hyperspectral remote sensing. *International Journal of Applied Earth Observation and Geoinformation* 4: 295–310.

Zhang M, Qin Z and Liu X. (2005). Remote Sensed Spectral Imagery to Detect Late Blight in Field Tomatoes. *Precision Agriculture* 6:489-508.

Zhang N, Wang M and Wang N. (2002). Precision agriculture-a worldwide overview. *Computers and Electronics in Agriculture* 36:113–132.

Zhu H, Ozkan HE, Fox RD, Brazee RD and Derksen RC. (1998). Mixture uniformity in supply lines and spray patterns of a laboratory injection sprayer. *Applied Engineering in Agriculture* 14: 223–230.

13

CHAPTER

Storage Fungi Infestation, their Detection and Management

☞ S.S. Jakhar[1], ☞ T. Ram[2] and ☞ M.S. Dalal[3]

INTRODUCTION

Seeds or grains are used for propagation of the species and also as food and feed.The storage of grain and other staple food and feed free from fungal contamination is of great importance to consumers of all nations. Process of seed to seed passes through a series of activities like planting, harvesting, threshing, processing or conditioning, storage and transportation. Deterioration of seed/grain can occur at time from planting to final use either through natural ageing process or due to damage caused by microorganisms, viruses, nematodes, insects, birds and rodents. The processing is done to improve the quality of seed or grain but sometime during processing damage is caused to the seed or grain. During processing, losses have been estimated to be as high as 30 per cent annually in some developing countries with an average between 8 to 15 per cent. Handling seed or grain involves many operations that can cause damage to seed or grain. Damage caused by breakage of the protective grain layers can provide sites for fungal infection or insect infestation. During storage interactions occur among insects, mites, fungi and the stored seeds. All of these factors must be considered in maintaining an efficient operation. Seed or grain deterioration by fungi has been studied by many workers in the world but deterioration by bacteria and yeasts has not been investigated extensively. Fungi that attack grains are classified into two categories *i.e.*, field or storage fungi on the basis of their ecological requirements (Agarwal and Sinclair, 1997).

FIELD FUNGI

These fungi invade grains either during development or after maturity but before harvest. Generally, field fungi cause damage in the field and no damage or little damage in storage. These fungi cause damage to grains, when the moisture content is in equilibrium with relative humidities above 95 per cent which gives a seed moisture content of 24 to 25 per cent on a wet weight basis *e.g.*, seeds of barley, maize, oat, soybean and wheat. This moisture content is far above that at which these grains are stored, hence field fungi rarely cause damage during storage.

[1] Assistant Seed Research Officer (Plant Pathology),Department of Seed Science & Technology, CCS Haryana Agricultural University, Hisar -125 004

[2] Sr. Scientist Agronomy PDFSR (ICAR) Modipuram, Meerut (UP) - 220 110

[2] Assistant Economic Botanist (Plant Breeding), Deptt. of Genetics & Plant Breeding, CCS-HAU, Hisar - 125 004

STORAGE FUNGI

Store fungi are those that grow in grains when the moisture content is in equilibrium, with 65 to 90 per cent relative humidities. Storage fungi normally do not play a role in disease development in the field but play a major role in seed deterioration in storage.

Main storage fungi are species of *Aspergillus*, such as *A. amstelodami, A. candidus, A. chevalieri, A. flavus, A. repens, A. restrictus, A. ruber* and *A. ochraceus and Penicillium* which usually invades seed with moisture content above 16 per cent and at low temperatures. *P. virdicatum* is a common storage fungus that invades dent corn kernels stored at 19 to 24 per cent moisture content between 8 to 24°C. *Penicillium* is more common in temperate than in tropical regions. Species of *Mucor, Rhizopus, Absidia, Byssochlamys*, and *Candida* and *Hansenula* have also been reported on stored seeds.

CHARACTERISTICS OF MAJOR STORAGE FUNGI

The common storage fungi are several species of *Aspergillus* and *Penicillium*. Each has a different relative humidity requirement, and thus its development is an indicator of the moisture content of the stored grain. *A. restrictus* is often the first fungus to appear, generally at grain moisture contents of 14.0 to 14.5 per cent. The fungus kills and discolours the germ, usually giving them a purplish-black cast. It grows too slowly to contribute to any detectable rise in temperature. It is responsible for a musty odor in grain. It is closely associated with granary and rice weevils. A high incidence of either of these weevils may indicate that the grain is heavily inoculated with *A. restrictus*..

A. glaucus actively grows at moisture contents of 14.5 to 15.0 per cent. It kills and discolors the germ; causes mustiness and caking at which time the fungi is apparent to naked eye. If it is growing vigorously, it can cause a rise in temperature up to 95^0 F. Most cases of fungus-caused spoilage, especially in the early stages involve *A. glaucus*. An increase in surface disinfected rice grains yielding this fungus when cultured between sampling periods, is an indication that spoilage is underway. Early stage of *A. glaucus* infestation can not be detected without a microscope.

A. candidus occurs at moisture contents of 15 to 15.5 per cent. *A. candidus* is major storage fungus and causes discoloration of the entire kernel while turning the germ black. The presence of this fungus should be an indication that serious grain deterioration is underway. Its growth results in heating of the total grain mass up to 130^0F at which point the grain is a total loss. Any increase in *A. candidus* from one sampling period to the next is cause for alarm. If heating is not already beginning, it soon will be.

A. flavus develops at moisture contents between 16 to 18 percent. It kills the germ, discolors the seed and causes rapid heating. If *A. flavus* is detected, major spoilage has already occurred. Under certain conditions, it can produce aflatoxin, a very potent poison. If *A. flavus* is discovered, immediate drying and cooling of the grain is essential. *Penicillium* is a fungus that develops at relatively high moisture content that is stored at low temperatures. *Penicillium* can develop at temperatures as low as 23° F. It causes kernel discolouration, caking and a musty odor. It can also produce mycotoxins.

INVASION BY STORAGE FUNGI

Species of *Aspergillus* and *Penicillium* normally do not infect seeds prior to storage but invade under favourable conditions prevailing in storage. A storage fungus enters seed tissues through injuries to the pericarp or testa. In addition, maturing seed can be damaged on the parent plant by birds or insects and fungi can readily infect these damaged tissues. *A. flavus* var. *columnaris* may enter to seed tissues via miropyle and peduncle or penetrate through microscopic cracks in seed surface of maize. The peduncle scar tissue is loose parenchyma which allows passage for advancing fungal mycelium. *A. flavus* penetrates maize seeds through pericarp or silk. *A. flavus* may enter the cotton seeds via the vascular tissues. It has also been reported in cotton that *A. flavus* invade bolls via the vascular system after inoculation of the nectaries and other natural openings. The systemic transmission of storage fungi can be one mode of seed infection. This type of infection has been observed in maize *A. flavus* var. *columnaris* was isolated from root, stems and leaves of plants produced from infected plants at all stages during the growth, as well as from male and female inflorescence. Infection by *A. flavus* was more on the basal than apical seeds of peanut.

LOSSES DUE TO STORAGE FUNGI

Storage fungi cause a reduction in germination, an increase in visible molds, discolouration, musty or sour odours, caking, chemical and nutritional changes, reduce processing quality and formation of mycotoxins.

REDUCTION IN GERMINATIONS

High germination is desired when seeds are to be used for planting and malting, or as edible sprouts. Reduction in germination can be caused by mechanical physiological reasons or by storage fungi. If caused by fungi, the effect is influenced by seed moisture content, storage period, storage conditions (temperature, humidity) and other factors. Decrease in germination and seedling vigour results from ageing or infection by storage fungi. Seed deterioration occurs at faster rate when temperature and moisture contents are favourable for fungi. In pea *A. ruber* produces a toxin that kills the embryonic root shoot axes when its mycelium is primarily confined to the testa. Mitochondria in *A. ruber* infected seeds are damaged. The mitochondria respiration is low and not stimulated by ADP (adenosine 5-diphosphate) in aged infected peas.

The healthy seed samples of barley, maize, peas, sorghum, soybean and wheat retained germination 95 to 100 per cent when stored at moisture contents and temperatures favourable for storage fungi whereas the germination of similar samples inoculated with storage fungi reduced to near zero or zero. Rice seed samples stored at 20°C and 30°C with 14 per cent moisture content decreased in germinability and increase in storage fungi (*A. candidus*, *A. glaucus* and *A. restrictus*) in proportion to increasing storage time. Seed viability may be significantly reduced before the fungus is visible even with the aid of a microscope. Weakening or killing of the seed embryo precedes any discolouration. The seed will not germinate if discolouration is easily detected. The speed at which seed viability is lost depends on the storage fungi. *A. flavus* may kill the entire infected seed lot within three months of storage. In contrast, seed lots infected with *A. restrictus* may not loose all germinability for as long as eight months.

Sunflower seeds stored at moisture content of 10, 12 and 14 per cent at 3 to 5, 8 to 10 and 27 to 28°C and infested with *Alternaria, Aspergillus glaucus* and *Penicillium* decreased in germinability in proportion to increasing moisture content temperature and storage time. At 19 to 20 per cent moisture contents and 20 to 25°C maize seeds invaded with *A. flavus* had 13 per cent germination, whereas non-inoculated seeds had 97 per cent germination after 74 days. In maize seeds free from storage fungi 98 per cent germination was observed after twelve weeks, whereas 6 per cent germination was recorded in seeds inoculated with storage fungi. Sorghum seeds with a moisture content of 14.5 per cent on a wet weight basis and invaded by storage fungi had a decreased germinability in proportion to the increased moisture content and storage period. Decrease in germination in soybean was correlated with the increase in *A. flavus* and *A. melleus, A. glaucus* and *A. niger* reduced germination on soybean.

A reduction in germination of wheat seeds was recorded due to storage fungi. Pea seeds inoculated with *A. flavus, A. candidus* and *A. ruber* when stored at 85 per cent RH were killed in 3, 6 and 8 months respectively, but uninoculated seeds maintained 97 per cent for 6 months. A toxin produced by *A. ruber* killed tissues in embryonic axes of pea seeds in advance of infection. A reduction in germination (61.11 per cent) of cauliflower seeds was also recorded by storage fungi.

DISCOLOURATION AND SHRINKAGE OF THE SEEDS

Tilletia indica, Karnal bunt of wheat cause black discolouration at the embryonal end and in groove (ventral side) of the seed. *Alternaria alternata,* black point of wheat cause brown to dark brown discolouration on part of the seed or whole seed. *Aspergillus candidus* especially, *A. restrictus* and *A. glaucus,* to a lesser extent, can cause a jet black discolouration in wheat seeds. Severe infection may result in discolouration of the whole seed. Such seeds produce musty odours. The official US standards for malting barley permit 4 per cent damaged kernels. Five fungi *viz. Alternaria brassicicola, A. alternata, Aspergillus* spp., *Penicillium* spp. and *Rhizopus stolonifer* have been reported to cause discolouration and shrivelling of cauliflower seeds in Haryana. *Aspergillus* and *Penicillium* cause discolouration and shrinkage of barley seeds. Wheat seeds having moisture content 16.0 to 16.4 per cent when inoculated with *A. candidus* and stored for three months at 25°C showed 79 per cent seed with dark germ ends and 6 per cent germination compared to 95 per cent germination of uninoculated seeds. This fungus in maize at 18 per cent moisture content and 25°C developed a dark brown discolouration at the germ ends after 4.5 months during storage.

PRODUCTION OF HEAT

A seed is a living organism that can respires throughout the life. During respiration, oxygen and starch are converted to carbon dioxide, water and energy. High temperature in storage increases the respiration rate. High respiration leads to loss in weight and quality of the seed. Cereal and legume seeds have low heat conductivity, so that the effects of temperature fluctuations are noticeable over short distances. This results in heat accumulation in pockets. Heat damage also may be due to fungal activity and not heat although both factors are present when stored seed/grain undergoes heating due to fungal growth. Respiration of moist grain was once thought to be responsible for the heating of stored grains. Research has subsequently proven that the metabolic processes of storage fungi are responsible. Heated grain does not become moldy, but just the opposite moldy grain becomes heated. Heating in stored rice begins, when either some of the grain is quite moist when stored or it acquires a high moisture content through bin leaks or insect activity or moisture transfer resulting from temperature gradients in the grain mass, so that fungi can grow. As the moisture increases due to fungal growth, the procession of different species (*A. candidus* and *A. flavus*) proceeds as described above raising the temperature up to 55°C and holding there for several weeks. The stored grain is blackened and a total loss at this point. The process may then subside or continue to the next step where thermophilic bacteria and fungi take over. Consequently, further raising the temperature to as high as 75°C the maximum attainable by microbiological heating. Under these conditions, pure chemical processes can occur raising the temperature to the point of spontaneous combustion. At 75°C temperature, short-chain hydrocarbons may be produced and when exposed to air, oxidize rapidly and result in fire. This heating to ignition occurs more commonly in maize and soybean, resulting in destruction of seeds and storage structures. A temperature increase from 50 to 55°C and a parallel increase in respiration were associated with proliferation of *A. flavus* and *A. glaucus* in soybean.

Temperature increased with the increase in water content in barley seeds from 15 to 24 per cent favoured development of storage fungi. It is quite difficult to separate increases in temperature due to storage fungi and insects. In temperate zones or tropics temperature rises in moist seeds due to fungi, however, in the dry tropics, insect induced dry grain heating is possible. Temperatures may rise to above 60°C when induced by storage fungi, whereas, insect induced hot spots never exceed 46°C. Hot zones generally rise vertically through the bulk of seed when induced by fungi and expand laterally when induced by insects because of insect mobility.

REDUCTION IN NUTRITIONAL VALUE

(a) Musty Odors, Caking and Decay

These characteristics are indicative of the advanced stages of spoilage, detectable by the unaided eye or nose. Substantial levels of fungal growth occur before it becomes readily apparent. There is generally visible growth on the grains before the musty odors are detected. Caking is a result of webbing of fine thread like mycelium between and within the kernels. The caking may sometimes be only a few inches thick, consisting of rotten kernels and fungal mycelium while the bulk of the grain underneath remains sound. Regardless of the depth, caking represents the final stages of decay. Stored seeds are used by storage fungi, which break down the nutrients of seeds chemically.

(*b*) Increase in fatty acid value

Seed or grain contains lipids and when these are damaged by improper storage conditions or by micro-organism, lipid generation occurs. Lipase and lipoxygenase are the principal enzymes involved in lipid generation in seeds or grains. Lipase may arise through activation of pre-existing materials, *de novo* synthesis or through production by microorganisms. In each case, the mechanism is similar and the end result is the same - an increase in free fatty acid, which is an important parameter of oilseed quality. According to the National Cottonseed Products Association Trading rules, grade No. 1, sunflower seed shall contain not more than 1.8 per cent free fatty acid in the oil and seeds and seeds containing over 3 per cent free fatty acid shall be sample grade subject to rejection. In seed deterioration, fatty acid value (FAV) is increased and the level of FAV is an indication of the stage of deterioration and seed storability. When the FAV increases, the flavour and odour of fatty acids make the seed rancid. *For example*, the FAV in sound maize seeds has been found low, but increased with increasing levels of damaged kernels and invasion by fungi. FAV increased more rapidly in *A. glaucus* and *A. restrictus* inoculated broken maize seeds than un-inoculated broken seeds when stored at 15.5 per cent moisture content and 30°C (Neergaard, 1998).

(*c*) Biochemical alteration in nutritional value

Seeds or grains with high concentration of nutrients are easily storable at low moisture content. Seeds or grains contain a large amount of oil, proteins and vitamins. Microorganisms grow luxuriantly on seeds or grains and decompose proteins to lower molecular weight components, resulting in formation of free amino acids. Moreover, there is depletion of some enzymes and an intensification of others production of new multiple molecular forms of enzymes. Many of these enzymes in seeds remain active during the infection period. Thus, the biochemical mechanisms operating in the saprophyte seed interaction vary efficiently and systemically enhance fungal growth at the expense of the seed reserves. Maximum endopolygalucturonose (endo-PG) was produced by *A. flavus*, followed by *Rhizoctonia solani*. Reduction in germinability and oil content and increase in FAV were more in seeds inoculated with *A. flavus*, *A. fumigatus* or *R. solani* than in those with *A. candidus* and *A. sydowii*. Endo-PG which randomly cleaves the α -1, 4-glucoside bond between the galacturonosyl moieties in the polymer of galacturonic acid, was considered one of the primary enzymes for maceration.

Peanut cotyledonary tissue contains 25-30 per cent protein (arachin, conarachin and albumins). Arachin and conarachin account for 78 per cent of peanut seed proteins. Microorganisms change these patterns in infected seed, because protein molecules are altered by microbial enzymatic action and new proteins of microbial origin are synthesized. These changes have been detected electrophoretically at 2 to

3 days after inoculation of seeds with *Aspergillus. A. niger* and *A. flavus* caused loss in oil content and unsaturated fatty acid content, namely linoleic, linolenic and oleic acids. Healthy Bengal gram seeds contained 14 amino acids whereas 12 were in *A. niger-* infected seeds. Proline and dihydroxyphenyl alaninl were absent. The remaining amino acids were in lower content, which indicated the deterioration of seed quality. In radish seeds, *A. flavus* increased the respiratory metabolism, when storage period was extended to 20 days and RH increased. The activity of starch phosphorylase, fructose diphosphate aldolase, pyruvic, α - Ketoglutaric, succinic acid oxidase, catalase and AT Pase was enhanced, in addition to increase in pyruvic acid and total keto acid but a decrease in α - Ketoglutaric acid.

A continuous loss in protein content was observed in *A. flavus* infested cowpea seeds. This might be due to hydrolysis of seed protein by hydrolytic enzymes.

PRODUCTION OF TOXINS

Storage fungi may produce mycotoxins that are injurious to man and animals on consumption.

(a) Aflatoxins

A. flavus produces aflatoxin B_1 and B_2 and *A. parasiticus* aflatoxin G_1, G_2, M_1 as well as B_1 and B_2. Aflatoxin B_1 causes liver damage to most animals and humans. A variety of animals such as cattle, chickens, dogs, ducklings, ferrets, guinea pigs, hamsters, horses, mice, monkeys, pigs, quail, rabbits, rainbow trout, rats, sheep and swine are affected by aflatoxins. Aflatoxins caused human fatalities in the third world, when heavily contaminated maize and rice were consumed under famine conditions. Diseases such as hepatitis, heptocarcinoma and Reye's Syndrome are caused by consumption of aflatoxin-contaminated food. Aflatoxins interfere with immune system in certain animals and are carcinogenic and mutagenic. The international agency for research on cancer placed aflatoxin on its list of human carcinogens.

(b) Ochratoxin A

Ochratoxin A is produced by a number of species of *Aspergillus* and *Penicillium*. In humans, it can cause degeneration of kidney, which can lead to death in extreme cases.

(c) Ergot alkaloids

Ergot alkaloids are produced by *Claviceps* in rye and triticale. Alkaloid consumption may result in gangrene, leading to necrosis of extremities, central nervous system effects ataxia, convulsions and paralysis, and gastrointestinal disorders.

(d) Citrinin

It is produced by *P. citrinum, P. verrucosum* and *Aspergillus terreus*. Citrinin is a nephrotoxin that occurs in apple juice, barley, oats, peanuts, rye, and wheat. Human kidney damage occurs with prolonged ingestion. It is toxic to birds, dogs and pigs and it results in watery diarrhoea, increased food consumption and reduced weight due to kidney degeneration in chickens, ducklings and turkeys.

(e) Zearalenone

Fusarium roseum and *F. tricinetum* produce it. It results in estrogenic and anabolic activity in animals and is associated with reddening and swelling of the vulva, uterine enlargement and vaginal and rectal prolapse. It occurs in cereals, primarily in maize and animal feeds.

(f) Citreovirdin

P. ochrosalmoneum and *P. citreonigrum* produce it. The disease beriberi (acute cardiac beriberi) has been associated with consumption of moldy "yellow rice". In Japan in 1910 yellow rice was banned due to this disease. In several animals species it causes vomiting, convulsions, ascending paralysis, and respiratory arrest.

FAVOURABLE CONDITIONS FOR FUNGI DEVELOPMENT

High temperature, relative humidity and moisture content of stored seed/grains are favourable for the development of micro-organisms.

The invasion and successful establishment of seed/grain deterioration by storage fungi are influenced by a number of factors in alone or in combination.

(a) Moisture content

Moisture content of seed/grain before storage is the most important factor for establishment, development and growth of storage fungi during storage. Different storage fungi require different moisture content below which they fail to develop or may remain dormant with seeds without causing any damage. In general, storage fungi grow at a moisture content in equilibrium with relative humidities of 65 to 90 per cent. Development of all types of microorganisms in seeds is retarded if the moisture content is lower than 13 per cent and below 10 per cent insect fail to develop. The moisture content within a seed lot may vary in different pockets and also may change from season to season. The invasion of *Aspergillus glaucus* and *A. restrictus* in sunflower seeds may occur at low moisture content *i.e.*, about 6 per cent but below 6.5 per cent, invasion is very low. *A. flavus* does not invade maize seeds below 17.5 per cent moisture content on wet weight basis but invades at 18.5 per cent and above. At 18.5 per cent moisture content, 25 and 35°C favour invasion by *A. ochraceus* and *A. candidus* respectively. With the increase in moisture content from 15 to 24 per cent and temperature from 20 to 30°C an increase in *Aspergillus* and *Penicillium* and decrease in germination in barley seeds have been observed.

It is generally accepted that rice can be stored for an extended period at a moisture content of 13.5 to 14.0 per cent. These values imply that nowhere in the entire storage mass is the moisture content higher. It is important to understand that while the moisture content of a small sample can be accurately measured, the moisture content of the entire storage bin can not be determined. For safe storage, it is important to know the highest moisture content in any portion of the bulk at any given location or time. Consider taking separate samples from different portions of the bin from time to time and analyze them separately. The more samples are better. For good results moisture contents could vary significantly from those that were initially recorded when the grain was first put into storage. This may need change your sampling locations based on the time of the year. During the winter months the grain will be warmer at the core and in the spring it will warmer next to the walls.

(b) Temperature

As with moisture content, uniformity of temperature is important to maintaining quality. A uniform temperature throughout the bulk greatly reduces moisture transfer from one area to another and thus reduces the likelihood of unexpected spoilage. Monitoring the temperature throughout the grain allows for the detection of potentially troublesome spots before damage spoilage becomes excessive.

Temperature monitoring devices for monitoring bulk stored grain were developed over 50 years ago. Most commonly, these devices consist of thermocouples attached to cables that extend from top to the bottom of the bin. Thermocouples are spaced about 6 feet apart on the cable and several cables are placed within the bin spaced at 20 feet intervals. The continuous output of the thermocouples can be monitored and stored using "off the shelf" data loggers. Temperatures should be recorded weekly. Any detectable temperature rise means that spoilage is at an advanced stage in the area where the heat is being generated. It indicates a need for immediate concern and corrective action. Storability and development of storage fungi in seeds are influenced by atmospheric, seed and inter-granular air temperature. Atmospheric heat slowly enters the bulk of seed. In addition, heat produced by fungi, insects and other organisms is considerably higher than the heat produced by seeds. Most storage fungi do not develop below 0°C, mites below 5°C and insects below 15°C. The minimum, optimum and maximum temperature requirement for the growth of most storage fungi is 0 to 5°C, 30 to 33°C and 50 to 55°C. Low temperature can be used as a substitute for control of high moisture content because at temperatures below 10°C storage fungi that invade seeds at moisture contents in equilibrium with relative humidities upto 85 per cent will grow very slowly.

Some *Penicillium* spp. grows slowly at -5°C when the moisture content is equilibrium with a RH of 90 per cent or more. It has been found that wheat seed with moisture of up to 10 per cent can be stored without deterioration for 1 year at 10°C or below and with moisture content of up to 18 per cent for 19 months at -5°C. Rice seeds with initial moisture content of 12 to 14 per cent were stored without deterioration for 465 days at 5 and 15°C. Germination of soybean seeds was maintained above 95 per cent after 24 weeks when stored at 12.1 to 16.5 per cent moisture content and at 15°C. Temperature, below 20°C tend to favour cold tolerant fungi like *Cladosporium* and *Penicillium* while higher storage temperatures favour *Aspergillus*. Store products seeds are more susceptible to *Aspergillus* than other fungi under tropical conditions. In maize the activity of *A. flavus* is enhanced between 30 to 40°C (high temperature).

(c) Aeration

One of the major functions of aeration is to reduce the temperature to a level where the growth of mold and insects is inhibited and to establish uniform temperatures through the grain mass so that moisture transfer is reduced.

(d) Physical damage of the seed

Mechanical damages during threshing, lifting, slow drying, improper storage facilities and rewetting of pods are ideal for *A. flavus* invasion. Damaged or injured seeds are more susceptible than whole seed to invasion by storage fungi and to deterioration by these fungi after invasion. Proper threshing and processing eliminate or reduce seed damage which helps in preserving germinability in wheat seeds stored under moist conditions. It has been found that germination of machine-threshed seeds decreased faster and resulted in fewer living seeds at 80, 85 and 90 per cent RH than hand-threshed seeds. A high level of injury on barley seeds reduced germination faster than a low level of injury.

(e) Seed infection infestation prior to storage

Seeds infected in the field are likely to deteriorate faster because storage fungi continue to develop at lower moisture and temperature conditions. Maize seeds invaded by storage fungi before storage deteriorate more rapidly under conditions favourable to fungi than seeds free from storage fungi.

(f) Admixtures with the seed

Admixture of plant parts, broken seeds, weed seeds, soil or field insects such as grasshoppers and crickets with seeds can serve as a substrate for development of storage fungi or insects. Admixtures usually are more moisture absorbent and retentive and thus more susceptible to storage fungi colonization. Weed seeds may have higher moisture content than cultivated crop seeds.

(g) Storage period

Length of storage influences by storage fungi. The chances of damage are less for seeds stored for a few weeks or months compared to seeds stored for years at a given moisture and temperature safe for storage.

(h) Insect and mite infestation

Infestation of seed or grain by insects and mite can accelerate deterioration by storage fungi. None infested seeds can be infested if insects and mites carry fungal spore. Moreover, they increase the moisture content of seeds through the release of water from their digestive process. Cereal seeds infested by the granary weevil, *Sitophilus granarium* and stored for three months at 75 per cent RH showed an increase in moisture content from 12.1 per cent to between 17.6 to 23 per cent compared to 14.6 and 14.8 per cent in weevil free seeds. Storage fungi increase was correlated with insect infestation. These mites, *Acarus siro* and *Tyrophagus castellanii* carried *Aspergillus glaucus* spores on their bodies and in their digestive tract and feces. An increase in moisture content by grain moth, *Sitotroga cerealella* was responsible for increased population of *A. amstelodam, A. repens,* and *A. ruber.* The incidence of the *A. flavus* group and aflatoxins was higher in maize seeds damaged by insects (*Sitophilus oryae* and *Tribolium castaneum*) than in nondamaged samples. *S. oryzae* may transmit *A. flavus* from infected to healthy rice seeds. The process of seed deterioration in storage is increased by carrying seed-borne fungal spores on *S. oryzae* in rice and *Rhizopertha dominica* in peanut. Fungi provide food for insects and mites or increase the susceptibility of seeds to attack by other microbiota

DETECTION METHODS FOR SEED BORNE FUNGI/DISEASES

Healthy and viable seed is the first pre-requisite for increasing seed production and to reduce possible seed crop failures. Until and unless we do not know the health status of the seed, it is not possible to manage the disease. To know health of the seed, different testing methods for different pathogens/diseases of different crops have been developed.

1. Examination of dry seeds

It is applied for detection of seed borne fungal pathogens which cause discoloration of the seed or change the shape and size of the seed. Also applicable for detecting fungal structures present in, on or with seed.

Procedure: Working sample-2000 seeds. All parts of seed sample are examined carefully by naked eye for the presence of discoloration and fungal structures and non-seed material are removed and identified.

Examples: Karnal bunt of wheat: *Neovossia indica*

Ergot of bajra: *Claviceps fusiformis*

2. Washing test

This method is used particularly for smut and bunt fungi in gramineous hosts except loose smut of wheat and barley. It can also be used for downy mildew (*Peronospora manshurica*) of soybean and tumour disease (*Protomyces macrosporus*) of coriander.

Procedure: Sample is taken by weight/number of seed and put in conical flask containing sufficient water. The flask is shaken for 5-10 minutes. Drops from the washing water are examined under microscope for identification of fungal spores.

Example: Flag smut of wheat (*Urocystis agropyri*), Smut of pearl millet (*Tolyposporium penicillariae*).

3. NaOH seed soak method

Applied for Karnal bunt of wheat and bunt of rice.

Procedure : Working sample - 2000 seeds.

Seeds are soaked in 0.2 per cent NaOH for 18-24th, at 20-25°C. After this swollen seeds are spread over blotter paper to remove excess water/moisture. Infected seeds giving jet black appearance can be separated from healthy seeds.

4. Blotter method

This method is widely used. All kinds of cereals, vegetables, crucifers, legumes, ornamentals and forests seeds.

Procedure: Seeds are placed on well water soaked filter paper and incubated at 20±2°C usually for 7 days in alternating cycles of 12 hr. light and 12 hr. darkness. Then individual seed is examined under stereo-microscope and fungi are identified based on habit characters. In fast germinating seeds 2, 4-D (2, 4-dichlorophenoxy acetic acid) @ 0.10 to 0.20 per cent solutions is used to check the growth of the seedlings. In case of cereals, this can be replaced by deep freeze blotter method (10°C for three days, then at 20°C for four days, then at -20°C for overnight and at 20°C for five days).

Examples: Black and gray leaf *(Alternaria brassicicola)*, Spot of crucifers *(A. brassicae)* and Ascochyta blight of gram *(Ascochyta rabiei)*.

5. Agar plate method

This method is used for detection of same type of pathogens as in blotter method. Those fungi which are not easily detectable in blotter method, can be detected by this method.

Procedure: Working sample - 400 seeds

Seeds are planted on specific medium after treating with 1 to 2 per cent sodium hypochlorite (NaOCl) and incubated in the same way as in blotter method. Fungi are identified based on colony characteristics. Colonies with doubtful identity should be examined under compound microscope *viz.* pathogen like, *Alternaria triticina* and *Fusarium oxysporum*.

6. Seedling symptom test

This test is applicable for those fungi, which are capable of producing symptoms on the root and shoot of the young seedlings. This test for certain pathogens, provides information pertaining to field performance of the seed lot.

Procedure: Seeds are sown in autoclaved soil or sand or any type of other media and incubated at 20°C for 14 days under 12h of alternating cycles of artificial light and darkness. After incubation, individual seedling is examined and per cent infection is calculated *viz.* *Alternaria* spp. in crucifers and wheat, *Fusarium* spp. in a number of hosts.

7. Embryo count method

This method is specifically used to detect loose smut of wheat and barley. Downy mildew *(Seclerospora graminicola)* of bajra can also be detected by this method.

Procedure: 2000 seeds or 100g seed are soaked in 5 per cent solution of NaOH and 0.01 per cent (100 ppm) trypan blue solution for 24h at 25-30 °C. Pass the material through different sieves of 3.5, 2.0 and 1.0 mm size along with showers of tap water. Dehydrate the embryos with methylated spirit or 95 per cent ethyl alcohol for 2- 3 minutes. Transfer the embryos in 200 ml of lactic acid + glycerol + water mixture

(1:2:1). After that transfer the embryos into a 250 ml beaker containing 75 ml of lactic acid + 150 ml glycerol (1:2) and then embryos are boiled for 2 minutes. Then the mixture is allowed to cool down. Observe the embryos under stereomicroscope for the presence of mycelium. Calculate the per cent infection.

8. Non-destructive seed health test

This test is conducted on high valued germplasm that can not be sacrificed as in conventional method. This test is easily applicable in large seeded crops such as corn, soybean and common bean, however, it can also be applied in small seeded crop like alfalfa, cabbage and lettuce. It consists of extracting tissue from dry seed with a metallic drill or cork borer (1 to 3 mm) and testing extracted tissue for the disease. This test does not decrease germination rate and also help in detection of the disease *e.g. Ustilago segetum* in wheat and *Phoma betae* in beet.

9. Fluorescence method

The fungus to produce a fluorescent substance under NUV light, *e.g. Ascochyta pisi* in pea seeds exhibits yellow green fluorescence.

MANAGEMENT

Deterioration of seeds by storage fungi can be reduced by various methods given as under:

(a) **Avoid damage to seeds during harvesting, threshing and processing:** After maturing to processing stage, storage fungi may enter into the seeds. Any damage to seeds results in the entry of storage fungi into the seeds and these seeds carry a heavy invasion of fungi. Proper precautions at harvest and during post-harvest operations reduce occurrence of storage fungi. Dried, sound, cleaned and undamaged seeds are ideal for storage.

(b) **Storage conditions:** The important factors in seed storage are seed type, type of container or storage method, storage period, seed temperature, moisture content, protection from physical damage and spoilage and relative humidity of the atmosphere. Among these factors, moisture is the most important. If this is higher than the equilibrium moisture content at 70 per cent RH, deterioration is certain to take place even after 2 or 3 months and at 5°C, under natural storage conditions moisture content is the dominant factor determining the storage life of seed or commodities.

Pre storage drying of seed and keep it dry is the most effective method in controlling the storage fungi because at low moisture content these fungi do not invade the stored seeds or grains. Phosphine fumigation is used in tropical countries for controlling storage insect pests because it is cheap and effective and application is easy. Generally, it is used @ 3 g/m³. Phosphine, even at low levels (0.1 g/m³) can slow the development of storage fungi in seeds where the moisture content is above the levels normally accepted for safe storage. Fumigation with phosphine also can reduce mycotoxin production. However, phosphine is effective only against growing fungi and has little effect on dormant conidia. The most effective means of avoiding damage to seeds by storage fungi are drying of seeds to a safer moisture level before storage and then frequently aerating during storage to maintain moisture and temperature below which the storage fungi do not develop. Aeration is the most important method to uniformly reduce seed temperature to 5 to 10°C throughout the storage bin. At this temperature range storage fungi grow slowly and insects and mites are dormant. It has been found that wheat seeds maintained 90 per cent germination after 22 months in a metal bin equipped with aeration system.

(c) **Reduction in seed moisture content to safe limits:** Fungi's growth initiate at a relative humidity above 65 to 70 per cent. The safe moisture contents for foodstuffs for long-term storage, therefore, are those that provide equilibrium at a RH of 65 to 70 pe rcent. The values vary with differences in chemical composition of various types of stored produce. Seeds with high lipid content will have higher equilibrium moisture content than cereals which are composed largely of starch.

Drying of seeds to moisture content at which no storage fungi develop is the most effective means to avoid damage to seeds. Safe storage moisture content for seeds of different crops are13 per cent for cereals like barley, wheat, rice, maize and sorghum, below 12 per cent for soybean and below 10 per cent for flax and at these moisture levels fungi do not occur in storage bins. At moisture content of 15 per cent and at temperature 20°C barley retains germination up to 41 months.

(d) **Seed treatment:** For storage seed treatment an ideal chemical must have low mammalian toxicity and long lasting microbial inhibiting properties and should not have an adverse effect on germination. *Aspergillus* and *Penicillium* has been controlled by iprodione or thiabendazole in soybean in mineral oil rather than in water. These treatments maintained the quality of seeds. Pre-storage spraying of soybean oil at 200 µg/ml on maize and soybean seeds has proved an effective mean of reducing development of storage fungi in stores.

Propionic acid salts are used by bakers to prevent fungal development by preventing development of bacteria, fungi and yeast. Also, they block the activity of enzymes that decompose carbohydrates and they have no adverse toxicological effects. They also check the formation of aflatoxin and overheating of wet cereals. Propionic acid at 9.5 g/kg eliminated *Aspergillus* and *Penicillium* up to 120 days in maize at 30 per cent moisture content. The rate of infection of *A. candidus* was reduced significantly by sodium propionate at 5000 µg/ml in rice seeds at 12.77 to 13 per cent moisture, stored for 6 months at 75 per cent and 4 months at 85 per cent RH. Application of propionic acid at 2 ml/kg to maize seeds stored in cloth bags inhibited fungal growth for 12 months without any loss of either germination or nutritional components. A combination of two or more chemical may be more useful because some fungi and bacteria have varying degree of tolerance to propionic acid. Application of benomyl or thiourea decreased *A. flavus* seed infection in maize during storage and maintained high seed quality.

(e) **Resistance:** Disease control by using plant resistance is an ideal, simple and practical method. The seed coat is the most important component in resistance to seeds to deterioration. The seeds of most wild cottons are impermeable to water through chalaza, whereas, cultivated cottons are permeable. The impermeable seed character imparts immunity to deterioration as well as to infection by *A. flavus* and elaboration of aflatoxin. It has been reported that phenolic compounds inhibit the growth of fungi and restrict the availability of nutrients to organism outside the seeds. The localized tannin in the nucleus may serve to protect the embryo from attack by micro-organism. It has been found that a protease from barley embryos is inhibitory to *Aspergillus* proteases. It has also been observed that trypsin inhibitor from maize inhibited fungal growth and it may serve to protect seeds against fungal attack. Seed containing high contents of calcium, potash and total phosphate both in testa and germs are resistant to *A. flavus* in cowpea. For identification of resistant cultivars and practical screening techniques are needed.

REFERENCES

Agarwal VK and Sinclair JB. (1997). Principles of Seed Pathology. Second Edition. CRC Lewis Publishers, New York. pp. 539.

Neergaard P. (1988). Seed Pathology. Vol. 1. The MacMillan Press Ltd. London: pp. 839.

■

<table>
<tr><td>

14

CHAPTER
</td><td>

Utility, Scope and Implementation of System of Rice Intensification (SRI) Under Temperate Conditions

☞ Parmeet Singh[1], ☞ Purshotam Singh[1] and ☞ Lal Singh[1]
</td></tr>
</table>

INTRODUCTION TO SRI SYSTEM

"SRI is a method of rice cultivation that involves efficient utilization of natural resources in conjunction with judicious use of external inputs to produce optimum rice yields"

The system of rice intensification (SRI) is a method of agronomic management of rice cultivation for increasing the yield of rice per unit area per unit time with special and mechanical arrangement, reduced seed and water requirement and modified soil (field) ecosystem". Although SRI is best explained operationally in terms of making certain changes in conventional rice-growing practices, as listed below, it is not best defined in terms of practices. SRI is better understood by focusing on its objectives than on its means. *SRI is a strategy of irrigated rice production, adapted to local conditions, that alters plant, soil, water and nutrient management practices* with the purpose of : (a) inducing larger and better-functioning root systems, and (b) more abundant, diverse and active communities of soil biota that live in association with the root systems. These organisms include both flora and fauna. This improved method of rice cultivation was developed in 1983 in Madagascar by Fr. Henri de Lau Lanie in association with non government organization - Association Tefy Saina (ATS) and in many small farmers in the 80"s is spreading in the many countries. SRI cultivation is a system rather than a technology. It is based on the insights that rice has the potential to produce more tillers and grains than now observed and that early transplanting and optimal growth conditions spacing, humidity, biological active and healthy soil and aerobic soil conditions during vegetative growth can fulfill this potential. Rice has been well adapted to inundation and half-hydrophytic environment. However, part of water requirement in rice is only to meet the ecological demand of improving nutrient uptake, soil process and thus is of certain plasticity. During 1999, SRI introduced in China and Indonesia. In 2002-International Conference on SRI was held in China (15 countries participated including India). Then, 2002 onwards-trials and experiments begun throughout the world. In India pioneer work was initiated at Tamilnadu Agricultural University (TNAU), Coimbatore, through the communication of Dr. H.F.M Ten Berge of PRI (Plant Research International) in the Netherland. Presently sizeable farmers are practicing SRI in Tamil Nadu. Andhra Pradesh, West Bengal, Punjab and Orissha work has been also initiated in Gujarat, Uttar Pradesh, Maharashtra, Jammu and Kashmir, Bihar, Madhya Pradesh etc.

[1] Division of Agronomy, SKUAST-K. Shalimar, Srinagar- 191 121 (J&K)

BASIC PRINCIPLES OF SRI SYSTEM

- Take very young age seedlings (12-14 days old or 16 days old in temperate)
- Well drained, raised nursery bed
- Only single seedling per hill
- Check-row planting (Wider spacing)
- Maintaining aerated soil conditions throughout the vegetative growth phase
- More Incorporative weeding (Use conoweeder)
- Using more organic amendments rather than chemical fertilizers alone

ADVANTAGES OF SRI SYSTEM

- Reduced requirement of seed
- Water saving
- Less requirement of purchased inputs
- SRI encourages rice plant to grow healthy with increased root volume, profuse tillering, restricted lodging.
- Higher yields due to registration of higher yield attributes
- Environmental benefits (Eco-friendly)

PROBLEMS ASSOCIATED WITH SRI SYSTEM

- Need to replace any damaged seedlings after transplanting.
- Only land that can be drained freely should be used for SRI.
- Land preparation should be done carefully to ensure that seedlings are transplanted into a well leveled field.
- Clay soil is less suitable for SRI, because of drying creates cracking problems.
- Vigorous weed growth increases the cost of production; the last weeding should be done by hand to avoid elimination of old roots.
- Uprooting the seedlings from the seedbed, without injuring roots, is difficult.
- It is labour intensive.

MANAGEMENT PRACTICES OF SRI SYSTEM

A. Nursery Management

- Select healthy seed and follow all the necessary seed treatments measures.
- Use pre-sprouted seeds (but under temperate conditions sprouted seeds).
- Prepare nursery on raised beds

- Seed rate remains 5kg/ha (50g/m^2 or 5 kg/100m^2)
- After sowing seeds, cover the surface with layer of manure.
- Keep the surface well saturated with water and protect from birds.

B. Field preparation

- Land preparation is similar to traditional method.
- After every 5-10 m distance (depending upon the convenience), prepare a channel to facilitate drainage.
- Level the field properly so that water can be applied very evenly.
- Shallow transplanting should be done in well puddled & leveled field.
- Seedling is ready for transplanting at 2-3 leaf stage (14-18 days after sowing).
- Take seedlings along with soil intact to root mass.
- Transplant only one seedling/hill without disturbing the root mud boll at a space of 25x25 cm.
- Use the recommended dose of fertilizer as under the traditional system.

C. Water management

- Maintain aerated soil condition.
- Maintain a soil condition between field capacity to saturation (no submergence) during vegetative stage (up to panicle initiation stage).
- Maintain a thin water layer (2 cm) after panicle initiation up to one fortnight before harvesting.

D. Nutrient management

- Use organics amendments to improve soil health.
- Reduce usage of chemical fertilizers.
- Biofertilizers perform more efficiently under aerobic conditions.
- Incorporation of weeds *in situ.*

E. Weed management

- Use pre-emergence herbicides .
- Use conoweeder.
- Do hand weeding at later stages.

Table 14.1: Difference between SRI and traditional method of rice cultivation

S.No.	System of rice intensification	Traditional method
1.	Nursery on raised beds with old gunny bag at the bottom and manured heavily without NPK.	Wet beds with less FYM in nursery beds and NPK applied as recommended (1:0.5: 0.5 kg NPK/ 100 m^2)

(Contd...)

S.No.	System of rice intensification	Traditional method
2.	Thin nurseries (50 g/m^2) @ 5 kg/ha in 100 m area.	Thick nurseries (150 g/sq m) @ 30 kg/ha in 500 m^2 area. In Kashmir conditions 60 kg/ha in 500 m^2.
3.	Roots are not washed in water, transplanted along with soil at 2.5 leaf stage immediately.	Roots are washed in water, transplanted at 4-5 leaf stage.
4.	Only shallow planting with a spacing 25 × 25 cm.	Shallow to deep planting with a spacing 20 × 10cm.
5.	Seedlings planted at 12-18 days and only one seedling /hill.	Planted after 30 days with 3 seedlings/hill.
6.	Gap filling compulsory, Gap filling may done at 10th day after trans-planting.	Gap filling not necessary.
7.	No transplanting shock observed	Transplanting shock may be observed.
8.	Use of cono weeder in both the directions (3 times) to control weeds and weeds incorporated.	Cono weeding was not practiced, only hand weeding twice (30 and 45 DAT) and weeds were removed
9.	Saturation is maintained up to PI stage and later alternate wetting is required up to 15 days before harvesting.	Submergence of 2-3 cm is maintained throughout crop growth.

Basis for higher yields

- Phyllochron effect
- Photosynthesis
- Soil nutrient availability
- Nitrate ammonium balance
- Phosphorus solubilisation
- Better soil environment
- Root functioning
- Birch effect
- Biological nitrogen fixation

Table 14.2: Soil environment under SRI and traditional method of rice cultivation

S.No.	System of rice intensification	Traditional method
1.	No observable root degeneration	Hypo-oxic conditions necessitate roots to degenerate cortex quickly from air pockets inlarge numbers.
2.	Energy saving (which otherwise wasted for adoption process).	Energy is diverted for adoption.
3.	Roots grow deeper and wider.	Roots become shallow due to lack of oxygen.
4.	Optimum growth conditions for roots.	Decrease redox potential, thus decreased root metabolism.

Research work on SRI

Wang *et al.* (2002) from Nanjing Agricultural University, China reported that increasing the seedling number per hill and nitrogen fertilization can promote the population development in SRI. They indicated that a combination of two seedlings per hill and application of 150 kg N/ha could promote population tiller number in SRI rice up to that in conventional rice. They also observed that water saving efficiency under SRI was remarkable with a 75 per cent increase in irrigation water-to-grain ratio. The SRI technique with 100 per cent of recommended NPK through chemical fertilizer produced significantly higher paddy yield over the other treatments (Singh and Chand, 2008).

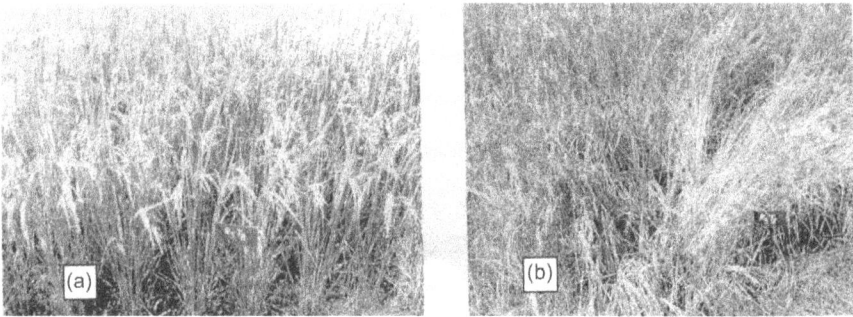

Fig. 14.1: *(a) SRI plot at experimental farm of SKUAST-Kashmir with good standing rice plants cultivar Shalimar-Rice-1, (b) A farmer practice trial with severe lodging responsible for yield and quality reduction compared with SRI.*

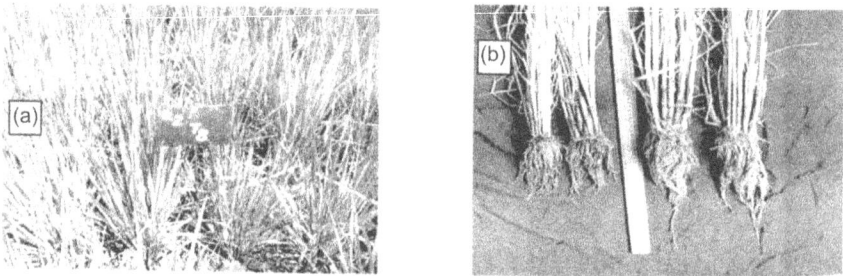

Fig. 14.2: *(a) SRI plot at experimental farm of SKUAST-Kashmir with profuse tillering of rice cultivar Shalimar-Rice-1, (b) Difference between rhizoshpere and root length of SRI and traditional rice*

Table 14.3: Effect of system of rice intensification on plant height, Number of effective tillers panicle length and number of filled grains / panicle of rice.

Treatments	Plant height (cm)		Effective tillers/ hill (No.)		Panicle length (cm)		Filled grains/panicle (No.)	
	2006	2007	2006	2007	2006	2007	2006	2007
T_1: Farmers practice	98.27	105.71	7.9	8.4	21.9	21.4	110.1	118.3
T_2: Recommended package of practices	99.53	105.21	8.8	9.2	22.2	ʳ22JT	124.4	128.5
T_3: SRI practices	96.67	108.84	13.8	15.8	22.8	24.1	134.6	138.2
T_4: SRI practices with no fertility	99.40	105.06	13.3	15.2	22.1	23.3	133.2	136.3
T_5: SRI practices + 50 % N through FYM	100.27	109.29	14.1	15.7	22.5	23.6	136.6	145.3
T_6: SRI practices + 50 % N through chemical fertilizer	100.23	110.23	14.4	15.9	22.8	23.8	137.2	144.6
T_7: SRI practices + 50 % N through FYM + 50 % N through chemical fertilizer	103.00	111.48	15.0	16.4	23.5	24.4	137.8	146.2
T_8: SRI practices + 100% N through chemical fertilizer	107.67	115.23	15.3	17.2	23.7	24.8	138.4	146.4
CD at 5 %	3.24	3.56	0.7	0.9	0.61	0.72	3.86	4.12

Source: RCM *Kharif*, 2008

The treatment T_8 = SRI + 100 per cent of RDF of NPK through chemical fertilizer produced tallest plants over all the treatments during both the years of study. Whereas, it remained at par with T_7 = SRI + 50 per cent of RDF of N through FYM + 50% of RDF of N through chemical fertilizer and T6 = SRI + 50 per cent of RDF of N through chemical fertilizer in case of number of effective Villers/hill, panicle length (cm), test weight (g) and number of grains per panicle. It gave significantly higher grain yield (q/ha) over all the other treatments and remained at par with T7 = SRI + 50% of RDF of N through FYM + 50% of N through chemical fertilizer in case straw yield (Table 14.3 and 14.4).

Table 14.4: Effect of system of rice intensification on grain (q/ha), straw yield (q/ha) and test weight (g) of rice.

Treatments	Grain yield (q/ha)		Straw yield (q/ha)		Test weight (g)	
	2006	2007	2006	2007	2006	2007
T_1: Farmers practice	51.38	53.11	65.25	68.25	25.74	26.12
T_2: Recommended package of practices	52.80	54.30	66.53	72.11	26.65	27.23

(Contd...)

Treatments	Grain yield (q/ha)		Straw yield (q/ha)		Test weight (g)	
	2006	2007	2006	2007	2006	2007
T_3: SRI practices	57.52	62.20	72.42	78.24	27.34	28.10
T_4: SRI practices with no fertility	55.24	59.05	68.26	74.05	26.88	27.67
T_5: SRI practices + 50 % N through FYM	58.90	62.22	73.44	78.42	27.19	28.36
T_6: SRI practsices + 50 % N through chemical fertilizer	60.65	64.26	73.99	80.08	27.70	28.31
T_7: SRI practices + 50 % N through FYM + 50 % N through chemical fertilizer	61.42	66.36	74.32	82.66	27.62	28.45
T_8: SRI practices + 100% N through chemical fertilizer	64.77	68.22	77.72	84.20	27.92	28.80
CD at 5 %	2.28	3.06	4.08	4.31	0.46	0.58

Source: RCM *Kharif,* 2008

Table 14.5: Gain yield rice as influnced by age of seedings and biofertilizers under SRI at different locations.

Treatments	Shalimar			Khuwani					
	2008	2009	Pooled	2008	2009	Pooled	2008	2009	Pooled
Seediling age (days)									
A_1 (12)	76.4	77.0	76.7	68.72	69.56	69.14	63.88	65.03	64.45
A_2 (16)	82.1	84.1	83.1	71.77	72.03	71.9	65.02	66.45	65.73
A_3 (20)	76.9	76.2	78.2	69.9	71.62	70.76	57.02	57.92	57.42
A_4 (24)	75.7	79.5	76	66.09	67.25	66.67	53.55	54.21	53.88
CD at 5%	3.6	4.8	2.0	3.66	20.5		5.92	6.06	3.77
No biofertilizer	73.5	76.2	74.9	68.16	69.14	68.65	46.56	47.16	46.86
Azospirillium	78.0	77.4	77.7	68.23	69.33	68.78	64.35	65.60	64.97
PSB	79.4	79.5	79.4	69.21	70.40	69.81	58.55	60.10	59.32
Azospirillium+PSB	80.2	83.7	81.9	71.25	72.39	71.82	70.01	70.75	70.38
CD at 5 %	4.70	4.7	3.20	9.37	NS		5.29	5.34	3.66

Source: RCM *Kharif,* 2010

The experiment was conducted at three locations viz., Shalimar, Khudwani, and Wadura during *kharij* 2008 and 2009. Averaged over two years, the results indicated that at Wadura 16 days old seedlings are par with 12 days old seedlings recorded significantly higher rice yield over 20 and 24 days old seedlings; at Shalimar 16 days old seedling significantly excelled in the rice yield over 12, 20 and 24 days old seedlings. However, at Khudwani 16 days old seedlings at par with 20 and 12 days seedling during 2008 and at par with 20 days old seedlings during 2009 recorded significantly higher rice grain yield as against other seedling age. As regards biofertilizers, application of Azospirillium + PSB registered significantly higher rice yield over Azospirillium and control, whereas at Khudwani, all biofertilizers treatments proved similar in affecting the rice grain yield during 2008 and 2009. (Table 14.5)

Table 14.6 : Periodic number of tillers per m^2 as affected by different treatments

Treatment	Maximum tillering	Booting	Anthesis	Maturity
T_1	331.61	319.63	314.74	308.84
T_2	352.33	341.35	336.46	331.93
T_3	375.26	365.23	361.73	355.91
T_4	393.61	381.61	376.61	371.72
T_5	275.47	266.47	261.28	257.63
T_6	283.39	272.49	268.92	263.90
T_7	410.90	399.75	394.58	389.36
T_8	443.33	433.72	429.81	424.18
T_9	463.75	454.16	449.10	443.59
T_{10}	475.56	464.35	460.39	456.27
SE ± (m)	3.58	3.46	3.39	3.31
C.D.(p=0.05)	10.81	c, -74	9.13	8.92

T_1 = 30 days seedlings; 03 seedling hill^{-1}; 15 × 15cm; RFD + FYM 5 t ha^{-1}; Butachlor + 1 hand weeding; Submergence with 3-5 cm water

T_2 = 16 days seedlings; 03 seedling hill^{-1}; 15 × 15cm; RFD + FYM 5 t ha^{-1}; Butachlor + 1 hand weeding; Submergence with 3-5 cm water

T_3 = 16 days seedlings; 01 seedling hill^{-1}; 15 × 15cm; RFD + FYM 5 t ha^{-1}; Butachlor + 1 hand weeding; Submergence with 3-5 cm water

T_4 = 16 days seedlings; 01 seedling-hill^{-1}; 25 × 25 cm; RFD + FYM 5 t ha^{-1}; Butachlor + 1 hand weeding; Submergence with 3-5 cm water

T_5 = 16 days seedlings; 01 seedling hill$^{-1'}$; 25 × 25 cm; RFD + FYM 10 t ha^{-1}; Butachlor + 1 hand weeding; Submergence with 3-5 cm water

T_6 = 16 days seedlings; 01 seedling hill^{-1}; 25 × 25 cm; RFD + FYM 20 t ha^{-1}; Butachlor + 1 hand weeding; Submergence with 3-5 cm water

T_7 = 16 days seedlings; 01 seedling hill^{-1}; 25 × 25 cm; RFD + FYM 10 t ha^{-1}; Butachlor + 1 hand weeding; Submergence with 3-5 cm water

T_8 = 16 days seedlings; 01 seedling hill^{-1}; 25 × 25 cm; RFD + FYM 5 t ha^{-1}; Butachlor + 1 hand weeding; Submergence with 3-5 cm water

T_9 = 16 days seedlings; 01 seedling hill$^{-1'}$; 25 × 25 cm; RFD + FYM 10 t ha^{-1}; Butachlor + 1 hand weeding; Submergence with 3-5 cm water

T_{10} = 16 days seedlings; 01 seedling hill^{-1}; 25 × 25 cm; RFD + FYM 10 t ha^{-1}; chemical + Rotary weeder, Alternate wetting and drying (AWD)

Source: *Nadia Nazir, 2010*

Table 14.7: Yield attributes of rice as affected by different treatments

Treatment	Panicles m^{-2}	Paniscle weight (g)	Number of grains panicle^{-1}	1000 grain weight (g)
T,	307.30	2.30	104.70	24.93
T$_2$	322.38	2.70	106.40	25.16
T$_3$	345.46	2.93	108.30	25.24
T$_4$	363.62	3.20	110.80	25.33
T$_5$	293.38	2.13	100.40	23.83
T$_6$	301.18	2.26	101.80	24.10
T$_7$	380.42	3.56	114.30	25.47
T$_8$	413.73	3.94	119.40	25.53
T$_9$	435.44	4.45	122.70	25.64
T$_{10}$	447.88	4.80	130.30	25.71
S.E ± (m)	**4.36**	**0.32**	**4.71**	**0.79**
C.D.(p=0.05)	**13.04**	**0.96 m**	**14.01**	**NS**

T$_1$ = 30 days seedlings; 03 seedling hill^{-1}; 15 × 15 cm;; RFD + FYM 5 t ha^{-1}; Butachlor + 1 hand weeding; Submergence with 3-5 cm water

T$_2$ = 16 days seedlings; 03 seedling hill^{-1}; 15 × 15 cm; RFD + FYM 5 t ha^{-1}; Butachlor + 1 hand weeding; Submergence with 3-5 cm water

T$_3$ = 16 days seedlings; 01 seedling hill^{-1}; 15 × 15 cm; RFD + FYM 5 t ha^{-1}; Butachlor + 1 hand weeding; Submergence with 3-5 cm water

T$_4$ = 16 days seedlings; 01 seedling hill^{-1}; 25 x 25 cm; RFD + FYM 5 t ha^{-1}; Butachlor + 1 hand weeding; Submergence with 3-5 cm water

T$_5$ = 16 days seedlings; 01 seedling hill^{-1}; 25 x 25 cm; FYM 10 t ha^{-1}; Butachlor + 1 hand weeding ; Submergence with 3-5 cm water

T$_6$ = 16 days seedlings; 01 seedling hill^{-1}; 25 × 25 cm; FYM 201 ha^{-1}; Butachlor + 1 hand weeding; Submergence with 3-5 cm water

T$_7$ = 16 days seedlings; 01 seedling hill^{-1}; 25 x 25 cm; RFD + FYM 10 t ha^{-1}; Butachlor + 1 hand weeding; Submergence with 3-5 cm water

T$_8$ = 16 days seedlings; 01 seedling hill^{-1}; 25 x 25 cm; RFD + FYM 10 t ha^{-1}; Rotary weeder; Submergence with 3-5 cm water

T$_9$ = 16 days seedlings; 01 seedling hill^{-1}; 25 x 25 cm; RFD + FYM 10 t ha^{-1}; chemical + Rotary weeder; Submergence with 3-5 cm water

T$_{10}$ = 16 days seedlings; 01 seedling hill^{-1}; 25 x 25 cm; RFD + FYM 10 t ha^{-1}; chemical + Rotary weeder; Alternate wetting and drying (AWD)

Source: *Nadia Nazir, 2010*

Table 14.8: Grain yield, straw yield (q ha^{-1}) and harvest index (%) of rice as affected by different treatments

Treatment	Grain yield (q ha^{-1})	Straw yield (q ha^{-1})	Harvest index (%)
T$_1$	63.23	88.52	44.86
T$_2$	66.36	93.51	44.23
T$_3$	67.53	94.54	43.67
1	68.56	95.90	42.79
T$_5$	50.33	70.62	41.62

(Contd...)

Treatment	Grain yield (q ha⁻¹)	Straw yield (q ha⁻¹)	Harvest index (%)
T_6	51.36	72.01	41.89
T_7	70.86	96.20	43.11
T_8	72.63	101.66	42.72
T_9	74.76	104.64	42.57
T_{10}	77.61	108.68	42.36
SE ± (m)	1.18	1.42	0.95
C.D.(p=0.05)	3.15	4.22	NS

T_1 = 30 days seedlings; 03 seedling hill⁻¹; 15 × 15 cm; RFD + FYM 5 t ha⁻¹; Butachlor + 1 hand weeding; Submergence with 3-5 cm water

T_2 = 16 days seedlings; 03 seedling hill⁻¹; 15 × 15 cm; RFD + FYM 5 t ha"¹; Butachlor + 1 hand weeding; Submergence with 3-5 cm water

T_3 = 16 days seedlings; 01 seedling hill⁻¹; 25 × 25 cm; RFD + FYM 5 t ha⁻¹; Butachlor + 1 hand weeding; Submergence with 3-5 cm water

T_4 = 16 days seedlings; 01 seedling hill⁻¹; 25 × 25 cm; FYM 10 t ha⁻¹; Butachlor + 1 hand weeding; Submergence with 3-5 cm water

T_5 = 16 days seedlings; 01 seedling hill⁻¹; 25 × 25 cm; FYM 10 t ha⁻¹; Butachlor + 1 hand weeding; Submergence with 3-5 cm water

T_6 = 16 days seedlings; 01 seedling hill⁻¹; 25 × 25 cm; RFD + FYM 10 t ha⁻¹; Butachlor + 1 hand weeding; Submergence with 3-5 cm water

T_7 = 16 days seedlings; 01 seedling hill⁻¹; 25 × 25 cm; RFD + FYM 10 t ha⁻¹; Rotary weeder; Submergence with 3-5 cm water

T_8 = 16 days seedlings; 01 seedling hill⁻¹; 25 × 25 cm; RFD + FYM 10 t ha⁻¹; chemical + Rotary weeder; Submergence with 3-5 cm water

T_9 = 16 days seedlings; 01 seedling hill⁻¹; 25 × 25 cm; RFD + FYM 10 t ha⁻¹; chemical + Rotary weeder; Submergence with 3-5 cm water

T_{10} = 16 days seedlings; 01 seeding hill⁻¹, 25 × 25 cm; RFD + FYM 10 t ha⁻¹; chemical + Rotary weeder, Alternate weeting and drying (AWD)

Source: *Nadia Nazir, 2010*

Table 14.9 : Relative economics as affected by different treatments

Treatment	Cost of cultivation (Rs. ha⁻¹)	Gross returns (Rs. ha⁻¹)	Net returns (Rs. ha⁻¹)	B:C Ratio
T_1	26994	79617	52623	1.94
T_2	26994	83895	56901	2.10
T_3	27018	85033	58015	2.14
T_4	27018	86262	59244	2.16
T_5	28168	63533	35385	1.28
T_6	36168	70168	38000	1.05
T_7	31018	91775	60757	1.95
T_8	32018	94234	62216	1.94
T_9	32218	97035	64817	2.01
T^{10}	32518	104058	71540	2.20

T_1 = 30 days seedlings; 03 seedling hill⁻¹; 15x15 cm; RFD + FYM 5 t ha⁻¹; Butachlor + 1 hand weeding; Submergence with 3-5 cm water

T_2	=	16 days seedlings; 03 seedling hill^{-1}; 15x15 cm; RFD + FYM 5 t ha^{-1}; Butachlor + 1 hand weeding; Submergence with 3-5 cm water
T_3	=	16 days seedlings; 01 seedling hill^{-1}; 15x 15 cm; RFD + FYM 5 t ha^{-1}; Butachlor + 1 hand weeding; Submergence with 3-5 cm water
T_4	=	16 days seedlings; 01 seedling hill^{-1}; 25 x 25 cm; RFD + FYM 5 t ha^{-1}; Butachlor + I hand weeding; Suhmergcnce with 3-5 cm water
T_5	=	16 days seedlings; 01 seedling hill^{-1}; 25 x 25 cm; FYM 10 t ha^{-1}; Butachlor + 1 hand weeding; Submergence with 3-5 cm water
T_6	=	16 days seedlings; 01 seedling hill^{-1}; 25 x 25 cm; FYM 201 ha^{-1}; Butachlor + 1 hand weeding; Submergence with 3-5 cm water
T_7	=	16 days seedlings; 01 seedling hill^{-1}; 25 x 25 cm; RFD + FYM 10 t ha^{-1}; Butachlor + I hand weeding; Submergence with 3-5 cm water
T_8	=	16 days seedlings; 01 seedling hill^{-1}; 25 x 25 cm; RFD + FYM 10 t ha^{-1}; Rotary weeder; Submergence with 3-5 cm water
T_9	=	16 days seedlings; 01 seedling hill^{-1}; 25 x 25 cm; RFD + FYM 10 t ha^{-1}; chemical + Rotary weeder; Submergence with 3-5 cm water
T_{10}	=	16 days seedlings; 01 seedling hill^{-1}; 25 x 25 cm; RFD + FYM 10 t ha^{-1}; chemical + Rotary weeder; Alternate wetting and drying (AWD)

Source: *Nadia Nazir, 2010*

SRI recorded significantly higher tillers and consequently panicle weight and number of grains/panicle, test weight which in turn recorded significantly superior grain and straw yield as compared to recommended practices of rice cultivation in Kashmir valley. Further, System of rice intensification proved economically efficient in terms of B:C ratio.

Future Needs/work to be done

- Agro meteorological aspects in SRI (Rainfall)
- Nursery management
- Water management
- Spacings in different agroclimates/soils/for varieties
- Nutrient management
- Tillage practices

CONCLUSIONS

SRI is a system to which improve rice production manifolds by exploring full potential of this semi aquatic plant. Besides, most dwindling resource *i.e.*, water could be utilized more efficiently and economically.

REFERENCES

Nadia Nazir. (2010). Studies on the effect of agronomic manipulation of system of rice intensification (SRI) practices on growth and yield of rice *(Oryza saliva* L.) under temperate valley conditions. M.Sc. Thesis submitted in Division of Agronomy, SKUAST of Kashmir, Srinagar.

RCM, *Kharif.* (2008). Evaluation of system of rice intensification (SRI) under temperate conditions of Kashmir. RCM report, Division of Agronomy, SKUAST of Kashmir, Shalimar.

RCM, *Kharif.* (2010). Evaluation of System of rice intensification (SRI) with biofertilizers and varying age of seedling under temperate condition. RCM report, Division of Agronomy, SKUAST of Kashmir, Shalimar.

Singh, KN and Chand, LC (2008). Evaluation of system of rice intensification (SRI) under temperate conditions of Kashmir. Report, 41st Research Council Meeting, Division of Agronomy, SKUAST-K, Shalimar, Srinagar; pp. 11-13.

Wang Shao-bua,Cao Weixing, Jiang Dong, Dai Tingbo and Zbu Yan (2002). Physiological Charateristics and High yielding Techniques with SRI. Cornell International Institute for Food, Agriculture and Development; pp. 116-124.

■

INTRODUCTION

Over the past four decades, crop management in India has been driven by increasing use of external inputs. Among them fertilizer nutrients have played a major role in improving crop productivity. During the period 1969-2007, food grain production was more than doubled from about 98 million tonnes (Mt) in 1969 to a record 212 Mt in 2001-02, while fertilizer use increased by nearly 12 times from 1.95 Mt in 1969 to more than 23 Mt in 2007-08 (Rao, 2009). Not with standing these impressive developments, food grain demand is estimated to increase about 300 Mt yr^{-1} by 2025 for which the country would require 45 Mt of nutrients (ICAR, 2008). With almost no opportunity to increase the area under cultivation, over 142 M ha, much of the desired increase in food grain production has to be attained through yield enhancement, in particular those of major staple food crops like rice, wheat and maize, which incidentally responded considerably to the introduction of green revolution technologies, contributing to more than 80 per cent of total food grain production. To sustain production demands, the productivity of major crops has to increase annually by 3.0 to 7.5 per cent (NAAS , 2006). Much of this has to be met by increasing genetic potential and improved production efficiency of the resources and inputs like water and nutrients. In addition, the growing concern about poor soil health and declining factor productivity or nutrient use efficiency, has raised concern on the productive capacity of agricultural systems in India. *For example*, research on farmer's fields has revealed that there is no compelling evidence of significant increases in fertilizer N efficiency in the rice-wheat system during the past 30 years (Dobermann and Cassman, 2002). The average plant recovery efficiency of fertilizer N is still only about 30 per cent (Dobermann, 2000).

Major factors contributing to the low and declining crop responses to fertilizer nutrients are (a) continuous nutrient mining due to imbalanced nutrient use, which is leading to depletion of some of the major, secondary, and micro nutrients like P, K, S, Zn, Mn, Fe and B, and (b) mismanagement of irrigation systems which leads to

[1, 2&3]Associate Professor, Faculty of Agriculture, Sher-e-Kashmir University of Agricultural Sciences and Technology of Jammu, Main Campus, Chatha-180 009, Jammu, J&K

[4] Scientist, PDFSR, Modipuram, Meerut (UP)

[5] FCLA, Faculty of Agriculture, Sher-e-Kashmir University of Agricultural Sciences and Technology of Jammu ,Main Campus, Chatha-180 009, Jammu, J&K

serious soil quality degradation. Furthermore, such low efficiency of resources and fertilizer inputs has impacted the production cost with serious environmental consequences.

Recent research conducted in various countries, including India (Dobermann *et al.*, 2002; Wang *et al.*, 2001; Wopereis *et al.*, 1999; Angus *et al.*, 1990) has indicated limitations of the blanket fertilizer recommendations practice across Asia. Cassman *et al.*, (1996 a, b) observed that indigenous N supply of soils was variable among fields and seasons, and it was not related to soil organic matter content. On-farm research has clearly demonstrated the existence of large field variability in terms of soil nutrient supply, nutrient use efficiency and crop responses. Thus, it was hypothesized that crop productivity and input use efficiency will require soil and crop management technologies that are knowledge intensive and are tailored to specific characteristics of individual farms or fields to manage the variability that exists between and within them (Tiwari, 2007).

NUTRIENT MANAGEMENT

Nutrient management is a major component of soil and crop management system. Knowing the required nutrients for all stages of growth and understanding the soil's ability to supply those needed nutrients are critical to profitable crop production. The importance of fertilizer nutrient application to crops for achieving higher yield and to sustain the soil fertility on long term basis are well known. Nutrient management implies managing all nutrient sources, fertilizers, organic manures, waste materials suitable for recycling nutrients, soil reserve and bio-fertilizers in such a way that yield is not knowingly jeopardized whilst every effort is made to minimize losses of nutrients to environment. The challenges of nutrient management are to maintain and increase sustainable crop productivity to meet demands for food and raw materials to enhance the quality of land and water resources.

Improving nutrient management strategies is equally important as breeding for high yielding. Several approaches have been used for fertilizer nutrient management in the past such as recommending a fixed rate based on nutrient response curve or recommendations based on balance sheet method. Fertilizer nutrient management based on soil test value *i.e.*, soil testing is one of the most widely used method in India, where fertilizer recommendations are made based on initial soil test value. Among the other approaches, the target yield approach has got popularity in India. This method not only estimates soil test based fertilizer dose but also the level of yield the farmer can achieve with a particular dose.

The success of any nutrient management strategy depends on the spatial scale for which decisions are made, hence, nutrient management concept can distinctly defined for two broad domains: (i) Package approach based on blanket recommendation over wider area and (ii) site specific nutrient management. Conventionally blanket fertilizer recommendations are often applied over larger areas without taking into account the wide variability in soil nutrient supplies, fertilizer efficiency and site-specific crop nutrient requirements within each recommendation domain. Package approach based on blanket recommendation including fixed rate nutrient application, soil test based approaches and balanced sheet method assume that crop needs for nutrients is constant among years and over large areas. But crop growth and demand for nutrient are strongly influenced by climate and crop growing conditions which can vary greatly among locations, seasons and years. Hence, current nutrient management strategies suffer from a number of limitations:

- They do not assist farmers in dynamic nutrient management
- Do not consider need of crop grown in different seasons and places
- Lead to a very low nutrient use efficiency *e.g.* N-0.3-0.5; P-0.15-0.2,K-0.80
- Give poor yields because of low nutrient use efficiency
- Poor returns
- Soil and water pollution due to excess nutrient losses

To eliminate wastage of fertilizer and increase farmers income, need was felt to make adjustments in fertilizer amount and timing with respect to location and season specific needs of crop. It is argued that a sustainable nutrient management strategy must ensure high and stable overall productivity with optimum economic returns and sufficient nutrient supply for potential yield increases with minimum wastage of nutrients. This can be achieved if exogenous nutrient supply is matched with nutrient supplies from soil and crop demands. Therefore, a site specific nutrient management (SSNM) strategy is required to increase efficiency of nutrients and crop productivity (Janssen *et al.*, 1990; Cassman *et al.*, 1996). Nutrient management strategies based on more site specific approaches can offer opportunities for efficient utilization of nutrients applied through fertilizers or organic sources.

What is SSNM

Site specific nutrient management is an approach of feeding crops with nutrients as and when needed. The application and management of nutrients are dynamically adjusted to crop needs of the location and season. It is a low technology plant based approach for managing the nutrient requirement of crops. It provides principles and tools for supplying nutrients to crops as and when needed to achieve high yields while optimizing use of nutrients from indigenous sources. This concept can be applied to any field and any crop. While most often, thought in relation to use of computer and satellite technology, the site-specific nutrient management does not require special equipment and does not require a large farming operation. The technology tools certainly expand the capabilities for using site specific management. Site-specific crop and soil management is really a repackaging of management concepts that have been promoted for many years. It is basically taking a systematic approach to applying sound agronomic management to small areas of a field that can be identified as needing special treatment.

The components of site-specific management may not be new, but we have the capability with new technology to use them more effectively. Site-specific management includes practices that have been previously associated with maximum economic yield (MEY) management, best management practices (BMPs), as well as general agronomic principles. The systematic implementation of these practices into site-specific systems is probably our best opportunity to develop a truly sustainable agriculture system.

CONCEPT AND APPROACHES OF NUTRIENT MANAGEMENT

Site-specific nutrient management is a widely used term all over the world with reference to addressing nutrient differences which exist within fields and making adjustments in nutrient application to match those locations or soil differences. Our search indicates that the earliest use of the term 'site-specific' comes from the late 1920's in the USA, when scientists at the University of Illinois were providing recommendations on the application of lime to acidic soils (Jones, 1993). In the mid to late 1980's, the development of GIS and GPS technology and the associated use of this technology in agriculture to map field variability, fostered the use of 'site-specific management' approaches in the developed world. International Plant Nutrition Institute (IPNI) started a site-specific soil testing and assessment programme in India, which can be best described as a systematic approach to soil fertility evaluation. It outlined a local approach for evaluating soil fertility and making fertilizer recommendations which could be implemented by any institute or state research group (Portch and Hunter, 2002). This method involves two major steps in the laboratory, prior to any field research being carried out. The first is to conduct a series of fixation reaction studies on nutrient elements with soil, determining the ration of nutrient supply versus nutrient applied with a specific soil. This information helps to indicate how much nutrient is required to be added to a soil to bring it to a level which is more than adequate for maximum growth, but less than that which could be toxic or out of balance with other plant nutrients. The second step involves a greenhouse study where sorghum is grown as an indicator plant with the complete nutrient treatment, including all nutrients required for growth provided in the fertigation system and deletion treatments for each of the individual nutrients. The result is biomass sample data which will provide an indication of when a potential positive or negative response is to be expected. This information is then used to proceed to field trials with the specific crop or crop's providing the necessary insight as to what nutrients need to be considered in the evaluation. The absence of any response to a nutrient in the greenhouse trial helps to eliminate it from further assessment in the field trials, bringing the focus on the deficient nutrients. Using yield goals and the soil analysis, the lab is in a position to provide fertilizer recommendations covering the full range of nutrients required, including macro, secondary, and micronutrients (Portch and Stauffer, 2005). This procedure was introduced in both India and China, with a national soil fertility lab, using the methods now operating in Beijing.

Many field trials by IPNI, conducted over the years in India, fertilizer rates were established based on the concept of crop removal, with an adjustment for soil residual nutrients. While this approach actually fits most production systems in India quite well, given that most of the crop biomass is removed from harvested fields, the role that residual soil nutrients play in meeting crop nutrient requirements becomes a challenge. *For example,* if a soil tests medium or low in most of the plant nutrients then application of nutrients based on target yield crop removal is going to address these nutrient demands. However, on soils where the soil nutrient analysis indicates a high level of nutrient supply, issue of whether to apply the nutrient at removal rates becomes a challenge to the researcher. The issue is one of the balanced nutrition for the crop. Addition of high rates of N, P and K as part of the treatment actually stimulates a deficiency of a secondary or micro-nutrient which according to soil test considered adequate. The best example of this is K use in many production systems where soil test shows that K levels should be more than adequate to meet crop demand, but at what yield level. Many of the recommendations and soil test , used for K guidelines are associated with much lower yields than are currently being targeted by growers. Research conducted on IPNI in India clearly show that many of these guidelines are inadequate for current yield targets, as a result the soil test K level once considered adequate turns out to be insufficient to balance the high rates of N and P being applied (Tiwari, 2005). As a result, best option is to apply all macro and secondary nutrients which are required to meet crop yield removal and those micro-nutrients which soil testing shows to be marginal or deficient. This then provides the environment for full yield expression in the absence of any deficiency. Once yield potential of this site has been determined, the next step is to refine nutrient application rates with further field trials.

A plant based SSNM approach was developed in the early 1990s by the International Rice Research Institute (IRRI) in collaboration with IPNI in India. IRRI working with the rural poor, meant that soil testing was not an option to consider, due to lack of service or high cost of the service. This approach focused on managing field specific spatial variation in indigenous NPK supply, temporal variability in plant N status occurring within a growing season and medium term changes in soil P and K supply resulting from actual nutrient balance. The approach required a data management option to predict soil nutrient supply and plant uptake in absolute terms in the high yielding irrigated rice systems in Asia. A modified QUEFTS model (Janssen *et al.*, 1990 and Witt *et al.*, 1999) was used for this purpose. It described the relationship between grain yield and nutrient accumulation as a function of climatic yield potential and the supply of the three macronutrients. In this situation of balanced nutrition, the QUEFTS model assumed a linear relationship between grain yield and plant nutrient uptake or constant internal efficiencies until yield targets reach about 70-80 per cent of yield potential. As yields approach the potential yield, the internal nutrient efficiency declines as the relationship between grain yield and nutrient uptake enters a non linear phase. To model this in a generic sense, it required an empirical determination of two boundary lines describing the minimum and maximum internal efficiencies of N, P and K in the plant across a wide range of yields and nutrient status. A database containing more than 2000 entries on the relationship between grain yield and nutrient uptake was used to derive the generic boundary lines of internal efficiencies (Dobermann and Witt, 2004). For rice the balanced N, P and K uptake requirements for 1000 kg of grain yield were estimated from the respective envelope functions as 14.7 kg N, 2.6 kg P and 14.5 kg K which is valid for the linear phase of the relationship between yield and nutrient uptake. The corresponding borderlines for describing the minimum and maximum internal efficiencies were estimated at 42 and 96 kg grain kg^{-1} N, 206 and 622 kg grain kg^{-1} P and 36 and 115 kg grain kg^{-1} K, respectively (Witt *et al.*, 1999). The parameters were found to be valid for any site in Asia for modern rice varieties with a harvest index of around 0.45 – 0.55. Principles of SSNM were compiled into a practical guidebook to nutrient management for rice. This popular guidebook, which provides guidelines on optimal rates of N, P and K adjusted to field specific yield levels and indigenous supply of nutrients was subsequently updated (Fairhurst *et al.*, 2007).

Initial Concept of SSNM

The concept of SSNM was developed for rice in the late 1990s by IRRI in collaboration with National Agricultural Research and Extension System (NAREC) and then evaluated from 1997 to 2000 in about 200 irrigated rice farms at eight sites in six Asian countries (Dobermann *et al.*, 2002). SSNM aimed at dynamic field-specific management of N, P, and K fertilizers to optimize the supply and demand of crop nutrients. The crop's need for nutrient N, P, or K was determined from the gap between the crop demand to achieve a target yield and the nutrient supply from indigenous sources.

(A) Establish a yield target for average climatic conditions

This yield target could be based either on a percentage (*for example* 70–80 per cent) of the potential yield estimated with crop growth model or on yields currently achievable by farmers practice under good crop management.

(B) Estimate crop demand for N, P, and K for a target yield

Based on a large database of modern rice varieties with harvest index of 0.45 to 0.55, the balanced plant nutrient requirement to produce one metric ton of un-milled rice was estimated 15 kg N, 2.6 kg P (6 kg P_2O_5) and 15 kg K (18 kg K_2O) for the linear portion of the relationship between grain yield and nutrient accumulation in mature crop.

(C) Estimate field-specific indigenous supply of N, P and K

The indigenous supply is the cumulative crop uptake of a nutrient from various sources other than fertilizer (that includes soil, crop residues, manures, irrigation water, rainfall, and atmospheric deposition). It is determined by the nutrient omission plot technique, whereby the indigenous supply of a nutrient is estimated by its accumulation in a crop not fertilized with the nutrient of interest but fertilized with sufficient amounts of other nutrients to ensure they do not limit yield. Indigenous K supply was determined in a K omission plot receiving no K fertilizer but sufficient N and P to ensure they do not limit yield.

(D) Create the recovery efficiencies of N, P and K

Crop recovery efficiencies of 0.4 to 0.6 kg^{-1} for fertilizer N, 0.2 to 0.3 kg kg^{-1} for fertilizer P, and 0.4 to 0.5 kg kg^{-1} for fertilizer K were used as targets.

(E) Optimal estimate of N, P and K rates

The estimated crop demand for N, P, and K to optimally achieve the yield target, the estimated indigenous supply of N, P, and K, and targeted recovery efficiencies for fertilizer N, P, and K were used to determine optimized fertilizer N, P, and K rates for filling the gap between crop demand for a yield target and indigenous supply.

EVOLUTIONARY DEVELOPMENT OF SSNM (2001 TO PRESENT)

Since 2001 to 2004 reach toward optimal productivity (RTOP) workgroup of the irrigated rice research consortium (IRRC) has collaborated with national agriculture research and extension systems in eight Asian countries to systematically transform the initial SSNM concept into an inclusive, simplified framework for the dynamic plant-need-based management of N, P, and K.

The SSNM approach now enables

- Dynamic adjustments in fertilizer N, P, and K management to accommodate field- and season-specific conditions.
- Effective use of indigenous nutrients.
- Efficient fertilizer N management through the use of the leaf colour chart (LCC), which helps ensure that N is applied at the time and in the amount needed by the rice crop.
- Use of the nutrient omission plot technique to determine the requirements for P and K fertilizers.
- Use of micro-nutrients based on local recommendations.

Since 2001 on the farm evaluation and promotion of SSNM have markedly increased. In 2002-2004 SSNM was evaluated and promoted with farmers at 20 locations in tropical and sub-tropical Asia, each representing an area of intensive rice farming on more than 100,000 ha with similar soils and cropping systems. The countries involved were Bangladesh, China, India, Indonesia, Myanmar, Philippines, Thailand and Vietnam.

REACHING TOWARD OPTIMUM PRODUCTIVITY WORKGROUP

The RTOP workgroup was instrumental in the formulation and promotion of site-specific nutrient management including the use of leaf colour chart as an approach for increasing farmer's profit through more efficient use of nutrients. The suggestions of this group in 2004 are as:

- Developed an improved and less expensive 4-panel leaf colour chart (LCC) in response to the recommendation of 2003 review of IRRC.
- Distributed about 100,000 new LCCs across Asia.
- Motivated local governments and extension agencies in Asia to emphasize the concept of SSNM to promote nutrient management in rice. The government of Hunan Province in China committed to extend SSNM with its own funds as a provincial initiative in 2005.

- Formulated principles of SSNM and simple guidelines of locally adapted nutrient management practice of 20 major rice producing area in Asia.

- Strengthened partnerships among research, extension, local government and the fertilizer sector for delivery of SSNM through six interactive workshops with stakeholders in India, Vietnam, and Philippines.

- Conducted ten seminars in various part of Asia to promoted and disseminated of SSNM.

- Trained around 30 participants in a two-week programme at IRRI to developing and disseminating SSNM.

During the four years period from 2001 to 2004 more than 25,000 farmers and extension workers attended training courses on SSNM organized by NARES and around 30,000 farmers/ extension workers and government officials participated in fields days and field visits at NARES sites across Asia.

Present Approach for SSNM

Research on the development and evaluation of SSNM, data were obtained on the relationship between yield of un-milled rice and total N, P, and K in the mature rice crop. As a result, uptake of N, P, and K by a mature rice crop with harvest index of 0.45 to 0.55 can be estimated from grain yield with sufficient reliability. Measurement of N, P, and K in grain and straw was no longer required for determining fertilizer N, P, and K rates by the SSNM approach. Grain yield targets and yield in nutrient omission plots can now be directly used in a simplified manner by estimating fertilizer N, P, and K requirements as follows.

(A) *Establish an attainable yield target for farmers' fields*

The yield target in the SSNM approach is directly used to calculate fertilizer rates for farmer's field. It must be reasonably attainable by farmers. A yield target higher than one attained by farmers would lead to recommendations of more fertilizer than required for high use efficiency and profit. A yield target below that realistically attained by farmers could result in suboptimal yield and profit. Grain yield from a fully fertilized plot with no nutrient limitations and good management (like NPK plot or NPK plus micro-nutrient plot in the nutrient omission plot technique) can be used to estimate the yield target.

(B) *Approximate a fertilizer N rate and formulate dynamic N management*

The difference between the yield target and N-limited yield (*i.e.*, yield with no N fertilizer and no limitation of other nutrients) provides an estimate of anticipated crop response to fertilizer N. The estimated yield response to fertilizer N and a targeted efficiency for fertilizer N use are used to approximate the total requirement of crop for fertilizer N which is dynamically apportioned among multiple times of application to best match the crop's need for N.

(C) Estimate field-specific nutrient-limited yields

Nutrient limited yields were determined by the nutrient omission plot technique. The K-limited yield is determined in a K omission plot receiving no K fertilizer but sufficient N and P to ensure they do not limit yield. The P-limited yield is determined in a plot receiving no P fertilizer but sufficient supply of other nutrients.

(D) Regulation of P and K rate

Crops need for fertilizer P based on a comparison of yield target and P-limited yield, whereas the crops need for fertilizer K is based on a comparison of the yield target and K-limited yield. The SSNM approach advocates sufficient use of fertilizer P and K to both overcome P and K deficiencies and maintain soil P and K fertility.

GENERAL SITE SPECIFIC NUTRIENT MANAGEMENT STRATEGY

A general site specific nutrient management strategy has been presented in Fig. 16.1. It requires an estimate of the climatic yield potential and current yield to decide the target yield. Prices for harvested grain and fertilizer inputs also should be considered for deciding the target yield. Once the target yield is set, a farm or field specific estimate of the indigenous N, P and K supply is required. This is the most sensitive input in the SSNM and can be estimated using a variety of methods, depending on available facilities. Using QUEFTs model nutrient requirement was estimated using the relationship between grain yield and nutrient uptake *i.e.*, crop nutrient demand for a specified target yield. Finally the recovery efficiency of applied nutrients is considered to calculate the amounts of fertilizer. It should be noted that soil moisture availability may not be a constraint because one of the major assumptions for the model is that water availability does not limiting growth.

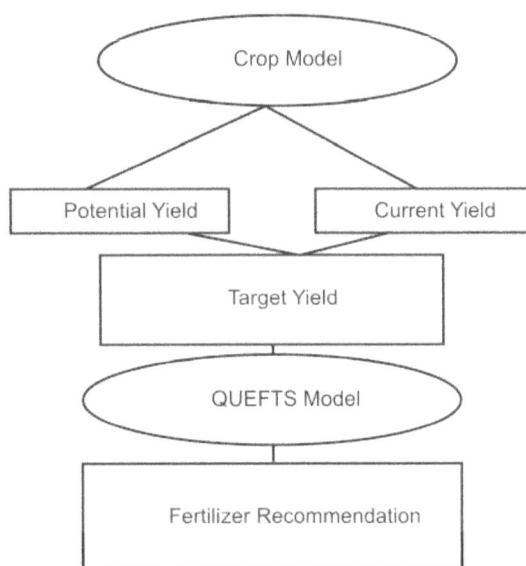

Fig. 15.1: *QUEFTS Model*

The following steps adopted by Dobermann and Fairhurst, 2000

- Identify and alleviate nutritional constraints other than N, P and K (*i.e.,* improved crop management to prevent toxicities or deficiencies of nutrients other than N, P and K).

- Estimate farm or field specific potential for indigenous supply of N (INS), P (IPS), and K (IKS, all in kg ha $^{-1}$).

- To developed a farm or field specific recommendation for NPK use and to achieve a defined target yield by optimizing the nutritional balance of N, P and K in the plant. Fertilizer rate = (crop nutrient requirement – indigenous nutrient supply)/ first crop recovery of fertilizer.

- Optimize the timing and amount of N fertilizer applied based on plant growth. Decision about the timing of N application and number of splits required can be based on (1) two or four split applications (*i.e.,* following basic agronomic principles). (2) regular monitoring of plant N status up to the flowering stage, using tools such as chlorophyll meters or green colour charts or (3) predictions of N split applications using simplified simulation models.

- Measure the grain yield and amount of straw and stubbles return to field, and calculate the amount of fertilizers nutrients applied. These data are again used to predict the change in INS, IPS and IKS during previous crop cycle based on the estimated nutrient budget. For nutrients such as P and K reasonable estimates of the nutrient budget can be obtained by estimating nutrient inputs from manures, fertilizers and crop residues and nutrients removed with grain and straw.

- Specify a fertilizer recommendation (repeat calculation as in step 3) for the subsequent crop cycle and modified INS, IPS, and IKS values resulting from step 5 are used for this.

- Continue using this procedure (Step 5 to 6) for a succession of crops. After about 3 to 5 years a new measurement of INS, IPS, IKS and other constraints may be necessary (Step 1 and 2) to restart the whole recommendation cycle.

Management of N, P and K require different strategies. To optimize N use efficiency for each season a dynamic N management strategy is required with the adjustment of quantity of N applied in relation to the variation in indigenous N supply. This is as important as timing, placement and sources of N applied. N management therefore includes the following measures.

- An estimate of crop N demand, potential supply from indigenous sources (soil and BNF), and recovery from inorganic and organic sources applied. These factors are used to estimate the total N requirement.

- An estimate for the need of a basal N application according to soil N release patterns, crop variety and crop establishment methods.

- Plant N starts monitoring to optimize the timing of split application of mineral fertilizer in relation to crop demand and soil N supply and

- Long term soil and crop management practices to manipulate the INS.

On the other hand P and K require a long term management strategy. It is more important to predict the need for the application of P and K and amount required than to maximize recovery efficiency for fertilizer P and K, this is because these nutrients are not readily lost or added to the root zone by biological and chemical processes affecting N Management and must be geared towards maintaining available soil nutrient supply to ensure that P and K do not limit crop growth and thus reduce N use efficiency. Changes in potential indigenous P and K supply can be predicted as a function of the overall nutrient balance. Key components of P and K management are follows.

- An estimate of crop P and K demand and potential indigenous P and K supply.

- Recovery of P and K from applied inorganic and organic sources to predict P and K inputs required to maintain a targeted yield level.

- A schedule for timing of K application depending on soil K buffering characteristics and an understanding of the relationship between K nutrition and pest incidence, and

- Knowledge of relationship between P and K budget and residual effects of P and K fertilizers and changes in soil supply over time.

MANAGEMENT OF OTHER NUTRIENTS

Prevention, diagnosis and treatment are the key management tools for nutrients (*e.g.*, Ca, Mg and S), micro-nutrients and minerals toxicity. Over the longer term, prevention through general crop management (*e.g.*, using adopted cultivars), water management and fertilizer management (*e.g.*, choice of fertilizer containing secondary minerals) is important. Deficiencies can be alleviated by regular or one time measure as a part of general recommendations. Similarly, diagnostic tools can be used to identify other nutritional disorder such as salinity, Fe and B toxicity to make adjustments in N, P and K management. In some cases, it may be necessary to alter soil management practices to reduce the severity of minerals toxicity.

Nitrogen Management

It is an essential element for plant growth. Rice plants can obtain much of their required N from soil and organic amendments, but supply of N from these naturally occurring indigenous sources is seldom sufficient for high rice yield. Supplemental N from fertilizers is typically essential for higher yields and profit from irrigated and favorable rainfed rice fields.

The demand of rice for N is strongly related to growth stage. Rice plants require N at early and mid-tillering (branching) stages to ensure sufficient number of panicles. Nitrogen absorbed at panicle initiation stage increases spikelet per panicle. Nitrogen absorbed during the ripening phase, in presence of adequate solar radiation, enhances the grain filling process. For best results, farmers should apply fertilizer N several times during the growing season to ensure that the N supply matches the crop need for N at the critical growth stages of tillering, panicle initiation and grain filling. The SSNM approach for managing fertilizer N aims to increase profit for farmers by achieving high rice yield and high efficiency of N use.

With this approach, the recommended use of N could be higher or lower than the current farmer's practice. Managing N by the SSNM approach typically involves a change in the farmer's practices on distributing fertilizer N during a crop growing season.

Estimating fertilizer N required for rice

The fertilizer N required by a rice crop is estimated through three steps in the SSNM approach.

Step 1: Establish an attainable yield target

The yield target provides an estimate of the total amount of N needed by rice crop because the amount of N taken up by a rice crop is directly related to crop yield (Fig. 16.2). The yield target depends upon the location specific climate, rice cultivar and crop management.

Step 2: Effectively use existing nutrients

Much of the N taken up by rice crop comes from naturally occurring (indigenous) sources. Like soil, organic amendments, crop residue, manure, and irrigation water. This indigenous supply of N can be estimated from the total amount of N taken by mature rice crop that does not receive fertilizer N and is not limited by other nutrients. Because the amount of N taken by rice is directly related to yield, indigenous N supply can then be estimated from N-limited yield, which is the grain yield for a crop not fertilized with N but fertilized with other nutrients to ensure they do not limit yield (Fig. 15.2).

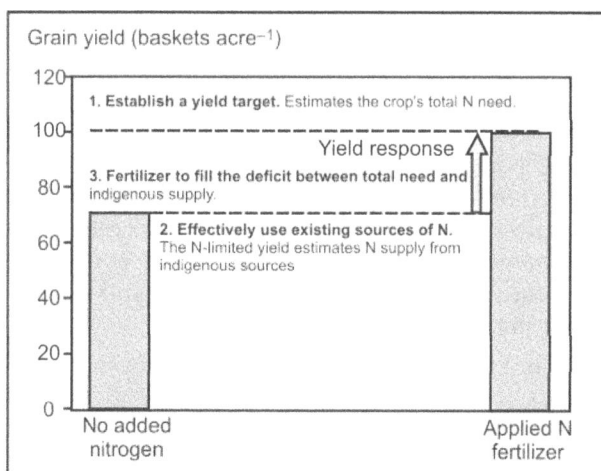

Fig. 15.2: Steps in determining N required for rice.

The N-limited yield can be determined by the nutrient omission plot technique. When results from the omission plot technique are unavailable, information on use of organic amendments, soil texture, soil testing, or previous measurements of N-limited yield on similar soils can often be used to suitably estimate N-limited yield. The direct measurement of N-limited yield by the omission plot technique is not required when N-limited yield can be estimated from other information within accuracy of ± 10 baskets acre^{-1}.

Step 3: Apply fertilizer to fill the deficit between crop need and indigenous supply

Fertilizer N is required to supplement the nutrients from indigenous sources and achieve yield target. The total fertilizer N requirement depends upon deficit between the crop's total N need to achieve a yield target and the N supply from indigenous sources, as determined by the N-limited yield. This deficit in N that must be filled by fertilizer N is directly related to estimated yield response to fertilizer N which is the difference between yield target and N-limited yield (Fig. 15.2).

Table 15.1 shows the simplified guidelines in estimating fertilizer N required by rice based on grain yield response to fertilizer N and efficient use of fertilizer N. As a general rule, round 37 lbs urea acre^{-1} should be applied for each 10 baskets acre^{-1} increase in yield anticipated from use of fertilizer N.

Table 15.1. Estimation of fertilizer N required for rice based on yield response to fertilizer N and efficiency of fertilizer N use.

Yield response (kg ha^{-1})	Urea required (lbs acre^{-1})
10	37
20	74
30	111
40	148

P and K Management

Principles of P and K management

Phosphorus and potassium are essential elements for plant growth. Phosphorus is particularly important in early growth stages. It promotes root development, tillering and early flowering. Potassium strengthens plant cell walls and contributes to greater canopy photosynthesis and crop growth. It does not have a pronounced effect on tillering, but it can increase the number of spikelet's per panicle and percentage of filled grain.

Rice plants obtains much of their required P and K from the soil, crop residues, organic amendments, and irrigation water, but the supply of P and K from these naturally occurring, indigenous sources are typically insufficient to sustain high rice yields. Supplemental P and K from fertilizers are thus essential for sustaining high and profitable yields of rice without depleting the fertility of soil.

Potential of SSNM

For any SSNM program, the principle approach should be to look for opportunities to increase crop production, especially where as imbalance of nutrient exists. This was an easy task in India when IPNI initiated its work on SSNM. Preliminary evaluation of production practices, soils and crop yields revealed that there was a large potential to show improved yields with balanced fertilization. Early on in the programme the focus was on the balance between N:P:K, which was highlighted in a couple of publications focused on P and K in particular (Hasan, 2002). Most obvious in these early years was the widening ratio of N:P:K in the use of fertilizers, especially after the wide spread nature of P and K deficiencies had been identified. Work with balancing P and K deficiencies resulted in significant yield increases in many trials with multiple crops. In addition, use of K in many areas of India, where K was considered sufficient, increased yield potential and deficiency of other secondary and micronutrients in crops. This is a characteristic response which we have observed in countries around the world, where N and P use forms basis of crop fertilization. Introduction of K results in moderate to significant yield increases along with quality improvement in crops. However, it also results in deficiency of secondary and micro-nutrients becoming obvious in field- under crops. In most of these instances, it is most limiting nutrients and once corrected opens the door to further deficiency in the cropping system (Tandon, 1997).

The full effect of inappropriate nutrients recommendation ratios and rapidly growing evidence of secondary and micro-nutrients deficiency, lead to the conclusion that India had entered the era of multiple nutrient deficiency and future food production goals were going to be largely limited by the ability to meet these multiple nutrient shortage. This impact of secondary and micro-nutrients was clearly shown in a report of research conducted with sugarcane in Maharashtra (Phonde *et al.*, 2005). A SSNM treatment, which reduced N rates by 30 per cent increased P and K rates and included secondary and micronutrients, increased cane yield by 30 t ha^{-1}. The addition of S, Zn, Fe and Mn all had a major impact in the SSNM treatment increasing cane yields. Deletion plots for these nutrients showed a yield loss of 5 t ha^{-1} for Mn, 18 t ha^{-1} for Zn or S, and 25 t ha^{-1} for Fe when left out of the SSNM treatment schedule. The results of this study provided overwhelming evidence that we are now well beyond the realm of NPK in terms of nutrient deficiency and existing practices are providing sub-optimal fertilizer recommendations for sugarcane yield potential. Similarly, a balanced application of macro and micro-nutrients in pulse crop provided 40 per cent increase in both yield and economic benefit (Bhattacharya *et al.*, 2004). Results of on-farm demonstrations (Singh, 2001) across crops and soils showed that S application increased grain yield in cereals by 650 kg ha^{-1} (+24 per cent over NPK), 570 kg ha^{-1} (+32per cent over NPK) in oilseeds and 375 kg ha^{-1} (+20 per cent over NPK) in pulses. Large number of trials on Zn application (over optimum NPK) in wheat increased grain production of 200-500 kg ha^{-1} in 35 per cent cases and 500 to 1000 kg ha^{-1} in 60 per cent cases. A similar number of trials in rice revealed that Zn application increased paddy yield by 200-500 kg ha^{-1} in 39 per cent cases and by more than 1000 kg ha^{-1} in 11 per cent cases (Tiwari 2008).

DELIVERING THE SSNM APPROACH TO FARMERS IN INDIA

A major aspect of any research programme in plant nutrition is determining how this information will ultimately be used to positively impact the farmer's practices. India has a vast population of farmers, many of which are poorly educated and have very limited ability to interpret new technologies. In fact, an NSS survey reported in the Hindu Business Line on 2 August 2005 reported that most farmers get their information from their progressive neighbours (16.7 per cent), input dealers (13.1 per cent) or from the radio (13.0 per cent). Government extension workers were noted by only 5.7 per cent of farmers as a source for accessing information on modern technology. This dependence on progressive neighbours is not uncommon, and indicates that many farmers are making business decisions based on the experiences and judgment of their neighbours. Radio program's provide much more general information, requiring a certain degree of interpretation relative to an individual farm. However, the role that input dealers play indicates that these farmers are seeking out a person who has contact with their location in the region, understands the soils, crops grown, yield potentials and the emerging issues.

Another area in which IPNI in collaboration with IRRI and fertilizer industry is making good strides is to consolidate the complex and knowledge intensive SSNM information into a simple delivery system enabling farmers to rapidly implement SSNM. Such delivery system developed in 2008 by IRRI and called 'Nutrient Manager' is an easy to use, interactive computer based decision tool (Buresh, 2008). The tool endeavours to acquire necessary information required for decision making on nutrient management through a series of easy to answer questions which essentially mask the rigor of SSNM principles from end users while maintaining robustness of the process. These decisions tools consist of about 10 multiple choice questions that can easily be answered by an extension worker or farmer. Based on response to the questions a fertilizer guide line with amounts of fertilizer by crop growth rate is provided for the rice field. They accommodate transplanted and direct-seeded rice, inbred or hybrid varieties with a range of growth durations as well as rice growing seasons to provide a range of options to wide spectrum of farmers.

Recently, IPNI has successfully tried Geographical Information System (GIS) mapping approach to measure spatial variability in nutrient status (Sen *et al.,* 2007) and used such maps as a site specific fertilizer recommendation tool to positively impact rice yields in farmer's fields (Sen *et al.,* 2008). This mapping is based on two factors (a) Nutrient content of cultivated soils varies spatially due to variation in genesis, topography, cropping history, fertilization history, and resource availability and (b) Soil testing of all holdings to estimate native fertility levels to ensure appropriate recommendation is a logical step but we do not have adequate infrastructure to accomplish this task. The process of soil fertility mapping involved geo-referenced soil sampling and using the soil analysis data in a GIS platform to develop surface maps of analyzed soil parameters across the study area. The spatial variability maps created by combining the location information of sampling points and analyzed soil parameter are capable of predicting soil parameter values of un-sampled points. This is possible because the interpolation technique used in the GIS platform creates smooth surface map of the study area utilizing point information (geographic location and corresponding soil parameters), where each point on the map has soil parameter value associated with it. The possibility of using such maps as fertilizer decision support tool to guide nutrient application in site specific mode is being explored in several studies under IPNI search initiative. Besides, delineating the fertility management zones these maps can give a clear visual indication within study area, these maps can give a clear visual indication of changing fertility scenario at village level with time, which is important for nutrient management planning (Sen *et al.*, 2008). Besides the logistical and economic advantages of implementing such a system, once established the technique can create an effective extension tool where field agents work more directly with farmers. Thus, farmers become more aware of how their fields rank within the landscape in terms of basic soil fertility which in turn enables a system of more rational use of fertilizer application.

FUTURE SSNM RESEARCH AND EXTENSION IN INDIA

Major challenges for SSNM research in future will be twofold (1) Retain the demonstrated potential of the approach and (2) To build upon what has already achieved using this approach while reducing the complexity of the technology as it is disseminated to farmers and their advisors. Thus, there is a need to further refine location-specific N management strategies and test them against other forms of N management. The nature of SSNM approach will need to be tailored to specific circumstance in different situations. In some areas, SSNM may be site or farm-specific, but in many areas, it is likely to be just region or season-specific. Thus, a simplified future SSNM approach should combine decisions that are made on a site-specific basis as well as decisions that are valid for somewhat larger regions with similar agro-climatic conditions. Estimates that allow placing a site into one of several broad categories of indigenous nutrients supply are probably sufficient for most SSNM applications and are easier to follow.

For extension of SSNM, need to development a decision support system (DSS) programmes like the nutrient manager mentioned above is a good example of the type of technology industry and extension agronomist can use to support nutrient management decisions at field level in villages. Developing nutrient manager for wheat and maize are logically goals considering the high acreage under these crops and their importance for food and nutritional security in the country.

REFERENCES

Angus JF, St Groth CFD and Tasic RC. (1990). Between-farm variability in yield responses to inputs of fertilizers and herbicide applied to rainfed lowland rice in the Phillipines. *Agriculture, Ecosystems and Environment* 30: 219-234.

Bhattacharya SS, Mandal M, Chattopadhyay GN and Majumdar K. (2004). Effect of balanced fertilization on pulse crop production in red and laterite soils. *Better Crops* 88(4): 25-27.

Buresh RJ. (2008). Management made easy: A new decision-making tool is helping rice farmers optimize their use of nutrient inputs. *Rice Today*, October-December 2008. International Rice Research Institute.

Cassman KG, Dobermann A, Sta Cruz PC, Gines HC, Samson MI, Descalsota JP, Alcantara, JM, Dizon MA and Olk DC. (1996a). Soil organic matter and the indigenous nitrogen supply of intensive irrigated rice systems in the tropics. *Plant and Soil* **182**: 267-278.

Cassman KG, Gines HC, Dizon M, Samson MI and Alcantara JM. (1996b). Nitrogen-use efficiency in tropical lowland rice systems: contributions from indigenous and applied nitrogen. *Filed Crops Research* **47**: 1-12.

Dobermann A (2000). Future intensification of irrigated rice systems. In: *Redesigning Rice Photosynthesis to Increase Yield* (Sheehy J E, Mitchell P L and Hardy B Eds.) pp. 229-247. IRRI/Elsevier, Makati City, Philippines/ Amsterdam.

Dobermann A and Cassman KG. (2002). Plant nutrient management for enhanced productivity in intensive grain production systems of the United   States and Asia. *Plant and Soil* **247**: 153-175.

Dobermann A and Witt C. (2004). The evolution of site-specific nutrient management in irrigated rice systems of Asia. In: *Increasing productivity of intensive rice systems through site-specific nutrient management* (A. Dobermann, C. Witt and D. Dawe, Eds.). pp. 410. Science Publishers Inc. and International Rice Research Institute (IRRI) Enfield N. H. (U. S. A.) and Los Banos (Phillipines).

Dobermann A, Witt C, Dawe D, Abdulrachman S, Gines HC, Nagarajan R, Satawathananont S, Son TT, Tan PS, Wang GH, Chien NV, Thoa VTK, Phung CV, Stalin P, Muthukrishnan P, Ravi V, Babu M, Chatuporn S, Sookthongsa J, Sun Q, Fu R, Simbahan GC and Adviento MAA. (2002). Site-specific nutrient management for intensive rice cropping systems in Asia. *Field Crops Research* **74**: 37-66.

Fairhurst TH, Witt C, Buresh RJ and Dobermann A (Editors). (2007). Rice: A Practical Guide to Nutrient Management. Los Banos (Philippines) and Singapore: International Rice Research Institute (IRRI), International Plant Nutrition Institute (IPNI) and International Potash Institute (IPI).

ICAR. (2008). From the DG's desk. *News Reporter* (April-June 2008), p. 2.

Janssen BH Guikin FCT, Van der Eijk D, Smaling EMA, Wolf J and Van Reuler H. (1990). A system for quantative evaluation of the fertility of tropical soils (QUEFTS). *Geoderma* **46**: 299-318.

Jones JB (1993). Review of perspective, issues, and trends in soil and plant testing in the United States of America. *Australian Journal of Experimental Agriculture* **33**: 973-981.

Khurana HS, Phillips SB, Bijay Singh, Alley MM, Dobermann A, Sidhu AS, Yadvinder Singh and Peng S. (2008). Agronomic and economic evaluation of site-specific nutrient management for irrigated wheat in northwest India. *Nutrient Cycling in Agroecosystems.* **82**: 15-31.

Khurana HS, Phillips SB, Bijay Singh, Dobermann A, Sidhu AS, Yadvinder Singh and Peng S. (2007). Performance of site specific nutrient management for irrigated, transplanted rice in northwest India. *Agronomy Journal* **99**: 1436-1 447.

NAAS. (2006). Low and declining crop responses to fertilizers. *Policy Paper* **35**: 1-8.

Phonde D.B, Nerkar YS, Zende NA, Chavan RV and Tiwari KN. (2005). Most profitable sugarcane production in Maharashtra. *Better Crops* **89**: (3) 21-23.

Portch S and Hunter A. (2002). A systematic approach to soil fertility evaluation and improvement. *Modern Agriculture and Fertilizers, Special publication No. 5.* IPNI China programme, Beijing, China.

Portch S and Stauffer MD (2005). Soil testing: A proven diagnostic tool. *Better Crops with Plant Food* **89** (1): 28-32.

Rao KV. (2009). Site-specific integrated nutrient management – principles and strategies. In: *Proceedings of the workshop on SSNM in Rice and Rice based cropping systems* held in Hyderabad, Feb. 4-13, 2009 pp. 22-29.

Sen P, Majumdar K and Sulewsk G. (2007). Spatial-variability in available nutrient status in an intensively cultivated village. *Better Crops* **91** (2): 10-11.

Sen P, Majumdar K and Sulewski G. (2008). Importance of spatial nutrient variability mapping to facilitate SSNM in small land holding systems. *Indian Journal of Fertilizers* **4** (11): 43-50.

Singh MV. (2001). Evaluation of current micronutrient stocks in different agro-ecological zones of India for sustainable crop production. *Fertilizer News* **46** (2): 25-42.

Tandon HLS. (1997). Experiences with balanced fertilization in India. *Better Crops* **11** (1): 20-21.

Tiwari KN. (2005). Diagnosing potassium deficiency and maximizing fruit crop productivity. *Better Crops* **89** (4): 29-31.

Tiwari KN. (2007). Breaking yield barriers and stagnation through site-specific nutrient management. *Journal of the Indian Society of Soil Science* **55**: 444-454.

Tiwari KN. (2008). Future of plant nutrition research in India. *Journal of the Indian Society of Soil Science* **56**: 327-336.

Wang G, Dobermann A, Wirtt C, Sun Q and Fu R. (2001). Performance of site-specific nutrient management for irrigated rice in South-east China. *Agronomy Journal* **93**: 869-878.

Witt C, Dobermann A, Abdulrachman S, Gines HC, Wang GH, Nagarajan R, Satawathananont S Son TT, Tan PS, Tiem LV, Simbaham GC and Olk DC. (1999). Internal nutrient efficiencies of irrigated lowland rice in tropical and subtropical Asia. *Field Crops Research* **63**: 113-138.

Wopereis MCS, Donovan C, Nebie B, Guindo D and N Daiye MK. (1999). Soil fertility management in irrigated rice systems in the Sahel and Savanna regions of West Africa. Part I. Agronomic analysis. *Field Crops Research* **61**: 125-145.

■

16
CHAPTER

An Introduction of Plant Nutrients and Foliar Fertilization

J. Singh[1], M. Singh[2], A. Jain[3], S. Bhardwaj[4]
Ajeet Singh[5], D.K. Singh[6], B. Bhushan[7] and
S.K. Dubey[8]

INTRODUCTION

Abundant plant growth and development needs adequate supply of various nutrients, which are absorbed through the plant roots and leaves. Plants require sixteen nutrients for their growth and development, which are carbon, hydrogen, oxygen, nitrogen, phosphorus, potassium, calcium, magnesium, sulphur, iron, boron, copper, manganese, molybdenum, zinc and chlorine. First nine are regarded as macro-nutrients and other seven signified as micro-nutrients, these terms are denoted in concern of plant's requirement *i.e.,* macro-nutrients are required in more quantities, whereas micro-nutrients are needed in minor amounts. Besides above elements, some trace elements such as silicon (Si), aluminum (Al), sodium (Na), cobalt (Co), and nickel (Ni) are also required in micro levels. Other elements, vanadium (V), lanthanum (La), cerium (Ce) and selenium (Se) have left evidences of their presence in some lower plants. Not all required for all plants, but all have been found to be essential or beneficial to some plants. Each of the elements plays a role in growth and development of the plants, and when present in deficient quantities can reduce growth and yields (Tisdale *et al.*, 1993). Several researchers have suggested that some other nutrients are also beneficial for plant growth. These include sodium for halophytes and sugar beet (Rana and Mark, 2008), silicon for rice and sugarcane, cobalt for efficient N_2-fixation and synthesis for vitamin B_{12} in ruminants (Fertilizer Manual, 1998) and vanadium for growth of certain microorganisms (Rohilla and Gupta, 2012). In addition, uptake of Al by tea plants and subsequent release of protons from their roots may be an important mechanism by which they acidify soils in tea gardens (Qing *et al.*, 2012). Lanthanum and cerium are components of a fertilizer used in some parts of China (Guo, 1987).

Carbon, hydrogen, oxygen are obtained from the atmosphere during usual conditions. If proper conditions of aeration and moisture are maintained in the soil,

[1&7] Central Institute of Post Harvest Engineering and Technology, Ludhiana/Abohar, Punjab

[2] Division of Nematology, IARI, New Delhi

[3] National Bureau of Plant Genetic Resources, New Delhi

[4&6] Department of Zoology, University of Delhi

[5] Divyayan Krishi Vigyan Kendra, Ranchi, Jharkhand.

[8] Central Soil & Water Conservation Research & Trainning Institute, Research Centre, PO Chhaleser, Agra (UP)

there is no problem with these three nutrients and other thirteen elements are commonly absorbed from the soil through root system. Some of these may also be absorbed through the leaves, when sprayed over the plant canopy in an appropriate concentration. Thus, application of fertilizers on leaves of growing plants with suitable concentrations is termed as foliar application. Foliar feeding is a relatively new and controversial technique (Bernal *et al.*, 2007 and Baloch *et al.*, 2008). In many cases aerial spray of nutrients is preferred and gives quicker and better results than the soil application (Jamal *et al.*, 2006). Now-a-days, development of foliar fertilizers and their application techniques have been suggested by numerous agri-agencies. Recently foliar application of nutrients has become an important practice in crop production while soil application of fertilizers is the basic method (Alam *et al.*, 2010). Recently foliar fertilization is widely used practice to correct nutritional deficiencies in plants and has potential advantages over soil application and it may increase the efficacy of fertilizer use (Silberbush, 2002). Similarly, in terms of yield it is more efficient than soil fertilization for both macro and micronutrients in different soil types (Amberger, 1991; Arif, 2006 and Ali *et al.*, 2008). Many plant nutrients are needed in such great quantities that it is impractical to supply them through the foliage. However, when soil conditions are unfavorable and micro-nutrients are needed to the plants, foliar applications may be accomplished in terms of contemporary technique for correction of nutrient level in plants.

In the early 1950s the commercial production of mixed liquid fertilizers was started in the United States and now it has been popularized in many countries (Wiley-VCH and LASTWiley-VCH, 2007). Multi-nutrient foliar feeding products are often the most effective and may correct nutrient deficiencies giving increases in growth and development (Mona *et al.*, 2012). It is most economical way of fertilization to achieve quality production and yield, especially when sink competition for carbohydrates among plant organs take place, while nutrient uptake from the soil is restricted (Kannan, 1986 and Singh, 2007). In higher pH soils, it is well known that micro-nutrients as well as some macro-nutrients may hardly be absorbed by roots due to higher ion concentration, which lowers osmotic potential of soil water and consequently the availability of soil water to the plants became a limiting factor (Hirpara *et al.*, 2005), then foliar application is particularly useful (Swietlik and Faust, 1994). Therefore, foliar feeding of nutrients has become an established procedure in crop production to increase yield and quality of crop products (Roemheld and El-Fouly, 1999) and it also minimizes environmental pollution and improves nutrient utilization through reducing the amounts of fertilizers added to the soil (Abou-El-nour, 2002). On other hand, foliar feeding of a nutrient promotes root absorption of the same nutrient or other nutrients through improving root growth and nutrients movement from terminal leaves to depth roots (El-Fouly and El-Sayed, 1997, Oosterhuis, 1998). Leaf feeding enhances overall nutrient level in the plant and sugar production during times of stress (A&L Canada Laboratories, 2000) and it also increases biochemical activities in the leaf by increasing chlorophyll a, b and carotenoids, which presumably favour the photosynthesis. Similarly, Shitole and Dhumal, (2012) found that foliar applications of micro-nutrients increased the primary metabolites like photosynthetic pigments and organic constituents of *Cassia angustifolia*. Mona *et al.*, (2012) found that increasing NPK levels significantly increase pigments content in potato shoots. Increased leaf activities stimulate the need of water uptake by the plants vascular system and consequently, it increases the nutrients uptake from the soil. Mishra *et al.*, (2011) studied the foliar fertilization as a part of integrated nutrient management in ber (*Zizyphus mauritiana*).

It has examined that the availability of macro and micro-nutrients to the plants is affected by numerous soil biophysical factors, whereas foliar application in a particular way, could avoid these factors for their more rapid and efficient absorption. University of California of VA findings indicated that foliar feeding was more efficient than soil application in such a way; N, 6 times more efficient, B, 4 times more efficient, Mn, 30 times more efficient, Zn, 20 times more efficient, P, 20 times more efficient (Dixon, 2003) and Mo 14 times more efficient. Likewise, Liew (1988) advocated foliar application of micro-nutrients to be 6-20 time more efficient than soil application, depending on soil type.

Limitations of Foliar Fertilization

Those fertilizer materials suitable for foliar application must be soluble in water. Most of these are salts and when applied in too high concentration the solution will cause "burning" of the plant tissue. Often the safe concentration of the fertilizer material in the sulution is so low that repeated applications are required to supply the needs of the plants. This is especially true of nitrogen, phosphorus and potassium.

Mechanisms of foliar nutrients uptake in plant

The processes that regulate the uptake of foliar substances have been studied in detail as movement of nutrients into the plant through the leaf involves different mechanisms to nutrient transport into plant roots (Eichert and Burkhardt, 2001 Fernandez and Ebert, 2005). Most notably, plant leaves have thicker cuticles which are coated with a waxy layer making penetration of solutes difficult. For foliar fertilizers to be utilized by the plant for growth, the nutrient must first penetrate the leaf surface prior to entering the cytoplasm of a cell within the leaf. Penetration of foliar nutrients occurs through the cuticle the stomata, leaf hairs and other specialized epidermal cells. There is ongoing debate as to which of these penetration pathways plays most important role in nutrient uptake (Fernandez and Eichert, 2009 Oosterhuis, 2009).

Epicuticular waxes are a barrier to the retention and penetration of foliar fertilizers into plant organs (Jenks and Ashworth, 1999). Almost all plant surface waxes are hydrophobic and repel water-based sprays. The cuticle itself is a lipid layer making slightly permeable to both water and oils. This is one reason why the timing of foliar sprays is often targeted to stomatal opening (Jenks and Ashworth 1999). The epidermal layer is dotted with stomata which allow CO_2 exchange between the outside environment and photosynthetic cells (Marschner, 1995). The stomatal pore is enveloped by two guard cells which regulate the opening and closing of the pore. Stomata also provide the major pathways for evaporative loss of water, exchange of gases during photosynthesis and for controlling the transport of water across the epidermis (Raven and Johnson, 1999). Buick *et al.*, (1992) showed that pre-treatment of broad bean leaves with abscisic acid to close stomata resulted in diminished absorption of foliar-applied solution. This confirmed that stomatal penetration played a role in penetration and uptake of solutes. Similarly, other studies (Field and Bishop, 1988) have identified stomatal penetration and therefore it is important to consider when applying foliar fertilizers. However, some researchers believe that stomatal uptake plays a minor or negligible role compared to cuticle penetration of foliar-applied nutrients (Schonherr, 2006 and Oosterhuis, 2009). The cuticle is the first route available for penetration of solutes into leaves upon contact with the leaf, and the cuticle layer partially extends across the stomatal cavity forming cuticle ledges. These cuticle ledges will also interfere with the ability of the stomata to take up solutes (Dickison, 2000). The cuticle is a structurally complex waxy layer and, on the basis of urea and glucose absorption studies, Schonherr, (1976) concluded that liquids could penetrate through cuticular pores up to 0.9 nm in diameter. Permeability of cuticles to ions is dependent on similar environmental factors to stomatal uptake. Research to date suggests that both stomata and the cuticle play a role in nutrient uptake. However, stomatal penetration appears to be more complex and dependent on more environmental factors than cuticular penetration. But stomatal penetration is quick and rapid, when it occurs, whereas cuticular penetration is a slower process.

(A) FOLIAR APPLICATION OF MACRO-NUTRIENTS

Nitrogen (N)

Plant leaves can also absorb atmospheric nitrogen, Tisdale *et al.*, (1990) experienced that the field crops can take 10 per cent of their total N requirements by direct absorption of NH_3 from the air. Urea (N) can also be supplied to plants in liquid form through the foliage (Millard and Robinson, 1990), facilitating optimal N management. Most plants rapidly absorb foliar-applied urea and hydrolyze the urea in the cytosol (Nicoulaud *et al.*, 1996). Urea (46 per cent N) is most common foliar fertilizer, which is highly water soluble and reduce N loss from the soil through leaching, denitrification, volatilization, immobilization, or a combination of all these loss pathways. With few exceptions, it is impossible to feed crops their complete nitrogen requirements through the foliage. Foliar applied-nitrogen can be supplement of soil fertilization but cannot replace soil N application, in order to meet high N requirements of crops (Oosterhuis, 1999, Ling and Silberbush, 2002). Foliar-applied urea in olive plants gave significant increases in grain protein of about 22 per cent, starch 5 per cent and total yield 4 per cent (Tejada and Gonzalez, 2003). Similarly, Ashour and Thalooth, (2003) reported most effective treatment of 1 per cent urea as foliar application during pod development in soybean for better yield. Varga and Svecnjak (2006) concluded that late-season urea spraying consistently increased wheat grain yield at a low basal N rate and increase in protein content. Likewise, foliar spray of urea was applied along with soil application in durum wheat and obtained better quality (Abad *et al.*, 2004).

Parvez *et al.* (2009) advocated that, urea foliar spray has advantage to increase wheat grain yield over 25 per cent in soil moisture deficient areas. Similarly, foliar application of urea to wheat at tillering increased grain yield and grain protein content under soil moisture limiting conditions in the marginal areas (Njuguna *et al.*, 2009). Abad *et al.*, (2004) and Parvez *et al.*, (2009) were also observed that foliar application of urea increased yield components and grain yield by 19 per cent and grain protein (Borjian and Emam 2001). Grain protein content increased with higher fertilization level and improved the rheological properties of dough and baking quality (Salah, 2006). Whereas, Chauhan *et al.*, (2004) revealed that, foliar sprays of concentrate urea induced defoliation in wheat and increased mean yield by 29 per cent from 2.4 t ha^{-1} to 3.1 t ha^{-1}. Foliar fertilization also has limitations in wheat because high yielding varieties of wheat need 100-120 kg N ha^{-1}; whereas 10 kg N ha^{-1} can be applied in a single spray using high volume sprayer, hence scheduled and specified sprays are suggested. Foliar applications of urea in chickpea before terminal drought, at 50 per cent flowering increased yield and seed protein content. The increase in yield resulted from an increase in the number of pods with more than one seed rather than from increased pod number per plant or increased seed size, indicating greater seed survival under terminal drought (Jairo *et al.*, 2005).

Increased yield and fruit set in fuerte avocado trees have been reported after nitrogen foliar application (Lovatt, 1999, Lahav and Whiley, 2002). Foliar fertilization of urea (0.6 per cent) could be suggested to have better yield and growing properties in lettuce production (Guvenc, *et al.*, 2006). Thus foliar application in certain crops can secure the availability of plant nutrients to obtaining higher yield (Arif *et al.*, 2006). The nitrogen absorbed by the leaves appears to be translocated to the inflorescences and it also increases the number of lateral inflorescences. From commercial point of view this practice is applicable in apples and citrus for increasing more lateral inflorescences and maximum yield and it sometimes takes six or more applications in one season to provide all of the nitrogen needs Zilkah *et al.*, (1996) found a 26 per cent increase in leaf N when applying foliar low-biuret urea in Hass trees in Israel, which was maintained for one year. Additionally, the spray application was found to increase freezing tolerance by 2.5 times and reduced leaf senescence. Labelled ^{14}C-urea was applied onto Hass leaves to find out the ability of field grown plants to absorb N, using radioactivity tracking. Translocation of ^{15}N foliar-applied urea was also observed to vegetative and reproductive sinks of the plant. Maximum uptake was achieved after 2 days of application, but as only 2.1 per cent of applied urea entered into the leaf it was assumed to be physiologically insignificant (Nevin and Lovatt, 1990). A further study by the same authors failed to increase leaf N significantly after applying low-biuret urea to young 'Hass' trees.

Leslie *et al.*, (WTFRC Project AH-01-65) found that foliar-applied urea in autumn apple leaves indicated rapid absorption in both surfaces of a leaf, but the lower surface absorbed N four-times faster than the upper surface when the king bloom opened. The absorbed urea N was converted into amino-acids in leaves and then translocated into bark and roots for storage. Soil urea application during the growing season (summer) promoted shoot growth and aboveground biomass while foliar urea application increased root biomass in young potted apple trees. Soil urea application promoted extensive root initiation while foliar urea application promoted feeder root initiation.

Nitrogen applied to the foliage as urea is a commercial practice in apples and itrus. These crops apparently absorb nitrogen better than most other. It is mostly used to supplement soil treatments as it sometimes takes six or more applications in one season to provide all of the nitrogen needs (Weinbaum, 1978). Foliar applications also offer a means of overcoming temporary nitrogen deficiency when plant growth and root uptake of nutrients is limited by external factors *e.g.* waterlogging or post irrigation stress in flood irrigated crops on heavy clay soils. El-Otmani *et al.,* (2002) indicated that foliar-applied urea is an efficient and cost-effective way to supply N to Clementine trees and its efficiency was found greater at pH~6.5. Urea is the most common foliar N source for cotton, due to its relatively low toxicity, quick absorption, and low cost (McConnell *et al.,* 1998, Oosterhuis and Bondada, 2001). Stiegler *et al.,* (2009) suggested that foliar urea-N applications to putting green turf can be made to actively growing plant tissue throughout the season without concern for substantial N loss via this pathway. Shastri *et al.,* (2000) pointed out that foliar application of 2 per cent DAP increased seed germination and root length, while 0.5 per cent borax increased shoot length. Vigour index and dry matter production were also highest with 2 per cent DAP. A grain yield increase was documented for plants with foliar applications although, Alston (1979) attributed the response to the N applied and commented that increasing grain yield can be achieved after head emergence by keeping the soil wet to enable nutrient uptake or by applying foliar fertilizer directly to the plant.

It has been observed by many investigators that foliar-applied urea is the cause of leaf-burn in soybean during low humidity and high temperatures and by use of a too concentrated fertilizer solution and believed that the leaf burn depends upon the form of nitrogen fertilizer such as urea used, that is less likely to cause foliage bum than other nitrogen fertilizers because it has a lower salt index and is more rapidly absorbed into the leaf. However, leaf-burn has often been observed after foliar fertilization of plants with urea, and it has been reported that the leaf-bum observed with urea increases with leaf urease activity and is due to the ammonia produced from urea by this activity (Vasilas *et al.,* 1980). Leaf-tip necrosis observed after foliar fertilization of soybeans with urea is usually attributed to ammonia formed through the urea hydrolysis by plant urease (Harper, 1984). Thus, various crops have different tolerance of limit to urea or N concentration for better growth and in this concern, Wittwer, *et al.,* (1963) optimized following tolerance level to the different crops (Table 16.1).

Table 16.1: Tolerances of plant foliage to urea sprays in kg/100 L of water (Wittwer, *et al.,* 1963)

Vegetable crops		*Plantation of tropical crops*		*Deciduous tree and small fruit crops*		*Agronomic crops*	
Crop	Tolerance	Crop	Tolerance	Crop	Tolerance	Crop	Tolerance
Cucumber	0.36-0.60	Pineapple	2.40-6.00	Grape	0.48-0.72	Potatoes	2.40

Contd...

Vegetable crops		Plantation of tropical crops		Deciduous tree and small fruit crops		Agronomic crops	
Bean	0.48-0.72	Cacao	0.60-1.20	Raspberry	0.48-0.72	Sugar beets	2.40
Tomato	0.48-0.72	Sugar cane	1.20-2.40	Apple	0.48-0.72	Alfalfa	2.40
Pepper	0.48-0.72	Banana	0.60-1.20	Strawberry	0.48-0.72	Corn	0.60-2.40
Sweet corn	0.48-0.72	Cotton	2.40-6.00	Plum	0.60-1.80	Wheat	2.40
Lettuce	0.48-0.72	Tobacco	0.36-1.20	Peach	0.60-2.40		
Cabbage	0.72-1.44	Citrus	0.60-1.20	Cherry	0.60-2.40		
Carrots	2.40	Hops	4.80-6.00				
Celery	2.40						
Onions	2.40						

Phosphorus (P)

Available P in soil depends upon pH and soil microbial activities. In most cultivated soils the dominant species are H_2PO_4 and HPO_4^-. Soil P is found in two forms namely, organic P and inorganic P. Organic P may make up to 15.7 per cent to 46 per cent of total P in Indian soils, the rest of course is inorganic P. When water soluble P fertilizers are applied in acidic soils, these start reaction with Ca, Fe, Al, NH_4^+, Mg etc. to form insoluble compounds it called reversion or retention of applied P. Phosphorus involved in almost all biochemical pathways as a component part of energy carrier compounds, ATP and ADP (Khalil and Jan, 2003). Soil pH is important P fixation factor. In acidic soils, formation of insoluble Al and Fe phosphate increased. On the other side formation of insoluble Ca-phosphates takes place in alkaline soils. However, if pH is 9.0 and above or Na dominates in soil solution then the availability of P increases as the water soluble sodium phosphates. Fertilizer P does not move much from the point of placement. Hence P needs to be deep placed (5.7 cm below surface) and near to the point where seeding is done. However, for lowland transplanted rice and forage crops broadcasting and incorporation of P in surface soil is the only possible method. The diffusion coefficient of P in soil is very low, hence the root zone P is depleted and plants cannot get it when it is needed (Clarkson, 1981). Low soil moisture limits root access to P for uptake (dos Santos *et al.*, 2004). While water may still be available in the subsoil, mineral nutrition can become the growth limiting factor as nutrients are often stratified in the dry topsoil.

The soil pH hindrance for deficiency of P in crop can be overcome by the use of P fertilizers as foliar sprays. Many reports have indicated some significant beneficial approaches when yield levels were lower, likely due to moisture stress. Haloi, (1980) reported that when initial P deficiency symptoms appeared 25 days after sowing in wheat, higher doses of ammonium phosphate as a foliar spray gave highest yields. Similarly, Benbella and Paulsen, (1998) also showed that foliar applications of 5-10 kg, KH_2PO_4 ha[-1] (1.1-2.2 kg P ha[-1]) after anthesis increased the wheat grain yields. Sharaf and El-Naggar, (2003) conducted a field experiment to record the response of carnation plant to phosphorus and boron foliar fertilization. They reported that foliar application of P_2O_5 alone or in combination with different levels of B stimulated the length, diameter and dry weight of stem, number and dry weight of leaves per branch as well as enhanced flowering time, number, size and dry weight of flower per plant. The best results of vegetative growth and flowering characteristics were obtained at 200 mg P_2O_5 per litre plus 50 mg B per litre. Kefyalew *et al.*, (2007) applied foliar spray of P (0, 2, 4, and 8 kg ha^{-1}) in corn at three stages (i) collar of fourth leaf visible (ii) collar of eighth leaf visible and (iii) last branch of the tassel completely visible but silks not yet emerged. Foliar P applied at the growing stage at rate of 8 kg ha^{-1} improved grain and forage P concentration more than the smaller rates, which was reflected in increased grain yield. Thus, foliar P could be used as an efficient P-management tool in corn when applied at the appropriate growth stage and rate. The utilization of P as a foliar application becomes increasingly important. Increased yields in barley were obtained using dilute solutions of foliar P (Qaseem *et al.*, 1978).

FOLIAR APPLICATIONS OF PHOSPHORUS (P)

Foliar applications of P are the most effective way for a grower to supply P to crops grown late in-season. Foliar P potentially provides increased fertilizer use efficiency (Dixon, 2003 and Girma *et al.*, 2007) and this is controlled by the leaf area available to intercept the applied formulation. Early in the growing season the proportion of surface cover of the growing crop is often less than 50 per cent (Scotford and Miller, 2004). This halves the maximum possible efficiency of a foliar fertilizer, as the P that falls to the ground is not likely to contribute to plant P uptake due to both the low mobility of P (Hedley and McLaughlin, 2005) and the very small concentrations of P reaching to the soil surface.

A number of studies have investigated the physiology of foliar P fertilizer uptake under controlled conditions, often using single droplets (Dixon, 2003; Girma *et al.*, 2007). These studies provide valuable information about the mechanisms for foliar fertilizer uptake and the rate and amount of nutrient absorbed and translocated in the plant. Ahmed *et al.*, (2006) demonstrated that the salt load, pH and nutrient mixture all had important effects on the efficacy of foliar P fertilizer. Foliar applications of P have been tested on various agricultural crops such as soybeans (Mallarino *et al.*, 2001, 2005), clover, wheat (Arif *et al.*, 2006 and Mosali *et al.*, 2006) and corn (Girma *et al.*, 2007) to optimize best formulation, rate, timing, crop type and sites. Benbella and Paulsen (1998) applied 0, 2.2, 4.4 and 6.6 kg P ha^{-1} as KH_2PO_4 foliar P at late anthesis in wheat, in two separate growing seasons, and suggested that the most efficient P application rate was 2.2 kg P ha^{-1}. In wheat crops Mosali *et al.*, (2006) applied foliar potassium phosphate at 1, 2 and 4 kg P ha^{-1} comparative rate applications at different growth stages and found best response to foliar P during flowering at 2 kg P ha^{-1}. Girma *et al.*, (2007) had significant corn grain yield response to foliar P at 2 kg P ha^{-1} applied from eighth leaf through to tasseling growth stages. Most appropriate timing for P application was found during early-pod development in soybeans and from canopy closure to anthesis in cereal crops and early tasseling in maize (Girma *et al.*, 2007). The rate of application is dependent on crop requirement, crop type and number of applications, water volume and salt loading but in general 1.5-4 kg P ha^{-1} gave the best results (Mosali *et al.*, 2006 and Girma *et al.*, 2007). When bean plants were put under severe water stress their photosynthetic rate was found significantly reduced and there was only a small effect of applied P on yield (Dos Santos *et al.,* 2004). This study found that, addition of extra P as foliar application was able to alleviate the effects of mild water stress. Compared to P starvation and low foliar P additions, the higher foliar P application had greater pod numbers and seed dry weight under water stress (Dos Santos *et al.,* 2004). Results suggested increases in grain yield that resulted from foliar P application that generally took place in seasons of water stress (Mosali *et al.,* 2006).

The application of foliar nutrients under water stress conditions requires careful consideration of stomatal opening and rate of fertilizer drying on leaf before penetration is possible. Likewise, the best responses to foliar P were in soils with the lowest levels of soil P. Similarly in the studies of Girma *et al.,* (2007), only 50 per cent of trials showed significant yield response to foliar P corresponding with sites with the lowest levels of initial soil P. Phosphate foliar applications have been reported by induce local and systemic resistance to powdery mildew in cucumber (Mucharromah; Reuveni *et al.,* 1993). Sharma *et al.,* (1993) reported that, application of 15 kg P_2O_5 at first irrigation through 2 per cent diammonium phosphate spray gave higher seed cotton yield compared to untreated. The effect of diammonium phosphate as 2 per cent foliar application at squaring stage of crop improved the vegetative growth and yield significantly in cotton Cv. LRA 5166.

Potassium (K)

Potassium uptake from soil solution is regulated by several factors including texture, moisture condition, pH, aeration, and temperature. Clay soil can adsorb and hold more K than sandy soils and exchange it with the soil solution for plant uptake. Plants obtain K primarily from the soil in form of K^+, which is also strongly adsorbed by soil components. Particularly on clay particles and is therefore not readily mobile in most soil Both calcium and magnesium compete with K for root uptake, hence plants grown in these soils often exhibit K deficiencies even though soil analysis indicate adequate K.

Potassium is an important nutrient for plant meristematic growth and physiological functions, including regulation of water and gas exchange in plants, protein synthesis, enzyme activation, photosynthesis and carbohydrate translocation in plants. This attributed to the role of K in biochemical pathways in plants and it increases the photosynthetic rates, CO_2 assimilation and facilitates carbon movements (Sangakkara *et al.,* 2000). Potassium has favourable effects on metabolism of nucleic acids, proteins, vitamins and growth substances (Bisson *et al.,* 1994; Bednarz and Oosterhuis, 1999). Potassium is highly mobile in plants and its deficiency symptoms therefore first appear on older leaves. Potassium-deficient plants often have slow growth, poor drought resistance, and are more susceptible to plant disease (Sinclair, 1993). Most of K is taken up during the vegetative growth and developmental stages of plants, when roots are more active than in reproductively growth stages (Mills and Jones, 1996 and Marschner, 1998). Similarly, developing fruits are stronger sinks for photoassimilates than roots and vegetative tissues. This competition for photoassimilates reduces root growth energy supply for nutrient uptake including K (Marschner, 1995). Therefore during reproductive development soil K supply is seldom adequate to support crucial processes such as sugar transport to fruits, enzyme activation protein synthesis and cell extension that ultimately determine fruit yield and quality. Lu and Shi (1982) reported that 51.9 per cent of total K uptake in rice was during ear initiation to heading and 27.7 per cent during grain filling and maturity and similarly in wheat, 69.1per cent of total K uptake was during ear initiation and 23.8 per cent during flowering to maturity. This suggests a need for top dressing of K at later stages of growth of rice and wheat of K-deficient soils.

Foliar application of K has been recommended to overcome certain interference to K uptake in plants and it has been recommended for tea, banana (Kumar and Kumar, 2007), in flood prone rice areas and it has also been proved an efficient method for orchards. Under certain conditions this method offer an opportunity to correct the deficiency more quickly (within 20 hours) and efficiently, especially late in the season when soil application of K may not be effective. Thus, foliar K may be a supplemental option when climatic and soil conditions reduce nutrient uptake from the soil. Similarly, Suwanarit and Sestapukdee (1989) applied various concentrations of K sprays on sweet maize and found a stimulating effect at 2.5 per cent KNO_3 during appropriate development stage and increases sweetness and crude protein content of grains and also affected maize by stimulating chlorophyll synthesis and not by increasing leaf area.

Roberts *et al.*, (1997) pointed out that the soil applied K along with foliar sprayed K can increase cotton lint yields for fast-fruiting cotton cultivars and also recorded that foliar K can be profitable on a low extractable soil K for at least two years. Outcome from foliar applications of K fertilizer, including tests in Virginia, have been inconsistent in showing an increase in cotton lint yield. It is important to recognize that it is generally cheaper way to supply K from soil-applied potash. Three to four foliar applications of K have been recommended during peak boll development at 7 to 10 day intervals; approximately 1.5 kg acre^{-1}, two weeks after flowering begins. The recommended source of K for foliar fertilization is potassium nitrate (KNO_3), although other sources such as potassium sulphate have also been used. Notable improvements in cotton yield and quality was found in foliar-applied K fields (Cassman *et al.*, 1992). These may be reflected as distinct changes in seed weight and quality. Pettigrew (1999) stated that the elevated carbohydrate concentrations remaining in source tissue, such as leaves, appear to be part of the overall effect of K deficiency in reducing the amount of photosynthate available for reproductive sinks and thereby producing changes in yield and quality seen in cotton.

Lester *et al.*, (2006) studied foliar K feeding in muskmelon and found that during fruiting of muskmelon the soil K fertilization alone is often inadequate due to poor root uptake and competitive uptake inhibition from calcium and magnesium. Soil supplemented K with foliar K applications during muskmelon fruit development and maturation, improved fresh fruit quality by significantly increasing firmness (26 per cent), sugar content (20 per cent), ascorbic acid (18 per cent), carotene (17 per cent), and K (14 per cent) compared to non-foliar treated fruits. Such supplemental property of K foliar fertilization is also found as supplemental option in soybean to mitigate K deficiency during the growing season and minimize yield loss when climatic and soil conditions reduce nutrient uptake from the soil (Kelly *et al.*, 2005). Foliar-applied potassium during fruit development and maturation of cantaloupe (*Cucumis melo*) improved fruit market quality by increasing firmness, sugar content, and nutritional value through increased beta-carotene, ascorbic acid and K concentrations in the edible flesh (Lester *et al.*, 2007) and apparent K deficiency during fruit development and maturation can be mitigated by foliar glycine aminoacid complexed K application (Lester *et al.*, 2005). Foliar K fertilization in tomatoes (*Lycopersicon esculentum*) has also been indicated increased fruit firmness (Chapagain and Wiesman, 2004). However, it is unclear whether the beneficial effects of foliar K fertilization on firmness and other quality parameter differ according to the source of K used, including the most common fertilizer K source-potassium chloride (KCl).

Foliar potassium nitrate has been recommended in prune orchards as an interim corrective measure until soil applications take effect (Swietlik and Faust, 1984). It has been observed that 2-year-old prune trees would require 140 applications of 1.2 per cent potassium nitrate per year to meet their requirement. Winter prebloom foliar applications of low-biuret urea or potassium phosphite (a form of P, HPO_3^{-2} readily taken up by leaves and translocated to the roots, (Lovatt and Mikkelsen, 2006) have been shown to increase in commercially valuable total yield such as, large size fruit and total soluble solids (TSS) of sweet oranges (*Citrus sinensis*). When mixed low-biuret urea and potassium phosphite were applied, the yield benefits were additive (Albrigo, 1999). ICL Fertilizers Co., applied ferti-K (0.2 per cent KCl) in olives during fruit set and observed that forti-K supplied the available potassium to the leaves and improves even fruit development, weight, size, quality and yield. Forti-K was also found compatible with the major pesticides. Soluble potassium silicate has been found correct for powdery mildew diseases control in cucurbits (Menzies *et al.*, 1992) and grapes (Bowen *et al.*, 1992).

The potassium phosphite formulation Nutri-Phite (0-28-26) has been used on perennial tree crops in US research trials as a winter prebloom spray resulted in an annual net increase in total yield by 7 metric tons ha^{-1}. Whereas the winter prebloom urea treatment resulted in a net increase of only 4 metric tons ha^{-1} (Albrigo, 1999), both treatments significantly increased the total soluble solids concentration of the fruit. Use of urea and potassium phosphite in *Clementine mandarin* production in Morocco produced similar beneficial yield results (El-Otmani *et al.*, 2003a, b). Foliar application of potassium sulphate (K_2SO_4) at the post-shooting stage of banana (*Musa* spp.) increased yield, fruit quality and post-harvest shelf-life (Kumar and Kumar, 2007).

Oosterhuis *et al.*, (1993) observed that cotton is most sensitive to low potassium availability than most other crops. Cotton sensitivity to low potassium availability is intensified by widespread potassium deficiencies in more recent years across the Cotton Belt, primarily associated with the use of higher yielding and faster fruiting cotton varieties and increased use of nitrogen. Foliar potassium applications have been successfully integrated with soil applications to attain a much better potassium response in cotton, primarily because the foliar application can be timed to apply potassium later in the growing season when it is most critical for fibre growth and strength. Similarly, an Arkansas study has shown that foliar-applied potassium (potassium nitrate gave the biggest yield increase, followed closely by potassium thiosulphate and potassium sulphate) increased the distribution of potassium in the fibre, seed, and capsule wall of the developing boll. The development of the boll load occurs late in the season, which coincides with a decrease in root activity and potassium supply to the boll, explaining the need for a timely and supplemental foliar application of potassium on cotton. The optimum timing of foliar-potassium application was found during the boll development. Initial application should be made at the first sign of bloom to help improve the set. A second application of nutrients will aid the later developing bolls, and should be made approximately three weeks after the first. Petiole analysis prior to boll development can determine the requirement of K for foliar-potassium rates. But generally, optimum foliar-potassium rate has been shown about 2.3 kg potassium per acre per application.

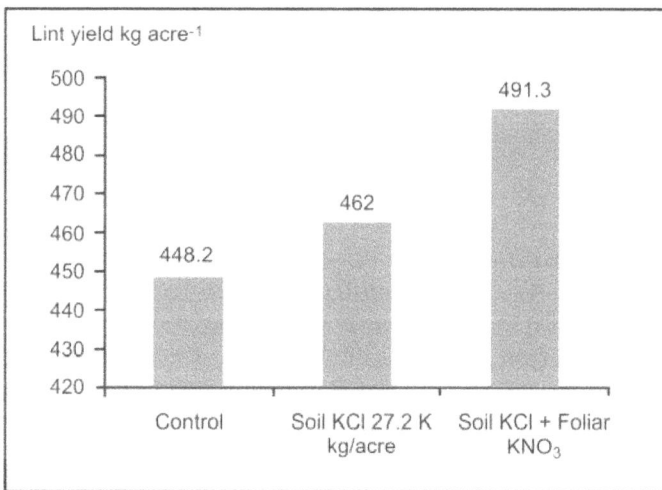

Fig. 16.1: *Cotton lint yield, soil-applied potassium and supplementary foliar-applied potassium*

Above results of K foliar feeding has shown that supplemental feeding can enhance production and quality at critical stages of growth. Foliar-applied K in form of KNO_3 has indicated more yield compared to both K-applied as KCl and control in terms of improvement in fibre quality, as well as more uniform maturity (Figure 16.1). As in any other treatment, the timing of application is of great importance, as are the type of formulation and ingredients.

Sulphur (S)

In recent years S-deficiency has become an increasing problem for agriculture resulting in decreased crop quality parameters and yields (Zhao *et al.*, 1999). Plants vary considerably in their S requirements. Oilseed rape, as with most *Brassicaceae*, has greater S requirements than other large crop species such as wheat or maize. *For example*, the production of 1 ton of rape seeds requires 16 kg of S (McGrath, 1996 and

Blake-Kalff *et al.*, 2001), compared with 2-3 kg for each ton of grain in wheat. The agronomic consequences of insufficient S are well documented with decreased yields and a substantial impact on S-content under extreme deficiency (Zhao *et al.*, 1999). In many cases of mild S-deficiency stress there may be little impact on yield but important consequences resulting for quality, with substantially modified N: S ratio (Zhao *et al.*, 1996). The sulphur requirements of cereal crops are lower in comparison to root crops, rapes and legumes (Motowicka-Terelak and Terelak, 2000). Following common sulphur containing fertilizers are used in various crops (Table 16.2).

Table 16.2: Common fertilizer sources of sulphur

Material		S (per cent)
Ammonium sulphate	$(NH_4)_2SO_4$	24
Potassium sulphate	K_2SO_4	18
Potassium magnesium sulphate	$K_2SO_4.2MgSO_4$	23
Calcium sulphate (gypsum)	$CaSO_4.2H_2O$	17
Magnesium sulphate (Epsom salts)	$MgSO_4.7H_2O$	14
Ammonium thiosulphate solution	$(NH_4)_2S_2O_3+H_2O$	26
Ordinary superphosphate	$Ca(H_2PO_4)_2+CaSO_4$	14
Zinc sulphate	$ZnSO_4$	18
Elemental sulphur	S	88-98

However, due to decreasing sulphur emission to the atmosphere and application of sulphur-free mineral fertilizer the deficiency of this element in cereal crops has been observed lately (Grzebisz and Przygocka-Cyna, 2003, Stern, 2005). Plants absorb sulphur from the soil through their root system in the sulphate form and transport it to the chloroplasts of leaf cells, where sulphate are reduced to sulphide and built into organic compounds (Hell and Rennenberg, 1998). Elemental form of sulphur is not directly available to the plants and has to be transformed in the soil into sulphate form. The transformation of elemental sulphur to sulphate depends on many factors *viz.* the size and activity of the microbial populations, soil temperature and moisture, and the degree of sulphur crumbling (Eriksen, 2009; Wen *et al.*, 2001). Elemental sulphur in the finely ground form can also be used for foliar fertilization because it undergoes oxidation on the leaf surface and is included in the processes of plant metabolism (Legris-Delaporte *et al.*, 1987).

In soil, sulphur is found as suphates of Na, K, Ca, and Mg and it has been recognized as the fourth major plant nutrient after nitrogen, phosphorus, and potassium. (Zaman, *et al.*, 2011) and its deficiencies and crop responses have been reported from all over the world. Sulphur requirements and its maximum benefits were reported in oil seed crops followed in pulses and least in cereals (Singh, 2001). However, good responses to sulphur application have been reported in both rice and wheat in rice-wheat cropping system (Yadav *et al.*, 2002 and Prasad, 2005). Sulphur is a structural component of amino-acids, proteins, vitamins and enzymes and is essential to produce chlorophyll. It imparts flavour to many vegetables and its deficiencies shows light green leaves. Sulphur is readily lost by leaching from soils and should be applied with a nutrient formula. Some water supplies may contain sulphur. Tea *et al.* (1990) reported a synergistic effect of the foliar applied N and S fertilizers that appears to increase their assimilation in grain and may improve bread-making qualities. Scott *et al.* (1984) have shown that foliar S treatment of barley increased the number of grains per ear, suggesting the possibility of advantageous influence on metabolism with higher plants.

Elemental sulphur has long been used for control of spider mites and certain plant diseases, and is approved for organic production. Organic gardeners primarily use microfine sulphur dust formulations for liquid spray. Sulphur can cause skin and eye irritation and can cause plant injury if offensively used, especially if applied in combination with oils, or if applied within several weeks of an oil treatment. Sulphur sprayed on leaves is readily absorbed by the plants. This fact was demonstrated, however, in connection with the study of the influence of certain sulphur sprays when used as a fungicide. Although there have been no reports of a sulphur deficiency being relieved by sulphur sprays, the practice may become established because it is physiologically sound.

Orlovius (2002) evaluated sulphur foliar fertilization on spring wheat in pot trials along with same dose of nitrogen soil application. Sulphur was used in the form of epsom salt, which resulted in considerable yield responses under S deficient conditions. The yield increase depended on the available sulphur reserves of the soil. Yield increase was also found in oil seed-rape seed and sugar beet, an early application at stem elongation was more effective than at heading. In field trials a foliar application of epsom salt with the addition of Mn and B resulted in an average yield increase of 3-7 per cent for oilseed rape and 3–5 per cent for sugar beets.

Malhi (2001) evaluated relative effectiveness of sulphate-sulphur fertilizer, applied at different growth stages of canola. Sulphate-sulphur was separately applied as soil and foliar applications along with similar dosage of nitrogen and found that foliar spray was more effective than soil application in terms of seed yield of canola. Ramos *et al.*, (2008) studied the leaf area development, biomass production and yield of four spring barley varieties grown in a Mediterranean environment (southern Spain) in response to an early application of foliar sulphur or etephon and observed similar results in both sulphur and etephon compared to untreated one. Whereas, the maximum leaf area index was found in untreated plots at the beginning of shooting and the growth of foliar area in the treated ones was found extended until anthesis. This led to a significant improvement in the biomass at anthesis, which was closely correlated with grain yield as well as in the leaf area duration during grain filling. An application of sulphur or etephon at tillering increases grain yield by raising both the number of ears per plant and per plot, without modifying the number of grains per ear or 1000-grain weight. The similarity between the effects of sulphur and etephon may be due to the fact that sulphur absorbed by the leaves results in an increase in methionine. Several S products are reported to lower soil pH and act as reducing agents to convert ferric iron to more readily available ferrous form (Tisdale *et al.*, 1990). Similarly, positive effect on fruit yield of citrus was also observed after S application. Salwa *et al.*, (2010) pointed out a combined application of amino-acid with micro-nutrient Fe, Zn, Mn in the presence of elemental sulphur significantly increased the sesame yield and improved nutritional quality. Kulczycki (2010) conducted a study to evaluate the effect of sulphur fertilization on the yield and chemical composition of winter wheat and found no difference in efficacy of the soil or foliar fertilization. Whereas, Randall *et al.*, (1981) conducted a study in winter wheat and found that combined soil and foliar application of sulphur S was most effective for grain yield. Similarly, Tea *et al.*, (2004) observed that foliar application with sulphur increases not only grain yield but also the content and polymerization of grain proteins. This in turn significantly improves dough mixing properties. Wheat straw yield increases significantly after foliar application with a dose of 10 dm^3 S$^-$ ha^{-1} along with elemental sulphur soil fertilization with a dose of 80 kg S$^-$ ha^{-1}. Application of elemental sulphur to the soil significantly decreases soil pH while foliar application did not influence this soil characteristic and foliar application of sulphur influenced neither the total, nor the sulphate content of sulphur in the soil (Kulczycki, 2010). Following timing and rates of S foliar application has been recommended for different crops (Table 16.3).

Table 16.3: Midwest Laboratories (1994) optimized following timing and rates of foliar S application in different crops (kg acre^{-1})

Crops	Growth stages	kg acre^{-1}
Tomato	First sign of bloom	0.18-0.36
	21 days after first spray	018-0.36
	14-21 days after second spray	0.18-0.36
Soybean	From R3 and two times at 7-10 days apart	0.68-0.68
Small grains	Just before tillering	0.23-0.45
	At tillering	0.23-0.45
	Post tillering	0.23-0.45
	Boot stage	0.23-0.45
Potatoes	6-10 inch plant height	0.34-0.45
	10-14 after first spray	0.34-0.45
	10-14 after second spray	0.34-0.45
Peppers, Eggplant and Okra	Bud formation	0.11-0.23
	10-14 after first spray	0.11-0.23
	10-14 after second spray	0.11-0.23
Grain Sorghum	Cold wet conditions (3-5 leaves)	0.09-0.18
	Normal conditions (5-leaves stage)	0.09-0.18
	7-10 days after first spray	0.09-0.18
	Reproductive growth stage	0.23-0.45
	Repeat 7-10 days	0.23-0.45
Corn	Cold wet conditions (3-4 leaf stage)	0.05-0.11
	Normal conditions (6-8 leaf stage)	0.05-0.11

(Contd...)

Crops	Growth stages	kg acre⁻¹
	7-10 days of first spray	0.05-0.11
	Early silk	0.23-0.45
Citrus	Early spring regrowth	0.34-0.45
	Late spray early summer	0.34-0.45
Avocados	Pre-bloom	0.23-0.23
	Post-bloom	0.23-0.23
	Post-bloom	0.23-0.23
Alfalfa	After each cutting - new growth reappears	0.23-0.34
	When alfalfa is of 6-8" height	0.23-0.34

Calcium (Ca)

Calcium plays an important role in maintaining quality of fruits and vegetables. Calcium treatment helps to retain fruit firmness, increase vitamin C content, decreased storage reserve breakdown and rotting and also decrease browning (Shukla, 2011). It performs important function in activation of a number of enzymes including cycle nucleotide phosphodiesterase, adenylate cyclase, membrane bound Ca^{+2}–ATPase and NAD-kinase. As the symptoms of calcium deficiency develop in plants, there is often a stage in which the tissues are water-soaked and one involving cell breakdown with loss of turgor, as in internal breakdown of apples. Eventually the tissue may become desiccated yielding a dry, more or less extensive area of necrosis. There is evidence that calcium deficiency renders membranes permeable which would account for a loss of turgor and permit cell fluids to invade intercellular spaces. An alternative situation may develop in soft, succulent fruits, the cells of which burst under hypotonic conditions *in vitro*. It is suggested that exogenous water may enter into fruit from the atmosphere or (in apple) through the phloem. Such exogenous water in the intercellular spaces of the fruit may cause cells to swell, so cracking the fruit or it may result in a bursting of the cells.

Foliar application of calcium is recommended for correction of fruit disorders of some crops. Calcium foliar spray applications were evaluated for reducing fruit disorder in waterberries. Calcium administered to the plant through the foliar feed has been reported to be important for resistance to bacterial wilt (Yamazaki and Hoshima, 1995) and fusarium crown rot resistance in tomato (Woltz *et al.*, 1992). Sprays containing calcium can also be used to prevent blossom end rot in tomatoes and other fruits, or tip burn in lettuce and cabbage which often occurs during dry periods. Ca and B which are phloem immobile should be applied in small amounts at frequent intervals rather than single application for correcting temporary deficiencies in vegetables (Maynard and Hochmuth, 1996). Kundu and Sarkar (2009) conducted a study on Gangetic alluvial soil to evaluate the effect of foliar calcium nitrate, $Ca(NO_3)_2$ and potassium nitrate (KNO_3) on growth and yield of rice (*Oryza sativa* L.) and observed that, foliar application of $Ca(NO_3)_2$ followed by KNO_3 during 50 per cent flowering stage increased the growth parameters and yield attributes, which ultimately resulted in higher grain yield.

Shukla (2011) applied three concentrations, 0.2 per cent, 0.4 per cent and 0.6 per cent of $Ca(CO_3)$ alone as foliar spray in gooseberry (*Emblica officinalis*). Other treatments were made by mixing of same concentrations of B (borax) and untreated one. Fruits under Ca+B, 0.4 per cent foliar spray indicated lesser incidence of fruit deformation, blossom and fruit drop than untreated plants and maximum yield was also found in this treatment. Although over all treatments increased the fruit weight, fruit size, number of fruit, shoot, and quality of fruits as compared to control, yet calcium carbonate + borax 0.4 per cent treatment produced superior results. Foliar application of calcium carbonate (0.4 per cent) significantly reduced the fruit drop and increased the retention of blossom and deformed fruit. In South Africa a number of foliar calcium formulations had been tested on the 'Pinkerton' variety in an effort to reduce internal physiological disorders associated with that cultivar (Penter and Stassen, 2000). An unexpected beneficial effect of Calcimax' on anthracnose disease reduction was found compared to the untreated one. Calcimax is described by the South African manufacturers 'Plaaskem', is a liquid organic calcium, with the calcium chelated to a carbohydrate and containing $100g^{-1}$ calcium and $5g^{-1}$ boron. Phytotoxicity was tested locally on avocados in New Zealand prior to implementing a grower trail. No phytotoxic symptoms were observed on trees sprayed 10 times the recommended concentration and at high water rates. Generally, following calcium containing fertilizers are used for fertilization (Table 16.4).

Table 16.4: Sources of calcium fertilizers

Material		*Ca (per cent)*
Calcitic lime	$CaCO_3$	40
Dolomitic lime	$CaCO_3 + MgCO_3$	22
Gypsum	$CaSO_4. 2H_2O$	22
Slaked lime	$Ca(OH)_2$	54
Ordinary superphosphate	$Ca(H_2PO_4)_2 + CaSO_4$	20
Triple superphosphate	$Ca(H_2PO_4)_2$	14
Calcium chloride	$CaCl_2$	36
Calcium nitrate	$Ca(NO_3)_2$	24

Peryea and Willemsen (2000) observed that calcium sprays may reduce the incidence of certain fruit disorders and may improve fruit quality. However, the relationships are not very precise. Physiological disorders such as bitter pit of apples, cork spot, alfalfa greening of Anjou pears, and cracking and firmness of cherries are often related to calcium content of the fruit. The most commonly used calcium spray material is calcium chloride, available as either food-grade product or specifically formulated for use as a foliar spray. Construction-grade calcium chloride contains impurities that can severely damage the fruit. Calcium chloride can cause leaf burn and fruit injury, and has limited compatibility with pesticides. Calcium nitrate has also been successfully used to reduce bitter pit of apple; however, it is more likely to cause fruit injury than calcium chloride. Calcium nitrate sprays applied at the rates and frequencies for bitter pit control will not improve green color of green apple varieties and may produce a duller red colour in red apple varieties. Foliar sprays of calcium sulphate may actually increase bitter pit and should not be used. Calcium-containing chelates and organic complexes have not been found to be more effective than calcium chloride. Only specifically labeled chelated organic complexes may be used for foliar application. Peryea and Willemsen, (2000) optimized following rates of $CaCl_2$ for foliar application (Table 16.5)

Table 16.5: Optimized rates of $CaCl_2$ for foliar application (Peryea and Willemsen, 2000)

Malakouti and Afkhami (2001) evaluated the impact of continuous calcium chloride (0.5 per cent) foliar application in apple to reduce diazinon and phosalon pesticide residues. Foliar sprays were carried out from early stage of apple fruit

Nutrient	Combinations	Amount per acre	Dilution (in 100 L)	Remarks and restrictions
Calcium (alfalfa greening of pear; cork spot of Anjou pear) Calcium (cherry fruit firmness and reduced cracking)	$CaCl_2$, dry, 34-36% Ca $CaCl_2$ liquid, 6-11% Ca $CaCl_2$, dry, 34-36% Ca $CaCl_2$ liquid, 12% Ca $CaCl_2$, dry, 34-36% Ca $CaCl_2$ liquid, 12% Ca	2.72-3.624 kg 3785.6 ml 3785.6 ml 2.72-3.624 kg 3785.6 ml 3.624-5.44 kg 3785.6 ml	0.68-0.91 kg 946.4 ml 946.4 ml 0.68-0.91 kg 946.4 ml 0.91-1.36 kg 946.4 ml	All products - apply five to eight applications from early June to late August. Dilute sprays are most effective. Can cause fruit injury. See text. Both products - apply four applications from early June to August. Dilute sprays are most effective and may be optimized. Limited effect and can reduce fruit size. Three or more applications are needed at weekly intervals before anticipated harvest.

formation and two weeks prior to harvest. The diazinon concentration reduced from 4.0 microgram per kilogram (ppb) to 3.0 ppb, and similarly, phosalon residues decreased from 1.24 ppb to 1.10 ppb after calcium application, which revealed that calcium chloride application left dual benefits in improving the quality of apples as well as reducing pesticides residues in an eco-friendly way. Calcium has a well-established role in strengthening the cell wall and thus in firmness of the fruits. The benefits of direct calcium application in reducing the incidence of physiological disorders have been reviewed by Poovaiah *et al.,* (1988). Post-harvest treatment of apple by dipping or infiltrating with calcium was found to maintain firmness during storage and also found a sigmoid relationship between concentration of calcium and increase in tissue strength (Abbott *et al.,* 1989). Conway and Sams (1987) found that Ca^{2+} was more effective than magnesium (Mg^{2+}) or strontium (Sr^{2+}) in increasing firmness. Pre-harvest calcium sprays also resulted in firmer fruit (Raese and Drake, 1993), which further confirms the involvement of calcium in maintaining texture. Sidiqui and Bangerth (1995) reported that pre-harvest calcium application in leaves might not always lead to firmer fruit at harvest, but may result in a better retention of firmness during storage. Mango fruits were sprayed with calcium chloride during pre-harvest at 0, 2.5, and 5.0 per cent to verify the influence of calcium on texture and activity of some enzymes such as polygalacturonase, methylpectinesterase, β-galactosidase and found that the fruits treatment with calcium chloride at 5.0 per cent presented firmer texture and less activity of the enzymes polygalacturonase and β-galactosidase (Evangelista *et al.*, 2000).

Magnesium (Mg)

Soil magnesium is derived from mineral biotite, phlogopite, hornoblende, olivine and serpentine. Mg is absorbed by plants as Mg^+ and absorption depend on the available Mg in soil and soil pH, CEC, K and Ca also determine the Mg uptake. Mg is a primary constituent of chlorophyll and accounts for 15 to 20 per cent of the total Mg content of plants Bybordi and Jasarat (2010). It has been observed that growth depressions and visible Mg deficiency symptoms occur if the Mg in the chlorophyll exceeds 20-25 per cent of the total Mg content in plants. This means that synthesis of plant components or metabolic processes which require Mg, includ-ing the formation of chloroplasts, is retarded or cannot fully operate. Vegetables such as beans, peas, lettuce and spinach can grow and produce good yields in soils with low magnesium levels, but plants such as tomatoes, peppers, roses etc. need high levels of magnesium for optimal growth. However, plants may not show the effects of magnesium deficiency until it is severe. Some common deficiency symptoms are yellowing of the leaves between the veins, leaf curling, stunted growth, and lack of sweetness in the fruit. Calcium and potassium compete with magnesium for uptake by plant roots, and magnesium often loses. Generally, first visible sign of Mg deficiencies appears in old leaves. Sometimes, a soil test will show adequate magnesium levels in soil, but a plant grown in that soil may still be deficient because of such ionic competition in those conditions foliar spray of epsom salts exhibited positive results in garden plants (Nardozzi, 2012). Flora Hydroponics Company, suggested foliar application of magnesium sulphate. Advantage of the practice is that soil applications of magnesium commonly take three years to correct magnesium-deficiency symptoms of such perennials as apple trees, whereas foliar sprays are effective within a few days after application. A foliar application of a 2 per cent solution of $MgSO_4$ to tomatoes, oranges, and apples has relieved magnesium deficiency and has increased crop yields. Following timing and rates of Mg foliar application has been recommended for different crops (Table 16.6).

Table 16.6: Midwest Laboratories (1994) optimized following timing and rates of foliar Mg application in different crops (kg acre^{-1})

Crops	Growth stages	kg acre^{-1}
Alfalfa (Hay)	After each cutting just as new growth reappear	0.50-0.90
	When alfalfa is 6-8 is high	0.50-0.90
	7 to 10 after spray no.2	0.50-0.90
Alfalfa (Seed)	Late bud stage	0.32-0.41
	10 to 20% bloom	0.32-0.41
	14-20 days interval starting at seed formation	0.32-0.41
Apples	Fruit development	0.23-0.45
Edible beans	Just after bloom	0.45-0.90
(Green, Lima,	2, 3 weeks after spray no. 1	0.45-0.90
Snapbeans Seed)	Pod set	
	10-14 after second spray	0.34-0.45
Kiwi fruit	Extension shoot growth	0.23-0.45
	Flower buds visible	0.23-0.45
Grain Sorghum	Cold wet conditions (3-5 leaves)	0.23-0.45
	Normal conditions (5-leaves stage)	0.23-0.45
	7-10 days after first spray	
	Reproductive growth stage	0.23-0.45
	Repeat 7-10 days	0.23-0.45
Corn	Cold wet conditions (3-4 leaf stage)	0.02-0.06
	Normal conditions (6-8 leaf stage)	0.02-0.06
	7-10 days of first spray	0.02-0.06
	Early silk	0.23-0.45
	7-10 apart	0.23-0.45
	Late spray early summer	0.34-0.45
Peas, Lentil	Just after bloom	0.45-0.90
	2-3 weeks after spray 1	0.45-0.90

Foliar spray rates are much less than soil application rates and it has been optimized that foliar application of 35 kg epson salt ha^{-1} was as good as soil application of 500 kg kieserite ha^{-1}. Magnesium sulphate sprays are recommended on some crops to correct magnesium deficiency (Jensen *et al.*, 1980) and it used in grapes as an interim corrective measure along with soil applications. It has also been used in some pulses. Sztuder and Swierczewska (1998) found that, foliar application of magnesium to peas at 6 - 8 leaf stage and just before flowering stage and increased the number of branches per plant and seed yield was observed. Muhammad *et al.* (2002) studied that, foliar application of magnesium as $MgSO_4$

(300 ml/m³) on lentil Cv-Mazoor 93 recorded the highest number of pods per plant, number of seeds per pod and more seed weight of 1000 seeds were recorded. Klacan (1997) studied the combined effect of seed treatment and foliar application of magnesium on seed yield of two garden pea Cvs Bohtyr and Smaragd and observed the marked increase in the number of pods per plant, number of seeds per pod and seed yield ha⁻¹ under combined treatments. Similarly, increase in seed yield due to magnesium application in garden pea was also observed (Rubes, 1984).

Similarly, Hanumanthareddy (1999) applied 10 ppm NAA in combination with MgSO$_4$ (1per cent) in cotton and revealed increased number of monopodial and sympodial branches of cotton, number of good bolls per plant, average boll weight, seed weight per plant, number of seeds per boll and ultimately highest lint yield. Shastri *et al.*, (2000) studied the effect of foliar nutrition on non aged and aged seeds of cotton Cv. MCU5 with 2 per cent MgSO$_4$, 2 per cent DAP and 0.5 per cent borax at 75 and 90 days after sowing. Seed cotton yield and seed yield of both aged and non-aged were significantly higher with 2 per cent DAP over control and other treatments. Namdeo *et al.*, (1992) reported that, the foliar application of Micnelf (1 per cent) at 30, 65 and 90 days after sowing increased the number bolls per plant but boll weight was increased by magnesium sulphate (2 per cent) and murate of potash (2 per cent) at 90 DAS. Micnelf (1 per cent) at 30, 65 and 90 DAS gave higher seed cotton yield, which followed by magnesium sulphate (2 per cent) at 30 and 60 DAS. Likewise, Chitdeshwari *et al.* (1997) observed that, foliar application of magnesium sulphate increased the seed cotton yield and decreased leaf reddening. Yallappa *et al.*, (2006) recorded more number of sympodial branches in cotton plant after foliar applications of MgSO$_4$ (1 per cent) at 60 DAS and boron (0.1 per cent) at 75 DAS. Similarly, highest number of flowers per plant, number of bolls per plant, average boll weight, seed cotton yield and seed yield were recorded with foliar application of MgSO$_4$ (per cent) at 60 DAS in combination with boron (0.1 per cent) at 75 DAS. Rajakumar and Gurumurthy (2008) recorded higher boll production, boll setting, boll weight and ultimately the seed cotton yield applying foliar application of magnesium and micronutrients under all plant densities. Likewise, synergistic effect of magnesium, zinc, copper and boron on seed cotton and lint yield was also observed by Namdeo *et al.* (1992).

Leaching of Mg is often a severe problem (35 kg per ha per year) and its deficiency occurs in the north-western soils of Iran, which are under the cultivation of grape (Havlin *et al.*, 2005). Bybordi and Jasarat (2010) evaluated the effects of the foliar applications of magnesium (Mg) and zinc (Zn) on the yield and quality of three grape cultivars. The highest yield was obtained with the combined foliar applications of Mg and Zn-fertilizers. Ghezel variety produced the highest yield among the three cultivars. The highest pH values of the fruits were obtained with the foliar application of Mg sulphate and it did not decrease the K/Mg ratio to the desirable level. For maximum yield and quality, soil application is necessary along with foliar application of Mg and Zn and also for lowering K/Mg ratio under Mg deficient soils. Nannette (2011) suggested foliar spray of magnesium sulphate hydrate (epsomite) in tomatoes, peppers and roses for maximum growth, yield and quality of the produce. Trolove (2007) suggested that foliar spray of Mg is a critical strategy to increase leaf Mg content and revealed that Mg nitrate at 1-2per cent increased leaf Mg concentration and did not cause leaf burn. Ceccarelli, (1987) pointed out that foliar application of Zn, K or Mg increased drought tolerance of mungbean plants and lowers the S value in the plants, which is an indication to drought tolerance. Similarly, fertilizer application of NPK, farmyard manure, magnesium sulphate and zinc sulphate treatment recorded the highest seed yield in niger (Paikray *et al.*, 2001).

(B) FOLIAR APPLICATION OF MICRO-NUTRIENTS

The term micronutrient also signifies that micronutrients are needed in small quantities for adequate plants' growth and production, often just 1-100 g ha^{-1}; however, their deficiencies cause a great disturbance in the physiological and metabolic processes in the plant (Bacha *et al.*, 1997). These are iron (Fe), manganese (Mn), copper (Cu), zinc (Zn), boron (B), molybdenum (Mo) and chlorine (Cl). Throughout the world microelements as Fe, Zn, Mn and Cu are added to foliar fertilizers, in order to compensate their deficiency especially in arid and semi arid regions (Kaya *et al.*, 2005). Micro-nutrients, especially Fe and Zn, act either as metal components of various enzymes or as functional, structural, or regulatory cofactors. Thus, they are associated with saccharide metabolism, photosynthesis, and protein synthesis (Marschner, 1995). Iron is mainly present in the form of insoluble Fe(III), therefore, unavailable to higher plants, particularly in neutral and alkaline soils (Shao *et al.*, 2007).

Micro-nutrients are sprayed about four weeks after emergence or transplanting. Because many micro-nutrients are not readily translocated within the plant, a second spray is required two weeks later to cover the new foliage and it is required in mid-season to correct deficiency (Vitosh *et al.*, 1994). Foliar spraying of micro-elements is very helpful when the roots cannot provide necessary nutrients (Kinaci and Gulmezoglu, 2007; Babaeian *et al.*, 2011). Now is known that plant leaves uptake some nutrients better than soil application and foliar spraying came in practice (Bozorgi *et al.*, 2011).

Sarkar *et al.*, (2007) showed that a small amount of nutrients, particularly Zn, Fe and Mn applied by foliar spraying increases significantly the yield of crops. Foliar spraying with suspension, micronutrient induced stimulatory effects on growth parameters and nutrients uptake either before or after the salinization treatments in wheat and it has been suggested that foliar spray with micro-nutrient may have a potential role for increasing wheat tolerance to salinity stress (El-Fouly *et al.*, 2011). Foliar spray of zinc, boron and copper micro-nutrients has been reported to be equally or even more effective as soil application to overcome micronutrients deficiency in subsoil (Ali *et al.*, 2009 and Hussanin *et al.*, 2012). Foliar application lead to increase in grain yield components and protein percentage in seed; for instance wheat, maize, rice, barley and sorghum showed increase in yield components by application of micronutrients (Boorboori *et al.*, 2012), especially in wheat, improved yield and yield components were found after foliar application of zinc, boron and copper. Micro-nutrients are absorbed into the plant roots in the following forms (Table 16.7).

Table 16.7: Forms of micro-nutrients taken up by plant

Element	Ionic form(s) taken up by plant
Boron(B)	$B(OH)_3$
Chlorine(Cl)	Cl^-
Copper(Cu)	Cu^{2+}
Iron(Fe)	Fe^{2+}, Fe^{3+}
Manganese(Mn)	Mn^{2+}
Molybdenum(Mo)	MoO_4^{2-}
Zinc (Zn)	Zn^{2+}, $Zn(OH)_2$

Treatment of micronutrient in combinations can also be done in most cases. Once a potential deficiency has been diagnosed, a single compound spray can be used to prevent or correct deficiency (Christensen, 1986). Sometimes diagnose in deficiency symptoms is confused then mixture of nutrients may be applied. Suggested micronutrient sources and spray rates are as under (Table 16.8).

Table 16.8: Sources, rates, and final concentration of the micro-nutrient for single foliar sprays for correcting micronutrient deficiencies.

Source	kg acre⁻¹	gm/100 L	Conc. (ppm)	per cent
Iron sulphate	0.5-1.0	30	62 Fe	0.2
Manganese sulfate	0.5-1.0	15	40 Mn	0.2
Zinc sulphate	0.15-0.35	15	56 Zn	0.2
Tri basic copper sulphate (53 per cent copper)	0.25-0.5	30	159 Cu	0.2
Sodium molybdate or	0.03-0.05	15	57 Mo	0.05
Ammonium molybdate	0.03-0.05	15	57 Mo	0.05
borax -11 per cent boron or	—	5.7	0.25 B	0.1
Soluble borate-20 per cent boron	—	0.13	0.25 B	0.1

Minimum 114 litre of water is used acre⁻¹

Do not apply combinations without first testing on a small number of plants. Use the same spreader- sticker product and rate with the above foliar sprays as used with insecticide and fungicide sprays

Source: North Carolina Cooperative Extension Service

Iron (Fe)

Iron is very essential nutrient for all the organisms and its deficiency in crops has been reported as problem world-wide, which depends on soil types (Welch, 1995). Arid, alkaline and calcareous soils are more likely to show iron-deficiency and it has been reported from India, Bangladesh and several African countries, Central and South America. Under anaerobic conditions excess availability of iron could lead to Fe-toxicity (Singh *et al.*, 2003). Fe is largely stored as ferric phosphoproteins called phytoproteins. Chloroplasts are richest organelle of Fe and may contain as much as 80 per cent of the total Fe in plants. Ferredoxins are another group of important Fe compounds that act in redox systems of photosynthesis, nitrate reduction, sulphate reduction and N_2 assimilation. It also involved in metabolism of protein, nucleic acid and lipids. Plants having Fe deficiency show accumulation of nitrate, amino acids and amides due to retarded protein synthesis. Iron has ability to form complexes with organic molecules and several anions. In soil solution Fe is surrounded by six molecules of water as $Fe(H_2O)_6^{3+}$. Increasing pH removes H^+ from $Fe(H_2O)_6^{3+}$ and gives, $Fe(OH)_2^{4+}$, $Fe(OH)_4^{5+}$, $Fe(OH)_2^+$, $Fe(OH)_3^{\circ}$ and $Fe(OH)_4^-$ ions. Thus pH has an important role in determining the solubility of Fe, which is quite low in the pH range 7.5 to 8.5, characteristic of calcareous soil. This explains why Fe-deficiency is most prevalent in calcareous alkaline soils. Soil application of iron-containing compounds to eliminate iron deficiency has not been economically possible in high pH fields. Because the problem of iron deficiency is associated with high pH and excessive calcium carbonate in the soils. Iron chlorosis of plant is essentially prevalent in alkaline soils in arid and semi-arid areas as some sandy soils having low organic matter (Mortvedt *et al.*, 1991). Synthetic chelates are generally most effective Fe sources for soil and foliar applications (Mortvedt *et al.*, 1991). Following iron containing compounds are used for the correction iron deficiency (Table 16.9).

Table 16.9: Common sources of iron fertilizer

Material		Fe (per cent)
Chelates	FeEDDHA	6
	FeDTPA	10
	FeEDTA	9-12
Ferrous sulphate	$FeSO_4.7H_2O$	20
Ferrous ammonium sulphate	$(NH_4)_2Fe(SO_4)_2.6H_2O$	20
Iron first	Glass (silicate) carrier	Varies
Organic iron complexes	Organic	6-11

Foliar application of water soluble iron containing compounds may be applied in alkaline soil conditions to meet out iron requirement to the crops. Foliar spray of ferrous sulphate (0.5 per cent) solution is required repeatedly if chlorosis symptoms persist. It is most effective when applied to young plants and repeated at 10-day to two-week intervals. After foliar application the highest re-greening was observed in plants treated with synthetic chelates and amino acid complexes. Fe-EDDS and Fe-EDTA performed in a similar way when applied in nutrient solution or as foliar sprays (Patricia *et al.*, 2010). Following timing and rates of Fe foliar application has been recommended for different crops (Table 16.10).

Table 16.10: Midwest Laboratories (1994) optimized following timing and rates of foliar Fe application in different crops (kg acre⁻¹)

Crops	Growth stages	kg acre⁻¹
Alfalfa (Hay)	After each cutting just as new growth reappear	0.05-0.09
	When alfalfa is 6- 8 high	0.05-0.09
	7 -10 days after spray no. 2	0.05-0.09
Alfalfa (Seed)	Late bud stage	0.05-0.14
	10-20per cent bloom	0.05-0.14
	14- 20 days intervals starting seed formation	0.05-0.14
Small Grains	Application in combination with2-4 days at or just before tillering	0.23-0.05
	At or before tillering	0.23-0.05
	Post tillering	0.23-0.05
	Boot stage	0.23-0.05
	At or before tillering	0.11-0.23
	10 -15 days after spray no.2	0.11-0.23
	Boot stage	0.11-0.23
Grain Sorghum	Cold wet conditions (3-5 leaves)	0.45-0.90
	Normal conditions (5-leaves stage)	0.45-0.90
	7-10 days after first spray	0.45-0.90
	Reproductive growth stage	0.23-0.45
	Repeat 7-10 days	0.23-0.45
Corn	Cold wet conditions (3-4 leaf stage)	0.11-0.23
	Normal conditions (6-8 leaf stage)	0.11-0.23
	7-10 days of first spray	0.11-0.23
	Early silk	0.11-0.23
	7- 10 days apart	0.34-0.45
Soybeans	For correction of chlorosis (stage v3-v4)whenfirst sign of chlorosis appear	0.68-0.90
	7 – 10 days spray no.1	0.68-0.90
	Growth stage R3-R5 two application and 7-10 days apart	
	Starting at growth stage R3 three application and 7-10 days apart	

One per cent solution of ferrous sulphate is prepared by adding 3.63 kg iron sulphate (20 per cent Fe) in 189.3 litre of water, including a surfactant. Thus any desired strength of solution may be prepared. However, ferric sulphate and ferrous sulphate are not effective as soil applications. Iron fertilization is recommended only for those field crops that are sensitive to low soil iron levels such as corn, sorghum, sudan, sorghum-sudan hybrids, beans and potatoes. Although a foliar spray produces quick results, the improvement is temporary because iron will not move into the tree beyond the tissue that was sprayed (Iron chlorosis in Trees, 2012). Basavarajappa *et al.*, (1997) reported that, foliar application of $FeSO_4$ (5 kg ha^{-1}) recorded significantly superior cotton yield followed by boron (soil), $CuSO_4$ (soil), $ZnSO_4$ (foliar), $CuSO_4$ (foliar) as compared to untreated field.

Foliar spray of iron is effective in restoring green color in plants, but may not restore top yields. Iron chlorosis in most of plants is essentially prevalent on alkaline soils in arid and semi-arid areas and also as in some sandy soils having low organic matter (Mortvedt *et al.*, 1991). Iron chlorosis of citrus could be controlled by foliar spray of ferrous sulphate or iron chelates and sulphur. Patel (2006) revealed that total iron content in leaves increased due to foliar application of iron which recovered iron chlorosis in citrus. The results were in agreement with those reported for citrus (Fernandez-Lopez *et al.*, 1993). It is the fact that the Fe foliar treatments increased the contents of total Fe by 3-4 times more than that of untreated, suggesting a decrease in total Fe content in leaves as a possible cause for chlorosis. The spray of 0.1 per cent iron as ferrous sulphate in 0.05 per cent citric acid solution proved more effective compared to other treatments in controlling these iron chlorosis in citrus. Several reports indicated the positive effect on fruit yield in citrus due to foliar application of iron (El-Kassas, 1984). Kansas State University, Agricultural Experiment Station and Co-operative Extension Service (2012, MF-718) suggested that if a rapid response is needed to correct a chlorotic condition in trees, a foliar spray with iron sulphate or iron chelate solution may be applied when the tree is in full leaf. A rate of 2.27 kg iron sulphate in 378.5 L of water is recommended for tree crops. Farahat *et al.*, (2007) reported positive effect of iron on chlorophyll content in *Cupressus sempervirens*. Total chlorophyll content of sugar cane leaf increased due to ferrous sulphate treatment (Rakkiyappan *et al.*, 2002). When iron becomes limiting, the chlorophyll synthesis slows down and the chlorophyll gets diluted due to continuous leaf expansion (Miller *et al.*, 1982). Terry and Low (1982) also reported close correlation between chlorophyll content of leaves and iron content.

Manganese (Mn)

Plant absorbs Mn in the form of Mn^{2+}, but it is easily oxidized to Mn^{3+} and Mn^{4+} forms. It is a constituent of water splitting enzyme associated with O_2 evolution and electron transfer in photosystem II of photosysnthesis. Magnesium deficiency can impair export of photosynthate from leaves (Hermans, 2004), and reduces photosynthetic efficiency (Skinner and Matthews, 1990). It influences auxin levels and maintains chloroplast membrane structure. Being a constituent of super oxide dismutase, it detoxifies or protects cells against the ill effects of superoxide-free radicals.

Camberato (2004) suggested fertilization strategies for overcoming Mn deficiency, which are dependent on soil pH and available methods of fertilizer application. In soils where pH is acidic Mn fertilizers can be applied as broadcast, banded, or foliar and residual Mn will be available in future seasons. At higher pH levels soil applications lose effectiveness, particularly when broadcast and residual value will be negligible. In high pH soils banded and foliar applications are preferred and any soil applications should be made as close to planting time as possible.

Rates of Mn application are highly dependent on the application method 4.5-6.8 kg Mn acre^{-1} broadcast to the soil, 1.4-2.3 kg Mn acre^{-1} banded near the crop row, or 0.45-0.9 kg Mn acre^{-1} applied to the foliage. Potential mineral sources are manganese sulfates, manganese chloride, or manganese oxides. It is also used in chelated forms such as manganese ethylenediamine tetraacetate, manganese diethylene triamine pentaacetic acid, and Mn-lignosulfonate for foliar application (Table 16.11).

Table 16.11: Fertilizer sources of Manganese

Material		Mn (per cent)
Manganese chelate	MnEDTA	12
Manganese sulphate	$MnSO_4. 3H_2O$	26-28
Manganese oxide	MnO	41-68
Manganese dioxide	MnO_2	63
Manganese chloride	$MnCl_2$	17
Manganese carbonate	$MnCO_3$	31
Manganese frits	Glass(silicate) carrier	10-25
Manganese complexes	Organic	7-9

Foliar application rates of Mn were found effective from 0.3-5.4 kg Mn ha^{-1} as $MnSO_4$ in 31-230 L water for correction of Mn deficiency in barley, corn, oats, onion, peanut, safflower, soybean and fruit trees. In most cases multiple foliar sprays were found better than a single application (Nayyar *et al.*, 1985). The choice of Mn fertilizer, its dose, stages of application is differ from crop to crop and these should be carefully determined by experimentation before making a recommendation to the farmers. Mishra *et al.*, (1985) recorded that the addition of Mn and B along with recommended doses of N, P, K increased seed yield and thousand seed weight in sunflower.

FFTC Taiwan, reported that manganese deficiency is common in leached tropical soils, particularly in calcareous soils derived from limestone. Legumes are particularly sensitive and the main symptom of manganese deficiency is chlorosis or yellowing between the veins of new leaves. Manganese deficiency limits legume production on black calcareous soils in Asian countries. Corn and sorghum are not sensitive to manganese deficiency in calcareous soils. Peanut, on the other hand, showed chlorotic symptoms in calcareous soils. Chlorotic peanut plants with severe manganese deficiency failed to form nodules until foliar applications of manganese were given. Porro *et al.*, (2002) pointed out that normally leaf drop is linked to Mn deficiency and examined the effect of foliar application of Mn in order to correct Mn deficiency and to prevent leaf drop in Golden Delicious apple. The effects of Mg and Mn on apple leaf drop are not well documented in literature and it is known that Mn is involved as cofactor in many reactions of plant metabolism. Compartmentation of Mn^{++} in cells could compete with Mg uptake; therefore, it can be strongly depressed by Mn cations and vice versa. The disorder is characterized by the development of necrotic blotches or irregular areas of dead tissue (brown blotches) in mature leaves. Foliar treatment of Mn in spring had higher leaf Mn concentration than autumn treated trees and untreated ones. Leaf drop incidence, leaf greenness and percentage of blotched leaves were statistically affected by foliar application. The spring combination was more effective in reducing leaf drop than autumn treatment. The greatest disorders were associated with low values of both Mn and Mg. The role of Mn seems to be more important in determining of leaf drop. Perveen (2000) observed that application of Mn in sweet orange significantly increased leaf Mn content, fruit yield and intensified the red colour of skin and juice.

Moosavi and Ronaghi, (2011) conducted a greenhouse study to find out the effect of soil and foliar applications of Fe and Mn on yield and Fe-Mn status of soybean plant. Both soil and foliar applications of Fe significantly increased shoot Fe concentration and uptake. However, foliar application was found more effective. Foliar spray of 1per cent Fe sulphate improved plant Fe content and had no effect on shoot Mn concentration. Soil addition of Fe decreased root Mn uptake probably due to the well-known antagonistic effect of Fe on Mn absorption. Whereas, foliar Fe application had no adverse effect on shoot Mn status. Similarly, also high level of soil applied Mn (*i.e.*, 30 mg Mn kg^{-1}) decreased Fe translocation from root to shoot. Foliar Mn applications caused greater increase in shoot Mn concentration and uptake than soil applications indicating that foliar Mn application is more effective in improving Mn nutrition status of soybean plants. Conclusively, foliar Fe/ Mn applications are appropriate methods of applying these nutrients for preventing yield reduction and nutrient imbalance in soybean grown on calcareous soils.

Likewise, Heckman, (2010) observed that Mn fertilizers applied to the soil are generally not cost effective and suggested Mn foliar spray during the growing season. Multiple applications of Mn sprays are often needed, especially in cases of severe Mn deficiency. The first Mn spray should be applied soon after emergence of the first tri-foliate soybean leaf, the second application at early bloom growth stage and the third during early pod growth stage. Delays in Mn application can result in substantial losses in crop yield. It was observed that both $MnSO_4$ (32 per cent Mn) and chelated Mn (5-12 per cent) were about equally effective in increasing seed yield of soybean, which exhibited severe Mn deficiency symptoms. Researchers in Delaware, Virginia, and North Carolina have shown that soybeans are very responsive to foliar Mn especially when applied before soybeans begin to bloom. Taylor, (2010) pointed out that the yield reductions of soybean can be avoided to a large degree by early diagnosis and treatment with foliar spray of Mn. Multiple applications of foliar Mn may be needed especially when Mn deficiency is severe. If enough leaf area is present to absorb adequate Mn, a single application higher rate (0.45-0.9 kg Mn acre^{-1}) was shown to be effective by Virginia and North Carolina researchers and if the crop is in the vegetative stage, mild to moderate symptoms can be alleviated with a 0.23 kg Mn per acre foliar spray.

Camberato (2004) recommended Mn foliar application in soybean before first bloom. Earlier foliar feeding to limited leaf area and mixing Mn fertilizers with glyphosate should be avoided. Foliar applications should be made immediately if deficiency symptoms appear and again if symptoms reappear. The most common foliar rate recommended is 0.045-0.9 kg Mn acre^{-1}, however, research with soybeans has shown the optimum rate to be much lower 0.45-0.9 kg Mn acre^{-1}. The lowest effective rate is preferred because of lower cost, less likely leaf burn, and ease in dissolving the fertilizer. The soil and foliar application of Mn significantly increased the yields, but the rates of soil applied Mn (40-50 kg ha[-1]) are uneconomical than its foliar sprays due to more reversion of soil applied Mn to higher oxide in alkaline soils (Manal *et al.*, 2010).

Likewise, Hong (2010) observed that when deficiency symptoms are visible in corn and soybean, foliar application of Mn is only way to quick correction of deficiency rather than band-application and Mn application can be combined with regular fungicide and insecticide sprays. Dehnavi (2008) studied the effects of foliar application of zinc and manganese on oil content and fatty acid profiles in fall-sown safflower cultivars under drought conditions and found that the zinc and manganese foliar applications significantly increased palmitic and oleic acids, whereas these foliar applications decreased linoleic acid per cent. Generally, drought stress during flowering stage imposed the most damage to oil per cent and yield and also decreased linoleic acid content and increased stearic and oleic acid content. Hence, flowering stage is more sensitive to drought than vegetative or grain filling stages and foliar application of zinc and manganese can compensate the adverse effects of drought on safflower.

Bohner (2007) evaluated various sources of foliar Mn, for beans but the most effective and economical was found manganese sulphate and recommended 2 kg of actual manganese ha[-1] from manganese sulphate (8 kg of manganese sulphate ha[-1]) in at least 200 L of water. Ontario research also suggested that equal rate of manganese sulphate, while Michigan trials have recommended half rate to get similar results. Bameri (2012) revealed that the application of Mn+Fe had the highest positive effect on yield components and grain yield.

AgTactics (2012) reported that foliar application of Mn at the 4 kg ha[-1] of Mn sulphate in 75-100 L of water in lupin plant seed production was found effective in increasing Mn concentrations in seeds above critical limits in most cases, but not all. The development stage of the seed at the time of spraying is critical. Spraying when the pods on the main stem were 2-3 cm long and the secondary stems nearly finished flowering have produced good results.

Copper (Cu)

Deep sandy and light loamy soils are much more likely to be Cu deficient than medium or heavy textured soils it is due to the parent material forming these soils may contain low Cu concentrations inherently. Some soil nutritional components also determine the availability of Cu in soil, such as organic matter, pH, high soil N and P. Copper strongly bound to organic matter and as organic matter increases, the probability of Cu deficiency increases. Similarly, soil pH increases so does the probability of Cu deficiency in such a way, soil pH increase of 1 unit (between pH 7 and 8) then 100-fold reduction in Cu availability to the crop. High soil test values for P and N are often associated with Cu deficient soils and high N levels delay the translocation of Cu from older leaves to the growing points or may tie up the Cu in protein-like compounds in the roots, thus contributing to Cu deficiency. Cu soil applications are more common since its single application can be effective for many years. In addition a higher concentration of Cu adversely affects plant growth characterized by fewer leaves, lesser chlorophyll content and shorter roots (Xiong *et al.*, 2006).

Copper exists in plants as cuprous Cu(I) and cupric(II) oxidation states and is taken by plants as Cu^{2+}. The plants contain concentration ranging between 5 and 20 mg kg^{-1}. The critical copper concentration in the shoot tips usually ranges from 1-1.5 mg kg^{-1} (dry weight), depending on the crop. It is constituent of several proteins or enzymes that perform important biochemical functions. Since the copper is immobile in plants, hence its deficiency symptoms appear on the young mature leaf and even on the emerging one. In Cu deficient soil, plants show stunted growth and foliage turn pale green with interveinal pale yellow chlorosis at the base of the blade. In cereals, Cu deficient condition is popularly known as white tip or reclamation disease and deficiency causes male sterility, delay in flowering and loss of grain yield. *For example*, complete fertilizer treatment of peanut in Thailand depressed copper concentrations in the leaves. Following some Cu containing fertilizers are used as foliar as well as soil applications (Table 16.12).

Table 16.12: Common sources of copper fertilizer

Material		Cu (per cent)
Copper chelate	Na_2Cu EDTA	13
Copper sulphate	$CuSO_4.5H_2O$	25
Cupric oxide	CuO	75
Cuprous oxide	Cu_2O	89
Copper complexes	Organic	Variable

Copper has been used as a fungicide even before it was recognized as an essential plant nutrient (Graham and Webb, 1991). Copper fertilization is reported to decrease the severity of some fungal and bacterial diseases. However, as a fungicide its concentrations required are 10-100 times greater than those normally needed as foliar spray (0.1-0.2 kg ha^{-1}). (a) Suppression of air borne diseases like; *Pseudomonas eichori* on ginsenz, *Puccinia triticiana* on wheat, Ergot on rye and barley, *Pyricularia oryzae* on rice and Septoria on wheat. (b) Suppression of soil borne diseases such as; *Heterodera* on sugar beet, *Vertcillinum alb-atrum* on tomato, *Vertcillinum dahliae* on cotton, *Strptomyces scabies* on potato and take-all (*Gaeumannomyces graminis* var. tritici (Ggt) on wheat.

Foliar application of Cu is an emergency measure and rates may vary from 0.09-4 kg Cu ha^{-1} as copper sulphate and its reasonable rate has been 1.1 kg Cu ha^{-1} in 280 liters of water. When copper sulphate is foliarly applied, it is recommended to make it basic by buffering with $CaCO_3$ (0.5 kg ha^{-1}) to reduce scorching of leaves. In severe deficiency conditions two spray of Cu have been recommended. Varvel (1983) suggested 2 foliar applications of 0.3 kg Cu ha^{-1} as $CuSO_4$ the first when the symptoms of deficiency initially appear and the second about 2 weeks later. In wheat two foliar applications of 2 kg Cu ha^{-1} as $CuSO_4$ is recommended, the first at mid-tillering and the second at booting (Grundon, 1980). Generally lower levels are required when Cu-chelates are used as foliar sprays. Haq *et al.*, (1994) found increased cotton seed yield in the order Cu>Zn>Fe>Mn>B trace elements applied together. Hosny *et al.*, (1984) observed that, foliar application of boron and copper with different concentrations increased the cotton seed yield and number of bolls per plant but boll weight was unaffected by boron or copper application. Similarly, AAFRD (1999) postulated that a foliar application corrects Cu deficiency if applied during the late tillering stage. Following responses of various crops to copper fertilizer in copper deficient soils were observed (Table 16.13).

Table 16.13: Responses of various crops to copper foliar fertilization in copper deficient soils

Crop	Response
Wheat	High
Flax	High
Barley	Medium-high
Oats	Medium
Corn	Medium
Peas	Low-medium
Clovers	Low-medium
Canola	Low
Rye	Low
Forage grasses (hay)	Low

Source: Department of Soil Science, Faculty of Agriculture, University of Manitoba

Agrichem (2007) observed that as the copper supply is increased; grain yield rises sharply, whereas the straw yield is only slightly enhanced. Oxides have a much lower salt index than sulphates and therefore are safer to the plant if being used as a foliar spray. The lower salt index also means oxides are much less corrosive than sulphates and therefore less harsh on spray equipment. Korzeniowska and Ewa (2011) pointed out that foliar application of copper in Jelcz-Laskowice did not cause any significant increase in Cu content in the wheat grain. But in some varieties almost 20-23 per cent increases in yield was found after Cu foliar application. Flora hydroponics Company postulated that copper deficiency can be controlled by spraying the leaves with a mixture of eight pounds of $CuSO_4$ plus eight pounds of $Ca(OH)_2$, in 100 gallons of water. Without the calcium hydroxide, the copper sulphate injures the foliage. Copper oxide has also been used successfully as a spray. A higher concentration of Cu adversely affects plant growth characterized by fewer leaves, lesser chlorophyll content and shorter roots (Xiong *et al.*, 2006). Similarly, Brennan (2005) studied the efficacy of Cu foliar application in wheat and found the separate spraying of one kg ha[-1] of copper sulphate, copper oxychloride and copper chelated at the 6th leaf stage can enhance maximum yield. The trials confirmed that the recommended rates of 1-2 kg ha[-1] of copper sulphate spray are highly effective and cause negligible leaf damage. When wheat was sprayed at the 6th leaf stage, yields were higher than if spraying was delayed until the late boot stage. This is because the pollen in wheat heads can be infertile if copper is deficient. Pollen fertility is important in determining the number of grains in each head, which means, it is critical to the final yields. Likewise, findings were also reported as foliar applied Cu chelates or Cu sulphate to correct Cu deficiencies in the growing season of cereal crops. They are best applied after elongation starts (first node visible) to flag leaf fully emerged stage to be effective in restoring seed yield. Foliar rates should be no less than 0.081 kg and no greater than 0.14 kg actual Cu acre^{-1}. Copper applied at the four leaf stage or at heading was ineffective in restoring yield (Micro-nutrients in Crop Production, 2006).

The chelated copper always resulted in the highest yield per quantity of copper used and copper oxychloride always resulted in the lowest yield. *For example,* 200 g ha^{-1} of copper as copper chelate is required to enhance maximum yield and to compensate same yield or 400 g ha^{-1} of copper sulphate will be required. Copper oxychloride appears to cause less leaf scorching than copper sulphate. Since similar quantities (about 1 kg spray ha^{-1}) of each of the compounds were needed for maximum grain yields. Chelated copper costs four to five times as much as copper sulphate and copper oxychloride costs about twice as much as copper sulphate.

Peryea and Willemsen, (2000) suggested postharvest foliar application of 0.45 kg of Cu acre^{-1} in fruit trees as copper sulphate or basic copper sulphate usually sufficient for the correction of symptoms. If symptoms are severe, mid-season sprays of copper chelate or basic copper sulfate, copper sulphate products can be applied but may cause foliage and fruit injury. Majid and Ballard 1990) optimized (0.1per cent) safe foliar application rate of copper sulphate for lodgepole pine trees and the application of 1per cent $CuSO_4$ was extremely toxic to the trees. It was also indicated that the critical value for the Cu was found to be 4 ppm and Cu toxicity in lodgepole pine might occur whenever foliar Cu concentration exceeds 17 ppm. Foliar feeding of urea did not seem to have any physiological interaction with foliar Cu.

Zinc (Zn)

Plants absorb Zn as Zn^{2+}, which occurs only as Zn(II) oxidation state. Zinc is constituent of carbonic anhydrase, alcoholic dehydragenase, superoxide, dismutase, lactase dehydrogenase, aldolase, phosphatase, DNA-RNA polymerase etc. Zinc is also involved in the synthesis of RNA, IAA, and metabolism of gibberelic acid. Zinc deficiency is observed in crops growing in high soil pH, and $CaCO_3$ or low in organic matter or waterlogged/ submerged rice soils. The solubility of zinc in soils and its uptake by plants falls rapidly as the soil pH increases. Sometimes zinc may be present in the soil, but not available to the plants due to high pH. Because zinc being less soluble in calcareous or in high soil pH conditions. Consequently, crops under these soil conditions may suffer from zinc deficiency. Fruit trees deficient in zinc may have a growth at the end of their shoot tips which looks rather like a rosette. Citrus trees often show chlorosis between the veins of the leaves, a condition sometimes known as "mottle-leaf" (Chang, 1999). Repeated applications of phosphate fertilizer and high levels of phosphorus in soils have been known to make zinc deficiency in a number of crops (Forth and Ellis, 1997, Chang, 1999). Following zinc bearing fertilizers are used to correction of Zn deficiency in soil as well as in foliar applications (Table 16.14).

Table 16.14: Fertilizer sources of zinc

Material		Zn (per cent)
Zinc sulphate	$ZnSO_4.H_2O$	36
Zinc oxide	ZnO	77-80
Zinc oxysulphate	oxide plus sulfate	varies
Zinc chelate	Na_2Zn EDTA	14
Zinc polyflavonoid	organic	10
Zinc lignisulphonate	organic	5
Zinc frits	glass(silicates) carrier	varies
Zinc-ammonia complexes		10

To overcome Zn deficiency in various soils, foliar spray of zinc is the most common because it is the most widely deficient micronutrient in various parts of the world. Treatment can also be quite effective if the correct methods of application are used (Christensen *et al.*, 1982). Neutral zinc (52 per cent Zn) and zinc oxide (75 per cent Zn) are the most economical and effective compounds on a recommended label basis (Christensen, 1986). There is no advantage in using chelated zinc products in sprays. They were originally intended for soil application, are more expensive, and less effective than neutral zinc and zinc oxide on a label recommended basis. Since zinc uptake is greatest with dilute application as compared to concentrate one. The optimum timing to influence fruit set is three weeks before bloom up to bloom (Christensen, 1980). Patil and Somawanshi (1983) had observed increased chlorophyll in leaf as a result of foliar feeding of Zn; it may be due to its role in increasing the rate of phytochemical reduction (Kumar *et al.*, 1988).

Essential oil content of *Mentha piperita* was found increased by 28.2per cent by foliar application of 3 ppm zinc chloride (Akhtar *et al.*, 2009). Similarly, essential oil biosynthesis in basil *(Ocimum sanctum* L.) is strongly influenced by Zn foliar application (Misra *et al.*, 2006). Misra and Sharma (1991) reported that zinc foliar application stimulated the fresh and dry matter production, essential oil and menthol concentration of Japanese mint. Foliar spraying with zinc (100 ppm) in blue sage (*Salvia farinacea* L.) enhanced the length of peduncle, length of main inflorescence, number of inflorescence and florets, and fresh and dry weight of inflorescences/ plant (Nahed and Balbaa, 2007). Yousef *et al.*, (2010) concluded that foliar application of Fe and Zn can considerably improve the flower and essential oil yields of chamomile, particularly if these micronutrients were applied together at both stages of stem elongation and flowering.

Pandey *et al.*, (2010) studied growth parameters and yield attributes of mungbean under induced salinity conditions and observed minimized adverse effects of zinc deficiency after Zn foliar application as zinc sulphate. Bashir (2012) reported foliar application of zinc at 300 ppm in optimum water availability increased maize grain yield. Potarzycki and Grzebisz (2009) conducted a study to evaluate foliar fertilization of zinc in maize and found that maize crop responded significantly to zinc foliar applications at the 5th leaf stage. The optimal rate of zinc foliar spray to achieve increased grain yield response was in the range from 1.0-1.5 kg Zn ha^{-1}.

Grain yield increase was found circa 18 per cent as compared to the NPK fertilizers applied. It was also observed that plants fertilized with 1.0 kg Zn ha^{-1} increased both total N uptake and grain yield. The number of kernels per plant showed the highest increase (17.8 per cent) as compared to the NPK plot and zinc foliar application also increased the length of cob and shape.

Ozturk *et al.* (2006) observed that large increases in loading of zinc into wheat grain can be achieved when foliar Zn fertilizers are applied to the plants. Similarly, Wheat is grown in Brazil, mostly without tillage under such system zinc can become potentially deficient, due to excessive application of acidity corrective and phosphate fertilizers in surface and, or at shallow depths. Zoz (2012) evaluated the effect of foliar application of zinc in agronomic characteristics and yield of wheat and found foliar application of zinc increased the number of fertile tillers and wheat yield. It has been documented that zinc foliar application is a simple way for making quick correction of plant nutritional status, similar has been reported for wheat (Erenoglu *et al.*, 2002) and maize (Grzebisz *et al.*, 2008).

Significant increases in grain yield with foliar Zn application has been reported in other crops such as rice (Cakmak, 2008) and common beans (Teixeira *et al.*, 2004). The nutritional requirement of zinc is specific to each culture (Motta *et al.*, 2007), but cereals are more responsive. Zinc application to leaves increased the number of fertile tillers and wheat yield; however, it had no effect on the agronomic characteristics of crop in no-till system with high nutrient content in soil (Zoz, 2012). Significant increases in wheat grains yield to foliar Zn application were also reported in other countries including Egypt (Seadh *et al.*, 2009; Zeidan *et al.*, 2010) and China (Zhao *et al.*, 2011).

Bybordi and Malakouti (2003) reported that wheat is sensitive to zinc deficiency, but less sensitive to Iron and copper deficiencies. The most effective method for increasing Zn in grain was the soil + foliar application method that resulted in about 3.5-fold increase in the grain Zn concentration. The highest increase in grain yield was obtained with soil + foliar and seed+foliar applications (Yilmaz *et al.*, 1997). Timing of foliar Zn application is an important factor determining the effectiveness of the foliar applied Zn fertilizers in increasing grain Zn concentration. It is expected that large increases in loading of Zn into grain can be achieved when foliar Zn fertilizers are applied to plants at a late growth stage. Ozturk *et al.*, (2006) studied changes in grain concentration of Zn in wheat during the reproductive stage and found that the highest concentration of Zn in grain occurs during the milk stage of development. In practical agriculture, it is known that foliar uptake of Zn is stimulated when Zn fertilizer is mixed with urea (Mortvedt and Gilkes, 1993). Habib (2009) pointed out that foliar application of Zn and Fe increased seed yield and its quality compared with untreated and among treatments, application of (Fe+Zn) obtained highest seed yield and quality.

Agrichem (2007) shown that the combination of zinc as seed dressing followed by a foliar application in cereals increased approximately 40 per cent yield, where zinc availability was limited. Because the primordial cells of the inflorescence head are set by the 4-5 leaf stage in cereals, zinc needs to apply by this stage in order for a yield response to occur. Application at a later growth stage will result in greening of the crop, but not yield enhancement. Where the crop shows signs of deficiency after the 6 leaf stage, application of zinc will prevent further yield loss. Zn foliar application at early stages has also been recommended in case of wheat to get maximum yield.

Oliver *et al.*, (1997) examined the effectiveness of foliar applications of zinc sulphate to decrease cadmium (Cd) concentration in wheat grain at 3 field sites in South Australia, Tumby Bay, Cummins and Keppoch. Foliar zinc treatments were found to decrease Cd concentrations in grain at only 1 site, Tumby Bay. At this site the highest foliar Zn treatment (0.67 kg Zn ha^{-1}) was given, which consisted of two applications of 0.33 kg Zn ha^{-1} during early and late crop season, then decreased the mean Cd concentration was observed *i.e.* 0.017 mg kg^{-1} in treated compared to untreated (0.025 mg kg^{-1}). Timing of application of foliar Zn had no significant effect on Cd concentration in wheat grain. Whereas, the effect of soil applications of zinc sulphate on grain Cd concentration was assessed at Tumby Bay only. There was no significant difference in grain Cd concentration between the soil Zn treatments. This is most likely due to the recommended foliar rate of 0.33 kg Zn ha^{-1} not providing excess Zn to the plant such that there is enough Zn to be translocated to the root, which is the site of Cd uptake by the plant. The results suggested that the benefits of foliar Zn to minimize Cd concentration in grain are variable or that the rates used to correct Zn deficiency under field conditions are too low to decrease Cd uptake.

Ullagaddi (2000) indicated that, the foliar spray of $ZnSO_4$ (0.1per cent) plus boron (0.1per cent) in combination with gibberellic acid-3 (GA3) (50 ppm) significantly increased the number of squares, flowers and matured bolls per plant and highest seed yield, shoot length, root length and vigour index in cotton. Similarly, Rathinavel *et al.*, (1999) reported that foliar application of zinc sulphate (0.5 per cent) on 90 and 110 days after sowing increased 100 seed weight, root length, shoot length and vigour index of cotton. The values for the electrical conductivity of seed leachate did not show consistency for treatments. Similarly Rashid and Rafiq (2000) obtained 14 per cent increase in cotton yield using B and Zn foliar application in 15 cultivation sites, while in another study (Rashid, 1996) cotton crop responded quite favourably to Zn fertilization in the fields with low native Zn. Soomro *et al.*, (2001) observed the highest seed cotton yield, number of bolls and boll weight when B and Zn were applied alone or in combination over untreated. Ali *et al.*, (2011) applied Zn and B as foliar application in cotton along with recommended dose and found higher cotton seed yield. Combined foliar treatment of Zn, K and P could bring about better impact on cotton productivity and quality in comparison with the ordinary cultural practices adopted by Egyptian cotton producers. The impact in cotton productivity and quality due to the application of K, Zn and P are believed to be sufficient enough to cover the cost of using those chemicals and to attain an economical profit. However, it is worthy to mention that about 40 per cent increased yield of lint and seed may secure an increase of gross income (Zakaria, 2008).

Boron (B)

Plants absorb boron in the form of boric acid (H_3BO_3) and under high pH conditions also as H_2BO_3 boron affects at least 16 functions in plants such as, flowering, fruiting, cell division, water relationships and the movement of hormones, transport of sugars, synthesis of cell wall and its structures, lignification, metabolism of phenols, carbohydrates, IAA, RNA, pollen germination and the growth of the pollen tube, regulates opening of stomata and imparts drought resistance to crops.

Generally, B-sufficient plants contain 10-200 mg B kg^{-1} dry matter. Its critical limit ranges between 20-70 mg kg^{-1} in dicotyledons and 5-10 mg kg^{-1} in graminaceuos plants. Boron is immobile in plants and hence its deficiency symptoms develop on new tissues or young leaves as white or transparent lesions in the interveinal mid-areas of the leaf. With the severity of its deficiency, the lesions join together and extend beyond the mid-section while the veins remain green. Under acute deficiency the leaves become thick, curl become quite brittle, the internodes become short and shoot tips die resulting in the growth of lateral shoots, the tips of which also die and the plants or trees form bushy appearance. Also plants develop poor root growth, defective inflorescence and even flowers do not form. The deficiency symptoms are popularly known as brown heart or heart rot in sugarbeet and turnip, browning of curds and hollow stem in cauliflower and external and internal cork in apples. The amount of boron needed to correct a deficiency is fairly small. Boron deficiency in crops is more critical in highly calcareous soils, sandy leached soils, limed acid soils or reclaimed yellow or lateritic soils. In a soil of Thailand with a boron level of only 0.12 mg kg^{-1} of hot water soluble boron (HWS-B), the application of 4 kg ha^{-1} borax was enough to prevent deficiency in early crops of black gram grown during the rainy season. However, if the boron-deficient fields were cropped continuously, the residual effect of this small application was found short-lived. By the third successive crop, it could no longer prevent boron deficiency. If higher borax rates of 10-20 kg ha^{-1} are applied, the residual effect lasted for ten crops. Thus, Boron must be available entire life of the plant. It is not translocated and is easily leached from soils. Hence, foliar application is an appropriate approach for its deficiency correction. It is most effective when applied as a foliar spray. Boron is important for ovule development, pollen tube growth and fruit set. Following boron bearing fertilizers are used to correction of B deficiency in soil as well as foliar applications (Table 16.15).

Table 16.15: Sources of boron fertilizers

Material	B (%)
Borax	11
Boric acid	17
Sodium pentaborate	18
Sodium tetrabrate	
Fertilizer borate-48	
Fertilizer borate-65	1420
Solubor	20

Foliar applications may sometimes be more effective than applying boron to the soil. Two foliar sprays of borax of only 50 g ha^{-1} each, applied at the strategic times of flower development and pod set, were as effective in correcting boron deficiency in black gram in the Philippines as a higher rate applied to the soil. Boron deficiency symptoms may sometimes appear for the first time when crops are suffering from water stress. When boron deficiency symptoms of citrus in Taiwan were observed in a drought year, the boron content was less than 10 mg kg^{-1} in the fruit peel and leaves. However, the boron content in the same orchard, when precipitation was abundant was 20 mg kg^{-1} in leaves and 14 mg kg^{-1} in peel, and no boron deficiency symptoms were observed. In Taiwan, a foliar spray of 0.3 per cent boric acid solution was also effective in citrus. However, a considerable amount of boron accumulated in the soil when trees were sprayed each year. For this reason, it is recommended that a foliar spray should not be applied every year. A water-soluble boron content of 1-1.5 mg kg^{-1} in the soil is regarded as the critical level for pineapple in Taiwan. The soil boron level should be checked after the first good rain.

Kuruppaiah (2005) stated that, the foliar application of borax (0.5 per cent) at 35 and 65 DAT was found to be best in terms of number of flowers per plant, number of productive flowers per plant, number of fruits per plant, individual fruit weight and yield, flowered by copper sulphate (0.5 per cent) and zinc sulphate (0.5 per cent) sprayed at 35 and 65 DAT in brinjal cv. Annamali. Dongre *et al.*, (2000) recorded the maximum fruit yield per plant and ha^{-1} in chilli, when sprayed 0.25 per cent boron as compared to untreated further recorded the highest average fruit length and fruit diameter when sprayed with 0.1per cent H$_3$BO$_3$. Sangale *et al.*, (1981) observed that, the foliar application of 0.2 per cent boron on safflower at 60 and 90 days of crop age increased the 1000 kernel weight, seed and oil yield. Singh *et al.*, (1990) reported that, in potato var. Kufri Chamatkar, foliar spray of 0.3 per cent boric acid along with 2per cent urea in conjunction with top dressing of 30 kg N ha^{-1} resulted in higher tuber yield compared to untreated. The 1000-seed weight and germination percentage were found to be higher with the foliar spray of boron in radish crop compared to untreated (Sharma *et al.*, 1999).

Roberts *et al.*, (2000) observed that boron deficiency in cotton may be corrected with foliar or soil boron applications. Foliar application of boron at the 0.11 kg ha^{-1} and soil application of boron at the 0.56 kg ha^{-1} provided about the same net returns. Sinha *et al.*, (2001) studied that, the spraying of boron concentration of (0.33 mg L^{-1}) significantly increased the seed yield, weight of 100 seeds and fibre within cotton. Whereas, low (0.0033 mg L^{-1}) and excess (3.3 mg L^{-1}) boron reduced the weight of fibres, biomass, seed yield and the content of starch, proteins and oils. Silva *et al.*, (1982) reported that, application of boron at the 1.23 kg ha^{-1} increased seed cotton yield, boll weight and induced earliness. Khodzhaev and Stesnyagina (1983) reported the spraying of plants at the flowering stage with a mixture of 0.02 per cent boric acid and 0.1 per cent zinc sulphate markedly increased heat and drought resistance during the period of moisture stress in cotton. Stoyanow and Gikov (1990) reported that the foliar application of B, Zn and Mn at different growth stage increased seed cotton yield. Similarly in tomato, increased number of fruits and fruit yield per plant were recorded with foliar application of boron at the rate of 4 ppm compared to untreated (Umajyothi and Shanmugavelu, 1985). Malik *et al.*, (1990) reported significant increase in seed cotton yield, boll weight and seeds per boll with B fertilization in Multan.

Malewar *et al.*, (1992) concluded that, the foliar application of phosphorous through bronated single super phosphate was beneficial in increasing yield and uptake of phosphorus and boron in the cotton and groundnut. Patil and Malewar (1994) reported that, foliar application of 0.2 per cent boron in combination with Fe and Zn in cotton proved its superiority in the production of chlorophyll 'a' and 'b'. Wankhade *et al.*, (1994) revealed that, foliar spray of 0.1 per cent borax at peak square and peak flowering stages recorded significantly more seed cotton yield over control. Similar results were also recorded with respect to bolls per plant and seed cotton yield. Jinfeng (1995) reported that, spraying boron as borax or boric acid with 0.2 per cent at the seedling stage, early flowering and boll formation stage increased the yield of 16.1 per centover control.

Rajeswari (1996) reported that, foliar application of 0.5 per cent boron increased the number of bolls per plant, mean boll weight and lint yield. Similarly, Carvalho *et al.*, (1996a) observed that, boron applied to the cotton Cv. IAC-17 side dressing by 0.75 kg and 0.15 kg or by foliar application of boron at the early growth stages increased yield and fibre length. Similarly, application of boron at planting time and foliar spray at 45-80 days increased yield and fibre length (Carvalho *et al.*, 1996b). Zhu *et al.*, (1996) concluded that, the foliar application of boron at seedling and internode elongation stages gave better results than seed treatment or basal application. Concentration of boron in the spray solution in the range of 0.1-0.25 per cent increased seed yield significantly in rape, 0.2 per cent being the optimum concentration with a 17.8 per cent yield increase over the untreated. Bowszys (1996) reported that, foliar spray of boron (0.4 kg ha^{-1}) at the bud stage significantly increased the seed yield and seed oil content in rape. Application of boron as foliar spray significantly increased number of seeds per siliqua, 1000 seed weight, seed yield, seed-oil content and oil yield in toria (Bora and Hazarika, 1997). McConnel *et al.*, (1992) found that, foliar application of boron at different growth stages increased the seed cotton yield and lint yield significantly. Anter *et al.*, (1978) reported that, foliar application of boron at 10, 20 or 40 ppm applied at various growth stages and increased fibre strength and fibre maturity were found but it decreased the fibre length. Howard *et al.*, (2000) found that foliar application of boric acid increased the cotton yield up to 10.3 per cent compared with check.

Molybdenum (Mo)

Plants absorb Mo in the form of MoO_4^{2-}, it exists as Mo (VI) in plants under oxidative conditions and undergoes reduction to Mo(V) and Mo(IV) forms. Molybdenum is a structural component of the enzyme that reduces nitrates to ammonia. Proteins synthesis is blocked in absence of Mo. It is an essential constituent of many enzymes such as nitrogenase, nitrate reductase, xanthine oxidase and dehydrogenase and it is also important element for nitrogen fixing bacteria in root nodules. Thus Mo performs biochemical role in plants. In case of legumes, a shortage of molybdenum affects the nitrogen-fixing activities of soil microorganisms; so that the plants are looked like N-deficient. Bell *et al.*, (1990) observed a correlation of Mo concentrations in leaves and nodules with the shoot dry weight and nitrogen content in several legume crops, including peanut, soybean, green gram and black gram. The correlation may be used to establish critical concentrations for the diagnosis of molybdenum deficiency. The critical values for nodules were much higher than the critical values for leaves. Symptoms of molybdenum deficiency resemble those of nitrogen deficiency, *i.e.* overall chlorosis, stunted growth, and low yield, yellow-spot of citrus, "blue-chaff" of oats, and "whiptail" of cauliflower are some examples of Mo deficiency. Depending on the kind of crop, critical deficiency levels of molybdenum range from 0.1-1 mg kg^{-1} (Chang, 1999). Molybdenum deficiency in peanut has been observed in the acidic soils of many Asian countries, including Thailand, Japan, Korea, Philippines, and Taiwan. Following molybdenum bearing fertilizers are used to correction of Mo deficiency in soil as well as foliar applications (Table 16.16).

Table 16.16: Fertilizer sources of Molybdenum

Material		*Mo (%)*
Sodium molybdate	$Na_2MoO_4.2H_2O$	39
Ammonium molybdate	$(NH_4)_6Mo_7O_{24}. 2H_2O$	54
Molybdenum trioxide	MoO_3	66
Molybdenum sulphide	MoS_2	60
Molybdenum frits	silicate	2-3

Foliar application of molybdenum are most effective if applied at early stages of plant development, and generally a 0.025-0.1 per cent solution of sodium or ammonium molybdate (~200g Mo ha^{-1}), is recommended for spray. Common foliar spray rates range from 0.01-0.07 kg acre^{-1}. Since a lower amount of Mo is required, this eliminates larger soil application that may lead to soil pollution (Kotur, 1995). Foliar feeding of molybdenum are often more effective than soil application, particularly for acid soils or under dry conditions (Hamlin, 2006). Ziolek and Ziolek (1987) found that foliar application of boron, manganese and molybdenum and their combination increased number of pods, seed per plant, hundred seed weight and seed yield of soybean.

Kotur, (1995) conducted a study to standardize foliar application of ammonium molybdate in cauliflower in Mo deficient fields and a quadratic relationship was observed between curd yield and Mo content of leaf. Maximum curd yield and quality was found when Mo concentration in the leaves was 6.73 µg g^{-1}. Similarly, Wojciechowska, (2011) studied the effect of foliar treatment of urea (1per cent)+Mo (1mg dm^{-3}) + BA (benzyladenine, (5 mg dm^{-3}) on lettuce crop, during summer-autumn growing season, when lettuce leaves accumulated more nitrates as compared to the spring cycle. A combined foliar treatment of urea, Mo and BA, simultaneously decreased nitrates and increased soluble sugars in leaves. Hristozkova *et al.* (2006) studied the effects of foliar absorbed nutrients on root processes related to assimilation of nitrogen under presence or absence of molybdenum (Mo) in pea plants (*Pisum sativum* L.) and observed that foliar application of Mo had a positive effect on the activities of nitrate reductase and glutamine synthetase enzymes in shoots. It was found that foliar application of nutrients reduced the inhibitory effect of Mo shortage on root nodulation, plant dry biomass and protein content.

In Australia, Mo was first reported to increase bunch yield of grapevines when two applications of sodium molybdate were applied pre-flowering to the canopy of own rooted Merlot grapevines in the Mt Lofty Ranges, South Australia. Foliar feeding of molybdenum to deficient vines has previously been shown to increase yield and bunch weight of Merlot grapevines (Williams *et al.*, 2004). Effects of molybdenum foliar sprays on the growth, fruit set, bunch yield and berry characteristics were examined in *Vitis vinifera* L. Cv Merlot on own roots and 5 rootstocks and improved fruit set, increased bunch yield, reduced numbers of green berries and increased numbers of coloured berries per bunch were found. But there were no major effects on the vegetative growth of grapevines. Two foliar sprays of Mo (118 g ha^{-1}) were applied before flowering, and yield results were recorded at harvest. The concentration of Mo in petioles from sprayed vines was much higher than that from unsprayed vines at all sites. Robinson and Burne (2000) showed that Mo deficiency may be a factor associated with the 'Merlot' problem. They reported that foliar Mo sprays increased in vine growth and accumulated nitrate returned to normal after approximately 4 weeks. Gridley (2003), Williams *et al.*, (2003) and Longbottom *et al.*, (2004) have reported yield increases in response to foliar Mo sprays, applied before flowering to Merlot vines on own roots. Williams *et al.*, (2004) reported no further benefits of the higher dose on bunch yields, and no any detrimental effects on yield. However, in the spring following Mo application grapevines exhibited delayed budburst compared with untreated vines. A response to foliar applications of Mo was found when the petiolar Mo levels were less than 0.1 mg kg^{-1} (Williams *et al.*, 2004). Foliar applications of molybdenum to Mo deficient grapevines have been shown to reduce the effects of fruit set disorders in Merlot grapevines, resulting in increased yields (Gridley, 2003, Williams *et al.*, 2003 Longbottom *et al.*, 2004, Williams *et al.*, 2004).

Chlorine (Cl)

Plants absorb chlorine in the form of chloride (Cl$^-$) both through roots and leaves. Normal plants generally contain 100-500 mg Cl kg^{-1}. It regulates water potential of plant through osmotic adjust, *viz.* stomata opening, cell elongation, turgour pressure, water potential of plants and osmotic pressure. Deficiency symptoms include wilting, stubby roots, chlorosis and bronzing. Chlorine provides resistance to plants against several diseases. Chloride (Cl$^-$) is found beneficial to the wheat culture for several reasons. It is an anion, required in evolution of oxygen in photosynthesis and has a major role in ion transport or movement of nutrients from the soil solution into the cellular solution and movement between cells, as well as being a significant ion in maintaining turgor, or preventing wilting.

Deficiency of Cl$^-$ is corrected along with soil and foliar application of chloride salts such as, potassium chloride (47 per cent Cl$^-$), ammonium chloride (66 per cent Cl$^-$), calcium chloride (65 per cent Cl$^-$), magnesium chloride (74 per cent Cl$^-$) and sodium chloride (66 per cent Cl$^-$). Chloride (Cl$^-$) bearing fertilizers have been widely used in crop production for numerous years, but primarily Cl$^-$ has been applied as a salt associated with K, with little specific recognition of the potential of Cl$^-$ as an essential nutrient.

Miller (1994) studied the comparative effect of NH_4Cl and $MgCl_2$ on wheat yield and fungal disease infestation. Results revealed that foliar application of Cl^- as NH_4Cl or $MgCl_2$ responded well. But no significant difference was found in 9-18 kg acre^{-1} rate of Cl^- as foliar NH_4Cl and the 18 kg acre^{-1} rate of Cl^- as $MgCl_2$. But wheat yield was significantly greater than the untreated fields. Both the foliar applications indicated less injury from prevalent fungal diseases. Some phytotoxicity was noted with foliar Cl^- applications, but wheat grew well after application, and no difference in plant size or biomass was noted at harvest. In view of available literature, chloride deficiency may be successfully corrected by foliar application of chloride bearing compounds at recommended doses.

Nickel (Ni)

Plant absorbs nickel as Ni^{2+} and normal plants contain 0.1-10 mg Ni kg^{-1}. It is essential for dehydrogenase, methyl reductase and urease activation, which regulate and catalysis of urea into two moles of NH_3 and one mole of CO_2 (Witte *et al.*, 2002) and lack of urease activation makes metabolically N deficient plant (Gerendas and Sattelmacher, 1997a). Though growth of plants with ammonium nitrate (NH_4NO_3) was found not affected by Ni supply, urease activity was significantly reduced irrespective of N source (NH_4NO_3 or urea) in plants grown without supplementary Ni (Gerendas and Sattelmacher, 1999). Leaf urease activity was found suppressed in certain plant species, wheat, soybean, rape, zucchini, and sunflower grown on urea-based nutrient media without supplementary Ni thereby causing accumulation of urea and reduction in dry matter and total N (Gerendas and Sattelmacher, 1997). Brown *et al.* (1987) noted 15-20 times higher levels of urea in leaf tips of Ni deficient wheat, barley, and oat plants. Nickel deficiency symptoms were observed in apical tips of leaves, leaflets and catkins due to the accumulation of urea in rapidly growing cells and tissues (Wood *et al.*, 2006). Nickel affected N metabolism *via* disruption of ureide catabolism, amino acid metabolism and ornithine cycle intermediates, and also affected the respiration process via disruption of citric acid cycle (Bai *et al.*, 2006). These results further support the hypothesis that Ni deficiency disrupts normal N-cycling via disruption of ureide metabolism. Increasing Ni in the nutrient solution reduced translocation of Cu and Fe from roots to tops but the translocation of Mn and Zn was unaffected (Rahman *et al.*, 2005). Absorption of Ni^{2+}, Mg^{2+}, Fe^{2+}, Mn^{2+}, Cu^{2+} and Zn^{2+} appears to be competitive, thus an excess of one can result in a shortage of the other with the lowest availability (Wood *et al.*, 2004).

Soluble salts like nicke sulphate ($NiSO_4$), which contains the Ni_2^+ ion, are suitable fertilizers to prevent or correct plant Ni deficiency. Applying a foliar spray at a concentration of 0.03–0.06 ppm Ni is sufficient for crop plants. An application of 0.23 kg Ni acre^{-1} is required in most the crops (NIPAN LLC, 2011). Following Ni bearing fertilizers are used to correction of Ni deficiency in soil as well as foliar applications (Table 16.17). Municipal biosolids fertilizer is also a good source of Ni.

Table 16.17: Some Ni-containing fertilizers

Fertilizers	Materials	Ni (per cent)
Nickel sulphate (also called nickelous sulfate)	$NiSO_4.6H_2O$	32.1
Anhydrous nickel sulphate	$NiSO_4$	37.5
Nickel nitrate	$Ni(NO_3)_2.6H_2O$	20.2
Nickel chloride	$NiCl_2.6H_2O$	37.2
Nickel(II) EDTA[1] complex	$NiC_{10}H_{16}N_2O_8$	16.7
Ni^{++}	Complex	5.4
Sewage sludge	Composite	2.4-5.3

[1]EDTA = Ethylene diamine tetra-acetate

[2]Nickel Plus also contains N (5per cent) and S (3 per cent)

Source: IAFS extension nickel nutrition in plants (HS1191), University of Florida

Betula nigra (black birch, river birch, water birch) is a species of birch native to the Eastern United States from New Hampshire west to southern Minnesota, and south to northern Florida and west Texas. It is commonly found in flood plains and or swamps. Mouse-ear is a disorder on container-grown river birch (*Betula nigra* L.) that is a national problem, caused by nickel deficiency. Symptomatic river birch trees were treated with foliar applications of nickel sulphate and a substrate drench. Topdress applications of superphosphate and miloroganite, products known to contain nickel, were also applied. At 16 days after treatment, new growth occurred on plants sprayed with nickel sulphate and foliar concentrations of nickel in the new growth increased fivefold compared to untreated one. Similarly at 30 DAT, shoot length increased 60per cent, leaf area increased 83per cent, and leaf dry mass increased 81per cent for trees receiving a Ni foliar application compared to non-treated plants (Ruter, 2004). Wood (2004) observed successful correction of Ni deficiency after foliar application of 25-100 ppm Ni solution in most U.S. pecan orchards depending on degree of Ni deficiency. Sprays can be applied in early spring within about 10 days after bud break or, in the case of severe deficiency, in the autumn.

Orchard and greenhouse studies on trees treated with either Cu or Ni indicated that foliar applied Ni corrects mouse-ear disease. Disease symptoms were prevented, in both orchard and greenhouse trees, by a single mid October foliar spray of Ni (nickel sulfate), whereas non-treated control trees exhibited severe disease. Foliar application of Cu in mid-October to greenhouse seedling trees increased disease severity the following spring. Post budbreak application of Ni to these Cu treated seedling trees prevented mouse-ear disease symptoms in post Ni application growth, but did not alter morphology of foliage exhibiting disease prior to Ni treatment. Nickel was combined with standard fungicide sprays and applied to foliage and fruit of Wichita, Apache, Desirable, Oconee, and Kiowa Pecan. Treatments reduced scab severity on all varieties, with those best most susceptible to scab being most influenced. The preliminary data indicates that foliar application of nickel can be used as a management tool with combined other disease management tools to reduce tree susceptibility to scab disease.

Ranjbar *et al.* (2011) applied foliar spray of nickel sulphate in strawberry at

4-5-leaf stage with concentrations of 0, 150, 300 and 450 mg L^{-1} and urea at 0 and 2 g L^{-1}, and found that 300 mg L^{-1} nickel sulphate without urea significantly increased the yield, primary and secondary fruits' weight and inflorescence number as compared to untreated. Also, the 300 mg L^{-1} nickel sulphate treatment with 2g L^{-1} urea had the highest rate of total acid (0.75per cent). In general, spraying 300 mg L^{-1} nickel sulphate with 2 g L^{-1} urea is recommended for increasing the strawberry yield.

The effect of nickel foliar application on *Cucurbita ficifolia* (tropical squash) plants was evaluated at 1.0, 2.5 and 5.0 mg L^{-1} concentrations in terms of later extraction of crystalline urease from the seeds. Nickel applications were found to cause phytotoxicity on all plants and fruit and seed growth were indirectly proportional to the concentrations used. The 1.0 and 2.5 mg L^{-1} concentrations were tolerated by the plants, but intervein chlorosis was observed. Contrarily, extracted urease amounts were directly proportional to applied Ni concentrations. In all cases, urease catalytic response to Ni concentration was found maximum at 1.0 mg L^{-1}, minimum at 2.5 mg L^{-1} and medium in untreated conducted a long-term field study to determine the effect foliar Ni applications on ring nematode (*Mesocriconema xenoplax*) populations at a site having a history of peach tree short life tree death and found increased plant mortality in Ni treated plots as compared to the untreated. These data provide useful insights into foliar Ni application in peach that enhanced peach tree short life tree mortality and which may be useful in defining the mode of action responsible for tree death in this disease complex.

PRECAUTIONS FOR FOLIAR FERTILIZATION

Do obtain an accurate plant tissue analysis before foliar spraying. Don't foliar spray plants under drought conditions, or when plants are under other forms of stress, such as disease or insect damage. Don't foliar spray during hot, dry mid-day temperatures. The best time is late evening; however, early morning or overcast days are acceptable. Avoid foliar spray under windy conditions if drift into adjacent fields is a factor. Use narrow tires and foliage deflectors on ground applicators and adjust wheels to accommodate row widths to avoid damage to plant foliage. Clean application equipment thoroughly before spraying. Use wide angle hollow cone spray nozzles in ground applicators. Select the proper nozzle spray tip that will operate within the desired pressure range of your pump capacity. Calibrate each spray nozzle for proper volume using pressure setting and ground speed desired. Adjust nozzles on boom to spray at a 90 angle straight down from spray boom. Adjust boom height above plant canopy for uniform spray pattern. Make sure all hoses, pumps or pipelines are thoroughly drained of any other plant food products and flushed with water before use.

Flush residual amounts of fertilizer solution from aluminum equipment with water after use to avoid possible pitting of equipment from prolonged exposure.

Foliage should be sprayed in the evening or on a cool, cloudy day to prevent leaf burning and few drops of liquid soap or wetting agents are added as surfactant. Consequently, for best results apply spray when leaves will remain moist for a period of time.

Copper sulphate is highly corrosive when it comes into contact with metals. Stainless steel and plastic components are required on fertilizer applicators and sprayers. Cool weather is essential, particularly for copper sulphate, because spraying later in the season during warm sunny weather can result in leaf scorching.

The concentrations of spraying solutions must be standardize by spraying a few healthy leaves to see if there is any damage such as, the leaves curl or look unhealthy in any way two days after spraying, it is diluted again and pH is adjusted to neutral range.

REFERENCES

A&L Canada Laboratories Small Fruit News Letter Vol. 6 May 7, 2000.

Abad, A., Lloveras, J., and Michelena, A. (2004). Nitrogen fertilization and foliar urea effects on durum wheat yield and quality and on residual soil in irrigated Mediterranean conditions. *Field Crops Research*, **87**:2 pp 257-269.

Abbott, JA, Conway, WS and Sams, CE (1989). Post-harvest calcium chloride infiltration affects textural attributes of apples. *J. Am. Soc. Hort. Sc.* **114**: 932–936.

Abou El-nour, EAA (2002). Can supplemented potassium foliar feeding reduce the recommended soil potassium? *Pak. J. Biol. Sci.* **5**:259–262.

Agrichem (2007). Liquid logics, agronomy update, no. 0007 and 0008

AgTactics (2012). Will your lupin seed areas be caught short of Manganese. A timely and tactical news letter for the Northern Agricultural Region, ISSN 1836 9294.

Ahmed, AG, Hassanein, MS, El-Gazzar, MM (2006). Growth and yield response of two wheat cultivars to complete foliar fertilizer compound "Dogoplus". *J. Appl. Sc. Res.*, **2**: 20-26.

Akhtar, N, Abdul, MSM, Akhter, H, Katrun, NM (2009). Effect of planting time and micro-nutrient as zinc chloride on the growth, yield and oil content of Mentha piperita. *Bangladesh J. Sci. Ind. Res.*, **44**(1): 125-130.

Alam, SS, Moslehuddin, AZM, Islam, MR and Kamal, AM (2010). Soil and foliar application of nitrogen for Boro rice (BRRIdhan 29) *J. Bangladesh Agril. Univ.* 8(2): 199–202.

Alberta Agriculture, Food and Rural Development (AAFRD) (1999). Copper deficiency: diagnosis and correction, Agdex 532-3.

Albrigo, LG (1999). Effects of foliar applications of urea or Nutri-Phite on flowering and yields of Valencia orange trees. *Proc. Fla. State Hort. Soc.* **112**:1-4.

Ali, Liaqat., Mushtaq Ali, and Qamar Mohyuddin (2011). Effect of foliar application of zinc and boron on seed cotton yield and economics in cotton-wheat cropping pattern, *J. Agric. Res.*, **49**(2): 173-180.

Ali, S, Khan, AR, Mairaj, G, Arif, M, Fida, M, and Bibi, S. (2008). Assessment of different crop nutrients management practices for improvement of crop yield. *Aust. J. Crop Sci.* **2**(3):150-157.

Ali, S, Shah, A, Arif, M, Mirja, G, Ali, I, Sajjad Khan, MFA., Moula, N (2009) *J. Agric.*, **25**, 1.

Alston, AM (1979). Effects of soil water content and foliar fertilization with nitrogen and phosphorus in late season on the yield and composition of wheat. *Aus. J. Agric. Res.* **30**: 577-585.

Amberger, AA (1991). Importance of micro-nutrients for crop production under semi-arid conditions of North Africa and Middle East. In Proceedings of the 4th Micro-nutrients Workshop, Amman, Jordan, 5–30. Cairo: NRC and Fertilizer 16: 130-152. (In Thai with English abstract).

Anter, F, Rasheed, MA, El-salam, MA and Metvally, A.I. (1978). Effect of foliar application of copper, molybdenum, zinc and boron on the fibre qualities of cotton plants grown on calcareous soil. *An. Agric. Sc.*, **6**: 303-311.

Arif, M, Chohan, MA, Ali, S and Khan, S (2006). Response of wheat to foliar application of nutrients. *J. Agri. Bio. Science*, **1**: 30-34.

Ashour, NI and Thalooth, AT (2003). Effect of soil and foliar application of nitrogen during pod development on the yield of soybean (*Glycine max* (L.) Merry) Plants. *Field Crops Research*, **6**: 261-266.

Babaeian, M, Tavassoli, A, Ghanbari, A, Esmaeilian, Y, Fahimifard, M (2011). Effects of foliar micro-nutrient application on osmotic adjustments, grain yield and yield components in sunflower (*Alstar cultivar*) under water stress at three stages. *Afr J Agric Res*. 6(5): 1204-1208.

Bacha, MA, Sabbah, AM, and Hamady, MA (1997). Effect of foliar application of iron, zinc and manganese on yield, berry quality and leaf mineral composition of Thompson seedless and roomy red grape cultivars. J. King Saud Univ. *Agric. Sci.*, **1**: 127-140.

Bai, C, Reilly, CC and Wood. BW (2006). Nickel Defi ciency Disrupts Metabolism of Ureides, Amino Acids, and Organic Acids of Young Pecan Foliage. *Plant Physiology*, **140**:433–443.

Baloch, QB, Chachar, QI and Tareen, MN (2008). Effect of foliar application of macro and micro nutrients on production of green chilies (*Capsicum annuum* L.). *J. Agric. Tech.*, 4(2): 177-184.

Bameri, M, Abdolshahi, R, Mohammadi-Nejad, G, Yousefi, K, and Tabatabaie, SM (2012). Effect of different microelement treatment on wheat (*Triticum aestivum*) growth and yield, Intl. *Res. J. Appl. Basic. Sci.*, 3(1): 219-223.

Basavarajappa, R, Koraddi, VR, Kamath, KS and Doddamani, MB (1997). Response cotton cv. Abadhita (*Gossypium hirsutum*) to soil and Foliar Application of micro-nutrients under Rainfed condition. *Karnataka J. Agric. Sci.*, **10** (2): 287-291.

Bashir, F, Muhammad, M, Sarwar, N, Ali, HM, Muhammad K, and Shehzad, A (2012). Effect of foliar application of zinc on yield and radiation use efficiency (RUE) of maize (*Zea mays* L.) under reduced irrigation conditions, *Asian J Pharm Biol Res* **2**(1): 33-39.

Bednarz, CW and Oosterhuis, DM (1999). Physiological changes associated with potassium deficiency in cotton. *J. Plant Nutr.*, **22**: 303-313.

Bell, PW, Rerkasem, B, Keerati-Kasikorn, P, Phetchawee, S, Hiunburana, N, Ratanarat, S, Pongsakul P and Loneragan, JF (1990). Mineral Nutrition of Food Legumes in

Thailand, with Particular Reference to Micro-nutrients. Australian Centre for International Agricultural Research (ACIAR). Technical Report No. 16. 52 pp. Bohatyr and Smaragd. Rosthlinna Vyroba, 30: 505-514.

Benbella, M, Paulsen, GM (1998). Efficacy of treatments for delaying senescence of wheat leaves: II. Senescence and grain yield under field conditions. *Agron. J.*, **90:** 332-338.

Bernal, M, Cases, R, Picorel, R, Yruela, I (2007). Foliar and root Cu supply affect differently Fe and Zn-uptake and photosynthetic activity in soybean plants. *Environ. Exp. Botany*, **60:** 145-150.

Bisson, P, Cretenet M and E Jallas, (1994). Nitrogen, phosphorus and potassium availability in the soil physiology of the assimilation and use of these nutrients by the plant, Challenging the Future: Proceedings of the World Cotton Research Conference, Brisbane Australia, February 14-17, Eds., Constable, G.A. and N.W. Forrester, CSIRO, Melbourne, pp: 115-124.

Blake-Kalff, MMA, Zhao, FJ, Hawkesford, MJ and McGrath, SP (2001). Using plant analysis to predict yield losses by sulphur deficiency. *Annals Applied of Biol.*, **138:** 123-127.

Bohner, H and Reid, K (2007). Manganese Deficiency, crop pests Ontario, http://www.omafra.gov.on.ca/english/crops/field/news/croppest/2007/12cpo07a2.htm

Boorboori, MR, Eradatmand Asli, D and Tehrani, M (2012). *J. Advances in Environmental Biology.*, **6**(2): 740.

Bora, PC and Hazarika, U (1997). Effect of lime and boron on rainfed toria (*Brassica campestrtis* subsp *Oleifera* var. Toria). *I. J. Agron.*, 42 : 361-364.

Borjian, AR and Emam, Y (2001). Effect of urea foliar feeding on grain protein content and quality in two winter wheat cultivars. *Iran Agricultural Research* **20**, 37-52.

Bowen, P, Menzies, J, Ehret, D., (1992). Soluble silicon sprays inhibit powdery mildew development on grape leaves. *J. Am. Soc. Hort. Sc.* **117:** 906–12.

Bowszys, T (1996), Response of winter rape to foliar application of boron fertilizer. *Zesyty Problemowe Portepow Nauk Rolniczych*, **434** (1): 71-76.

Bozorgi, HA, Azarpour, E, Moradi, M (2011). The effects of bio, mineral nitrogen fertilization and foliar zinc spraying on yield and yield components of faba bean. *J. World Appl. Sci.* **13**(6): 1409-1414.

Brennan, R (2005). Overcoming copper deficiency in wheat with foliar sprays, Department of agriculture and food, Govt. of Australia, http://www.agric.wa.gov.au/PC_92056.html

Brown, PH, Welch RM, Cary, EE and Checkai, RT (1987). Beneficial effects of nickel on plant growth. *J. Pl. Nutr.* **10:** 2125-2135.

Buick, RD, Robson, B, Field, RJ (1992). A mechanistic model to describe organosilicone surfactant promotion of Triclopyr uptake. *Pesti. Sci.* **36:** 127-133.

Bybordi, A Jasarat AS (2010). Effects of the foliar application of magnesium and zinc on the yield and quality of three grape cultivars grown in the calcareous soils of iran, *Not. Sci. Biol.*, **2**(1): 81-86.

Bybordi, A, Malakouti, MJ (2003). Effects of Iron, Manganese, Zinc and Copper on Wheat Yield and Quality under Saline Condition. *Olom-e-Ab Va Khak*, **17**(2): 48-59.

Cakmak, I (2008). Enrichment of cereal grains with zinc: agronomic or genetic biofortification, *Pl. and Soil*, **302**(1-2): 1-17.

Camberato, JJ (2004). Manganese deficiency and fertilization of cotton, *Soil Fertility* Series no.1, Clemson, University.

Carvalho, LH, Silva, NM, Brasil Sobrinho, MOC, Kondo, JI and Chiavegato, EJ (1996a). Application of boron to cotton by side dressing and foliar spray. *Revista Brasilevia de Ciencia do Solo*, **20**: 265-269.

Carvalho, LH, Silva, NM, Brasil Sobrinho, MOC, Kondo, JI and Chiavegato, EJ (1996b). Methods of boron application to cotton. *Revista Brasileria de Ciencia do Solo*, **20**: 271-275.

Cassman, KG, Roberts, BA and Bryant, DC (1992). Cotton response to residual fertilizer potassium on vermiculitic soil, organic matter and sodium effects. *Soil Sci. Soc. Am. J.*, **56**: 823-830.

Ceccarelli, S (1987). Yield potential and drought of segregation population of barley in contrasting environment. *Euphytica*, **36**: 265-273.

Chang, SS (1999). Micronutrients in crop production of Taiwan. In: Proceedings of International Workshop on Micro-nutrient in Crop Production, held Nov. 8-13, National Taiwan University, Taipei, Taiwan ROC.

Chapagain, BP and Wiesman. Z. (2004). Effect of Nutri-Vant-Pea K foliar spray on plant develop-ment, yield, and fruit quality in green house tomatoes. *Sci. Hort.* 102:177-188.

Chauhan, YS Apphun, A, Singh, VK, Dwivedi, BS (2004). Foliar sprays of concentrated urea at maturity of pigeonpea toinduce defoliation and increase its residual benet to wheat, *Field Crops Research* **89**: 17–25.

Chitdeshwari, T, Sankaran, K and Krishna, DD (1997). Role of magnesium sulphate in controlling seedling of leaves in cotton. *Madras Agric. J.*, **84**: 386-387.

Christensen, LP, Kasimatis, AN and Jensen, FL (1982). Grapevine nutrition and fertilization in the San Joaquin Valley. University of California Div. Agric. Sci. Publication 4087.

Christensen, P (1986). Additives don't improve zinc uptake in grapevines. California Agriculture. 40(1 and 2) and Boron application in vineyard. *California Agriculture*. **40**(3 and 4).

Clarkson, DT (1981). Nutrient interception and transport by root systems. In Physiological Processes Limiting Plant Productivity; Johnson, C.B., ed; Butterworths: London, 307-330.

Conway, WS, and Sams. CE (1987). The effects of post-harvest infiltration of calcium, magnesium, or strontium on decay, firmness, respiration, and ethylene production in apples. *Journal of the American Society for Horticultural Science* **112**: 300–303.

Dehnavi, MM, Mohammad, SA, and Sanavy, M (2008). Effects of withholding irrigation and foliar application of zinc and manganese on fatty acid composition and seed oil content in winter safflower, 7th International safflower conference, Wagga Wagga Australia.

Dickison, WC (2000) 'Integrative Plant Anatomy' (Academic Press: USA).

Dixon, RC (2003). Foliar fertilization improves nutrient use efficiency, *Fluid J.*, pp. 1-2.

Dongre, SM, Mahorkar, VK, Joshi, PS and Deo, DD (2000). Effect of micro-nutrients spray on yield and quality of chilli (*Capsicum annum* L.) cv. *Jayanti. Agric. Sc. Digest*, **20**: 106-107.

Dos Santos, MG, Ribeiro, RV, de Oliveira, RF, Pimentel, C (2004). Gas exchange and yield response to foliar phosphorus application in *Phaseolus vulgaris* L. under drought. *Brazilian J. Plant Physiol.* 16, 171-179.

Eichert, T, Burkhardt J (2001). Quantification of stomatal uptake of ionic solutes using a new model system. *Journal of Experimental Botany*, **52**: 771-781.

El-Fouly, MM and El-Sayed AA (1997). Foliar fertilization: An environmentally friendly application of fertilizers. Dahlia Greidinger International Symposium on "Fertilization and Environment" 24-27 March, Haifa, Israel, John I (ed.). pp. 346-357.

El-Fouly, MM, Mobarak, ZM and Salama, ZA (2011). Micro-nutrients (Fe, Mn, Zn) foliar spray for increasing salinity tolerance in wheat *Triticum aestivum* L., *Afr. J. Pl. Sc.*, 5(5): 314–322.

El-Kassas, SE (1984). Effect of iron nutrition on the growth, yield, fruit quality, and leaf composition of ceded balady lime trees grown on sandy calcareous soils. *J. Pl Nutr..* **7**: 301-311.

El-Otmani, M, Ait-Oubahou, A, Gousrire, H, Hamza, Y, Mazih, A, and Lovatt, CJ (2003b). Effect of potassium phosphite on flowering, yield, and tree health of 'Clementine' mandarin. *Proc. Intl. Soc. Citriculture* 1:428–432.

El-Otmani, M, Ait-Oubahou, A, Taibi, FZ, Lmfoufid, B, El-Hila, M and Lovatt, CJ (2003a). Prebloom foliar urea application increases fruit set, size, and yield of Clementine mandarin. *Proc. Intl. Soc. Citriculture* 1:559–562.

El-Otmani, M, Ait-Oubahou, A, Zahra, F and Lovatt, CJ (2002). Efficacy of foliar urea as an n source in substainable citrus production systems, International Symposium on Foliar Nutrition of Perennial Fruit Plants, ISHS Acta Horticulturae 594.

Erenoglu B, Nikolic, M, Römhold, V, Cakmak, I (2002). Uptake and transport of foliar applied zinc (65Zn) in bread and durum wheat cultivars differing in zinc efficiency. *Plant and Soil*, **241**: 251–257.

Eriksen, J (2009). Chapter 2 Soil sulphur Cycling in Temperate Agricultural Systems. Advances in Agronomy. **102**: 55–89.

Evangelista, RM, Bosco, A and Fernandez. MI (2000). Influence of the application pre-harvest of the calcium in the poligalacturonase, pectinmetilesterase and γgalactosidase activity and texture of the mangos 'Tommy Atkins' stored under refrigeration. *Ciencia Agrotecnology* **24**: 174–181.

Farahat, MM, Ibrahim, SS Lobna, T and Fatma, EM El-Quesni (2007). *World, J. of Agric. Sci.*, **3**(3).

Fernandez, V and Ebert, G (2005). Foliar iron fertilization : a critical review. *J. pl. Nutr.*, **28**: 2113-2124.

Fernandez, V, and Eichert, T (2009). Uptake of hydrophilic solutes through plant leaves: current state of knowledge and perspectives of foliar fertilization. *Crit. Rev. Pl. Sc.* **28**: 36-68.

Fernandez-Lopez, JA eta/. (1993). *I. Pl. Nutr.*, **16**(8): 1395-1407.

Fertilizer Manual (1998). UN Industrial development organization, Int'l Fertilizer Development Center. Springer, ISBN, 0792350324, 9780792350323.

Field, RJ and Bishop, NG (1988). Promotion of stomatal infiltration of glyphosate by an organosilicone surfactant reduces the critical rainfall period. *Pesti. Sc.* **24**: 55-62.

Forth, HD and Ellis, BG (1997). Soil Fertility. 2nd edition. Lewis Publishers, New York, USA.

Gerendas, J, and Sattelmacher, B (1997). Significance of Ni supply for growth, urease activity and the contents of urea, amino acids and mineral nutrients of urea-grown plants. *Pl. Soil* **190**: 153–62.

Girma, K, Martin, KL, Freeman, KW, Mosali, J, Teal, RK, Raun, WR, Moges, SM, and Arnall, B (2007). Determination of the optimum rate and growth stage for foliar applied phosphorus in corn. *Commun Soil Sc. Pl. Anal.*, **38**: 1137-1154.

Graham, RD and Webb. MJ (1991). Micro-nutrients and disease resistance and tolerance in plants. Pp. 329-370. In J.J. Mortvedt, F.R. Cox, L.M. Shuman, and R.M. Welch eds.) Micro-nutrients in Agriculture. 2nd edition. Soil.Sc.Soc. Am.,

Gridley, KL (2003). The effects of molybdenum as a foliar spray on fruit set and berry size in *Vitis vinifera* cv. Merlot. Honours Thesis, The University of Adelaide. June.

Grundon, NJ (1980). Effectiveness of soil dressings and foliar sprays of copper sulphate in correcting copper deficiency of wheat (*Triticum aestivum*) in Queensland. *Aus. J. Exp. Agric. An. Hus.* **20**(107): 717-723.

Grzebisz, W, and Przygocka-Cyna, K (2003). Aktualne problemy gospodarowania siark[1]w rolnictwie Polskim. Nawozy i Nawozenie. 4(17): 64–77.

Grzebisz, W, Wronska, M, Diatta, JB, Dullin, P (2008). Effect of zinc foliar application at early stages of maize growth on patterns of nutrients and dry matter accumulation by the canopy. Part I. Zinc uptake patterns and its redistribution among maize organs. *J. Elementolgy*, **13**:17–28.

Guo, B (1987). A new application of raze earth's-agriculture. P. 237-246. In Race Earth Horizons, Aust. Department of Industry and commerce, Australia.

Guvenc, I, Karatas, A and Kaymak, HC (2006). Effect of foliar applications of urea, ethanol and putrecine on growth and yield of lettuce (*Lactuca sativa*), I. J. Agric. Sc., **79**:1.

Habib, M (2009). Effect of foliar application of Zn and Fe on wheat yield and quality, *African Journal of Biotechnology* **8**(24): 6795-6798.

Haloi, B (1980). Effect of foliar application of phosphorus salt on yellowing of wheat seedlings. *J. Res., Assam Agricultural University*, **1**:108-109.

Hamlin, RL (2006). Handbook of plant nutrition, Molybdenum, pp 375 -394.

Hanumanthareddy, LP (1999). Impact of nutrients and growth regulators on drying of reproductive structures in cotton (*Gossypium barbadense*). Thesis, University of Agricultural Sciences, Dharwad.

Haq Nawaz, Saeed, M, Iqbal, MM, Shah, SM and Wasil Mohammad (1994). Growth and yield of seed cotton as influenced by micronutrients. *Sarhad J. Agric.*, **10**:21-25.

Harper, JE (1984). in Nitrogen in Crop Production, ed. Hauck, R. D. (Am. Soc. Agron., Madison, WI), pp. 165-182.

Havlin, JL, Beaton, JD, Tisdale, SL and Nelson, WL (2005). Soil fertility and fertilizers: An introduction to nutrient management. 7th edition. Pearson Prentice Hall. Upper Saddle River, New Jersey, USA.

Heckman, J (2010). Effective Treatment of Mn deficiency in soybean: Comparing Fertilizer Sources, Plant & pest advisory, *A Rutgers Cooperative Extension Publication*, **16**(7): 3.

Hedley, MJ and McLaughlin, MJ (2005). Reactions of phosphate fertilizers and by-products in soils. In 'Phosphorus: Agriculture and the Environment'. (Ed. AN Sharpley) pp. 181-252. (American Society of Agronomy, Crop Science Society of America, Soil Science Society of America: Madison, WI).

Hell, R, Rennenberg, H (1998). The plant sulphur cycle. Sulphur in agroecosystem, Schnug E. redakcja, Kluwer Academic Publishers: 221.

Hermans, C, Johnson, GN, Strasser, RJ, and Verbruggen, N (2004). Physiological characterization of magnesium deficiency in sugar beet: acclimation to low magnesium differentially affects photosytems I and II. Planta, 220, 344-355. http://dx.doi.org/10.1007/s00425-004-1340-4.

Hirpara, KD, Ramoliya, PJ, Patel, AD and Pandey, AN (2005). Effect of salinisation of soil on growth and macro- and micro-nutrient accumulation in seedlings of *Butea monosperma* (Fabaceae) *Anales de Biología* **27**: 3-14.

Hong, E, Ketterings, Q, McBride, M (2010). Nutrient management spear program, agronomy fact sheet series, Fact Sheet 49, Cornell University, Cooperative Extension http://nmsp.cals.cornell.edu

Hosny, AA, Kadry, W and Mohamad, HMH (1984). Effect of foliar application of boron and copper on growth, yield and yield components of Giza 75 cotton variety. *An. Agric. Sc.*, **21**: 25-35.

Howard, DD, Essington, ME, Gwathmey, CO and Percell, WM (2000). Buffering of foliar potassium and boron solutions for no- tillage cotton production. *The Cotton Sci.* **4**: 237-244.

Hristozkova, M Geneva, M Stancheva I (2006). Response of inoculated pea plants (*Pisum sativum l.*) to root and foliar fertilizer application with reduced molybdenum concentration, *Gen. Appl. Plant Physiol, Special Issue*, 73-79.

Hussanin, M, Ayaz Khani, M, Khan, MB., Farooq, M, Farooq, S, (2012). *J. Rice Sc.*, **19**: 2.

Iron Chlorosis in Trees (2012). Kansas State Universtiy, Kansas Forest Service Publication, MF-718. http://www.hfrr.ksu.edu/doc1649.ashx

Jairo A Palta, Ajit S, Nandwal, Sunita Kumari, Neil, C, Turner (2005). Foliar nitrogen applications increase the seed yield and protein content in chickpea (*Cicer arietinum* L.) subject to terminal drought, *Aus. J. Agric. Res.* **56**(2): 105–112.

Jamal, Z, Hamayun, M, Ahmad, N and Chaudhary, MF (2006). Effect of soil and foliar application of different concentrations of NPK and foliar application of $(NH_4)_2SO_4$ on different parameters in wheat. *J. Agron.*, **5**(2): 251-256.

Jenks, MA, and Ashworth, EN (1999). Plant epicuticular waxes: Function, production and genetics. *Hort. Re.* **23**: 1-29.

Jensen, F, Luvisi, D, and Beede R (1980). The effects of adjuvants, pesticides, and mineral nutrients applied with the fruit set gibberellin treatments and growth regulators on fruit characteristics of table Thompson Seedless. San Joaquin Valley Agric. Res. and Extension Center Report, 2.

Jinfeng, D (1995). The yield increasing ability of spraying cotton with boron. Hevan Nangye Kexul, 3, 6. Puyang Agriculture and Animal Husbandry Bureau.

Kannan, S (1986). Foliar absorption and transport inorganic nutrients CRC *Crit. Rev. Plant Sci. J.*, 341-375.

Kaya, M, Atak, M, Mahmood, KK, Ciftci, CY, Ozcan, S (2005). Effect of pre-sowing seed treatment with zinc and foliar spray of humic acids on yield of common bean (*Phaseolus vulgaris* L.). *Int. J. Agri. Biol.*, **6**(7): 875–878.

Kefyalew, G, Martina, KL, Freemana, KW, Mosalib, J, Teala, RK, William, RR, Mogesa, SM, Arnalla, DB (2007). Determination of Optimum Rate and Growth Stage for Foliar-Applied Phosphorus in Corn. *Comm. Soil Sc. Pl. Anal.*, **38**: 1137-1154.

Kelly, A, Nelson, P, Motavalli, P and Manjula Nathan (2005). Response of no-till soybean (*Glycine max* L.) to timing of preplant and fliar potassium applications in a claypan soil. *Agron. J.*, **97**: 832-838.

Khalil, IA and Jan A (2003). Cropping Technology. National Book Foundation, Islamabad, Pakistan.

Khodzhaev, DKH and Stesnyagina, TYA (1983). Effect of top dressing with trace elements on heat and drought resistance of cotton. *Uzbekeskii Biologicheskii Zhurnal*, **6**: 19-23.

Kinaci, E, Gulmezoglu, N (2007). Grain yield and yield components of triticale upon application of different foliar fertilizers. Interciencia. **32**(9): 624-628.

Klacan, GR (1997). Combined effect of seed treatment and foliar dressing of magnesium on garden pea cultivars. Diss. Abstracts, **23**: 782.

Korzeniowska, J and Ewa Stanis³awska-Glubiak (2011). The effect of foliar application of copper on content of this element in winter wheat grain, *Polish J. Agron.*, **4**: 3-6.

Kotur, SC (1995). Standardization of foliar application of molybdenum for high yield of cauliflower (*Brassica oleracea* convar botrytis var botrytis) on Alfisol of Bihar plateau, *Indian Journal of Agricultural Sciences* **65**(6): 405-9.

Kulczycki, G, (2010). The effect of soil and foliar sulphur application on winter wheat yield and soil properties, Wroc³aw University of Environmental and Life Sciences. http://karnet.up.wroc.pl/~kulcz/images/Publikacjeper cent20Kulczycki/2011/43-Kulczycki-2011.pdf

Kumar, AR and Kumar, N (2007). Sulphate of potash foliar spray effects on yield, quality and post-harvest life of banana. Better crops **91**(2): 22-24.

Kundu, C, Sarkar, RK (2009). Effect of foliar application of potassium nitrate and calcium nitrate on performance of rainfed lowland rice (*Oryza sativa* L.), *I. J. Agron*, **54**(4): 428-432.

Kuruppaiah, P (2005). Foliar application of micro-nutrients on growth, flowering and yield characters of brinjal cv. Annamalai. *Plant Archives*, **5**(2): 605-608.

Lahav, E and Whiley, AW (2002). Irrigation and mineral nutrition. In The Avocado: Botany, Production and Uses. (Eds A.W. Whiley, B. Schaffer and B.N. Wolstenholme) CAB International Oxon, New York.

Legris-Delaporte, S, Ferron, F, Landry, J and Costes, C (1987). Metabolization of Elemental sulphur in Wheat Leaves Consecutive to Its Foliar Application. *Plant Physiology*. **85**(4): 1026–1030.

Leslie, H, Pinghai Ding, Guihong Bi, Minggang Cui, (WTFRC Project AH-01-65). The relationship of foliar and soil N applications to nitrogen use efficiency, growth and production of apple trees, Department of Horticulture, Oregon State University, Corvallis, OR 97331.

Lester, GE, Jifon, JL and Rogers. G (2005). Supplemental foliar potassium applications during muskmelon fruit development can improve fruit quality, ascorbic acid, and beta-carotene contents. *J. Am. Soc. Hort. Sci.* **130**:649-653.

Lester, GE, Jifon, JL and Stewart, WM (2007). Foliar potassium improves cantaloupe.

Lester, GE, Jifon, JL, Makus, DJ (2006). Supplemental foliar potassium application with or without a surfactant can enhance netted muskmelon quality. *Hort Science*. **41**(3):741-744.

Liew, CS (1988). Foliar fertilizers from Uniroyal and their potential in Pakistan. Proc. of Seminar on micro-nutrient in soils and crops in Pak. 277. (Abst.).

Ling, F, Silberbush, M (2002). Response of maize to foliar vs. soil application of nitrogen-phosphorus-potassium fertilizers. *J. Plant. Nutr.*, **11**: 2333-2342.

Longbottom, M, Dry, P and Sedgley, M (2004). Foliar application of molybdenum preflowering Effects on yield of Merlot. The Australian and New Zealand Grapegrower and Winemaker **491**: 36-39.

Lovatt, CJ (1999). Timing citrus and avocado foliar nutrient applications to increase fruit set and size *Hor Technology* **9**:607-612.

Lovatt, CJ and Mikkelsen, RL (2006). Phosphite fertilizers: What are they? Can you use them? What can they do? Better Crops., **90**: 11-13.

Lu, R and Shi, T (1982). Handbook of Agrochemistry, Science Press China.

Majid, NM and Ballard, TM (1990). Effects of foliar application of copper sulphate and urea on the growth of lodgepole pine, *Forest Ecol. Management*, **37**(1-3): 151-165.

Malakouti, MJ and Afkhami M (2001). Foliar application of calcium chloride for improving apple quality and reducing residual pesticides, *Acta Hort.* 564: IV International symposium on mineral nutrition of deciduous fruit crops.

Malewar, GU, Syed, II and Indulkar, BS (1992). Effect of phosphorus with and without Boron on the yield and uptake of Boron and phosphorus in cotton, groundnut cropping sequence. *An. Agric. Res.*, **13**: 269-270.

Malhi, SS (2001). Agriculture and Agri-Food Canada, Melfort, Saskatchewan.

Malik, MNA, Makhdoom, M.I., Shah S.I.H. and Chaudhry. F.I. (1990). Cotton response to boron nutrition in silt loam soils. *The Pak. Cottons.* **34**:133-140.

Mallarino, AP, Haq, MU and Murrell, TS (2005). Early season foliar fertilization of soybeans. *Better Crops*, **89**: 11-13.

Mallarino, AP, Haq, MU, Wittry, D (2001). Variation in soybean response to early season foliar fertilization among and within fields. *Agron. J.*, **93**: 1220-1226.

Manal, FM, Thalooth, AT and Khalifa, HM (2010). Effect of foliar spraying with uniconazole and micro-nutrients on yield and nutrients uptake of wheat plants grown under saline condition. *J. Amer Sci.* 6(8):398-404.

Marschner, H (1995). 'Mineral Nutrition of Higher Plants.' (Academic Press Ltd.: London).

Marschner, H (1995). Mineral Nutrition of Higher Plants. 2ed. New York: Academic Press, p. 889.

Marschner, H (1998). Mineral nutrition of higher plants. 2nd ed. Academic press gross, San diego, CA.

Maynard, D and Hochmuth, G (1996). Handbook for vegetable Growers. 4th ed. John Willey & Sons, inc

McConnell, JS, Baker, WH and Kirst Jr. RC (1998). Yield and petiole nitrate concentrations of cotton treated with soil-applied and foliar-applied nitrogen. *J. Cotton Sci.* **2**:143-152.

McConnell, JS, Baker, WH, Frizzel, BS and Varvil (1992). Response of cotton to nitrogen fertilization and early multiple application of mepiquate chloride. *J. Pl Nutr.*, **15**: 457-468.

McGrath, S and Zhao, J (1996). Sulphur uptake, yield responses and interaction between nitrogen and sulphur in winter oilseed rape (*Brassica napus*). *J. Agric. Sci.*, **126**: 53-62.

Menzies, J, Bowen, P, Ehret, D and Glass, ADM (1992). Foliar applications of potassium silicate reduce severity of powdery mildew on cucumber, muskmelon, and zucchini squash. *J. Am. Soc. Hort. Sc.*, **112**:902–5.

Micronutrients in Crop Production (2006). Ministry of agriculture. Govt. of Saskatchewan, http://www.agriculture.gov.sk.ca/default.aspx?dn=acbda80d-c76d-41a8-a0ea-871144bfee2f.

Midwest Laboratories (1994). Foliar nutrition. Inc., Omaha, NE

Millard, P and Robinson, D (1990). Effect of the timingand rate of nitrogen-fertilization on the growth andrecovery of fertilizer nitrogen within the potato (*Solanum tuberosum* L.) *Crop. Fert. Res.*, **21**: 133-140.

Miller, GW, Denney, A, Pushnik, . and Yu, MH (1982). J. *Plant Nutri.*, 5: 289- 300.

Miller, TD (1994). The effect of chloride fertilizers on yield and disease progress in Texas wheat, Intensive Wheat Management Conference Proceedings.

Mills, HA, and Jones, JB Jr. (1996). Plant analysis handbook: II. A practical sampling, preparation, analysis, and interpretation guide. Micro Macro Publ., Inc., Jefferson City, MO.

Mishra, NM, Ramkrishna Reddy, MG and Subramanyam, MVK (1985). Effect of micro-nutrients on sunflower. *The Andhra Agric. J.*, **32**: 149-159.

Mishra, S, Choudhary, MR, Yadav, BL, Singh, SP (2011). Studies on the response of integrated nutrient management on growth and yield of ber. *I.J. Hort.* **68**(3): 318-321.

Misra, A, Dwivedi, S, Srivastava, AK, Tewari, DK, Khan, A, Kumar, R (2006). Low iron stress nutrition for evaluation of Fe-efficient genotype physiology, photosynthesis, and essential monoterpene oil(s) yield of *Ocimum sanctum*. *Photosyntetica*, **44**(3): 474-477.

Misra, A, Sharma, S (1991). Zn concentration for essential oil yield and enthol concentration of Japanese mint. *Fertilizer Crit. Res.*, **29**: 261-265.

Mona, E Eleiwa, Ibrahim, SA and Manal, Mohamed, F (2012). Combined effect of NPK levels and foliar nutritional compounds on growth and yield parameters of potato plants (*Solanum tuberosum* L.) *Afr. J. Micro. Res.*, 6(24): 5100-5109.

Moosavi, AA, and Ronaghi, A (2011). Influence of foliar and soil applications of iron and manganese on soybean dry matter yield and iron-manganese relationship in a Calcareous soil, *Asian J. Crop Sc.*, **5**(12):1550-1556.

Mortvedt, JJ and Gilkes, RJ (1993). Zinc fertilizers. In: Robson AD (ed) Zinc in soils and plants. Kluwer, Dordrecht, *The Netherlands* 33-44.

Mortvedt, JJ, Cox, FR Shuman LM and Welch. RM (1991). Micronutrients in Agriculture, 2nd Edition. (Eds.): J.J. Mortvedt, F.R. Cox, L.M. Shuman and R.M. Welch. Soil Sc. Soc. Am.: Madison, WI.

Mosali, J, Girma, K, Teal, RK, Freeman, KW, Martin, KL, Lawles JW, and Raun, WR, (2006). Effect of foliar application of phosphorus on winter wheat grain yield, phosphorus uptake and use efficiency. *J. Plant Nutr.* **29**:2147-2163.

Motowicka-Terelak, T, Terelak, H (2000). Siarka w glebach iroslinach Polski. Fol. Univ. *Agric. Stetin., Agric.* **204**(81): 7–16.

Motta, ACV, Serrat, BM, Reissmann, CB and Dionisio, JA, eds. (2007). Micronutrientes na rocha, no solo e na planta. Curitiba, Universidade Federal do Parana, pp. 246.

Mucharromah, E and Ku, J (1991). Oxalate and phosphates induce systemic resistance against diseases caused by fungi, bacteria and viruses in cucumber. *Crop Prot.*, **10:** 266-270.

Muhammad Hussain, Shah, SH and Nazir, MS (2002). Effect of foliar application of calcium cum magnesium on different agronomic traits of three genotypes of Lentil (Lens culinaaris Medic). *Pakistan Journal of Agricultural Sciences*, **39:** 123-125.

Nahed, Abd El-Aziz G, Balbaa, LK (2007). Influence of tyrosine and zinc on growth, flowering and chemical constituents of *Salvia farinacea* plants. *J. Appl. Sci. Res.*, 3(11): 1479-1489.

Namdeo, KN, Sharma, JK and Mandoli, KC (1992), Effect of foliar feeding of micro-nutrients on rainfed hybrid cotton (JK Hy. 1). *Crop Research*, **5:** 451-455.

Nannette, R (2011). Magnesium sulphate Hydrate (Epsomite) Boosts Growth in Tomatoes, Peppers and Roses, How to use magnesium sulphate in the garden, http://voices.yahoo.com/magnesium-sulphate-hydrate-epsomite-boosts-growth-in-8638139.html?cat=6

Nardozzi, C (2012). Fertilize with epsom salts, National Gardening Association. http://www.garden.org/articles/articles.php?q= show&id=68&page=1.

Nayyar, VK, Sadanas, US and Takkar, PN (1985). Methods and rates of application of Mn and its critical levels for wheat following rice on coarse textured soils. *Fert. Res.* 8: 173-178.

Nevin, JM, and Lovatt, CJ (1990). Problems with urea-N foliar fertilization of avocado. *Summary of Avocado Research*, **1:** 15-16.

Nicoulaud, BAL and Bloom, AJ (1996). Absorption and assimilation of foliarly applied urea in tomato. *J. Am. Soc. Hort. Sci.*, **121:** 1117-1121.

NIPAN LLC, (2011). "Nickel Plus for Cotton" Accessed, March 22, HS1191 http://www.nickelplus.biz/id65.html.

Njuguna, MN, Macharia, M, Mwangi, HG, Kamwaga, JN, Ngari, CM, Ndembei, NW and Wasike, VW (2009). Effect of foliar fertilisation on wheat in marginal areas of Eastern Province of Kenya, African Crop Science Society, African Crop Science Conference Proceedings, 9: 209–211.

Oliver, DP, Tiller, KG Wilhelm, NS, McFarlane, JD and Cozens, GD (1997). Effect of soil and foliar applications of zinc on cadmium concentration in wheat grain, *Aus. J. Exp. Agric.*, 37(6): 677-681.

Oosterhuis, D (2009). Foliar fertilization: mechanisms and magnitude of nutrient uptake. In 'Proceedings of the fluid forum'. 15-17 February, 2009, Phoenix, Arizona.

Oosterhuis, DM (1998). Foliar fertilization of cotton with potassium in the USA. Proc. Symp. "Foliar Fertilization: Cairo Eds.

Oosterhuis, DM (1999). Foliar fertilization. In: Proc. Beltwide Cotton Conf., Orlando, pp: 26-29. National Cotton Council, Memphis, Tenn.

Oosterhuis, DM and Bondada. BR (2001). Yield response of cotton to foliar nitrogen as influenced by sink strength, petiole, and soil nitrogen. *J. Plant Nutr.* **24**:413-422.

Oosterhuis, DM, Miley, WN and Janes, LD (1993). A summary of foliar fertilization of cotton with potassium nitrate, Arkansas 1989-1992. Arkansas Fertility Studies. University of Arkansas, Arkansas Agriculture Experiment Station, Bulletin 425: 97-100.

Orlovius, K (2002). Effect of foliar fertilisation with magnesium, sulphur, manganese and boron to sugar beet, oilseed rape, and cereals, *Pl. Nutr. Developments Pl.Soil Sc.*, **92**: 788-789.

Ozturk, L, Yazici, MA, Yuce, IC, Torun, A, Cekic, C, Bagci, A, Ozkan, H, Braun, HJ, Sayers, Z, Cakmak, I (2006). Concentration and Nyczepir and Wood (2012) pp. 304 localization of zinc during seed development and germination in wheat. *Physiol. Plant* **128**: 144-152.

Paikray, RK, Mishra, PJ, Mohapatra, AK, Haldar, J and Panda, S (2001). Response of niger (*Guizotia abyssinica*) to secondary and micronutrients in acid red soil. *An. Agric. Res.*, **22**: 140-142.

Pandey, SK, Bahuguna, RN, Pal, M, Trivedi, AK, Srivastava, JP (2010). Effects of pre-treatment and foliar application of zinc on growth and yield components of mungbean (*Vigna radiata* L.) under induced salinity. *I.J. Pl. Physiol*, **15**(2): 164-167.

Parvez, K, Muhammad, YM, Muhammad, I and Muhammad, A (2009). Response of wheat to foliar and soil application of urea at different growth stages. *Pak. J. Bot.* **41**(3): 1197-1204.

Patel PC (2006). Effect of Foliar Application of Iron and Sulphur in Alleviation of Iron Chlorosis in Acid lime (*Citrus aurantifolia*, Swingle). 18th World congress of soil science, Philladelphia, Pennsylvania, USA.

Patil, UD and Malewar, GV, (1994). Yield and chlorophyll content of cotton as influenced by micronutrient spray. *Journal of Cotton Research and Development*, **8**: 189-192.

Patil, VM and Somawanshi, RB (1983). *Commun. Soil Sci. and Plant Anal.*, **14**: 471-480.

Patricia Rodriguez-Lucena, Lourdes Hernandez-Apaolaza, and Lucena, JJ (2010). Comparison of iron chelates and complexes supplied as foliar sprays and in nutrient solution to correct iron chlorosis of soybean, *J. Plant Nutr. Soil Sci.*, **173**: 120–126.

Penter, MG and Stassen, PJC (2000). The effect of pre- and post-harvest calcium applications on the post-harvest quality of Pinkerton avocados. South African Growers' Association Yearbook **23**: 1-7.

Perveen, S, and Hafeez-ur-Tahman (2000). Effect of foliar application of zinc, manganese and boron in combination with urea on the yield of sweet orange, *Pak. J. Agric. Res.*, **16**(2): 135-141.

Peryea, F and Willemsen. K, (2000). Nutrient sprays, WSU-Tree Fruit Research & Extension Center. http://www.tfrec.wsu.edu/horticulture nutspray.html# Calcium.

Pettigrew, WT (1999). Potassium deficiency increases specific leaf weights of leaf glucose levels in field- grown cotton. *Agron. J.*, **91**: 962-968.

Poovaiah, BW, Glenn, GM, and Reddy, ASN (1988). Calcium and fruit softening: physiology and biochemistry. *Horticultural Reviews* **10**: 107-152.

Porro, D, Comai, M, Dorigoni, A, Stefanini, M, Ceschini, A (2002). Manganese foliar application to prevent leaf drop, *Acta Horticulturae* 594: International Symposium on Foliar Nutrition of Perennial Fruit Plants.

Potarzycki, J and Grzebisz, W (2009). Effect of zinc foliar application on grain yield of maize and its yielding components. *Plant Soil Environ.*, **55**(12): 519-527.

Prasad, R (2005). Rice-wheat cropping system. *Adv. Agron.* **86**: 255-339.

Qaseem, SM, Afridi, MM and Samiullah, RK (1978). Effect of leaf applied phosphorus on the yield characteristics of ten barley varieties. *I. J. Agric. Sci.* **48**: 215-217.

Raese, JT and Drake., SR (1993). Effects of preharvest calcium sprays on apple and pear quality. *J. Pl. Nutr.*, **16**: 1807-1819.

Rahman, H, Sabreen, S, Alam, S, and Kawai. S (2005). Effects of nickel on growth and composition of metal micronutrients in barley plants grown in nutrient solution. *J. Pl. Nutr.*, **28**: 393-404.

Rajakumar, D and Gurumurthy S (2008). Effect of plant density and nutrient spray on the yield attributes and yield of direct sown and polybag seedling planted hybrid cotton. *Agric. Sci. Digest*, **28**(3): 174 – 177.

Rajeswari, RV (1996). Foliar application of growth regulators and nutrients on boll development and yield in cotton. *J. Ind. Soc. Cotton Improvement*, **21**:71.

Rakkiyappan, P Thangavelu S and Radhamani, R (2002) *I. J. Sugarcane Technol.*, 4(1&2): 33–37.

Ramos, JM, Garcia del-Moral, LF, Molina-Cano, JL, Salamanca, P, Roca de Togores F (2008). Effects of an Early Application of Sulphur or Etephon as Foliar Spray on the Growth and Yield of Spring Barley in a Mediterranean Environment, *J. Agron. Crop Sc.*, **163** (2): 129-137.

Rana Munns and Mark Tester (2008). *Annu. Rev. Plant Biol.* 2008.59:651-681. Downloaded from arjournals.annualreviews.org by CSIRO on 06/21/08.

Randall, P, Spencer, K and Freney, J (1981). Sulphur and Nitrogen fertilizer effects on wheat. I. Concentrations of sulphur and nitrogen and the nitrogen to sulphur ratio in grain, in relation to the yield response. *Aust. J. Agric. Res.* 32(2): 203–212.

Ranjbar, R, Eshghi, S and Rostami, M (2011). Effect of foliar application of nickel sulphate and urea on reproductive growth and quantitative and qualitative characteristics of strawberry fruit (*Fragaria ananassa* Duch. cv. Pajaro).

Rashid, A (1996). Secondary and micronutrients. In: Soil Science. A. Rashid and K. S. Memon (Managing Authors). National Book Foundation, Islamabad, Pakistan. p. 341-385.

Rashid, A and Rafiq. E (2000). Boron and zinc fertilizer use in cotton: importance and recommendation. PARC, Islamabad.

Rathinavel, K, Dharmalingam, C and Paneersewam, S (1999). Effect of micronutrient on the productivity and quality of cotton seed Cv. TCB 209 (*Gossypium barbadense* L.). *Madras Agric. J.*, **86**: 313-316.

Raven, PH and Johnson GB (1999). Evolutionary history of plants. In 'Biology'. pp 645-664. (WCB/McGraw-Hill: Boston, MA).

Reuveni, M, Agapov, V and Reuveni, R (1993). Induction of systemic resistance to powdery mildew and growth increase in cucumber by phosphates. *Biol. Agric. and Hortic.* **9**: 305-315.

Roberts, RK, Gerkman, JM and Howard, DD (2000). Soil and foliar applied boron in cotton production on economic analysis. *J. Cotton Sc.*, **4** (3): 171-177.

Roberts, RK, Gerloff, DC and Howard, DD (1997). Foliar Potassium on Cotton- a profitable supplement to broadcast potassium application on low testing soils. *Better Crops*, **81**(1).

Robinson, JB and Burne, P (2000). Another look at the Merlot problem: Could it be Molybdenum deficiency? In 'The Australian Grape grower and Winemaker' pp. 21-22.

Roemheld, V and El-Fouly, MM (1999). Foliar nutrient application Challenge and limits in crop production. Proceedings of the 2nd International Workshop on Foliar Fertilization, Bangkok, Thailand, 4-10.

Rohilla, R and Gupta, U (2012). Mean centering of ratio spectra as a new spectrophotometric method for the analysis of binary mixtures of Vanadium and Lead in water samples and alloys, *Res. J. Chem. Sci.* **2**(9): 22-29.

Rubes, L (1984). Effect of foliar application of magnesium on growth and yield of peas Cv. Rerksasem, (1989) pp. 297.

Ruter, JM (2004) Mouse ear disorder on river birch caused by nickel deficiency. (Abstr.). *HortScience* **39**: 892.

Salah, M (2006). The effect of nitrogen and manure fertiliser on grain quality, baking and rheological properties of wheat grown in sandy soil. *J. Sc. Fd Agric.*, **2**: 205-211.

Salwa, AI Eisa, M, Abass M and Behary SS (2010). Amelioration Productivity of Sandy Soil by using Amino Acids, Sulphur and Micro-nutrients for Sesame Production, *J. Am. Sc.*, **6**(11): 250-257.

Sangakkara, UR, Frehner M and Nosberger, J (2000). Effect of soil moisture and potassium fertilizer on shoot water potential, photosynthesis and partitioning of carbon in mungbean and cowpea. *J. Agron. Crop Sci.*, **185**: 201-207.

Sangale, PB, Patil, GD and Daftardar, SY (1981). Effect of foliar application of Zinc, Iron and Boron on yield of safflower. *J. Maharashtra Agric. Universities*, **6**: 65-66.

Sarkar, D, Mandal, B and Kundu, MC (2007). Increasing use efficiency of boron fertilisers by rescheduling the time and methods of application for crops in India. *Plant Soil*. **301**: 77-85.

Schonherr, J (1976). Water permeability of isolated cuticular membranes: the effect of pH and cations on diffusion, hydrodynamic permability and size of polar pores in cutin matrix. *Planta* **128:** 113-126.

Schonherr, J (2006). Characterization of aqueous pores in plant cuticles and permeation of ionic solutes. *J. Exp. Bot.* **57:** 2471-2491.

Scotford, IM, and Miller, PCH (2004). Combination of spectral reflectance and ultrasonic sensing to monitor the growth of wheat. *Biosystems Engineering* **87:**27-38.

Scott, NW, Dyson, PW, Ross, Y and Sharp, GS (1984). The effect of sulphur on the yield and chemical composition of winter barley, *J. Agric. Sci.* (Camb.) **103:** 699-702.

Seadh, SE, El-Abaday, MI, El-ghamry, AM, and Farouk, S (2009). Influence of micronutrients foliar application and nitrogen fertilization on wheat yield and quality of grain and seed. *Journal of Biological Sciences*, 9(8): 851-858.

Shao, G, Chen, M, Wang, W, Mou, R and Zhang, G (2007). Iron nutrition affects cadmium accumulation and toxicity in rice plants. *Plant Growth Reg.*, **53:** 33–42.

Sharaf, AI and El-Naggar. AH (2003). Response of carnation plant to phosphorus and boron foliar fertilization under greenhouse conditions. *Alexandria J. Agril. Res.*, **48**(1): 147-158

Sharma, AP, Taneja, AD, Madan, VK, Bishnoi, LK and Kairon, MS (1993). Studies on method and time of DAP application on yield and quality of *Gossypium hirsutum* cotton. *J. Ind Soc. Cotton Improvement*, **18:** 66-70.

Sharma, SK, Singh, H and Kohli, UK (1999). Influence of boron and zinc on seed yield and quality in radish. *Seed Res.*, **27:** 154-158.

Shastri, G, Thiagarajan, CP, Srimathi, P, Malarkodi, K and Venkatasalam, EP, (2000). Foliar application of nutrient on the seed yield and quality characters of non aged and aged seeds of cotton Cv. MCU5. *Madras Agric. J.*, **87:** 202-208.

Shitole, SM and Dhumal, KN (2012). Influence of foliar applications of micronutrients on photosynthetic pigments and organic constituents of medicinal plant *Cassia angustifolia* Vahl. *An. Bio. Res.*, 3(1):520-526.

Shukla, AK (2011). Effect of foliar application of calcium and boron on growth, productivity and quality of Indian gooseberry (*Emblica officinalis*), *The Indian Journal of Agricultural Sciences*, 81(7): 628–32.

Sidiqui, S, and F Bangerth. (1995). Differential effect of calcium and strontium on flesh firmness and properties of cell walls in apples. *Journal of Horticultural Science* **70:** 949–953.

Silberbush, M (2002). Simulation of ion uptake from the soil. In Plant Roots: The Hidden Half, 3rd ed.; Waisel, Y., Eshel, A., and Kafkafi, U. (eds.); Marcel, Dekker: New York, 651–661.

Silva, NMDA, Carvalo, LH, Chiavagaio, ET, Sabino, NP and Hiroce, R, (1982). Effect of rate of boron application at sowing time on cotton plants in different soils. Bragantia, 41: 181-191.

Sinclair, JB (1993). Soybeans. p. 99–103. In W.B. Bennett (ed.) Nutrient deficiencies & toxicities in crop plants. APS Press, St. Paul, MN.

Singh, AK, McIntyre, LM and Sherman, LA (2003). Microarray analysis of the genome-wide.

Singh, AR, Singh, B and Singh, A (1990). Efficacy of nitrogen and boron on the growth, yield and quality of potato (*Solanum tuberosum* L.) Var Kufri Chamtkar. *Progressive Horticulture*, **22:** 104-107.

Singh, MV (2001). Importance of sulphur balanced fertilizer use in India. *Fertilizer News*, **46:** 13-18.

Singh, MV (2007). Efficiency of seed treatment for ameliorating zinc deficiency in crops. Proceeding of Zinc Crop Conference, Istanbul, Turkey.

Sinha, P, Dube, BK and Chatterjee, C (2001). Boron stress induced changes in cotton. *An. Agric. Res.*, **22** (3): 365-370.

Skinner, PW, and Matthews, MA (1990). A novel interaction of magnesium translocation with the supply of phosphorus to roots of grapevine (*Vitis vinifera* L.). Plant, Cell and Environment, 13, 821-826. http://dx.doi.org/10.1111/j.1365-3040.1990.tb01098.x

Soomro, AW, Arain, AS, Soomro, AR, Tunio, GH, Chang, MS, Leghari, AB and Magsi. MR (2001). Evaluation of proper fertilizer application for higher cotton production in Sindh. OnLine *J. Biol. Sci.* 1(4): 295-297.

Stern, DI (2005). Global sulphur emissions from 1850 to 2000. Chemosphere. **58**(2): 163–175.

Stiegler, C, Richardson, J, McCalla, M, Summerford, J and Roberts, T (2009). Nitrogen rate and season influence ammonia volatilization following foliar application of urea to putting green turf. Arkansas Turfgrass Report 2008, *Ark. Ag. Exp. Stn. Res.* Ser. **568:**110-115.

Stoyanow, D and Gikov, GL (1990). Foliar application of trace elements to cotton. *Pochvozhanie*, **25:** 45-46.

Suwanarit, A and Sestapukdee, M (1989). Stimulating effects of foliar K-fertilizer applied at the appropriate stage of development of maize: A new way to increase yield and improve quality. *Plant and Soil*, **120** (1): 111-124.

Swietlik, D, and Faust. M (1984). Foliar nutrition of fruit crops. *Horticultural Reviews*, Vol. 6. Symptoms and causes. *Acta Horticulturae* **721:** 83-97.

Swietlik, D, Faust, M (1994). Foliar nutrition of fruit crops. *Hort. Rev.*, **6:**287-355.

Sztuder, H and Swierczewska, M, (1998). Efficiency of foliar application for leguminous crops with magnesium and magnesium combined with micro-nutrients. Proc. of the 5th Int. Symp., Wansaw, Poland, pp. 24-25.

Taylor, R (2010). Weekly crop update, check soybeans for Manganese (Mn) deficiency, http://agdev.anr.udel.edu/weeklycropupdate/?p=2091

Tea, I, Genter, T, Naulet, N, Boyer, V, Lummerzheim, M and Kleiber, D (2004). Effect of Foliar sulphurand Nitrogen Fertilization on Wheat Storage Protein Composition and Dough Mixing Properties. *Cereal Chemistry*. **81**(6): 759-766.

Tea, I, Thierry Genter, Norbert Naulet, Marie Lummerzheim, Didier Kleiber (1990).

Interaction between nitrogen and sulphur by foliar application and its effects on flour bread-making quality, *Soil Sci Soc Am J.*, **54**:257-262.

Teixeira, IR, Borem, A., Araujo, G,A.A., Lucio, R. and Fontes, F. (2004). Manganese and zinc leaf application on common bean grown on a "Cerrado" soil. *Scientia Agricola*, **61**(1): 77-81.

Tejada, M and Gonzalez, JL (2003). Effects of foliar application of a byproduct of the two-step olive oil mill process on rice yield. *Euro. J. Agron.*, **21**(1): 31-40.

Terry, N and Low, G. (1982) *J. Plant Nutr.*, **5**: 301-310.

Tisdale, SL, Nelson WL and Beaton, JD (1990). Soil fertility and fertilizer. Elements required in plant nutrition. 4ᵗʰ Ed. Maxwell McMillan Publishing, Singapore, 52-92.

Tisdale, SL, Nelson, WL Beaton, JD and Havlin. J.L. (1993). *Soil fertility and fertilizers*. 5ᵗʰ edition. Macmillan Publishing , Co., NY. 634 p.

Trolove, S (2007). Strategies to alleviate magnesium deficiency in perennial crops. Final Report, AGMARDT.

Ullagaddi, MS, (2000). Effect of dates of sowing and chemicals on seed yield and quality of Male parent SB (YF)-425 cotton (*Gossypium barbadonse* L.). Thesis, University of Agricultural Sciences, Dharwad.

Umajyothi, K and Shanmugavelu, KG (1985). Studies on the effect of triacontanol, 2,4- D and boron on the yield and quality of chilli fruit and seed. *Seed Res.*, **18**:114-116.

Varvel, GE (1983). Effect of banded and broadcast placement of Cu fertilizers on correction of Cu deficiency. *Agron. J.*, **75**: 99-101.

Vasilas, BL, Legg, JO and Wolf, DC (1980). *Agron. J.* **72**: 271-275.

Verga, B and Svecnjak, Z (2006). The effect of late-season urea spraying on grain yield and quality of winter wheat cultivars under low and high basal nitrogen fertilization. *Field Crops Res.* **96**:125-132.

Vitosh, ML Warncke, DD and Lucas, RE (1994). Secondary and Micronutrients for vegetables and field crops, Department of Crop and Soil Sciences, Michigan State University Extension , E-486. https://www.msu.edu/~warncke/E0486.pdf

Wankhade, ST, Meshram, LD, and Kene, HK (1994). Impact of foliar feeding of nutrients on hybrid seed production. *Punjabrao Krishi Vidya Peeth Research Journal*, **18**: 127-128.

Weinbaum, SA (1978). Feasibility of satisfying total nitrogen requirement of non-bearing prune trees with foliar nitrate. *Hort Science* **13**:1.

Welch, RM (1995). Micro-nutrient nutrition of plants. *Crit. Rev. Pl Sc.*, **14**(1): 49–82.

Wen, G, Schoenau, JJ, Yamamoto, T and Inoue, M (2001). A Model of Oxidation of An Elemental sulphur Fertilizer in Soils. *Soil Science.* **166**. 9: 607–613.

Wiley-VCH and LASTWiley-VCH (2007). "Ullmann's Agrochemicals," vol. 1, chap. 3.2 liquid fertilizers pp. 36.

Williams CMJ, Maier, NA and Bartlett, L (2004). Effect of molybdenum foliar sprays on yield, berry size, seed formation, and petiolar nutrient composition of "Merlot" grapevines. *J. Pl. Nutr.*, **27**: 1891-1916.

Williams CMJ, Maier NA, Bartlett L (2003) Nutrition, including molybdenum (Mo) for fruitfulness in grapevines. In 'Proceedings of the flower formation, flowering and berry set in grapevines workshop', May, 2003. (Eds GM Dunn, PA Lothian, T Clancy) pp. 92. (Grape and Wine Research and Development Corporation and Department of Primary Industries, Victoria).

Witte, CP, Tiller, SA, Taylor, MA and Davies. HV (2002). Leaf urea metabolism in potato. Urease activity profile and patterns of recovery and distribution of ^{15}N after foliar urea application in wild-type and urease-antisense transgenics. *Plant Physiology* **128**: 1129-1136.

Wittwer, SH, Bukovac, MJ and Tukey, HB (1963). Advances in foliar feeding of plant nutrients. In MHMcVickar, ed, Fertilizer Technology and Usage. *Soil Science Society of America*, Madison, WI, pp: 429-455.

Wojciechowska, R and Kowalska, I (2011). The effect of foliar application of urea, Mo and BA on nitrate metabolism in lettuce leaves in the spring and summer-autumn seasons, Folia Hort., 23(2): 119-123.

Woltz, SS Jones, JP and Scott, JW (1992). Sodium chloride, nitrogen source, and lime influence fusarium crown rot severity in tomato. *HortScience*, **27**(10):1087-1088.

Wood, BW (2004). Nickel nutrition of pecan. Pecan Grower **16**: 4-7.

Wood, BW, Nyczepir, AP and Reilly, CC (2006). Field deficiency of nickel in trees:

Wood, BW, Reilly, CC and Nyczepir, AP (2004). Mouse-ear of pecan: a nickel deficiency, *Hort Science* **39**: 1238-1242.

Xiong, ZT, Liu, C, and Geng. B (2006). Phytotoxic effects of copper on nitrogen metabolism and plant growth in *Brassica pekinensis* Rupr. Ecotoxicology and Environmental Safety **64**: 273-280.

Yadav, RL, Sen, NL, Fageria, MS and Dhaka, RS, (2002). Effect of nitrogen and potassium fertilization on quality bulb production of onion. *Haryana J. Hort. Sci.* **31**(3&4): 297-298.

Yallappa and Eshanna, MR (2006). Effect of foliar application of micronutrients on seed yield and quality of cotton hybrid DHH-11, Department of Seed Science and Technology, College of Agriculture, Dharwad, University of Agricultural Sciences, Dharwad.

Yamazaki, H and Hoshina,T (1995). Calcium nutrition affects resistance of tomato seedlings to bacterial wilt. *Hort Science*, **30**(1): 91-93.

Yilmaz, A, Ekiz, H, Torun, B, Gultekin, I, Karanlik, S, Bagci, SA, Cakmak, I (1997). Effect of different zinc application methods on grain yield and zinc concentration in wheat grown on zinc-deficient calcareous soils in Central Anatolia. *J. Plant Nutr.* **20**: 461-471.

Yousef, N, Saeid Zehtab-Salmasi, Safar Nasrullahzadeh, Nosratollah Najafi and Kazem Ghassemi-Golezani (2010). Effects of foliar application of micro-nutrients (Fe and Zn) on flower yield and essential oil of chamomile (*Matricaria chamomilla* L.) *Journal of Medicinal Plants Research*, **4**(17): 1733-1737.

Zakaria MS, Mahmoud H Mahmoud and Amal H El-Guibali. (2008). Influence of potassium fertilization and foliar application of zinc and phosphorus on growth, yield components, yield and fiber properties of Egyptian cotton (*Gossypium barbadense* L.) *Journal of Plant Ecology*, **1**(4): 259-270.

Zaman, MS Hashem, MA Jahiruddin M and Rahim MA (2011). Effect of sulphur fertilization on the growth and yield of garlic (*Allium sativum* L.). *Bangladesh J. Agril. Res.* **36**(4): 647-656.

Zeidan, M, Manal, S, Mohamed, F, and Hamouda, HA (2010). Effect of foliar fertilization of Fe, Mn and Zn on wheat yield and quality in low sandy soils fertility. *World Journal of Agricultural Sciences*, **6**: p.696-699.

Zhao, A, Lu, X, Chen, Z, Tian, X, and Yang, X (2011). Zinc fertilization methods on zinc absorption and translocation in wheat. *Journal of Agricultural Science*, **3**(1): 28-35.

Zhao, FJ, Hawkesford, MJ, Warrilow, AGS, McGrath SP, and Clarkson, DT (1996). Responses of two wheat varieties to sulphur addition and diagnosis of sulphur deficiency. *Plant and Soil*, **181**: 317-327.

Zhao, FJ, MJ Hawkesford and SP McGrath, (1999). Sulphur assimilation and effects on yield and quality of wheat. *J. Cereal Sci.*, **30**: 1-17.

Zhu, H, Zhang X and Sun, C (1996). Characteristics of micro-nutrients uptake by rape plants and methods of B and Zn application. *Oil Crops of China*, **18**(2): 59-61.

Zilkah, S, Wiesmann, Z, Klein, I, and David, I (1996). Foliar applied urea improves freezing protection to avocado and peach. *Sc. Hort.*, **66**: 85-92.

Ziolek, E and Ziolek, W (1987). The effect of trace element application on yield and quality of soybean seeds. *Agraria*, **26**: 195-207.

Zoz, T, Fabio, S, Rubens, F, Deise, DC, and Edleusa, PS (2012). Response of wheat to foliar application of zinc, *Cienc. Rural* **42**(5).

■

INTRODUCTON

Indian economy is growing day by day and being a more than trillion dollar economy, now Indian economy is no more backward economy by any yardstick due to its inclusive development (growing with 8 plus per cent GDP) left behind *Hindu* rate of growth (below 5 per cent GDP), provided space for all round development. Information and communication technology (ICT) is one among the fastest growing service sector, neither any one nor agriculture is unaffected by this revolution. Increased accessibility of information and communication tools and techniques by common man and it's adventitious and multiple usages prove this boon for humanity. Like other arena of life, agricultural production system is also harnessing its (ICT) capabilities for its own benefits in particulars as well as for serving humanity with added responsibilities in general. Being an integral part of our cultural heritage and as a part and parcel of daily like, Agriculture in India is never treated as business because. Presently farming is facing bigger challenge than ever since civilization for its sustainability in profitable manner obviously farming activities are potentially threatened by global warming and impending climate change circumstances. Conceivably precision farming is befitting reply not only for improving production but also enhancing factor productivity. Blend of various precision farming techniques that can suit for site specific agricultural production coupled with the help of faster expanding information and communication technology can not only help in mitigating changing climate but also provide unique opportunity for profitable agricultural production system. Greater responsibilities are sensed by ICT by virtue of its potential utility for fragile Indian agricultural production system (Barnett et al., 1997).

This chapter is based on three basic component and due weighted has been given to (1) ICT, (2) Precision farming and only relevant introductory information has been narrated in the beginning about (3)climate change and its impact on agriculture in brief. It is because nature and scope of this chapter does not permit detailed discussion on climate chance.

[1] ICAR- Research Complex for Eastern Region Patna, Bihar-800 014

[2] Division of Agronomy, Sher-e-Kashmir University of Agricultural Sciences & Technology of Kashmir, Srinagar-191121

[3] Scientist, Plant Physiology, PDFSR (ICAR), Modipuram, Meerut (UP) - 250 110

CHANGING CLIMATE SCENARIO

Climate is usually the average weather or more rigorously as the statistical description in terms of the mean and variability of relevant quantities over a period of time ranging from months to thousands or millions of years. Climate change may be due to natural internal processes or external forces, or to persistent anthropogenic changes in the composition of the atmosphere or in land use, consequently the impact of rise in temperature and CO_2 concentration on crops yield may be negative. It is estimated that 2°C increase in air temperature could decrease rice yield by about 0.75 tons ha^{-1} in high yielding areas. It was further indicated that due to climate change, there is reduction in crop yield of 10 to 40 per cent at the present yield level by the turn of the century. This complex situation not only present great challenge to humanity but also provide unique opportunity to combats this natural cum man made reality. If not managed with sensitivity it may leads to havoc for human civilization. The green house gases (GHGs) are the main culprits of the global warming and climate change. The GHGs like carbon dioxide, methane and nitrous oxide are playing hazards in the present times. Why is there a need for agriculture to aid in the mitigation of climate change and concurrently better adapt itself for change? Despite of occupying 40-50 per cent of the total land surface agricultur contribution to GHGs is meager to the extent of 10-12 per cent of total global anthropogenic greenhouse gas (GHG) emissions. Further agriculture contributes 47 per cent and 58 per cent of total anthropogenic emissions of (non greenhouse gas) N_2O and CH_4, respectively. N_2O and CH_4 have 21 and 310 times the "global warming potential" of CO_2. However, CO_2 has large annual exchanges between the atmosphere and agricultural lands but the net flux is estimated to be approximately balanced, so accounts for less than 1 per cent of global anthropogenic CO_2 emissions. Agricultural N_2O and CH_4 emissions have increased by nearly 17 per cent from 1990 to 2005. Global warming potential is the ratio of the warming that would result from the emission of one kilogram of a greenhouse gas to that from the emission of one kilogram of carbon dioxide over a fixed period of time such as 100 years. The latest assessment report of the Intergovernmental Panel on Climate Change (IPCC) confirms that the global average temperature increased by 0.74°C over the last 100 years; and the projected increase in temperature by 2100 is about 1.8 to 4. 0°C. Global warming poses a potential threat to agricultural production and productivity throughout the world and this might affect the crop yields, incidence of weeds, pests and plant diseases; and the economic costs of agricultural production. Crop productivity is projected to decrease even by small rise in temperature (1-2°C) at the lower latitudes, especially in the seasonal dry and tropical regions (IPCC, 2007).

Temporal variability is driven to a large extent by inter-annual variations in local weather and how farmers respond to these variations. Our best estimate is that agriculture accounts for about 10-12 per cent of the total global anthropogenic emissions of GHGs or between 6 and 8 Gt CO_2e per annum. Non-CO_2 GHGs from management operations = 6.2 Gt CO_2e. Energy related CO_2 emissions (including emissions from manufacture of fertilizer) = 0.6 Gt CO_2e. Agriculture accounts for between 59 and 63 per cent of the world's non-CO_2 GHG emissions. This sector accounts for 84 per cent of the global N_2O emissions and 54 per cent of the global

CH_4 emissions. These emissions are principally from six sources (1) N_2O from soil, (2) N_2O from manure management (3) CH_4 from enteric fermentation CH_4 from manure management (4) CH_4 from rice cultivation(5) CH_4 from other sources(6) Savannah burning, Burning of agricultural residues, burning from forest clearing and agricultural soils (CH_4) (Yadav and Subba, 2000).

INFORMATION AND COMMUNICATION TECHNOLOGY (ICT) VIS - A-VIS E-AGRICULTURE

The agricultural sector in India is currently passing through a difficult phase. India is moving towards an agricultural emergency due to lack of attention, insufficient land reforms, defective land management, non-providing of fair prices to farmers for their crops, inadequate investment in irrigational and agricultural infrastructure in India, etc. Country's food production and productivity is declining while its food consumption is increasing. The position has further been worsened due to use of food grains to meet the demands of bio fuels. Even the solution of import of food grains would be troublesome, as India does not have ports and logistical systems for large scale food imports. The application of ICT in agriculture is increasingly important. E-Agriculture is an emerging field focusing on the enhancement of agricultural and rural development through improved information and communication processes. More specifically, e-Agriculture involves the conceptualization, design, development, evaluation and application of innovative ways to use ICT in the rural domain, with a primary focus on agriculture. E-Agriculture is a relatively new term and its scope to change and evolve as our understanding of the area grows.

Agriculture production system mainly consist of (i) Crop cultivation, (ii) Water management, (iii) Fertilizer Application, (iv) Fertigation, (v) Pest management, (vi) Harvesting, (vii) Post harvest, (viii) Handling, (ix) Transporting of food/food products, (x) Packaging, (xi) Food preservation, (xii) Food processing/value addition, (xiii) Food quality management, (xiv) Food safety, (xv) Food storage (xvi) Food marketing.

All stakeholders of agriculture production system need information and knowledge about these phases to manage them efficiently. Any system applied for getting information and knowledge for making decisions in any industry should deliver accurate, complete, concise information in time or on time. The information provided by the system must be in user-friendly form, easy to access, cost-effective and well protected from unauthorized accesses. ICT can play a significant role in maintaining the above mentioned properties of information as it consists of three main technologies, *viz.* computer technology, communication technology and information management technology. These technologies are applied for processing, exchanging and managing data, information and knowledge. The tools provided by ICT are having ability to

1. Record text, drawings, photographs, audio, video, process descriptions, and other information in digital formats,

2. Produce exact duplicates of such information at significantly lower cost,

3. Transfer information and knowledge rapidly over large distances through communications networks.

4. Develop standardized algorithms to large quantities of information relatively rapidly.

5. Achieve greater interactivity in communicating, evaluating, producing and sharing useful information and knowledge.

Applications of ICT in agriculture sector: The main focus of this article is to elaborate how the achievements of ICT can be applied in Agriculture sector and its development. The main applications of ICT in Agriculture sector are listed below.

Application of office automation: The office automation is application of computers, computer networks, telephone networks and other office automation tool such as photocopy machines, scanners, printers, cleaning equipments, electronic security systems to increase the productivity of organizations. There are many governments, private and non-government organizations are involving in agriculture sector and rural development. They all have to work together to give better service to farming community. Therefore, application of office automation is one of the solutions to enhance the efficiency and interconnectivity of the employees work in all above mentioned organizations. Many computer applications such as MS Office, Internet Explorer, Open Office and other tailor made office automation software packages are providing unlimited potential to organizations and individuals to fulfill their day to day data processing requirements to give an efficient service to their customers. Large agri-business organizations are now using SAP AG ERP software solutions to automate their business processes.

Application of Knowledge Management System : Knowledge increasingly has become a vital resource. Within our communities, institutions and organizations, practical insights are needed for optimizing its use. Knowledge management needs to become an object of study. This article deals with two issues. First, using both knowledge systems concepts and tools and insights gained from comparative research, it explores the vital qualities of agricultural knowledge systems. These qualities, like the multiplicity and relative autonomy of the actors, the level of integration reached through linkage mechanisms and the coordination needed to overcome default situations, might provide leverage points for effective knowledge management. Second, it probes into a more specific definition of the tasks and areas of attention of the knowledge manager. Knowledge management can focus on various levels of a system (*e.g.* the individual, organizational, or system level) and can make use of a variety of instruments and skills. Knowledge management is the management of the organization towards the continuous renewal of the organizational knowledge base, *e.g.* creation of supportive organizational structures, facilitation of organizational members, putting IT-instruments with emphasis on teamwork and diffusion of knowledge (as *e.g.* groupware) into place. Knowledge Management; as the word implies, the ability to manage knowledge. We are all familiar with the term Information Management. This term came about when people realized that information is a resource that can and needs to be managed to be useful in an organization. From this, the ideas of Information Analysis and Information Planning came about. Organizations are now starting to look at "knowledge" as a resource as well. This means that we need ways for managing the knowledge in an organization. We can use techniques and methods that were developed as part of Knowledge Technology to analyze the knowledge sources in an organization. Using these techniques we can perform Knowledge Analysis and Knowledge Planning.

Knowledge Management is the collection of processes that govern the creation, dissemination and utilization of knowledge. In one form or another, knowledge management has been around for a very long time. Practitioners have included philosophers, priests, teachers, politicians, scribes, liberians etc. So, if Knowledge Management is such an ageless and broad topic what role does it serves in today's information age? These processes exist whether we acknowledge them or not and they have a profound effect on the decisions we make and the actions we take, both of which are enabled by knowledge of some type. If this is the case and we agree that many of our decisions and actions have profound and long lasting effects, it makes sense to recognize and understand the processes that effect or actions and decision and, where possible, take steps to improve quality these processes and in turn improve quality of those actions and decisions for which we are responsible? Knowledge management is not a, "a technology thing" or a, "computer thing" If we accept the premise that knowledge management is concerned with the entire process of discovery and creation of knowledge, dissemination of knowledge and the utilization of knowledge then we are strongly driven to accept that knowledge management is much more than a "technology thing" and that elements of it exist in each of our jobs.

Knowledge increasingly has become a vital resource. Within our communities, institutions and organizations, practical insights are needed for optimizing its use. Knowledge management needs to become an object of study. This article deals with two issues. First, using both knowledge systems concepts and tools and insights gained from comparative research, it explores the vital qualities of agricultural knowledge systems. These qualities, like the multiplicity and relative autonomy of actors, the level of integration reached through linkage mechanisms and the coordination needed to overcome default situations, might provide leverage points for effective knowledge management. Second, it probes into a more specific definition of tasks and areas of attention of the knowledge manager. Knowledge management can focus on various levels of a system (*e.g.* the individual, organizational or system level) and can make use of a variety of instruments and skills.

Application of e-learning: E-Agriculture is an emerging field focusing on the enhancement of agricultural and rural development through improved information and communication processes. More specifically e-Agriculture involves the conceptualization, design, development, evaluation and application of innovative ways to use information and ICT in the rural domain with a primary focus on agriculture. E-Agriculture is a relatively new term and we fully expect its scope to change and evolve as our understanding of the area grows. E-Agriculture is one of the action lines identified in declaration and plan of action of the World Summit on the Information Society (WSIS). The "Tunis Agenda for the Information Society," published on 18 November, 2005, emphasizes the leading facilitating roles that UN agencies need to play in the implementation of the Geneva Plan of Action. The Food and Agriculture Organization of the United Nations (FAO) has been assigned the responsibility of organizing activities related to the action line under C.7, ICT Applications on E-Agriculture. Some of the benefits of ICT for the improvement and strengthening of agriculture sector in India include timely information on weather forecasts and calamities, better and spontaneous agricultural practices, better marketing exposure and pricing, reduction of agricultural risks and enhanced incomes, better awareness and information, improved networking and communication, facility of online trading and e-commerce, better representation at various forums, authorities and platform etc. E-agriculture can play a major role in the increased food production and productivity in India.

Application of e-commerce : E-commerce is business conducted via the Internet as a world-wide distribution channel for goods and services (Lange, 1996). Unlike previous forms of electronic commerce – via fax, telephone and telex, e-commerce represents a qualitative jump in the ability to bring buyers and sellers closer together. In the process, buyers and sellers can communicate more effectively about their requirements, product and price information is more readily available and just in time delivery of customised goods is more easily achieved. The development of new technology, such as e-commerce, is of major importance because of it's impacts on the relationship between efficiency, market structures and market outcomes. While most media attention has focused on the business to consumer side of e-commerce, there are more important and fundamental changes taking place in the business to business space. In this we are concerned with the impact of e-commerce in the business to business space. In particular, we are interested in e-commerce's impact on major agri-businesses in New Zealand. E-commerce draws on at least three fundamentally new developments.

- It is based on the existence of a ubiquitous telecommunications infrastructure and the hardware required moving large amounts of data cheaply.
- It relies on a simple to use and cheap browser.
- It allows a firm to gain information from purchasers of its products and suppliers of its inputs, which is critical to the running of its business. It also supplies that information to managers in an easily accessible way and in an understandable fashion.

Application of ICT for managing agricultural resources and services : The ultimate solution to this problem is a solid political will along with a competent bureaucracy, as without them all proposed reforms remain only paper works. India must also act at the grass-root and ground level. For instance, panchayats should encourage co-operative farming, power and irrigational facilities must be provided to the farmers, easy and effective financial access must be provided to the farmers, direct marketing and sale must be adopted by farmers, public investment in agricultural infrastructure must be enhanced, a minimum support price for food grains must be set etc. Finally, farmers in India must use ICT for agricultural purposes. India's food production and productivity may be increased by an effective use of ICT for agricultural purposes. The developed nations are using laser technology instead of tractors to plough lands. This helps in optimising the use of various inputs such as water, seeds, fertilisers etc. The problem is that Indian farmers cannot afford this technology and unless government comes in support for agricultural infrastructure, the same remains a dream only. Further, power and electricity also remains a major problem for Indian farmers and alternative means of power like solar energy panels, regulated and optimised by ICT can be a blessing for them. Thus, e-agriculture in India can put India on the higher pedestal of Green Resolution making India self-sufficient in the matters of food grains.

Application of Wireless Technologies: Wireless technology is the process of sending information through invisible waves in the air. Information such as data, voice and video are carried through the radio-frequency of the electromagnetic spectrum. The electromagnetic spectrum consists of different levels of energy waves including gamma rays, X-ray, ultraviolet light, visible light, infrared, microwave and radio. The equipment needed to send and receive information via wireless has to be accomplished through a modulator and demodulator. A modulator is used to send the information wirelessly into the air. A demodulator is used to convert it from air waves to another use. The use of wireless technologies is by no means a new concept. When people hear the word wireless they are probably thinking of cellular phones or pagers.

Methods of Wireless Connectivity

- Infrared Data Association (IrDA) comes with most laptops and PDAs and uses the infrared wavelength on the electromagnetic spectrum. It needs a clear line of site to operate. Typical examples of using IrDA is "beeming" electronic business cards between two PDAs, remote control for your TV or printing from a laptop to a printer equipped with a IrDA device. There is a limited use of IrDA compared to other wireless technologies today.

- **HF, SW, VHF and UHF Radio Frequencies:** Radio-frequencies use a radio transmitter in the device to send and receive signals. One example of using radio includes a wireless keyboard and mice for your computer. These devices can communicate up to 6 feet and operate at 27 MHz. CB radios can also operate at 27 MHz but can extend communication up to 5 miles due to the radio transmitter and tower capabilities. FM radios allowed closed networks of communication within short range distances. Most all FM frequencies are also governed by the FCC and you pay license fees to use specific frequency channels depending on the use. In agriculture, wireless meteorological stations, RFID readers for livestock identification, and sensors use frequencies between 20 MHz up to 900 MHz and can wirelessly communicate weather data. They have been economically feasible to install and easy to operate.

- **Wireless 3G and Cellular Technology:** Cellular technologies have been around for years and also have direct applications in agriculture. Today's third generation or 3G technology delivers the capabilities of video, music, voice, e-mail and data transmission through the use of cellular towers and specialized equipment. This technology requires high speed communications between 384 kbps and up to 2 Mbps which is much different from today's standard of cellular communication. Another difference of this technology is that you are always connected to the Internet and that you will not need to dial to connect to the Internet. However, several limitations still exist for rural areas and that is reluctance to several cellular phone towers. In addition, this technology is still being introduced so devices are not fully developed.

- **Wireless Networks:** A wireless network consists of several access points for connecting computers, PDA or other equipment with wireless network cards in a nearby area to network or get access to the Internet. There are two main standards for wireless networking, Bluetooth and IEEE 802.11. Both Bluetooth and IEEE 802.11 networks can coincide in the same environment but each have different approaches to connectivity.

- **Bluetooth:** It can be used in any device computers, printers, fax machines, GPS, cameras, and other devices. Bluetooth consists of a microchip and software within the chip called a link controller that works to identify other Bluetooth devices to send and receive data. It can send both voice and data. It operates at 2.4 GHz radio-spectrum and is designed for a small proximity. It can communicate up to 100 feet, however, the average standard distance for reliable use is typically within a 33 feet area. Devices connect with each other by examining each other's profiles that are coded into the devices. Profiles contain information about the device, what it can communicate with, and what it is used for. A connection of two or more Bluetooth devices is called piconet. If there are several devices near each other the radio-signals could have some interference but are usually very unnoticeable. Bluetooth networks are usually not very large.

- **Global Positioning Systems (GPS):** GPS units use a receiver chip and software to communicate to specific satellites in order to determine your location on earth. There is a network of 24 GPS satellites that orbit the earth. They are located in such a way that four should be visible from any spot on earth at one time. These satellites constantly transmit signals on two frequencies, 1575.42 MHz and 1227.60 MHz. Once the data of your location has been logged it can be further communicated through other wireless or non-wireless networks (Sanders *et al.*, 1996).

- **Satellite Internet and Video Connectivity:** This is broadcast connectivity between satellite dishes and specific subscription paid service satellites. Satellite television or Internet can be achieved directly by sending a signal from a large dish antenna to a geostationary satellite. Geostationary satellites remain in the same spot above the earth all times. This guarantees signals being sent and received without loosing connection to the satellite. Sending the data to the satellite is called uplink. The satellite receives the signal through a transponder that converts it to a different frequency and transmits back to earth called downlink. Downlink speeds are typically faster than uplink data speeds. In order for this to work properly the look angles of the satellite dishes need to be pointed toward the communication satellite. There is a great range of satellite data transmission rates between 1000 MHz to several hundred GHz for data connectivity.

Common Applications of Wireless in Agriculture and Natural Resources: There are many applications in agriculture that can use wireless technologies. Some examples includes

- Monitoring pesticide and herbicide applications.
- Animal tracking and identification.
- Monitoring water or flood levels.
- Indicate warning for frost events.
- Monitor crop health, rainfall, temperature and other meteorological data.
- Track shipments of perishable crops and crop inputs.
- Monitor equipment movement and performance.
- Web cameras to view hazardous or remote areas.
- Odour, gas and other environmental indicators for livestock housing facilities.
- Integrating GPS data into Geographic Information Systems.
- Precision agriculture applications in data collection and reporting.
- Food safety and security through continuous tracking capabilities from production (knowing what pesticides or other treatments have been used) and packing and shipping of products.
- Agro-security by reducing theft of farm products, vandalism of property, and detection of bio-chemicals.

Wireless Use in Crop Management and Precision Agriculture: Field sensors and other technologies and equipment are getting better defined for crop production use each day. Using principles and technologies of precision agriculture we are constantly collecting data along with GPS from various sources of sensors, controllers and other hardware for use as integrated into GIS management software. Precision agriculture applications have the greatest to gain for a comprehensive data collection and reporting system in order for all those involved to make timely decisions. Data sources typically gathered for crop management that can be shared through wireless technologies, *i.e.,* (i) Topo and elevation mapping, (ii) Soil sampling, (iii) Yield monitoring, (iv) Soil electromagnetic conductivity mapping, (v) Satellite and aerial imagery, (vi) Soil Moisture for irrigation needs, (vii) Crop input record keeping and tracking, (viii) Crop scouting of weeds, diseases, and insects, (ix) Meteorological data collection.

Some traditional limitations in collecting this data include timeliness of transferring data to the appropriate locations or central databases and loosing data from equipment malfunction or battery loss. Data can be collected in real-time or after a specific process has been completed. Information can also come in large sizes so the transfer rate of data needs to be handled through larger bandwidth and data storage capabilities. Another benefit of using wireless technology is that it can send several sources of data in real-time to one central location like a server database system. Allowing data to be pooled to a central location allows multiple users to utilize that data when they need it. For example, weed scouting data collected in the field from a consultant using a handheld PDA equipped to a cellular phone or local wireless network system relays treatment information to office server. Server-side software generates an appropriate application map to grower and sent on to commercial chemical applicator for application at the same time.

Application of Remote Sensing: It is the acquisition of information about an object from a distance, with precision, without coming into contact with the same. Although the use of RS is a decade old, its relevance to agriculture in spatial variability management is relatively new. RS measures visible and invisible properties of a field or a group of fields and converts point measurements into spatial information to monitor temporally dynamic plant and soil conditions. The visual observations are recorded through a digital notepad and geo-referenced to GIS database, most commonly used RS device but aerial photography and videography are also used in PF. Satellite RS has provided a tool for acreage estimation one month in advance with more than 95 per cent accuracy and in mono-croped area yield estimation with more than 90 per cent accuracy ten days in advance. These images allow mapping of crop, pest and soil properties for monitoring seasonally variable crop production, stress, weed infestation and extent within a field. RS can be used for PF in a number of ways for providing input supplies and variability management through decision support system. The point data of soil test results can be translated into spatial coverage based on geo-statistical interpolation which gives chemical properties of the soil, nutrient status, organic matter, salinity, moisture content etc. This information on spatial variability can be used with other geo-references to identify both seasonally stable and variable units, based on which management strategies can be developed. Space technology combined with satellite RS and communication provides valuable, accurate and timely information like early warning, occurrence, progressive dangers, damage assessment, quick dissemination of information regarding disaster and decision support to mitigate it. Recent developments in remote sensing technology have led to significant improvements of spatial and spectral resolution of commercially available aircraft or satellite based remote sensing imagery.

Application of GPS: The GPS technology enables precision agriculture because all phases of precision agriculture require positioning information. GPS is able to provide the positioning in a practical and efficient manner. GPS was developed by US military and later permitted for restricted civilian use. Expensive, high-precision differential GPS (DGPS) systems are available that achieve centimetre accuracies allow for automated machinery guidance and kinematic mapping of topography and are useful in the creation of digital elevation models needed for terrain analysis.

While the GPS signal is ubiquitous, there have been problems in making available GPS at the needed precision for agriculture. Additionally, some GPS receivers are susceptible to unwanted interfering signals from a variety of sources, including farm machinery, making the receiver useless in navigation or positioning. Some interference can be overcome in the design of the GPS receiver. The GPS can be used in two modes single receiver mode and differential mode (DGPS) using two receivers. Single receiver collects the timing information and processes it into position. This system is the cheapest and easiest but its accuracy suffers due to introduced positional errors. In the differential mode (DGPS), one receiver is mounted in a stationary position; usually at farm office while the other is on the machine implement. Regardless of problems, DGPS has greatly enabled precision agriculture. Of great importance for precision agriculture, particularly for guidance and for digital elevation modeling, position accuracies at the centimetre level are possible in DGPS receivers that use carrier phase in combination with DGPS. Accurate guidance and navigation systems will allow for farming operations not currently in use, including field operations at night when wind speeds are low and more suitable for spraying and the use of night tillage to reduce the light induced germination of certain weeds.

Application of GIS: A Geographical Information System has the capability to capture, store, manipulate, analyze and display spatial information and related attributes. The graphics data base in the GIS contains all of locational information relating to mapped features. The attribute data base contains all of descriptive information relating to the mapped features. Use of GIS in agriculture has increased because of misuse of resources like land, water etc. GIS is the principal technology used to integrate spatial data coming from various sources in a computer. GIS techniques deal with the management of spatial information of soil properties, cropping systems, pest infestations and weather conditions. This is primarily an intermediate step because it combines the data collected at different times based on sampling regimes to develop the subsequent decision technologies such as process models, expert systems etc. In the new millennium, GIS aided techniques are indeed needed for sustainable food production and resource utilization without further depletion of the environment. GIS technology will help the farmers and scientists in decision making, as precise information on field will be readily available. GIS techniques make weed control, pest control and fertilizer application site-specific, precise and effective, it would also be very useful for drought monitoring, yield estimation, pest infestation monitoring and forecasting. GIS coupled with GPS, microcomputers, RS and sensors is used for soil mapping, crop stress, yield mapping, estimation of soil organic matter and available nutrients. In combination these technologies have brought out rapid changes in data collection, storing, processing, analysis and developing models for input parameters.

Application of Modeling: Modeling is proposed as an important tool in precision agriculture to simulate spatial and temporal variation in soil properties, pests, crop yield and environmental performance of cropping systems. Models have been developed and calibrated for specific purposes but have not been used extensively in spatial prediction. A major problem of models is the availability of inputs needed to run them. A major advantage of models is their ability, once calibrated to simulate the temporal component of crop production. This capability should allow models designed to account for spatial variability to evaluate different precision management scenarios that would otherwise be prohibited by time and cost considerations. The application of models to the simulation of the space-time continuum of crop production is a critical research need.

External links: This type of association can be run without meeting each other in physical world, and business or activities can be perform with the help of linkage by one or several way. In 2007, a community of practice was set up to discuss and share lessons learned regarding the use of ICT to enhance sustainable agricultural development and food security. This Community, located at [www.e-agriculture.org,] today boasts over 6,000 members from over 150 countries, and include practitioners, policymakers, representatives of farmer organizations, researchers, and information and communication specialists involved in agriculture and rural development.

TECHNOLOGY DISSEMINATION

Extension System has crumbled - needs to be re-look

- Farming has become increasingly complex
- Small farm holder, with little education, finds it difficult to make right choices. *Therefore, Effective communication between technology developer-provider and technology user has becomes crucial.*

Initiatives on Transfer of Technology

- (i) One Krishi Vigyan Kendra (KVK) in each district of the country
- (ii) Institute-Village Linkage Programme for Technology Assessment & Refinement (IVLP)
- (iii) Agricultural Technology Information Centres (ATIC) at selected SAUs and ICAR Institutes
- (iv) ATMA in selected states of the country
- (v) Toll-free Kisan Call Centres.

Status of Applications of ICT in Agriculture Sector in India

Tata Kisan Kendra: The concept of precision farming being implemented by the TKKs has the potential to catapult rural India from the bullock-cart age into new era of satellites and IT. TCL's extension services, brought to farmers through the TKKs, use remote-sensing technology to analyze soil, inform about crop health, pest attacks and coverage of various crops predicting the final output. This helps farmers adapt quickly to changing conditions. The result: healthier crops, higher yields and enhanced incomes for farmers.

Government Organization: Space Applications Centre (ISRO), M.S. Swaminathan Research Foundation, Chennai, Indian Agricultural Research Institute, New Delhi, and Project Directorate of Cropping Systems Research, Modipuram had started working in this direction and in soon it will help the Indian farmers harvest the fruits of frontier technologies without compromising on the quality of land. ISRO has initiated Gramsat project in Orissa. Forecasting the yield of mono and multiple crops is being done at NRSA. Acreage estimates and crop inventory is being done during *Kharif* and *Rabi* seasons for Rice which is the major crop grown in our India. Other crops like Banana, Chillies, Cotton, Maize, Sugarcane and Tobacco are also being inventoried. Satellite data can also delineate different crops that are grown in the same area, and an inventory of each of the crops can be done.

Problems in adopting ICT in Agriculture Sector in India

(i)   Illiteracy of farmers, (ii) Knowledge and technological gaps, (iii), High cost of precision farming equipments, (iv) Lack of availability of such equipments, (v) Poor economic condition of farmers and (vi) Small land holdings.

Steps to be taken for implementing ICT in Agriculture Sector in India

Creation of multidisciplinary teams involving agricultural scientists in various fields, engineers, manufacturers and economists to study the overall scope of precision agriculture.

- Formation of farmer's co-operatives since many of the precision agriculture tools are costly (GIS, GPS, RS, etc.).
- Pilot study should be conducted on farmer's field to show the results of precision agriculture implementation.
- Government legislation restraining farmers using indiscriminate farm inputs and thereby causing ecological/environmental imbalance would induce the farmer to go for alternative approach.
- Creating awareness amongst farmers about consequences of applying imbalanced doses of farm inputs like irrigation, fertilizers, insecticides and pesticides.

Synonyms of Precision Farming

Precision farming, information-intensive agriculture, prescription farming, target farming, site specific crop management, variable rate management, farming by soil, grid soil sampling agriculture, grid farming, GPS farming , farming by the inch, farming by the foot, smart farming, farming by computer, farming by satellite, computer-assisted agriculture, automated agriculture, cyberfarm, etc.

Perspectives of Precision Farming

(i)   **Agronomical perspective** : Adjust cultural practices for real needs of the crop (*e.g.* better fertilization)

(ii)   **Technical perspective** : Better time management (*e.g.* planification of agricultural activity)

(iii) **Environmental perspective** : Reduction of agricultural impacts (better estimation of crop nitrogen needs implying limitation of nitrogen run off)

(iv)   **Economical perspective** : Increase of the output and reduction of the input, increase of efficiency (*e.g.* lower cost of nitrogen fertilization practice)

(v)   **Crop productivity perspective**

(vi)   Nutrient supply according to soil variation may be most significant aspect of PA which aims at improving the input output characteristics of the soil and crop system as they vary in space and time.

(vii)   Information and management technology are combined into a production systems to cater the variability that can increase production efficiency by allowing more efficient input use through efficient application.

(viii)   Precision instruments are able to complete in-field operations like cultivation, seed sowing, application of fertilizers & pesticides and harvest timely which enhance the crop productivity.

(ix)   Precision farming provides sufficient understanding of the processes involved to apply inputs in such a way as to be able to achieve a particular goal. The goal, however, might not necessarily mean maximum yield but may be to optimize financial advantage while operating within environmental constraints.

(x)   Precision farming involves the use of spatial asset allocation and management to distribute available time and money where it is most needed and will provide the best return.

(xi)   Precision agriculture provides tools for tailoring production inputs to specific plots within a field thus potentially reducing input costs, increasing yields and reducing environmental impacts by better matching inputs applied to crop needs.

(xii)   Precision agriculture technology is paving the way for agricultural producers by allowing for precise management of inputs. The appropriate management processes and information needs vary among different environments, but also among different decisions to be made.

SUMMARY

Information and Communication Technology (ICT) is unique blend of tools and technology for providing information in fastest way and correct manner without losing content and most importantly without any communication gap, where as precision farming is smart way of farming under changing climate scenario.

REFERENCES

Barnett V, Landau S, Colls JJ, Craigon J, Mitchell RAC and Payne RW (1997). Predicting wheat yields: The search for valid and precise models. In "Precision Agriculture: Spatial and Temporal Variability of Environmental Quality" (J.V. Lake, G.R. Bock and J.A. Goode. Eds.). pp. 79-92. Wiley, New York.

Lange AF (1996). Centimeter accuracy differential GPS foe precision agriculture applications. In "Proceedings of the Third International Conference on Precision Agriculture, Minneapolis, MN, 23-26 June 1996" (P.C. Robert, R.H. Rust and W.E. Larson, Eds.), ASA Miscellaneous Publication. pp. 675-680. ASA, CSSA and SSSA, Madison, WI.

O'Conner M, Bell T, Elkaim G and Parkinson B. (1996). Automatic steering of form vehicles using GPS. In "Proceedings of the Third International Conference on Precision Agriculture, Minneapolis, MN, 23-26 June 1996" (Robert PC, Rust RH and Larson WE, Eds.), ASA Miscellaneous Publication. pp. 767-777. ASA, CSSA and SSSA, Madison, WI.

Saunders WP, Larscheid G, Blackmore BS and Stafford JV. (1996). A method for direct comparison of differential global positioning systems suitable for precision farming. In "Proceedings of the Third International Conference on Precision Agriculture, Minneapolis, MN, 23-26 June 1996" (Robert PC, Rust RH and Larson WE, Eds.), ASA Miscellaneous Publication. pp. 663-647. ASA, CSSA and SSSA, Madison, WI.

Yadav RL and Rao Subba AVM. (2000). In: National Symposium "Agronomy: Challenges and Strategies for the New Millennium", November 15-18, 2000, GAU, Junagadh. Indian Society of Agronomy. pp. 3-4.

18

Precision Farming of Seed Spices

CHAPTER

☞ G. Lal[1], ☞ R. Singh[2] and ☞ Cherian Mathews[3]

INTRODUCTION

India's productivity of food grains per hectare is not more than three-fourths of the world average and less than half of that in agriculturally advanced countries, despite having all the natural advantages, per capita food grain availability even after the Green Revolution, it has been less than two thirds of the world average. Only five states in India, namely Himachal Pradesh, Punjab, Haryana, Uttar Pradesh and Madhya Pradesh produce more grain than their populations can consume. The combined population of the five states is less than one third of the total production of the country. More than two thirds of the population lives in states that are still food deficit. This requires transport of lakhs of tonnes of food grain, involving high costs and pilferage. Our effort should have been to make all the states self-sufficient with respect to food grains and if some disturbances occurred due to unnoticed natural calamities the nation must be in an ever ready position to mitigate such challenging tasks.

Spice farming is extensively done in the whole of India. Each spice is grown in an unique manner. Each plant needs a different climate soil condition, irrigation, manure, pest control methods etc. Every spice is grown in different parts of the country depending upon the climate of the particular place. Spices are grown with great care and many spices also go through various post agricultural treatment as well.

The Indian green revolution is also associated with negative ecological / environmental consequences. The status of Indian environment shows that, in India, about 182 million ha of the country's total geographical area of 328.7 million ha is affected by land degradation of this 141.33 million ha are due to water erosion, 11.50 million ha due to wind erosion and 12.63 and 13.24 million ha due to water logging and chemical deterioration (salinization and loss of nutrients) respectively. On the other end, India shares 17 per cent of world's population with 2.5 per cent of geographical area, 1 per cent of gross world product, 4 per cent of world carbon emission and hardly 2 per cent of world forest area. The Indian status on environment is though not alarming when compared to developed countries, it gives an early warning to take appropriate precautionary measures.

---

[1] Principal Scientist, NRCSS, Tabiji, Ajmer (Rajasthan)
[2&3] Senior Scientist, NRCSS, Tabiji, Ajmer (Rajasthan)

The growth rate of grain production during the ninth plan has been less than the population growth rate. The poor agricultural performance has not been because of the vagaries of monsoon. In India, per capita availability of grain and capita calorie intake, which were less than the minimum required for adequate nutrition, have further declined.

The decline in agricultural growth and increase in rural poverty have been due to the long persisting government indifference towards the farm sector, which is evident from plan outlays on agricultural and allied activities, Rural development and irrigation, which added up to 37.1 per cent of the total during the first plan were brought down to only 19.4 per cent during the ninth plan. However, the main reason, for poor performance of the farm sector has been the long persisting adverse terms of trade policies for agriculturists in addition to the mismanagement of natural resources leads to ever ending crisis.

THE NEED FOR PRECISION FARMING

The production of food grains in five decades, has increased more than threefold, the yield during this period has increased more than two folds. All this has been possible due to high input application, like increase in fertilization, irrigation, pesticides, higher use of HYV's, increase in cropping intensity and increase in mechanization of agriculture.

(i) Fatigue of Green Revolution

Green revolution of course contributed a lot. However, even with the spectacular growth in the agriculture, the productivity levels of many major crops are for below than expectation. We have not achieved even the lowest level of potential productivity of Indian high yielding varieties, whereas the worlds highest productive country have crop yield levels significantly higher than the upper limit of the potential of Indian HYV's. Even the crop yields of India's agriculturally rich state like Punjab is far below than the average yield of many high productive countries.

(ii) Natural resource degradation

In this situation, there is a need to convert this green revolution into an evergreen revolution, which will be triggered by farming systems approach that can help to produce more from the available land, water and labour resources, without either ecological or social harm. Since precision farming, proposes to prescribe tailor made management practices, it can help to serve this purpose.

WHAT IS PRECISION FARMING

It is a relatively new and highly technology driven farming practice which may potentially improve resource use and reduce environment impacts of arable crop production. Fundamentally, precision farming acknowledges that conditions for agricultural production as determined by soil resources, weather and previous management –vary across.

It is a management strategy that employs detailed site specific information to precisely manage production inputs. This concept is sometimes called precision agriculture, prescription farming or site specific management. The idea is to know the soil and crop characteristics unique to each part of the field and to optimize the production inputs within small portions of the field. The philosophy behind precision agriculture is that production inputs (seed, fertilizer, chemicals etc.) should be applied only as needed and where needed for the most economic production.

WHY IS PRECISION FARMING

It is a mangement strategy that employs detailed, site specific information to – precisely manage production imputs. This concept is sometimes called precision agriculture, prescription farming, or site-specific management. The idea is know the soil and crop characterisitcs unique to each part of the field and to optimize the production inputs within small portions of the field.

The philosophy behind precision farming is that production inputs (seeds, fertilizer, chemicals etc.) should be applied – only as needed and where needed for the most economic production. Why should producers be interested in precision farming? Precision farming techniques can improve the economic and environmental sustainability of croproduction. The only alternative left to enhance productivity in a sustainable manner from the limited natural resources at the disposal, without any adverse consequences, is by maximizing the resource input use efficiency. It is also certain that even in developing countries, availability for labour for agricultural activities is going to be in short supply in future. The time has now arrived to exploit all the modern tools (Mondal *et al.*, 2011).

PRECISION FARMING IN SEED SPICES

Seed spices are a group which includes the annual spices whose dried fruit or seeds are used as spices. In respect to all the other spices these are considered as underutilized and very less work has been done. India is not only home of seed spices but also having supreme position of the largest as producer, consumer and exporter in the world. Still the productivity level is very low mainly because the cultivation of these crops is still done traditionally on marginal lands with low fertility and improper use of inputs. Precision farming is the solution to increase the productivity level of seed spices as precision agriculture ensures better utilization of inputs and other resources based on agro climatic situations.

INTERVENTIONS OF PRECISION FARMING

1. Seed bed preparation

Levelled, well pulverized seed beds are required for seed spices. About 10t/ha of organic manures are incorporated during seedbed preparation in addition to basal dose of fertilizer. In light soils it is achieved by 3-4 ploughings and planking with traditional plough or tractor mounted tillers. In weedy field ploughing at least once with mould board plough is desirable. Rotavators are better option to create well pulverized seedbed.

For levelling bullock drawn plank or tractor mounted scraper are generally used. Precise land levelling on large farms is efficiently done with the help of motor grader. Laser land levellers have been introduced in the recent past which can do most precise levelling, which facilitates water application and yield enhancement.

2. Integrated Nutrient Management

Seed spices are mostly grown on marginal lands with low fertility. Present fertilizer recommendations are very old and developed based on agro climatic zones. The assumption is that agro-climatic zones are homogenous units. However, analyses of agro-climatic zones reveal variability in soil within each zone. Current agro-climatic zonal fertilizer recommendations are generalized for entire zone and not addressed to specific soil types. The agro-climatic zones vary widely in soils and in their potentials, behaviour and response to management. The fertilizer application efficiency varies within each zone and within the management units. These differences contribute to errors of both excess and insufficient applications. Besides, there is a continuous removal of secondary and micro-nutrients by crops in all farming situations resulting in inappropriate management practices. All these suggest that soil-based fertilizer recommendations should be preferred to achieve precision in farming and to maximize seed spice production as well as maintain the soil fertility.

3. Sowing and planting

Seed spices are generally directly soaked but in some cases like fennel and celery which are transplanted also. Leaving fenugreek seed spices are of size, shape and texture that are difficult to mechanically meter. It is desirable to adopt seed metering used for mustard or specially design metering mechanism. For efficient precision sowing specially for planting, it will be desirable to pellet them. Available seed drills/ seed cum fertilizer drill metering mechanism are designed for relatively larger and heavy seed without awns or fudge. Depths of sowing are very shallow 1-2cm or 3cm at the most. The furrow openers of commonly used seed drills are not very well suited for seed spices. The new concept emerging is that of Pneumatic planters.

4. Irrigation

Micro-irrigation system is irrigation system with high frequency application of water in and around the root zone of plant system. The micro-irrigation system consists of a network of pipes along with a suitable emitting device. Micro-irrigation has proved to be the most efficient method of irrigation with very high productivities per unit of resources employed. With the adoption of micro- irrigation technologies the use efficiencies of water, energy, nutrients, chemicals and human effort have all been found to improve substantially.

5. Plant protection measures

High incidence of diseases for example wilt, blight, powdery mildew and stem gall as well as aphids are one of the major constraints faced in getting the full potential of the seed spices production. Remote sensing and GPS etc are some of the tools by which we can forecast the disease in advance and adopt control measures. Plant health assessment and monitoring can be performed with the use of teledetection methods, among which satellite, air and infrared imaging are most common. Infected plants show signs of stress, which lowers their capability to absorb infrared waves or different temperature between healthy and infected parts of plants. Amended and variable application of plant protection agents is an inseparable part of precision farming and is commonly known as variable rate application (VRA) process. Different hi-tech sprayers with appropriate boom size and shape of nozzles can be used for the precise and required application of plant protection inputs.

6. Harvesting

Lack of proper adoption of harvesting and post-harvest technology adversely affects the quality of seed spices. Since seed spice crops are valued for their peculiar flavour and harvesting stage plays a major role detaining the flavour which ultimately adds to the quality. The essential oil content present in fruits of seed spices contribute towards the flavour and fragrances. Thus it becomes necessary to harvest the crop at right stage so that the high quality with reference to flavour could be maintained. Proper time of harvest by maturity index can be worked out using remote sensing application. Apart from the maturity index due care should be given to develop friendly tools and equipments for easy picking of the seeds. Combine harvester is also one of the best options to get the maximum produce with minimum loss.

7. Threshing

Post-harvest losses are up to the level of 25 per cent in seed spices. The maximum losses are during the post-harvest steps mainly threshing. Manual threshing practices is still practiced which is the main reason for deterioration and loss of the produce. Use of modern threshers is a best option to overcome this problem.

8. Farm processing

Processing mainly include drying, grading, packaging and storage. General practices of drying are not suitable as per the level of perfection required in case of seed spices. Moisture content is the main reason for deterioration in the quality of the material. Solar tunnels are more in practice as a best option for drying. Similarly mobile seed processing unit is a best option for grading. Packaging by controlled atmospheric and modified atmospheric packaging can also be worked out for longevity and keeping quality of seed spices.

9. Protected Cultivation

It is a competitive in today's marketplace, vegetable growers must strive for high-quality produce and extended production cycles. The use of structures enables vegetable producers to realize greater returns per unit of land. Yields under structures may be two or three times higher than those of field crops. Production under structures also offers other benefits, including early harvests which earn higher prices, cleaner and better quality produce, more efficient use of water and fertilizer, reduced leaching of fertilizers (especially on light, sandy soils), and better management of pests, weeds and diseases. Thus production under structures is well suited to Asia, because it can be used effectively by growers with either small or large farms (Mondal *et al.*, 2011).

CONCLUSION

The 21st century calls for the sustainability that is to transform green revolution to evergreen revolution. This is possible only through hi-tech agriculture comprising of precision farming and protected cultivation. Precision farming is doing right thing at right time in right amount. It involves studying and managing variations within fields that can affect crop yield. It identifies the critical factors where yield is limited by controllable factors. Finally it maximizes crop production and minimizes environmental pollution and degradation leading to sustainable development.

REFERENCE

Mondal P, Basu M and Bhadoria PBS. (2011). Ceritical review of precision agriculture technologies and score of adoption in India. American Journal Experimental Agriculture **1**:49–68.

19

CHAPTER

Remote Sensing Application for Crop Acreage and Production Estimation

☞ Karmal Singh[1], Sandeep Arya[2], Parveen Kumar[3], Meena Sewhag[4] and Bhagat Singh[5]

INTRODUCTION

There has been substantial boost in food grain production in post 'Green Revolution' era but it could not keep pace with the rate of increase in the population of country. Thus, increasing agricultural productivity has been main concern since scope of increasing area under agriculture is limited. Fulfilling this requirement, information on crop acreage, yield, production and conditions for strategic planning is required. The use of Remotely Sensed data is being investigated all over the world for crop production forecasting. Crop production consists of two components (i) area under the crop, and (ii) yield per unit area. Acreage estimation using RS data has been demonstrated in various parts of the world. A concentrated effort has been made under this programme to develop methodology applicable for larger areas. However, yield prediction is a subject of intensive research and is still at a developmental stage. Remote Sensing can provide information on crop growth variables on a regional scale. Vegetation Indices (VIs) derived from RS data are indicative of crop growth, vigour and potential grain yield. Numerous factors including weather parameters and agronomic practices, which vary from area to area affect crop yield in a given region. Over a decade, number of crop yield forecasting models using RS data have been developed.

Among various applications of remote sensing in natural resources management, its use for crop production forecasting is of great economic benefit. The successful launching of Earth Resources Technology Satellite (Landsat-1) in 1972 provided a new tool for gathering information on crops. The crop production forecasting system received a greater impetus with the launch and operationalisation of the state-of-out Indian Remote Sensing Satellites *viz.* IRS-1A (March 17, 1988), IRS-1B (August 21, 1991), IRS-P2 (October 15, 1994), IRS-P3 (March 21, 1996), IRS-1C (December 28, 1995) and IRS-1D (September 29, 1997).

[1] Department of Agronomy, CCS, Haryana Agricultural University, Hisar-125004
[2] KVK Yamunanagar, CCS, Haryana Agricultural University, Hisar-125004
[3,4&5] Department of Agronomy, CCS, Haryana Agricultural University, Hisar-125004

GLOBAL STATUS

The era of high quality observations of the earth surface through space began in 1972 the launch of ERTS-1 (later named as LANDSAT) in USA with the Multispectral Scanners (MSS) at 80-metre spatial resolution. Since 1982 Landsat Thematic Mapper (TM) has provided 30 meter spatial resolution data which are of better quality than the data received from MSS. The use of space borne remote sensing data for larger area crop surveys was explored in USA under Com Blight Watch Experiment (CBWE) in 1971, under Crop Identification Technology Assessment for Remote Sensing (CITARS) in 1973 followed by an attempt to forecast wheat crop production for major growing regions of the world under Large Area Crop Inventory Experiment (LACIE) during 1974-1977. Later a six year programme of research and development named Agriculture and Resource Inventory Survey through Aerospace Remote Sensing (AGRIS ARS) was taken up in 1988. Since then large scale methodology development-cum-demonstration studies for crop statistics have been carried out in Africa, Europe and a number of other countries (Argentina, Australia, Brazil, Canada, Japan etc). Currently major programmes are underway in Africa under Global Information and Early Warning System (GIEWS) and in Europe under Monitoring Agriculture through Remote Sensing (MARS). This project has developed rapid crop survey procedure for Crop Growth and Monitoring System (CGMS), which employ crop simulation models, agro-meteorological models and real time data for crop forecasting and assessment. USDA Statistical Reporting Service (SRS) has integrated Landsat data in domestic crop estimation programme. USDA makes use of the Landsat data for stratification (based on visual interpretation) and to classify the digital data into crop types and regress SRS ground-collected data results from the area sampling segments (0.7 sq. miles). A direct expansion estimator is used to expand the data to state, regional and national level. Studies have also been made to examine the relationship of crop growth parameters like leaf area index (LAI) and the spectral data in the form of several vegetation indices developed from the spectral data of various bands. Since the inception of civilian RS program in the U.S. in the early 1960s a major research and development thrust has been on agricultural crop identification and area estimation. Experiments such as Crop Identification Technology Assessments for Remote Sensing (CITARS) and Large Area Crop Inventory Experiment (LACIE) were conducted to demonstrate the capabilities of RS for crop inventory and forecasting. CITARS demonstrated the usefulness of automated data processing techniques and space-borne data for corn and soybean inventory in U.S. LACIE was the first worldwide experiment to demonstrate the operational capability of RS technology for wheat production forecasting. Currently major operational programme is under way in Europe under Monitoring Agriculture through Remote Sensing (MARS) project. The main objective of the project is a rapid crop survey procedure for crop area estimation. MARS can monitor and forecast crop information in three levels: the Community level, regional level and the national level.

INDIAN EXPERIENCE

The applications of remote sensing techniques for crop inventory has received attention right from the beginning in India. Identification of root wilt disease in coconut using aerial false colour photographs in 1970 was the pioneering experiment in India (Dakshinamurti *et al.*, 1971). A systematic study on crop inventory using CIR aerial data was carried out in the joint ISRO-ICAR Agricultural Resource Inventory and Survey Experiment (ARISE) project, which was conducted in Anantapur (Andhra Pradesh) and Patiala (Punjab). Early studies using space borne data employed visual mapping of crops such as wheat and rice and Under the IRS Utilization Programme initiated by Dept. of Space in 1984, different aspects of the problem of crop inventory using remotely sensed data were studied. First attempt in the country towards the use of satellite digital data for wheat acreage estimation was made in Karnal district of Haryana State using Landsat MSS data (Dadhwal and Parihar, 1985). A project on Crop Acreage and Production Estimation under the Remote Sensing Applications Mission (RSAM) with enlarged scope and objectives was formulated in 1986. A concentrated effort has been made under this programme to develop methodology applicable over large areas.

CROP ACREAGE AND PRODUCTION ESTIMATION (CAPE)

The satellite based studies graduated from visual to digital analysis and by launch of IRS-1A, large projects such as CAPE were in implementation covering large area crop inventory and yield modeling for important crops such as wheat, rice, cotton, groundnut, sorghum and mustard. This project was initially, undertaken for few districts/states for wheat and rice which was extended to more crops in their major producing areas at the request of the Department. of Agriculture and Cooperation. A large amount of experience has been gained in CAPE project on efficient sample design, factors affecting crop discrimination, spectral yield relationships and realization of timeliness and accuracy for pre-harvest crop forecasts. The experience gained over more than a decade of implementation of CAPE project a proposal called FASAL has been submitted to Department of Agriculture & Cooperation (DAC), Ministry of Agriculture (MOA). As Remote Sensing, weather and field observations provide complementary and supplementary information for making crop forecasts, FASAL proposes an approach, which integrates inputs from three types of observations to make forecasts of desired coverage, accuracy and timeliness. The concept of FASAL thus strengthens the current capabilities of early season crop estimation from econometric and weather-based techniques with RS applications. Mid-season assessments can be supplemented with multi-temporal coarse resolution data based analysis. In the later half of the crop growth period, the contribution of RS is direct in the form of acreage estimates and yield forecasts. India entered the space age by launching of the IRS-1A on March 17, 1988 which has opened new vistas for agriculture and natural resources in terms of speed and quality. IRS-IA provided comparable data to those obtained from LANDSAT, as the sensor LISS-I system was well comparable with MSS while LISS-II was comparable with TM as far as resolutions were concerned. As a follow up and towards ensuring continuity of data availability to the users the IRS-1B Satellite identical with the IRS-IA was launched in August 1991. Subsequently IRS-IC was launched in 1995 and IRS-1D in 1997. Data obtained from these satellites opened vast area of research and applications and several organizations are engaged in development of new and improved methodologies for the applications of satellite data.

Recently with a view to collect, collate and assimilate large data from different sources, a National Crop Forecasting Centre (NCFC) has been set up under the Ministry of Agriculture during 1998. Deptt. of Space has also recently launched a project-Forecasting Agricultural output using Space, Agrometeorology and Land based observations (FASAL) envisaging advance reliable assessment of crop acreage and production using remote sensing techniques. Very recently national wheat production forecast for (1998-99) using multi date WiFs and meteorological data have been developed under this project.

USE OF SATELLITE DATA IN CROP YIELD ESTIMATION

Remote sensing satellite data can also be used for improving the crop yield estimation through crop cutting experiments and also for developing models for crop yield forecasting using historical data, meteorological data and remotely sensed satellite data. Attempts to use the satellite data for improving the crop yield estimates based on GeES were initiated at IASRl during 1990. During 1990-93 a study was conducted at the Institute to examine the usefulness of satellite spectral data for stratification of crop area based on vegetation indices for improving crop yield estimation based on yield data from crop cutting experiments. The study pertained to wheat crop yield for district Sultanpur, UP for 1985-86 and the satellite data was used from the USA satellite Land Sat-4. This study showed that the efficiency of crop yield estimation could be increased considerably by using the satellite data along with the survey data. The results of this study are given by Singh *et al.*, Another similar study was undertaken during 1996-98 for improved estimation of wheat crop yield in district Rohtak, Haryana for 1995-96 using the IRS 1B -LISS II satellite and the crop yield data from crop yield estimation surveys showed that satellite data in the form of vegetation indices greatly improves the efficiency of crop yield estimator. From these studies it could also be observed that for obtaining crop yield estimates at district level, the use of satellite data can lead to a significant reduction in number of crop cutting experiments without affecting the efficiency. Alternatively, the crop yield estimates can be developed at smaller administrative units level (say tehsil block) using appropriate small area estimators.

ESTIMATION OF CROP YIELD USING DATA BASED ON CROP CUTTING EXPERIMENTS ALONG WITH THE SATELLITE SPECTRAL DATA

The factors like different soil types, agricultural inputs, adoption of improved technology etc. affect the crop yield and hence cause a lot of variability in the yield even within a stratum. Since the spectral reflectance is a manifestation of all important factors affecting the crop, hence a stratification of crop area on the basis of crop vigour as reflected by the spectral data is expected to result in a greater efficiency of the crop yield estimation.

(a) Identification of Crop Cutting experiment on the Satellite Imagery

Initially district boundary mask was generated using the topographic maps of scale I :250,000. False colour composite (FCC) using spectral data from band-2 (Green), band-3 (Red) and band 4 (Near infra red) was generated. For identifying the villages selected for crop surveys on the FCC, topographic maps of scale 1:50,000 which contain information on identifiable features like roads, canals, water bodies etc. were used. A Global Positioning System (GPS) was also used to locate the crop plots selected for crop cutting experiments. The GPS was taken to the plots and locations of the plots (longitude and latitude) were recorded. These locations were identified in the selected villages already earmarked on the FCCs.

(b) Stratification of Imageries using Density-slicing Technique

Spectral response characteristics of healthy vegetation, can easily be characterized in different parts of the electromagnetic spectrum. To further enhance the discrimination between different spectral vegetation classes, computation of different vegetation indices using infrared and red band data in the electromagnetic spectrum for describing the crop growth conditions are commonly used. Two most commonly used vegetation indices are:

The Normalized Difference Vegetation Index (NDVI) defined as

$$NDVI = \frac{1R - R}{1R + R}$$

The Ratio Vegetation Index (RVI) defined as

$$RVI = \frac{1R}{R}$$

Where, IR and R refer to radiance in infrared (band-4) and red (band-3) bands of the satellite. These two indices have been used in the present study to generate the index images for post-stratification of the study area on the basis of vegetation vigour.

The concept of density slicing was used to divide' the RVI and NDVI imageries into different vegetation classes. The RVI and NDVI grey level values were linearly stretched over the total range (0-255) of grey level values and were divided into 3 classes named as, (i) Non-vegetation class, (ii) Average vegetation class, and (iii) High vegetation class.

Assigning different colours to different class range values the stratified imageries were generated and area falling under different strata could be obtained which have been used as the strata weights.

POST STRATIFIED ESTIMATOR OF CROP YIELD

In case of yield estimation surveys, the original stratification is based on geographical considerations and may not be much effective in terms of making the strata more homogeneous. As such in the present discussion for simplification of results, the original stratification has not been taken into consideration. The general crop estimation surveys have been planned to provide reliable estimates of crop yield at district level. To improve the efficiency of these estimates satellite spectral data in the form of vegetation indices has been used to post-stratify the crop area into areas of homogeneous crop growth. Let the *n* villages selected in the sample be post-stratified into L'strata based on vegetation indices such that n_k villages fall in the k_{-th} post-stratum. Let Y_{kij} denote the yield for the j_{-th} field in the i_{-th} village of the k_{-th} post stratum. The sample mean for the k-th post stratum can be defined as

$$\overline{Y}_k = \frac{\sum_i \sum_j Y_{kij}}{m_k}$$

where, m_k is total number of field experiments falling in the k_{th} post-strata. Now the post stratified estimator of district average yield can be given by

$$\overline{y} = \frac{\sum_k^L A_k \overline{y}_k}{\sum_k^L A_k} = \sum_k^L W_k \overline{y}_k$$

where, A_k denote the area under crop in the k_{th} post-stratum.

$$W_k = \frac{A_k}{\sum A_k}$$

Ignoring the pre stratification and also ignoring the contribution to sampling error due to post stratification the variance and the estimator of variance of y can be obtained easily given by

$$V(\overline{y}) = \sum W_k^2 V(\overline{y}_k)$$

$$\hat{V}(\overline{y}) = \sum W_k^2 \hat{V}(\overline{y}_k)$$

SMALL AREA ESTIMATION

The issue of small area estimation has gained importance in view of growing needs of micro level planning. The advances in computer facilities have provided convenient tools for many theoretical developments for providing small area estimates. The small area estimation techniques make use of information from other available sources and borrow strength from related or similar areas through explicit and implicit models that connect the small area via supplementary data. Most of the small area estimation techniques in the early stages were developed in the context of demographic studies. Purcell and Kish (1979) categorized these areas under the general heading of Symptomatic Accounting Technique (SAT). Gonzales (1973) described a small area estimation technique well known as synthetic estimator. In this method an unbiased estimate is obtained from a sample survey for a larger area and this estimate is used to derive estimates for sub areas having the same characteristics as the larger area.

SMALL AREA ESTIMATION OF CROP YIELD AT TEHSIL BLOCK LEVEL

General crop estimation surveys have been designed to obtain crop yield estimates at the district level. However, with increasing emphasis on micro level planning, the estimates at lower administrative unit level like tehsil or block level are needed. Since it is not possible and desirable to further increase the number of crop cutting experiments it is desirable to make use of the satellite data and the small area statistics techniques to develop reliable crop yield estimates at tehsil and block level. Consider the population of a district consisting of T small areas of tehsil block. Let the district area be divided into V post strata representing crop condition like very good crop, average crop, poor crop, no crop etc. based on the vegetation indices derived from the satellite spectral data. The crop within these post strata is homogeneous in respect of the character under study (crop yield) and the boundaries of these post strata cut across the small areas. Hence, it can be easily assumed that the units within a small area belonging to particular post strata will have the same characteristics as the units belonging to those particular post strata irrespective of the small area. In order to develop crop yield estimates at tehsil block level from general crop yield estimation surveys based on crop cutting experiments we propose two estimators namely. (i) The Direct estimator, and (ii) The Synthetic estimator.

These estimators make use of available information on crop yield and also the information of crop acreage for all the post strata, which overlap the small area tehsil blocks. It has been seen earlier from the results that stratification based on NDVI provides more efficient estimates of crop yield for the district as a whole. Therefore, for small area estimation only NDVI has been used to develop post strata.

DIRECT ESTIMATOR

Let Y_{tvi} and Xi_{tvv} denote the crop yield and the crop acreage for the i-th plot in the v-th post strata of the t-th small area. Let Y_{tv} and XIV denote estimators of the character under study (crop yield) and the auxiliary character (crop acreage) from the t-th small area and v-th post strata (based on vegetation vigour). Further let SIV denote the set of sample observations belonging to the t-th small area in the v-th post strata. If all !ltv'S are non empty then an unbiased post stratified estimator known as the direct estimator for crop yield may be obtained as

$$\bar{y}_{dt} = \sum_Y W_{tv}\bar{y}_{tv}$$

Where $\bar{y}_{tv} = \dfrac{1}{n}\sum_1^{n_{tv}} y_{tvi}$ is the average yield for tv-th cell

$$W_{tv} = \frac{X_{tv}X_{oo}}{X_{to}X_{ov}}$$

X_{tv}= the crop acreage for tv-th cell

$$X_{to} = \sum_v X_{tv} \text{ is the crop area for the } t_{\text{-th}} \text{ small area}$$

$$X_{ov} = \sum_v X_{tv} \text{ is the crop area for the } v_{\text{-th}} \text{ post stratum, and}$$

$$X_{\infty} = \sum_1 \sum_v X_{tv} \text{ total crop acreage in the district}$$

The approximate estimate of variance of $\bar{y}_{dt}$ can be written as

$$\hat{V}(\bar{y}_{dt}) = \sum_v W_{tv}^2 \hat{V}(\bar{y}_{tv})$$

$$\hat{V}(\bar{y}_{tv}) = \frac{S_{y_{tv}}^2}{n_{tv}} \quad \text{and}$$

$$S_{y_{tv}}^2 = \frac{1}{n_{tv} - 1} \sum_i^{n_{tv}} (y_{tvi} - \bar{y}_{tv})^2$$

SYNTHETIC ESTIMATOR

The direct estimator is based on only the number of crop cutting experiments belonging to tv-th cell *i.e.* t-$_{\text{th}}$ tehsil in the post strata which is quite small and hence the estimator will not be quite efficient. To improve the efficiency of the direct estimator, a synthetic estimator is proposed which make use of the information from the whole sample given by y_{st}

$$\bar{y}_{st} = \sum_v W_{tv} \bar{Y}_{ov} \text{ where } W'_{tv} = \frac{X_{tv}}{X_{to}}$$

and $\bar{y}_{ov}$ is the average crop yield for the vth post stratum given by

$$\bar{Y}_{ov} = \sum_v W_{tv}^n \bar{y}_{ov} \text{ where } W'_{tv} = \frac{X_{tv}}{X_{to}}$$

The estimator of variance of Yst can be approximately written as

$$\hat{V}(\bar{y}_{st}) = \sum_v W_{tv}'^2 \hat{V}(\bar{y}_{ov})$$

$$= \sum_v W_{tv}'^2 \sum_t W_{tv}'^2 \hat{V}(\bar{y}_{tv})$$

Since, sample in each tehsil has been selected independently.

USE OF SATELLITE DATA IN CROP YIELD MODELLING

Forecasting of crop production is one of the most important aspect of agricultural statistics system. Yield forecasts at present are based on eye estimates and the final crop production estimates based on objective crop cutting surveys become available long after the harvests. This as such calls for the necessity of objectives methods for pre-harvest forecast of crop yields. The main factors affecting crop yield are inputs and weather. Use of these factors forms one class of models for forecasting crop yields. The other approach uses plant vigour measured through plant characters. It is assumed that plant characters are integrated effects of all the factors affecting yield. Yet another approach is measurement of crop vigour through remotely sensed data. These approaches are being tried by various organizations. *Box and Jenkins (1976)* used time series models for forecasting where the variation in yield during different years is explained using historical data through trend analysis and presented the well known technique of auto regressive integrated moving averages ARIMA.

The approach using weather parameters is normally based on time series data. The major work in this regard has been attempted at IMD. Their studies involve identification of significant weather parameters in different periods and utilizing these parameter, in the regression model along with trend. At IASRI, New Delhi studies have been carried out at district level using weekly weather parameters. Various composite weather variables were derived as weighted accumulations of weekly weather parameters or interactions up to the time of forecast and were used as regressors in the model along with trend. The main problem associated with meteorological model is assumption of same weather prevailing in a larger area, as observatories are sparsely located. These models also require long series of data, which are not available for most of the locations.

The other approach using plant characters collected at farmers' fields has also been attempted through pilot studies at IASRI, New Delhi. The data have been collected at different periodic intervals through suitable sampling design for 3 to 4 years. Mainly two types of models, between year and within year models have been used. Between year models are based on historic data and involve an assumption that present year is a part of the composite population of the previous years. These models utilize the plant characters at some suitable phenological stage of crop growth either as such or their suitable transformations through multiple regression technique.

In case of crop yield modelling using satellite data, several studies have been undertaken to establish relationship between spectral parameters through vegetation indices and the crop yield. Sridhar *et al.,* (1994) presented wheat production forecasting for a predominantly un-irrigated region in Madhya Pradesh. Singh and Ibrahim (1996) examined the use of multi date satellite spectral data for crop yield modelling using Markov Chain Model. Saha (1999) used satellite data and GIS for developing several crop yield models. However, most of these models have not been adopted in practices because of several problems affecting these models like non-availability of data in time, higher cost involved etc.

SPECTRAL YIELD MODEL

These are empirical models which directly relate the crop yield to the multi-spectral satellite data or derived parameters in the form of spectral vegetation indices (SVI). In this procedure SVI at the time of maximum vegetation growth stage of the crop is related to final crop yield through regression techniques and pre harvest crop yield is forecasted. In India district level yields of major crops like wheat, paddy, sorghum etc. have been developed under crop acreage and production estimation (CAPE) project undertaken by Department of Space. However, these models could explain about 60 per cent variation in yield and hence are not very efficient. In the situation of post-stratification based on vegetation indices an improvement in efficiency of the regression model is required. Regression coefficient of y (yield) on the spectral response parameter (x) can be expected if separate models are developed for different post-strata.

For this we define the regression model as

$$y_{hi} = y_{hi} = \beta_0 + \beta_h X_{hi} + e_{hi}$$

where y_{hi} and x_{hi} denotes the yield and the spectral response of the ith unit in the hth steatum and m_h is the number of sampling units in the hth stratum, h= 1.2.....L; i= 1,2,.....m_h

$$E(e_{hi}) = 0$$

$$V(e_{hi}) = \sigma_h^2$$

$$Cov(e_{hi} e_{hj}) = 0 \text{ for } i \neq j$$

Here β_h may be estimated separately for each stratum. Under the given assumption, the Ordinary Least Square (OLS) estimator of the regression coefficient β_h for the h_{th} stratum may be given by β_h as

$$\hat{\beta}_h = \frac{\sum_{i-1}^{m_h} (X_{hi} - \overline{X}_h)(y_{hi} - \overline{y}_h)}{\sum_{i-1}^{m_h} (x_{hi} - \overline{x}_h)^2}$$

With $\overline{X}_h = \sum_{i=1}^{m_h} x_{hi} / m_h$ and $\overline{y}_h = \sum_{i=1}^{m_h} y_{hi} / m_h$

The variance of $\hat{\beta}_h$ is given by

$$V(\hat{\beta}_h) = \frac{\sigma_h^2}{\sum_{i=1}^{m_h} (x_{hi} - \overline{x}_h)}$$

An unbiased estimator of h is given by

$$S_h^2 = \sum_{i=1}^{m_h}(y_{hi} - \hat{\beta}_0 - \hat{\beta}_h\, x_{hi})^2 / (m_h - 2)$$

The fitted regression equation can be used to predict corresponding to a chosen value x'_{hi} of the spectral response

$$\hat{y}_{hi} = \hat{\beta}_o + \hat{\beta}_h\, X'_{hi}$$

$$V(\hat{y}_{hi}) = V(\bar{y}_h) + (x'_{hi} - \bar{x}_h)^2 V(\hat{\beta}_h)$$

$$= \frac{\sigma_h^2}{m_h} + \frac{(x'_{hi} - \bar{x}_h)^2 \sigma_h^2}{\sum_{i=1}^{m_h}(x'_{hi} - \bar{x}_h)^2}$$

The regression coefficient estimator for the population *n*

$$\hat{\beta} = \sum_{h=1}^{L} w_h^2\, \hat{\beta}_h$$

and the variance of the estimator of the regression coefficient is given by

$$V(\hat{\beta}) = \sum_{h=1}^{L} w_h^2 V(\hat{\beta}_h)$$

where w_h is the weight of hth stratum m_h

INTEGRATED YIELD MODEL USING SPECTRAL DATA AND FARMERS EYE ESTIMATE OF CROP YIELD

Most of the crop yield models developed so far could not be adopted in practice either because of delay in the availability of data on different variables to be used in the model or the high cost in collecting the data and in analyzing the results. For any operational yield model to be successful for adoption, it is necessary that data should be available much before the harvest of the crop and it should be cost effective. Spectral data in the form of vegetation indices have proved to be very useful variable for explaining variability of the crop yield which can be early available for use in yield forecasting models. In a study for 'evaluation of crop cutting methods and farmers reports for estimating crop production' undertaken at Long acre Agricultural Development Centre, UK, Verma *et al.* (1988) have shown that farmers eye estimates are remarkably close to actual production figures. But, eye estimates being subjective and amenable to several non-sampling errors, it is desirable that these estimates are not used directly for estimation of crop yield. However, this information can be used as auxiliary variable along with the spectral vegetation indices to improve the efficiency of the crop yield models. An earlier such attempt on using eye appraisal of crop yield of a large number of sample fields as auxiliary information had been made by many research workers. Therefore in a study conducted by Singh (2003) suitable models using spectral vegetation indices in the form of NDVI and farmers eye estimate as explanatory variables have been developed for improved crop yield forecasting models. Both of these variables can be easily obtained at the time of maximum growth of crop and can prove very effective for developing suitable yield forecasting models.

Complete information of crop acreage and production estimation by remote sensing in tabular form in India are predicted in Table 19.1

Table 19.1 Information of Crop acreage and production estimation by remote sensing

Scale of Coverage	Local-regional to national
Crops covered	Wheat, Rice, Sugarcane, Mustard, Cotton, Sorghum, Groundnut, and Potato
Input Data	Spatial resolution IRS–/III/AWiFS 23/56 m Temporal Resolution 24/5 days
	Microwave Radarsat SAR
Sampling Design	Sampling Stratified sampling with 10-15% sample segments at National level
Image analysis	Supervised maximum likelihood at single date data for local to state level crop acreage estimation.
	Multi-stage classification using Mask (Nonagriculture areas) for multi-date AWiFS using decision rule based classification
Accuracy evaluation	> 95% compared to DoA estimates and production forecast >90 % compared to independent crop cutting experiments conducted by NSSO
User acceptance	User type: Govt / Industry/... Mainly Government, Centre as well as State
Timelines	Regular 4 forecasts – pre-season acreage to be sown, early season, mid-season and pre harvest
Software/package	ERDAS, PCI-GEOMATICA, ENVI & ARC-Map
Developed own module like	CAPEMAN /CAPEWORKS, SAR-CROPS & Auto-Fit programme.

(Source: Parihar and Markand, 2006).

REFERENCES

Box GEP and Jenkins GM (1976). Time Series Analysis: Forecasting and Control. Holden-Day, San Franscisco, 575.

Dadhwal VK and Parihar. (1985). Estimation of 1983-84 wheat acreage of Kamal district (Haryana) using Landsat MSS digital data. Technical note, *IRS-UP/SAC/CPFffN/09*, Space Application Centre, Ahmedabad.

Gonzalez ME. (1973). Use and evaluation of synthetic estimates. *Proc. of American Statistical Association, Social Statistics Section, 33-36.*

Purcell NJ and Kish L. (1979). Estimation of small domains. *Biometrics,* **35:** 365-384.

Saba SK. (1999). Crop yield modelling using satellite remote sensing and GIS-current status and future prospects. *Proc. Geoinformatics Beyond 2000,* an international conference on Geoinformatics for natural resources assessment, monitoring and management held at IIRS, Dehradun during 9-11 March, 1999.

Singh R and Ibrahim AE. (1996). Use of spectral data in Markov Chain model for crop yield forecasting. *Jr. Indian Socy. of Remote Sensing,* **24**(3):145-152.

Singh R, (2003). Use of satellite data and farmers eye estimate for crop yield modelling. *Jour. Ind. Soc. Agril. Stat.,* **56**(2),166-176.

Sridhar VN, Dadhwal VK, Chaudhary KN, Sharma R, Bairagi GD and Sharma AK. (1994). Wheat production forecasting for a predominantly unirrgated region in Madhya Pradesh. *Int. J. Rem. Sens.,* **15**(6): 1307-1316.

Verma V, Marchant T and Scott C. (1988). Evaluation of crop cut methods and farmers reports for estimating crop production. Longacre Agricultural Development Centre Ltd., London, U.K.

■

20

Yield Monitoring System for Grain Combine Harvester

☞ Manjeet Singh[1], ☞ Ankit Sharma[2], Karun Sharma[3], ☞ P.K. Mishra[4] and ☞ Rajnesh Kumar[5]

CHAPTER

INTRODUCTION

Yield monitoring is a recent development tool in precision farming and agricultural machinery that allows farmers to assess the yield variability in his field during harvesting of crop. In ordr to use yield maps in precision agriculture effectively, one needs to understand how grain flow is measured by a yield sensor and how this measured at specific location. When addressing "yield monitor accuracy" we shoud be aware of the interactions between yield sensors and how they relate to the accuracy of a yield map. This is logicaly first step for those who want to practice site-specific crop management or 'precision agriculture'. Yield monitering is used in conjunction with Global Positioning System (GPS) receiver for recording field and crop information during harvest and gives user an accurate assessment of how yields vary within a field. Yield maps provide feedback for determining the effects of managed inputs such as fertilizer and lime, seed and pesticides and cultural practices such as tillage, irrigation and drainage. Monitoring of crop yield is the most interesting operation to any farmer. Yield monitering for grain crops has been introduced in some developing as well as transitional countries. Although in advanced countries, high hp combines for large farms are available fitted with yield monitors as standard equipment or installed separately. However, it difficult to adopt on small combines directly, because the sensors and systems are usually design for those combines and are very costly. Consequently, indigenous yield monitor has also been designed and developed in India to keep various things in mind *i.e.,* small combine, low cost and easy working and installation. Indigenous Yield Monitor fills the wide gap in the yield monitoring technology for large farms in advanced countries and used in India for small and marginal farms.

YIELD MONITORING SYSTEMS

Yield monitors are a combination of several compo-nents (Fig. 20.1). They typically include a data stor-age device, user interface (display and keypad) and a console located in the combine cab, which controls the integration and interaction of these components. The sensors measure the mass or the volume of grain flow (grain

[1] Research Engineer, Department of Farm Machinery and Power Engineering, PAU, Ludhiana
[2] Assistant Professor (Agril. Engg.), Punjab Agricultural University, KVK, Mansa (Punjab)
[3&4] Research Associate, Department of Farm Machinery and Power Engineering, PAU, Ludhiana
[5] Senior Research Fellow, Department of Farm Machinery and Power Engineering, PAU, Ludhiana

flow sensors) separator speed, ground speed, grain moisture, and header height. Yield is determined as a product of the various parameters being sensed. One must understand the function of these components in order to understand the interaction of the yield monitor, combine operator and combine dynamics.

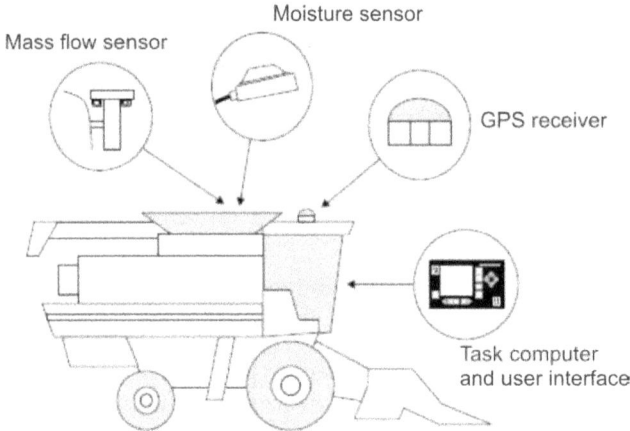

Fig. 20.1: *Components of a yield-monitoring system.*

YIELD AND MOISTURE SENSORS

Yield sensors measure the force of the grain hitting a plate, the attenuation of light passing through the grain stream, the weight of the grain collected for a period of time, or the volume of grain passing through a paddle. The most common method is to measure the force of the grain striking a plate located at the top of the clean grain elevator (Fig. 20.2). Yield and moisture data are collected simultaneously to obtain corrected yield. Moisture sensors are often located in the clean grain elevator or the clean grain auger. Most sensors measure the capacitance of the grain and can provide continuous moisture data (Fig. 20.3).

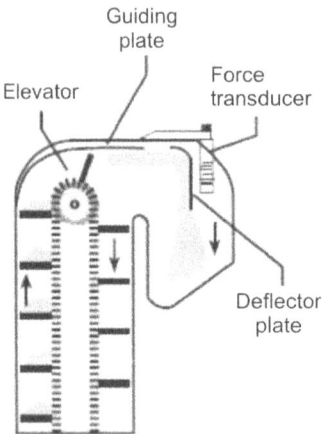

Fig. 20.2: *Mass flow sensor*

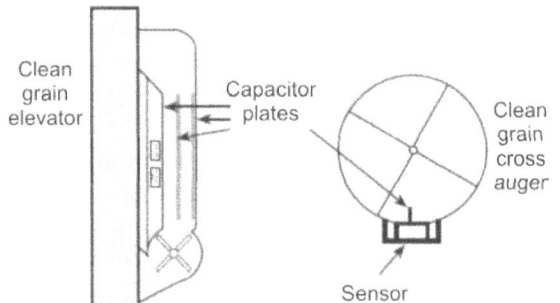

Fig. 20.3: *Moisture sensor installation.*

DATA COLLECTION AND STORAGE

Data is often recorded on removable memory cartridges which can be downloaded to a computer. Data should be downloaded daily to ensure that the yield monitor is working properly and to protect from accidental loss. Memory cards may store several megabytes of data. One megabyte of memory can store 15 to 45 hours of information for yield data collection intervals of 1 to 3 seconds.

YIELD MONITOR CALIBRATION AND ACCURACY

Yield monitor must be calibrated to provide accurate yield data. Calibration must be performed for each type of grain harvested at the beginning of the harvest season. Accuracy usually improves when several loads are used to perform the calibration. Re-calibration should be performed as necessary, especially later in the season as average moisture content drops or when there is a significant change in crop conditions. Calibration is usually as simple as weighing and recording the moisture of the first several loads collected under a variety of conditions such as various operating speeds or grain flow rates, consult the operator's manual for specific instructions. The accuracy of a yield monitor depends not only on its design but on how carefully the calibration procedure is followed. Some companies offer a training session or video tapes to teach calibration. These procedures vary considerably among manufacturers but all require carefully recording the weight of grain which can become a logistical problem.

OPERATING A COMBINE EQUIPPED WITH A YIELD MONITOR

The final appearance of a yield map depends on how the combine is operated. Frequently stopping or sudden changes in speed can cause erratic yield data due to delay and smoothing phenomena associated with the combine separating system. The combine must be operated on a uniform swath width to ensure accurate yield data. The width of the header must be manually entered into the monitor to accurately calculate yield. Yield will be underestimated if fewer rows are harvested. Many yield monitors allow the operator to change the number of rows or the per cent of width harvested to correct yield for point rows or field edges.

PRESENT TECHNOLOGIES FOR YIELD MONITORING

According to United States Department of Agricultural survey (Daberkow and McBride, 1998) of corn producing farms in 16 states of USA, about 9 per cent used some form of precision agriculture for corn production, this was equal to about 1/5th of the corn harvested in 1996. They also reported that 545 of the Precision Agriculture adopters useed grain monitors. In Argentina there were about 560 yield monitors in 2001 and about 4 per cent of the grain and oil seed area was harvested with headers equipped with yield monitors (Lowenberg, 2003). Informal reports indicate that in Australia about 800 yield monitors were used in 2000 harvest. Some fifteen farmers used the yield monitoring system in South Africa during the 1999–2000 crop season (Wilhem, 2000).

DIRECT METHODS

The grain yield monitor first became available in the early 1990's. Today, nearly every grain combine manufacturers of (USA) offers an optional, factory installed yield monitor. Most commercial sensors rely on impact sensing for mass-flow measurement. In these sensors the impact force or moment caused by the change in momentum of the grain flow, is measured. The plate can be flat or curved, or just a pair of fingers. These sensing devices are mounted at the top of the elevator. De Baerdemaeker *et al.* (1985) investigated an impact type flow meter which measured grain forces on a curved circular tube using a load cell, near the top of a clean grain elevator. The force developed by the change in momentum of the grain was found to be proportional to grain flow rate. The calibration of the sensor was influenced by the slope of the field and was dependent on material properties like moisture content, friction and kernel size. This work was a precursor to the development of most of the yield monitoring systems used today.

Volume of grain is measured when flowing through the sensor during a fixed time interval or time needed by a fixed volume of grain to flow through the sensor. To acquire accurate measurements, mass density has to be measured for each different field or even different measurements on one field. By means of optic sensors (light emitter and detector), height of the grain on the elevator paddles is measured. Together with the conversion of volume to mass, the conversion of height to volume is a second drawback of this sensor. This sensor has been studied in different forms, depending on the configuration of the light emitter and detector. This one-dimensional system has been studied by Strubbe *et al.* (1996). When tested at a transverse slope of 11 per cent, the difference between estimated and real volume approached 13 per cent at high flow rates. By using a two-dimensional system, placing two sensors at each side of the elevator, the results could be improved. To improve further, the grain should be spread more homogeneously on the paddles. By vibrating the machine, the grain surface is more flattened near the top of the elevator but placement of the sensor is more difficult. Reitz and Kutzbach (1992) describe a two-dimensional system with one emitter or detector pair aside of the elevator and two at the backside parallel to the driving direction. By introducing the latter pair of sensors, the accuracy while harvesting on a hillside could be strongly improved. Chosa *et al.,* (2002, 2003 and 2004) reported a grain yield monitoring system for head-feeding combines in Japan. This system uses optical sensors installed in a grain tank to measure the grain flow and a load cell that was settled below the grain tank to measure the weight of tank including the grains harvested.

Hindryckx and Missotten (1994) and Kutzbach and Schneider (1997) have given an overview of different devices for measuring grain yield. Mass or volumetric flow sensors can be divided in four groups, depending on the principle of measurement. The basic principle of these measurements is the combination of a weight and speed measurement. Frequently, the grain mass is measured by weighing machine components that transport grain. In general, problems are noted with dependence on the moisture content and combine operation on slopes. Moreover, these measuring devices are difficult to construct.

An intelligent yield monitor was developed for grain combine harvester by

Minzan *et al.*, (2005). The crop was wheat and the harvesting combine used to equip the monitor was JL1065, a typical machine with 4-meter swath width in northern China. The monitor can collect four analog signals, grain flow, grain moisture content, grain temperature and header up and down signal and two digital signals, ground speed and elevator speed. Two digital signals were sensed by Hall effect elements. The monitor can also synchronously receive Digital Global Positioning System (DGPS) signal. A liquid crystal display and a touch screen were integrated as an I/O interface. Field tests showed a linear relationship between actual yield and the output of the yield monitor. The error between measurement and prediction was less than 3 per cent. It is concluded that the developed intelligent yield monitor is practical. Michihisa *et al.*, (2005) developed a grain yield monitor for head-feeding combines. It consists of two grain flow sensors, a single kernel moisture sensor, Real Time Kinematic (RTK) GPS and so on. A method to correct the time lags in the separation and cleaning section and the circulation of grain flow using a return flow sensor is adapted to obtain the accurate grain yield. The yield monitor was installed on a 4 m combine and was tested in several hectares of paddy fields. In addition, a grain yield mapping software was developed for this monitor. A new crop mass flow sensing technology was proposed to reduce the effect of combine dynamic errors by Veal *et al.* (2010). This new sensor measures the flow of biomass through the feeder housing. Results were also compared with to yield data collected from an impact type mass flow sensor located in the clean grain elevator. The results indicated that biomass flow sensing at the feeder housing might complement existing technologies to improve yield monitor data quality.

INDIRECT METHODS

The dielectric constant of the mixture of air and grain increases as the mass flow increases. However, the dielectric constant is not only dependent of the mass flow but also on the moisture content and grain type. Separate calibrations must be executed for each grain type. The calibration curve is non-linear and partly dependent on the moisture content. The change of the dielectric properties of the material between two capacitor plates is measured by Stafford *et al.* (1991). It was proposed a less invasive method of determining the corn yield variation within the field through the use of remote sensing methods (Diker *et al.*, 2002). Aerial images of a field were taken with a multispectral digital camera and these images were analysed using the red, green and near infrared bands. Both traditional yield monitor data and the remote sensing data indicated that the yield variability within the field increased beyond the area covered by the center pivot irrigation system used in the field. Both methods indicated that the center of the field was less variable. Wild *et al.* (2003) proposed another method to determine crop yields that did not involve mass flow measurement within a grain combine. Instead, this method was based on radar pulses. In the test setup, a radar pulse was sent through the material once the pulse struck a metal sheet. The returning signal intensity was recorded. The results of the test were promising, as the coefficient of determination relating the amount of crop to the signal strength was between 94 and 99 per cent.

COMMERCIAL SYSTEMS

RDS Technology Limited produces a yield mapping system (Ceres) based on volumetric measurement. An emitter detector system is mounted on the side wall of the clean grain elevator. This system was patented earlier (1982) by the CLAAS Company. The CLAAS quantimeter II is similar to the RDS sensor. The Greenstar yield mapping system of the John Deere Company uses an impact style mass flow sensor with a curved plate. The deflection of the plate is measured via a linear potentiometer. Case IH (Advanced Farming Systems) utilizes the impact sensor developed by AgLeader. This impact type sensor has a flat plate on which the deflection is measured by aid of strain gauges. The Deutz-Fahr Teris system uses the same sensor. The Grain-Trak yield measuring system by Micro-Trak uses two fingers to measure the impact force. With the Fieldstar precision farming system of Massey Ferguson a radiometric yield meter is used. However, in countries where the use of the radioactive radiation source is not allowed, an impact system with two measuring fingers can be used.

DEVELOPMENT OF YIELD MONITOR IN INDIA

In Indian context yield monitor can help in various manner which have been describe below:

- In India, yield is measured during marketing of harvested crop and as a gross yield of the land owned by the farmer but with the help of yield monitor farmer can measure the exact yield during the harvesting which helps the farmers to get an exact assessment of crop production from their farm.

- The farmers having marginal and small farms, cannot enjoy the benefits of mechanization through individual ownership. Custom hiring of farm machinery and power is the only mean by which they can reach the benefits of farm mechanization. For instance, custom hiring of combine harvester is spreading very fast. By using the yield monitor in combine harvester, custom hiring charges may be charged on the actual basis *i.e.*, weight basis rather than area basis used nowadays to collect the charges from the farmers.

- Measure the yield variability in different fields and within the same field, provide tools to the researchers for performing experiments in the fields, helps in making strategies and planning to manage the yield variability, and helps the planners and decision makers to implement the food policies, price regulation and storage of food grains. Scientists and researchers can use yield monitors to measure the yield of experiment research plots of the crops.

- Yield data can be accumulated for a specific load or field, thereby facilitating the comparison of hybrids, varieties or treatments within test plots. *For example*, all yield monitors can measure grain mass and harvested area on a load-by-load or field by field basis. This feature allows an operator to get instantaneous readout in the field of accumulated grain weight, harvested area and average yield. With many yield monitors, these values can be exported to a personal computer and stored in nonvolatile memory for further analysis or printing via specialized software packages or more standard word-processing and spreadsheet software. Season summaries of harvested areas might then be used to settle custom harvesting charges or to keep track of production from individual fields when it is impractical to scale grain trucks.

- With a yield monitor, a producer also can conduct on-farm variety trials or weed control evaluations without the need of a weigh wagon. Such on-farm comparisons help producers fine-tune crop production practices to their soils.

- An understanding of yield variations within or between fields has been used to evaluate growth and management history of crops and this provides important information to determine site-specific management for the following year. Sites where lodging has been observed should have reduced amounts of fertilizer applied and sites where yield has been low should have increased amounts of fertilizer. Although, farmer can understand yield variations through routine farm work, the yield monitoring combine is expected to play an important role in establishing site specific crop management and spreading related technology to farmers.

NEED OF YIELD MONITORING SYSTEMS IN INDIA

- Mainly combine harvesters are used for the harvesting of wheat and rice in the northern part of India.

- In Punjab, the total area under combine harvesting for rice and wheat is 91 per cent and 82 per cent respectively.

- Custom hiring operators may charge from the farmers on weight basis rather than area basis.

- Yield monitor can also be used to measure the yield of experiment research plots of the crops.

- The yield monitoring combine is expected to play an important role in establishing site specific crop management and spreading related technology to farmers.

- To spread the use of yield monitoring technology in Indo-Gangetic plain and throughout India, it is necessary to develop an original yield monitoring technique for low hp combine.

Yield monitor developed in India at department of Farm Machinery and Power Machinery, Punjab Agricultural University, Ludhiana were of batch type and continuous type yield monitors which have been describe below:

TRAILER MOUNTED TYPE YIELD MONITOR

A trailer-mounted yield monitor was designed by keeping in view the potential yield of rice and wheat for 500 m² harvested area. The trailer-mounted yield monitor (Fig. 21.4), consisted of a single point parallel load cell with 700 kg capacity, offers total stainless steel construction, complete hermetic sealing, oil proof, water proof and anti-corrosion making it suitable for use in all kinds of environment, a mild steel drum of size 125×85×80 cm and a display unit EPS 301 with 24 bit sigma delta converter having 25.4 mm bright LED. The display unit was based on Switch Mode Power Supply (SMPS) that produces the desired voltage by rapidly switching between full power on and full power off. The average between these two voltages produces the power needed for a device. It works on the principle of switching regulations. The SMPS system is highly reliable, efficient, noiseless and compact because the switching is done at very high rate. Fig. 20.1 shows the working view of the trailer mounted yield monitor in the field. This type of yield monitor was of very low cost and mobile but the limitation was that the combine harvester had to be stopped after harvesting of each 500 m² plot to measure the yield.

Fig. 20.4: *View of trailer mounted yield monitor in the field*

COMBINE MOUNTED BATCH TYPE YIELD MONITOR

A combine-mounted batch type yield monitor was developed by using one S-type load cell, an auxiliary tank and display unit. An auxiliary tank of size

145×100×85 cm was mounted within the original tank of the combine harvester (Fig. 20. 5). Load cell was placed under the auxiliary tank with the help of fasteners. The capacity of the auxiliary tank was 400kg and it was made of steel sheet which has thickness of 3mm. The steel sheets of three sides of tank were welded and fixed with each other. One side of auxiliary tank was moveable and a handle was provided at the top of moveable plate.

The grain directly falls in the auxiliary tank instead of the main tank. The yield data is displayed over the display unit installed near the driver seat (Fig. 20.6). When the auxiliary tank is filled, grains are shifted to the main tank of the combine by opening the outlet fitted in the auxiliary tank of the combine. Opening of the outlet was done manually and it took approximately four to five minutes time for shifting of grains from auxiliary tank to main tank.

Fig. 20.5: *Auxiliary tank placed in the main tank*

Fig. 20.6: *Display unit attached on the combine*

COMBINE MOUNTED CONTINUOUS TYPE YIELD MONITOR

Auxiliary tank was fitted in the main tank of combine and cable of the load cell was connected with micro-controller's input pins. Inductive/speed sensor was fitted at the rear axle of combine (Fig. 20.7) and its signal wire was connected with Micro-controller's input pins. Display unit was placed in front of operator's seat. Harvested grains coming through the elevator fall in the auxiliary tank when combine harvested the crop in the field. Load cell fitted at the bottom of tank fluctuated when grains impacted on the load cell. Load cell sent an analogue signal to signal condition element. Similarly speed inductive sensor sent the signal to the conditioning element. In this element, analogue signal was converted into digital form and impurities of signal were removed. The digital data is sent to the micro-controller for processing the data. ICRO-controller stored and converted the data into understandable form. After processing the data, data presentation element helps to present the data in readable form. Display unit was is the data presentation element which shows yield data in kilogram and distance in meter form. Further development in the continuous type yield monitor have been done by replacing the auxiliary tank from the main tank and by fitting an impact plate with cantilever load cell below the grain elevator outlet (Fig. 20.8). This system will help in collecting the continuous yield data without making the auxiliary tank empty as in the previous developed yield monitor.

Fig. 20.7: *Rear axle of combine after mounting an inductive/speed sensor*

Fig. 20.8: *View of impact plate attached below the grain elevator outlet*

LIMITATIONS OF YIELD MONITORS

The combine operation is dynamic and the flow rate of material processed can vary depending on entering and exiting the crop. These varying flow rates can influence the results of the yield monitor data. Since the yield monitor measures the rate at which clean grain is enter-ing the grain tank, time delays between the time when grain enters the combine header and the time when it passes through the clean grain elevator can be significant. Combines also smooth abrupt changes in yield, hence, the yield monitor measures delayed averages of yield. The phenomena of time delays and smoothing are most obvious when a combine enters or leaves the crop at the ends of a field. The combine, in the example above has a delay of 15 seconds to process the entering crop and would travel 110 feet and harvest almost 0.04 of an acre before an accurate or stable yield is displayed on the yield monitor. Most yield mapping software compensates for equip-ment delays caused by the combine and corrects the yield data. The resulting yield map will not be perfect, but it will be very adequate for observing the magni-tude and location of yield variability.

CONCLUSION

Mainly yield monitors are mass, volumetric flow and impact type depending on the principle of measurement. While evaluating these sensors, the different critical points like ease of mounting on different types and models of combine harvesters and its calibration and precision or accuracy, there should be no obstruction to the normal threshing process, even when the sensor is damaged. There are six common sources of error, which affect the yield monitor's accuracy. The six errors are incorrect swath width, time lag through the threshing mechanism, GPS error, grain surging within the transport system, grain loss, and sensor calibration.

A batch type grain yield monitor having load cell of capacity 700 kg was developed for grain combines to measure the spatial variation of grains for use as single unit or by putting directly in trailer. Combine mounted batch type yield monitor was also developed by fitting an auxiliary tank having capacity 400 kg, in the main tank and load cell at the bottom of the auxiliary tank. A Continuous Type Yield Monitor was also developed with the speed/inductive sensor fitted at the rear axle of combine and its signal wire was connected with Micro-controller's input pins. Because of its low cost, the system is very useful for small and marginal fields and can be used for any crop or any combine. Yield monitor information can also be geographically referenced for the generation of yield maps to provide year-to-year comparisons of high- and low-yielding areas of a field throughout a crop rotation sequence for better management.

REFERENCES

Chosa T, Kobayashi K, Daikoku M, Shibata Y and Omine M. (2002). A study on yield monitoring system for head-feeding combines (Part 1) (in Japanese), J. Japanese Society of Agricultural Machinery, **64**(6): 145-153.

Chosa T, Shibata Y, Kobayashi K, Omine M and Daikoku M. (2003). A study on yield monitoring system for head-feeding combines (Part 2)(in Japanese), J. Japanese Society of Agricultural Machinery, **65**(2): 192-199.

Chosa T, Shibata Y, Omine M, Toriyama K and Araki K. (2004). A study on yield monitoring system for head-feeding combines (Part 3) (in Japanese), J. Japanese Society of Agricultural Machinery, **66**(2): 137-144.

Daberkow SG and McBride WD. (1998). Adoption of precision agriculture technologies by U.S. corn producers. J. Agribus. **16**(2): 151-168.

De Baerdemaeker J, Decroix R and Lindemans P. (1985). Monitoring the grain flow on combines. In proc. Agrimation I Conference and Exposition, 329-338. St. Joseph, Mich.: ASAE.

Diker K, Heermann DF, Bausch WC and Wright DK. (2002). Relationship between yield monitor and remotely sensed data for corn. ASAE Paper No. 021164. St. Joseph, Mich.: ASAE.

Hindryckx K and Missotten B. (1994). Krachten en snelheden van de graanstroom bij een impulsdebietmeter (Eindwerk KU Leuven).

Kutzbach HD and Schneider H. (1997). Scientific challenges of grain harvesting. ASAE Paper No. 97-1080.

Lowenberg D, Boer J.(2003). Precision farming or convenience agriculture. http:// www.regional.org.au/au/asa/2003/i/6/lowenberg.htm.

Michihis I, Yao Y, Kimura A and Umeda M. (2005). Development of Grain Yield Monitor for Head-feeding Combines. Presented at the ASAE Annual International Meeting, Florida 17-20 July 2005.

Minzan L, Peng L, Wang Q, Fang J and Wang M. (2005). Development of an Intelligent Yield Monitor for Grain Combine Harvester. Artificial Intelligence Applications and Innovations. Springer Boston. Volume 187, pp. 663-70.

Reitz P and Kutzbach HD. (1992). Technische Komponenten für die Erstellung von Ertragskarten während der Getreideernte mit dem Mähdrescher. VDI/MEG Kolloquiumsbericht 'Ortung und Navigation landwirtschaftlicher Fahrzeuge' Heft 14, Freising-Weihenstephan, March 5–6, pp. 91–105.

Stafford J V, Ambler B and Smith MP. (1991). Sensing and mapping grain yield variation. Automated Agriculture for the 21st Century, Proceedings of the 1991 Symposium, 16–17 December (ASAE, Chicago, IL), pp. 356–365.

Strubbe GB, Missotten and Baerdemaeker JD. (1996). Performance evaluation of a three-dimensional optical volume flow meter. Transactions of the American Society of Agricultural Engineers 12(4), 403–409.

Veal MW, Shearer SA and Fulton JP. (2010). Development and Performance Assessment of a Grain Combine Feeder House-Based Mass Flow Sensing Device. Trans. of the ASABE ISSN 2151-0032. Vol. 53(2): 339-348.

Wild K, Ruhland S and Haedicke S.(2003). Performance of Pulse Radar Systems for Crop Yield Monitoring. Paper No. 031038. (2003) ASAE Annual International Meeting Las Vegas, Nevada 27- 30 July, 2003.

Wilhem N. (2000). Personal communication. Centre for Agricultural Management, University of the Free State, Bloemfontein, South Africa.

■

21

Database Requirements for Precision Farming

☞ Afzal Ahmad[1] and Ramesh Kumar[2]

CHAPTER

INTRODUCTION

Precision agriculture is a systematic approach to managing fields. It involves multiple technologies and multiple disciplines. It should be first and foremost considered a tool to make better management decisions. It is not a silver bullet that will make any farm profitable but it will help in making more informed decisions that can lead to greater prodigality. Just as the seasons have a cycle, so does precision agriculture. Throughout the season, data is collected from various sources, analyzed and used for decision making. There are many types of information that can be collected and used in a precision agriculture system including yield, topography, prescriptions, imagery, electrical conductivity, soil types and as applied maps. Agricultural systems are continuing to change in response to economic, technological and social trends. Agriculture practices are being questioned not only by farmers but the public at large. Farmers adopt new technology that concerns the profitability and environmental impacts. Mechanization brought many types of equipment to use. Technology brought the genetically modified products. The hard work of researchers and agricultural experts has brought this new technique called precision farming, precision agriculture or site specific management of soils and crops.

Agricultural systems are continuing to change in response to economic, technological and social trends. Agriculture practices are being questioned not only by farmers but the public at large. Farmers adopt new technology that concerns the profitability and environmental impacts.

Precision farming is a new concept to this world, a great evolution in the agricultural fields. Increase in population has led to many disasters diseases and conflicts. Changes must take place in agriculture in order to increase the yield and reduce the demand. It is gaining popularity largely due to the introduction of technological tools which are effective in the agricultural community. These tools are cost effective, accurate and user friendly. These new inventions rely on computers, data collecting sensor, GIS and GPS techniques. Precision farming is really a systematic approach to managing fields.

---

[1] Subject Matter Specialist (Agronomy), Krishi Vigyan Kendra, Tepla, Ambala, Haryana

[2] Extension Specialist, Krishi Vigyan Kendra, Tepla, Ambala, Haryana

PRECISION AGRICULTURE

Precision Agriculture attempts to manage variability within fields. There are several reasons that precision farming has come about as a management method in the past decade

- High cost of crop inputs including seed, fertilizer, pesticides and fuel.
- Environmental concerns about fertilizers & pesticides near sensitive areas, run-off and de-nitrification.
- The technology has become available and economically feasible.

Precision agriculture goes by many names but they all refer to managing variability:

- Precision Farming
- Prescription Farming
- Spatially Variable Agriculture
- Site Specific Management
- GPS Farming
- Farming by Satellite
- Farming by the Foot
- Variable Rate Application
- Digital Farming

ENABLING TECHNOLOGIES OF PRECISION FARMING

Many technological developments occurred in 20th century mainly Information and Communication Technology (ICT) and Geoinformatics, which led to the development of the concept of precision farming. The success of the precision farming system relies on the integration of these technologies into a single system that can be operated at farm level with sustainable effort. These technological developments are as follows:

GEOINFORMATICS

It is a term that appears to have been independently coined by several groups around the world to describe a variety of efforts to promote collaboration between computer science and the geosciences to solve complex scientific questions, it is the science and technology of gathering, analyzing, interpreting, distributing and using geographic information. Geoinformatics otherwise called as geomatics encompasses a broad range of disciplines including surveying and mapping, remote sensing, geographic information system and global positioning system.

GLOBAL POSITIONING SYSTEM (GPS)

Without having a reliable method of locating equipment and items in a field, it is difficult to manage in-field variability. A crude method might be to stake out the field to show area that require different treatment, but this is not practical on large fields. A reliable positioning method is needed to accurately locate field features to make precision agriculture work. Some local positioning system were developed but not successfully commercialized. The advent of GPS allowed for low cost, reliable positioning of equipment in the field. The United States of America government initially as a way to locate military applications launched the GPS. It has grown into a commonplace utility, being used in everything from cell phones to cars to landing aircraft and guiding ships. There are two major satellites positioning systems circling the earth. The main system used in North America is the US military Navigation Satellite Timing And Range Global Positioning System or NAVSTAR GPS System. There is also the Russian developed GLONASS System. System under development includes the European Galileo System, Chinese Beidou and Compass, and India's Indian Regional Navigational Satellite System (IRNSS). The general term for all satellite positioning system is Global Navigation Satellite System (GNSS). The GPS system consists of three parts:

(i) **Space Segment**: 24 operating satellites in orbits about 20,200 Kilometer above the earth that pass by twice per day (12 hour orbits).

(ii) **Control Segment:** Monitor and control stations on the ground that maintain the satellites in their proper orbits and adjust the satellite clocks to maintain the health of the GPS.

(iii) **User Segment:** The user's GPS receiver on the ground. This can be in cab of the tractor, an airplane or handheld device. The GPS is a satellite based navigation system that can be used to locate position anywhere on the earth. GPS provides continuous (24 hours per day), real time, 3-dimensional positioning, navigation and timing worldwide in any weather condition. GPS was originally intended for military applications but in the 1980s, the government made the system available for civilian use. There are no subscription fees or set up charges to use GPS. Any person with a GPS receiver can access the system and it can be used for any application that requires location coordinates. The GPS positional accuracy when used in single receiver mode (autonomous navigation) can be degraded by various error sources. The positional (horizontal) accuracy of the GPS can be of the order of 20 meter. In order to achieve the required accuracies, especially needed for precision agriculture, the GPS has to be operated in a differentially corrected position mode *i.e.*, Differential Global Positioning System. In the DGPS, the errors computed by a reference station which is located in a known place, is transmitted to the mobile user and error correction is done to improve the accuracy. The most common use of GPS in agriculture is for yield mapping and variable rate fertilizer and pesticide applicator. GPS are important to find out the exact location in the field to assess the spatial variability and site-specific application of the inputs. GPS operating in differential mode are capable of providing location accuracy of 1 meter and also sub meter. The availability of GPS approaches to farming will allow field-based variable to be tied together. This tool has to be the unifying connection among field variables such as weeds, crop yield, soil moisture, and remote sensing data. The general uses of GPS in precision farming are yield mapping, variable rate control, field mapping, guidance, asset tracking, irrigation, tracking livestock, aerial spraying, rock picking, auto steering, automatic section control, drainage, data collection, crop scouting, topographic mapping, electrical conductivity mapping, soil sampling and new uses are being developed.

GEOGRAPHIC INFORMATION SYSTEM (GIS)

GIS are used in many industries around the world to collect and manage geographically referenced (geo-referenced) data. They serve a useful purpose in precision agriculture to manage and visualize the tremendous amounts of data involved. The most basic concept of a GIS is layers of information that can be compared to see trends in the data. A simple visualization is maps of a field printed on transparencies that show the field boundary, soil types, fertility, topography and yield. If these layers are overlaid on one another, one may be able to see trends. *For example*, there may be correlation between yield and topography or yield and fertility. GIS is a computerized data storage and retrieval system which can be used to manage and analyze spatial data relating crop productivity and agronomic factors. It can integrate all types of information and interface with other decision support tools. GIS can display analysed information in maps that allow (a) better understanding of interactions among yield, fertility, pests, weeds and other factors and (b) decision making based on such spatial relationships. Many types of GIS software with varying functionality and price are now available. GRAM⁺⁺ (GRAM Plus Plus) is an indigenously developed GIS software package for storage, analysis and retrieval of geographic information relevant to the task of local level planning. The package has been put to use in a variety of applications like watershed management, wasteland reclamation, land capability analysis, soil erosion assessment, energy budgeting, location/allocation of facilities and hazard zonation studies.

Many farm information system are available which are simple programmes to create a farm level database. One example of such FIS is LORIS. LORIS (Local Resource Information System) consists of several modules which enable the data import generation of raster files by different gridding methods; the storage of raster information in a database; the generation of digital agro-resource maps; the creation of operational maps etc. A comprehensive farm GIS contains base maps such as topography, soil type, N, P, K and other nutrient levels, soil moisture, pH etc. Data on crop rotations, tillage, nutrient and pesticide applications, and yields etc. can also be stored. GIS is useful to create fertility, weed and pest intensity maps, which can be used for making maps that show recommended application rates of nutrients and pesticides.

GIS A KEY COMPONENT OF PRECISION AGRICULTURE

The introduction of GIS and GPS is being utilized by most of the farmers in America and Europe for precision agriculture and they have been successful in achieving their goal. Most of the users and researchers confirm that GIS is the key in making precision farming to be a successful one in accomplishing their work to the full and correct extent (Kazuhik, 2001). Currently there are many software's entering the market to fulfill the requirements of users. But this GIS is one, which satisfies the users by making queries and getting the probable answers and allows the user to make decisions at each and every stage of the work. GIS plays a major role in decision making and monitoring with respect to agriculture. In addition to data storage and display, the GIS can be used to evaluate present and alternative management by combining and manipulating data layers to produce an analysis of management scenarios that will be benefiting the users and farmers in decision making (Deepak and Amin, 2003).

PRELIMINARY STEPS IN CREATING A GIS DATABASE

The prime steps to be considered before creating a GIS database are given below:

(i) Identifying the problem to be solved

(ii) Gathering data to solve this problem and creating a spatial database

(iii) Performing the analysis

(iv) Presenting the results of the analysis

Specifically, we need to create a vector data set by hand, and then create the digital version of this dataset in GIS – ARCVIEW/ARCINFO or any other GIS Software.

DEVELOPING A GIS DATABASE

There are two district stages involved in developing a GIS database. (i) Data acquisition and (ii) Data preparation. The data plays a major role in GIS application. Many potential applications for GIS exist in the agriculture field. A GIS database consists of a number of data layers, a few or many according to the application. Data layers often used in precision farming include soils, land use, topography, rainfall, evaporation, run-off, stream network and boundaries. The two main methods of data base development are obtaining existing digital data and converting existing paper maps to digital format through manual digitizing or by automated scanning.

DATA SOURCES

The potential source of data for GIS database includes the primary and secondary sources. These two sources of data should be considered well before achieving the outputs. The primary sources are the data that have to be collected directly from the field. Of course, data collection is accompanied by DGPS for location of the sampling sites. *For example*, crop yield, soil samples, water samples etc.

Secondary sources include the paper maps, aerial photographs and digital databases that exist already. The secondary data source can obtained from government departments or from survey departments and also from private organization.

DATA INPUT

The data input to the GIS database can be accomplished in many ways. Regardless the source of the data, still it has to be entered to the GIS database for getting output for precision farming. The method of entering data depends on factors such as the data source, format, required output, available hardware, software and type of use.

The primary data sources can be entered to the GIS database in many ways. Indirect methods include recording the values in the field manually on notebook and then entering them in the GIS software using a computer. Direct field measurements could be recorded through the data logger, resulting in a digital data record and that in turn could be transferred to the GIS format with any needed format conversions. The secondary data are from other digital sources and they must be converted to the required format for the GIS. Digital data can be obtained through electronic media, such as tapes, diskettes, CD etc. Conversion of paper maps to digital format can be achieved by manual digitizing or scanning. Techniques used to convert other analog materials include photogrammetric digitizing, coordinate geometry and key entry. If the data source were an image then scanning would be appropriate. If the data source were an image then scanning would be perfectly suitable. Digitizing is better for vector format and scanning is better for the raster format.

QUALITY OF DIGITAL DATA

The quality of the GIS results depends on the quality of the GIS database. A great quantity of data is available for developing GIS database for precision farming applications. However, more important is the quality rather than the quantity, regardless of the technique used to develop database. It is the responsibility of the user to confirm the quality of the database used in GIS analysis.

INTEGRATING DATA FROM OTHER SOURCES

The development of a quantitative and quality GIS database usually involves in the integration of data from multiple sources with varying accuracies, scales, geometric structures, spatial resolutions and other characteristics. The differences in these data layer characteristics must be considered in the integration of the data to ensure an acceptable quality database.

UPDATING & MANAGING DATABASE

Once the database is created, it is essential to know how to manage, maintain update. A data directory is very much essential for effective management of the GIS database. The directory defines the entities, their attributes, associated, domain values, convey the meaning to the database administrator and the user. The database management helps the user to manage the data and organize them for future analysis. A specific back up for all the data is very much required in order to have a record. There are many applications in precision farming where it is necessary to recover the previous versions of the information to determine changes that have occurred over time.

Globally Data Base Management System (DBMS) deals with the management of information that is:

(i) Information distribution

(ii) Information quality control

(iii) Information consistency maintenance

DBMS plays a major role in interface between data and the data users. Besides the fact that the whole set of data is centralized, DBMS offers query tools to the information. The development of communication network increased the accessibility to information (Collect et.al. 1996).

GIS CAPABILITIES

(i) The GIS software has common features and components as other software.

(ii) Data acquisition and preparation tools for the creation of the database

(iii) A geographical database management system will be derived

(iv) Data exploitation tools for information retrieval from the database

(v) Data analysis tools

Major difference in using GIS software is that depending on the purpose it offers complex capabilities for different processing stages. It is important to know the way it processes the spatial information. The object-processing mode considers spatial objects as units of description, while image-processing mode is based on arbitrary and regular units of description referred to as cells. These two approaches process the spatial information in a different manner and require a specific data structure, vector and rastor respectively.

BENEFITS OF GIS DATABASE

The evaluation and comparison of GIS software requires a clear understanding of the nature of GIS and its related concepts. There are comprehensive benefits that are achieved by using GIS, data is well organized and stored in a place such that the users have a constant access over them. Easy to retrieve the data and update in GIS database since it is user friendly and flexible maps, reports created with GIS have good graphic quality and more consistent. An important function of the GIS in agriculture is to store layers of information, such as yields, soil survey maps, remotely sensed data, crop scouting reports and soil nutrient levels. Geographically referenced data can be displayed in the GIS, adding a visual perspective for interpretation.

OPPORTUNITIES OF GIS DATABASE IN PRECISION AGRICULTURE FOR INDIA

Keeping in view the development of satellite, computer and communications technologies, the following opportunities and challenges exist in the application of GIS and precision farming technologies in India.

(i) Prioritization of macro and micro watersheds for implementation and impact assessment of watershed projects at national, state, district, taluka levels.

(ii) Forecasting of outbreak of pests and diseases based on soil water status and plant stress indicators in crops such as paddy, wheat, sugarcane, cotton, chilli and pigeon pea etc.

(iii) Development of decision support system for precise management of resources at farm level at least in commercially fruit and flower crops to begin with.

(iv) Airborne SAR data utilization for identification of *kharif* crops and development of procedures for canopy back scatter models for identification and yield prediction.

(v) Soil mapping at cadastral scale using high resolution spatial, spectral and radiometric resolutions.

(vi) Quantification of soil loss.

(vii) Detection of water logging due to rising ground water table.

(viii) Delineation of salt affected soils in black soil and sandy regions.

(ix) Soil moisture estimation and mapping using micro wave and optical thermal remote sensing techniques in surface and root zone.

(x) Land surface temperature estimation using thermal and microwave remote sensing techniques.

(xi) Hyper spectral studies on soils to establish quantitative relationship between spectral reflectance and soil properties.

(xii) Development of digital techniques for variety of applications using GIS techniques. *For example*, soil suitability and land irrigability assessment etc.

(xiii) Preparatory activities towards hyper spectral data utilization for understanding plant processes and development of spectral response models for stress detection.

(xiv) Improved yield models by integration of biophysical simulation and regional level crop models.

CHALLENGES OF DATABASE FOR PRECISION AGRICULTURE IN INDIA

The application of remote sensing, GIS, and precision farming techniques in the management of agricultural resources are increasing rapidly due to improvements in space science supported by computer and communication technologies in the application of these technologies-

(i) Identification of crops and estimation of area and production of short duration crops grown in fragmented land holdings, in particular during *kharif* season.

(ii) Forecasting of floods/droughts.

(iii) Detection of crop stress due to nutrients, pests, and diseases and quantification of their effects on crop yield.

(iv) Automation of land evaluation procedures for a variety of applications using GIS techniques.

(v) Information on sub-surface soil horizons.

(vi) Extending precision farming database to smaller farm size and/or diverse crops/cropping systems.

(vii) Developing decision support systems for management of biotic and abiotic stress at the farm level.

(viii) More accurate yield models.

(ix) Estimation of depth of water in reservoirs and quality assessment of ground water.

(x) Better than one meter contours for watershed development plan at the micro level.

(xi) Use of remote sensing and precision farming technologies in intercropping/multiple cropping situations.

(xii) Identifying ways and means of reducing the cost of remote sensing, GIS and precision farming technologies and time gap in collection, interpretation and dissemination of data to enable their usage on a large scale. A successful example in this direction is that of hand-held radiometre developed by Optomech Engineers, Hyderabad in collaboration with Space Application Centre (ISRO), Ahmedabad for standardizing the spectral signature *in situ* for inter pressing the remote sensing data.

(xiii) Convincing evidence to prove the utility and economic viability of these technologies so as to mobilize support for research and development work.

(xiv) Human resources development to hasten the process of large scale use of unexplained and cutting edge technologies that have tremendous scope and potential (Patil et al., 2001).

REMOTE SENSING TECHNIQUES

Remote Sensing refers to any non-contact method of getting information about a particular aspect of a field. This can mean plant health, soil quality, topography, weed pressure or anything that can be measured without contact, hence the remote part meaning "operating from a distance" and sensing meaning "detecting physical phenomena". This is different than physically taking a soil or plant sample for analysis. A remote sensing system consists of a sensor to collect the radiation and a platform- an aircraft, balloon, rocket, satellite or even a ground-based sensor supporting stand-on which sensor can be mounted. Currently a number of aircraft and spacecraft imaging systems are operating using remote sensing sensors. Some of the current image systems from spacecraft platform include Indian Remote Sensing Satellite, French National Earth Observation Satellite and IKONOS etc. Remote sensing holds great promise for precision agriculture because of its potential for monitoring spatial variability over time at high resolution. Using remote sensing data for mapping has many inherent limitations, which includes, requirement for instrument calibration, atmospheric correction, normalization of off-nadir effects on optical data, cloud screening for data specially during monsoon period, processing images from airborne video and digital cameras. Keeping in view the agricultural scenario in developing countries, the requirements for a marketable remote sensing technology for precision agriculture is the delivery of information with the following characteristics like low turn around time (24-48 hours), low data cost (100 Rs./acre/season), high spatial resolution (at least 2 meter multi-spectral), high spectral resolution (< 25 nano meter), high temporal resolution (at least 5-6 data per season) and delivery of analytical products in simpler format.

TYPES OF REMOTE SENSING

There are many non-contact method of gathering information from a field. They can be as far away as satellite or as close a sensor attached to equipment in the field that looks at individual plant.

SATELLITE IMAGERY

There are several providers of satellite imagery on a regular basis. One provision of satellite data is that cloud cover and darkness affect the image. Even though the satellite may have passed overhead, the image may not be usable if it is cloudy or dark. The timing of receiving satellite imagery has improved over the past few years. It is now possible to get images from a satellite pass within two or three days. This type of information can be useful to the grower if, *for example*, a decision on applying a fungicide needs to be made quickly. The resolution or size that each pixel in the image represents on the ground, varies from 30 meter down to sub meter level. While very high-resolution imagery is available to military via satellites, it is generally limited to 0.5-meter resolution for the public unless permission is granted for higher resolution. The cost generally goes up with the resolution as the data files get much larger for higher resolution and the satellite takes images of a smaller swath on ground when the resolution is higher, requiring more passes to cover the same ground area as a lower resolution image.

AERIAL IMAGERY

It refers to photos or digital imagery captured from an overhead elevation where the sensing device is not supported by ground based structure. It can be from a fixed wing aircraft-helicopter, balloon, blimp, drone, kite or parachute, cloud cover, darkness and shadows can affect the quality of aerial imagery. Aerial imagery can be used for highly detailed images down to centimeter resolution. They can be as simple as photos taken from a remote control plane with a digital camera or as complex a hyper spectoral-imaging device mounted on the belly of an aircraft. Aerial imagery has the benefit of being an "on-demand"service and timely imagery can be quickly acquired when needed. It can also be higher resolution than satellite data.

NON-CONTACT SENSORS

Remote sensing can also be done from ground level with non-contact sensors that measure reflected wavelengths from plants or soil to measure several properties. These sensors can be used for mapping of plant condition for later treatment or mounted on a sprayer or toolbar for the on-the-fly fertilizer or herbicide treatment. The models, which emit their own light source, are not affected by cloud or darkness so they can be effective in most conditions.

OTHER TECHNOLOGIES

Computer & Internet

The computer and internet are the most important component in enabling the precision farming possible as they are main source of information processing and gathering. The high-speed computer has made faster processing the data gathered during precise management of the land parcel. Internet which is a network of computer, is the most recent development among all these technologies. The internet has bridged the gap between the information provider and the user. In agriculture, like any other form of business, internet has the capability to supply timely data about changing conditions.

Spatial Decision Support Systems

These are designed to help growers to solve complex spatial problems and to make decision concerning to irrigation scheduling, fertilization, use of crop growth regulators and other chemicals. SDSS have evolved in parallel with decision support system (DSS) (*Sahoo 2010*). In addition, in order to effectively support decision-making for complex spatial problems, a SDSS will need to:

- Provide for spatial data input
- Allow storage of complex structures common in spatial data
- Include analytical techniques that are unique to spatial analysis
- Provide output in the form of maps and other spatial forms

GISs stand fall short of the goals of SDSS for a number of reasons

- Analytical modeling capabilities often are not part of GIS
- Many GIS databases have been designed solely for cartographic display of results. SDSS goals require flexibility in the way information is communicated to the user

- The set of variables or layers in the database may be insufficient for complex modelling
- Data may be at insufficient scale or resolution
- GIS designs are not flexible enough to accommodate variations in either the context or the process of spatial decision-making

SDSS provide a framework for integrating:

(i)   analytical modelling capabilities,

(ii)   database management system,

(iii) graphical display capabilities,

(iv)   tabular reporting capabilities and

(v)   the decision maker's expert knowledge

Yield Mapping

Yield mapping and soil sampling tend to be the first stage in implementing precision farming. Yield mapping is one of the principle uses of precision agriculture technologies. Yield maps can be useful tools to aid in understanding why crops performed as they did in variours areas of fields. Yield maps are produced by processing data from an adapted combine that has a vehicle positioning system integrated with a yield recording system. Massey Ferguson was the first company to produce a commercial yield mapping combine. This combine has a Differential Global Positioning System fitted to it that can be identified by the GPS receiver on the roof of the cab and the differential aerial above the engine. The out put from the combine is a data file that records every 1.2 seconds the position of the combine in longitude and latitude with the yield at that point. This data set can then be processed by various geo-statistical techniques into a yield map.

Crop Simulation Models

However, although significant technical advances have been made in measuring and displaying variation in crop yields across a field, it is not always clear how to determine the best management practice for each part of a field in order to achieve these goals. *For example*, does a farmer apply more fertilizer to the lower yielding areas to try and raise yield to the average or are there low-yielding areas at their potential yield already and would, be advised to apply more to the high yielding areas, believing they are able to make better use of it. Answer to these questions are not clear cut but can be found using optimization techniques based on knowledge of the marginal yield response which can be obtained from the curve describing yield response to the level of particular input for each homogenous regions are unlikely to be known. While mini-experiments on each homogenous region with a range of application levels of the input will provide this information, this approach is time consuming, labour-intensive and results are likely to be specific to that field only. An alternative and perhaps complementary, approach is to use crop simulation models to predict the likely yield response to different levels of particular input. Such models offer cost effective way in which agronomic knowledge accumulated from numerous previous experiments, usually with treatments of uniformly applied inputs on small and relatively homogenous plots, can be extended to larger spatially variable fields.

Crop simulation models are needed to help consultants, researchers and other farm advisors determine the pattern of field management that optimizes production or profit. However, the effective use of these tools require their evaluation in fields to be optimized, their integration with other information tools such as GIS, Geostatistics, remote sensing and optimization analysis. Crop simulation models like CERES (Maize, Wheat, Rice, Sorghum, Barley & Millet) CROPGRO (Soybean, Peanut, Dry bean and Tomato), SUBSTOR (Potato), CROPSIM (Cassava), and CANEGRO (Sugarcane) models have been developed by researchers from several countries. These models respond to weather, soil water holding and root growth characteristics, cultivars, water management, nitrogen management and row spacing/plant population. Also decision support system like, DSSAT incorporates crop/soil/weather models, data input and management software and analysis programmes for optimizing production or profit for homogenous fields. DSSAT also includes links to GIS and remote sensing information which allows mapping of spatially variable inputs across a field and mapping of predicted outputs from the models, such as yield, nitrogen leaching, water use etc. The site specific yield potentials can be estimated determining spatial pattern crop and land information and using it in above simulation models.

Variable Rate Technology

Variable Rate Technology, VRT, for short, combines, GIS, GPS and electronic controllers in the cab to change the rate of any product being applied in the field. In general terms, VRT is accomplished by developing a prescription map, transferring it to controller in the cab of the vehicle, driving the field with the controller changing the application rate based on the prescription map and recording how much was applied where. VRT can also be done on- the- fly with sensors that measure what is needed by the crop and adjust the rate accordingly in real time. The VRT is the most advanced component of the precision farming technologies, provides "on-the fly" delivery of field inputs. A GPS receiver is mounted on a truck so that a field location can easily be recognized. An in-vehicle computer, which contains the input recommendation maps, controls the distribution valves to provide a suitable input mix by comparing to the positional information received from the GPS receiver. Current commercial VRT systems are either map-based or sensor-based. The map-based VRT systems require a GPS and DGPS geo-referenced location and a command unit that stores a plan of the desired application rates for each field location. The sensor-based VRT system do not require a geo-referenced location but include a dynamic control unit, which specifies application through real time analysis of soil or crop sensor measurements for each field location. There are two methods of VRT-Map-based and Sensor-based.

Table 21.1: Major differences between Map based and Sensor based Precision farming systems (Patil and Shanwad, 2001)

S.No.	Parameter	Map-based	Sensor-based
1	Methodology	Grid sampling-lab analyses-Site Specific maps and use of variable rate applicator	Real time Sensors-feedback control measures and use of variable rate applicator
2	GPS/DGPS	Very much required	Not necessary
3	Laboratory Analyses (Plant & Soil)	Required	Not required
4	Mapping	Required	May not be required
5	Time consumption	More	Less
6	Limitations	Cost of soil testing and analyses limit the usage	Lack of sufficient sensors for getting crop and soil information
7	Operation	Difficult	Easy
8	Skills	Required	Required
9	Sampling Unit	2 to 3 acres	Individual spot
10	Relevance	Popular in developing countries	Popular in developed countries

INPUT DATA REQUIRED FOR PRECISION FARMING

Yield Data

It is recommended to collect few years of yield data to get a sense of the trends within the field. As water is typically the most limiting factor to crop production, the yield may vary widely in a dry year versus a wet year. Using only one year of yield data may not provide the necessary information to make informed decisions. It is recommended to view the trends year over year.

Fertility

An accurate assessment of the nutrients available in the field is important for varying fertilizer rates. Soil fertility can be determined by soil sampling, where to sample, and how many samples to take, requires some planning. Two different approaches to soil sampling are grid sampling and benchmark or landscape sampling. Grid sampling refers to taking soil samples at a regular spacing interval, such as every 2.5 acres or 5 acres. Benchmark or landscape sampling is done by determining beforehand where to sample based on imagery or zone map. Samples are taken from representative areas of the field, *for example* of the same soil type or slope. Several cores may be mixed to make a composite sample for each zone. Recording the GPS location of the soil sample allows a map to be produced of soil fertility. A branch of statistics called geostatistics can be applied to interpolate between the points. It is possible to return to the same sample points from one year to another if they are geo-referenced.

Topography

Field topography can have a significant impact on yields. Hilltops generally yield differently than low-lying depressions. Using topography as an input for determining zone maps can be beneficial. For drainage, accurate elevations are needed to ensure the water flows in the correct direction.

Imagery

Imagery required from satellites or aerial means can be used to determine where inputs should be applied. Imagery refers to film based aerial photos, satellite digital images or any other overhead view of a field generally taken looking directly down at the field.

Electrical Conductivity

An input with increasing popularity is electrical conductivity mapping of fields. This method involves injecting an electrical current into the ground and measuring the conductivity of the soil. This generally relates to soil texture as small particles such as clay tend to conduct more current than larger sand and silt particles. Another technology uses electromagnetic waves to measure soil conductivity without contacting the soil. This method can identify salinity and soil moisture content as well. Soil texture does not change much over time so EC mapping is a one-time expense for field. It can provide valuable information on soil texture and provides another layer of information in GIS.

Target Yield

To know if you have achieved your goals in a precision farming system, it is important to determine beforehand what you expect for yields. There are different approaches to managing yield. The first is to maximize production of each square foot of the field by maximizing inputs. The second is to make the yield more consistent across the entire field. The overall goal of using variable rate technology is to apply the correct product at the correct place at the correct time. Applying products where they are needed and not applying them where they are not, can maximize profitability.

Weed Mapping

A farmer can map weeds while combining, seeding, spraying or field scouting by using a keypad or buttons hooked up to a GPS receiver and data logger. These occurrences can then be mapped out on a computer with an electronic prescription map.

Salinity Mapping

GPS can be coupled to a salinity meter sled which is towed behind a pick up across fields affected by salinity. Salinity mapping is valuable in interpreting yield maps and weed maps as well as tracking the change in salinity over time.

Other Data

The precision farming requires other inputs such as crop characteristics like stage of the crop, crop health, nutrient requirements etc, Microclimate data (seasonal and daily) about the canopy temperature, wind direction and speed, humidity etc, Irrigation facilities, water availability and other planning inputs of interest.

CONCLUSION

Precision farming technology could revolutionise agriculture but most are in the early stages of development and will need extensive research before they are proven effective. It is clear that many farmers are at sufficient level of management that they can benefit from precision farming management. Questions remain about cost effective and the most effective ways to use the technological tools we now have, but the concept of "doing the right thing in the right place at the right time" has a strong instinct. Ultimately, the success of precision farming depends largely on how well and how quickly the knowledge needed to guide the new technologies can be found. There is a general concern for quality and access to data. For producers and support system to do an adequate job of analysis and research, data must be freely exchanged. Government agencies that collect data have made data available, though it may not be readily accessible for some people. This may be due to lack of computer skills, GIS skills, or telecommunications/internet access. Developing a database for precision farming is important and required with respect to the production and farmers. Database created using GIS is useful for researchers, farmers and for decision makers in the field. It is user friendly and can provide timely information to the users to take a firm decision. Thematic maps will be produced with the final reports that are useful and comprehensive to the new users. Future updates can be made with GIS to implement better technology.

REFERENCES

Collect C, Consuegera D and Joerin F. (1996). *GIS Needs and GIS Software*. Kluwer Academic Publishers. pp. 115-142.

Deepak TJ, Amin MS. (2003). Dominance of Geographic Information System in Developing a Database for Precision Farming. *Journal of Human Ecology*, **14**(5): 329-335.

Kazuhiko O. (2001). Precision Agriculture in Japan-Application of GPS in Agriculture, *Paper Presented to the DSMM/UN/USA/JUPEM WORKSHOP on the use of GNSS*, 20-24th August, 2001.

Patil VC, Shanwad UK. (2001). Relevance of Precision farming to Indian Agriculture. In: Proceedings of the First National Conference on Agro-Informatics (NCAI), INSAIT. Dharwad.

Patil VC, Maru A, Shashidhara GB, Shanwad UK. (2001). In: Proceedings of the First National Conference on Agro-Informatics (NCAI), INSAIT. Dharwad.

Sahoo RN. (2010). Geoinformatics in Precision Agriculture. Lecture delivered in FAI Workshop on 'Fertilizers Reform through ICT'. 14-17th June, 2010 at Kufri Holiday Resort, Kufri, Shimla.

■

22

CHAPTER

Precision Farming in Vegetables

☞ Raj Narayan[1], ☞ Sumati Narayan[2], ☞ Santosh Choudhary[3] and Anjum Ara[4]

INTRODUCTION

Vegetable crop production being relatively high risk, high input cost business requiring intensive management differs from other crop production enterprises. Successful vegetable growers manage capital and marketing intelligently and proficiently in designing and implementing systems of culture which include the selection of crop and variety, crop rotation, land tillage and seed bed preparation, seeding, transplanting, soil fertilization, irrigation, integrated pest management including weed control practices, use of wind breaks, pollination, harvesting and handling, packaging and sales of crop produce. Most of these crops are perishable in nature, have narrow market windows and must, therefore, be free from blemishes, punctures or scars for preventing further loses. Consequently, crop practice (crop rotation, soil, variety, fertilization, pest control, irrigation, etc) and post-harvest (handling, packaging, marketing etc.) operations must be accomplished wisely and timely in more precise manner for getting bumper produce. Further, growers are required to pay close and careful attention to these operations in order to compensate for the factors which are beyond their control, like weather, epidemics of pest infestation, market fluctuations and legislation, for increasing margin of profit and avoiding loss in the venture.

'Precision Farming' or 'Precision Agriculture' aims at increasing productivity, decreasing production costs and minimising the environmental impact on farming. In other words precision agriculture can be defined as a comprehensive system designed to choose risky agricultural production through the application of crop information, advanced technology and management practices. Precision agriculture is also known as precision farming, precision horticulture, site-specific farming, site-specific management, site-specific crop management, variable rate application, etc. The U.S. House of representatives, in 1977, have defined precusion agriculture as an integrated information and production based farming system designed to

[1] Programme Coordinator, senior scientist, K.V.K., ICAR Research Complex for Goa, Ela, Old Goa.

[2] Assistant Professor/Junior Agronomist (Vegetables), Shere-e-Kashmir University of Agricultural Science & Technology, Shalimar, Srinagar (J&K)

[3] Assistant Professor (Horticulture), College of Agriculture, Mandore, Jodhpur (Raj.)

[4] Research Associate, Division of Olericulture, SKUAST-K, Shalimar, Srinagar (J&K)

increase long term, site specific and whole farm production efficiency, productivity and profitability while minimizing unintended impacts on wild life and environment, and SSCM as a form of precision agriculture whereby decisions on resource application and agronomic practices are improved to better match soil and crop requirements as they vary in the field.

A more holistic approach uses information technology to bring data from multiple sources which have a bearing on decisions associated with agricultural production, logistics, marketing, finance and personnel (Arora, 2005).

TECHNOLOGICAL INTERVENTIONS IN PRECISION HORTICULTURE

Technological interventions in precision horticulture include genetic conservation, genetic engineering, integrated nutrient management, protected cultivation, micro irrigation and fertigation, post-harvest technology etc. Micro-irrigation and fertigation system save water upto 70 per cent and fertilizer upto 30 per cent besides preventing weed growth and increasing yield upto 100 per cent (Chadha, 2005). However, the technology related to precision farming needs refinement to realize benefits (Kalia, 2005). To be a truly comprehensive system, it must begin during the planning stage of crop, virtual crop care and continue through the post harvest management phase of production.

PRECISION FARMING *VIS-A- VIS* TRADITIONAL FARMING

Precision farming distinguishes itself from traditional farming basically by its level of management. Instead of managing whole field as a single unit in traditional agriculture, the management in precision agriculture is customized for small areas within fields. This increased level of management emphasizes the need for sound agronomic practices and unlike traditional crop management which assumes uniform field conditions and recommends average or thumb rule input application rates, the precision agriculture is information intensive and management here may be spot specific.

OBJECTIVES OF PRECISION FARMING

a) Increased profitability and sustainability

Precision farming aims at earning maximum profit and sustaining this profitability by using balanced precise amounts of inputs (variety, seed, fertilizer, herbicide, pesticide etc.) as per crop response need in each zone, site in a field as determined by weather, soil characteristics and historic crop performance.

b) Optimizing production efficiency

Precision farming aims to optimize returns across a field or optimize production quantity at each site or within each zone using differential management. The initial emphasis, in absence of any clear environmental benefits, should be on achieving production and cost benefits by differentially applying inputs as per requirement so that the marginal return equals marginal cost at each site or zone in the paddock.

c) Optimizing product quality

Precision farming also aims at optimization of product quality by ways of using sensors which detect the quality attributes of the crop and thus inputs are to be applied accordingly.

d) Most efficient use of chemicals and seed

Precision Agriculture involves efficient use of inputs like chemicals, seeds etc. in accordance with the yield potential of the soil and crop plants.

e) Effective and efficient pest management

Conventional farming methods apply herbicides, pesticides etc. to the entire field, whereas site specific variable rate application is made in precision farming with the goal of cutting down crop production inputs for cost and environmental savings to increase their efficacy and efficiency.

f) Energy, water and soil conservation

Precision agriculture begins from crop planning by including such tillage practices which conserve the soil or disturb the soil to its minimum, besides applying water efficiently by adopting techniques like drip irrigation, etc. in all these cases, very less energy is used leading to the conservation of energy also.

g) Surface and ground water preservation

Safeguarding the environment by way of efficient use of inputs like chemicals etc. prevents their loss by leaching through ground water or surface run-off and thus helps to preserve water and its quality.

h) Minimizing environmental impact and toxic inputs

Better management decisions made to preview inputs to meet production needs must eventually result in retarded loss of any input to the environment. This reduces the risks of environmental damage, although some actual or potential harm is always associated with production system.

i) Minimizing risk

Generally minimizing income risk is considered more important than environmental risk in traditional agriculture and thus a farmer may put an extra spray on, add extra fertilizer, buy more machinery or hire extra labour to ensure timely harvest and sale thereby guaranteeing a return but precision agriculture attempts to offer a solution that may allow both angles to be considered in risk management.

POTENTIAL BENEFITS OF PRECISION FARMING

Precision agriculture's effectiveness is highly dependent on how much variability exists within the fields and the ability of a producer to indentify and put into use the best management practices for each fields sub area. If the spatial data provided by precision agriculture is properly used, the following potential benefits can be realized:

- **Increased profits through increased efficiency**

  With PA, one can use farming inputs more efficiently, by applying more interventions where more are needed and less where less is needed.

- **Reduced agronomic inputs**

  Many producers who adopt precision agriculture technologies find their overall use of production inputs (fertilizer, lime and chemicals) decreased because they can better adjust to in accordance with the fields fertility level, the amount of inputs to be applied in accordance with the fertility level of a field.

- **Better record keeping**

  PA technology generates large amounts of data that are spatial records of inputs and outputs for fields. These data can help to create more accurate management or manageable plans.

- **Improved production decisions**

  PA can be used to make land use decisions. Profit maps which show the spatial distribution of a field's profitability, can help to make decisions about which cropping systems work best. We may also be able to identify areas that should be enrolled in government programmes such as the federal conservation reserve programme or removed from production, if low yields cannot be profitably corrected.

- **On-farm research**

  The ability to quantify the spatial performance of crops allows conducting more comparison trials within fields. *For example*, it is relatively simple to compare the performance of different varieties on different soil types when practicing precision agriculture.

- **Reduced environmental impact**

  If we apply and use production inputs more efficiently, less material will leave the field through surface water and ground water. Reduced environmental impact can only improve the public's perception and acceptance of producer's agricultural practices.

- **Property advantages**

  Many landlords are giving preference to farmers who can create yield maps and other files of spatial data for the fields they farm. This spatial history may also increase the value of cropland.

- **More ground farmed**

  The records generated can allow to effectively managing more crop land than we had in the past.

CONSTRAINTS INVOLVED IN PRECISION AGRICULTURE

- More than 60 per cent of operational land holdings in India have size less than one hectare, besides the fact that these land holdings are fragmented.
- PA implementation involves many cost effective technologies.
- Heterogeneity of cropping systems.
- Lack of local technical expertise and knowledge.
- Technological gaps.

MISCONCEPTIONS CARRIED BY PRECISION AGRICULTURE

Like many other new concepts, precision agriculture also carries with it some misconceptions:

- PA is often confused with yield mapping. Yield mapping is a tool that is one of the first steps towards implementing a SSCM strategy.
- PA is sometimes misinterpreted as sustainable agriculture. PA is a tool to help make agriculture more sustainable. It is not the total answer because PA primarily aims at maximum production efficiency, (productivity and profitability) with minimum environmental impact. In recent years the potential for this technology (SSCM as a form of PA) as a tool for environmental auditing of production systems has become more obvious. However, environmental auditing is not environmental management, but large amount of fine-scale data being collected in a SSCM system can be used for on-farm environmental risk assessment and incorporated into a whole farm plan to help viability in the long term.
- Finally, machinery guidance and auto steer systems needed for the successful adoption of new technology on farms. However, these are the tools that help with SSCM by themselves they are not PA.

ELEMENTS OF PRECISION FARMING

- A) Information
- B) Technology
- C) Decision Support (management)

A) Information includes data on

- Crop characteristics like stage of crop, crop health, nutrient and water requirement, etc.
- Detailed soil layer with physical and chemical properties, depth, texture, nutrient status, salinity and toxicity, soil temperature, productivity potential, etc.
- Micro-climate data (season and daily) about the canopy temperature, wind directions and speed, humidity, etc.

- Surface and sub-surface drainage conditions.
- Irrigation facilities, water availability and other planning inputs of interest.

B) Technology

Precision Agriculture is an integrated agricultural management system incorporating several technologies. Theses technological tools often include

a) Global positioning system (GPS)

GPS is widely available in agricultural community and its potential is growing. Global Positioning System is one of the cutting edge technologies of this information- age for vegetable farming. Which asks the basic question "where is". It is computer based and the key components in this technology are satellites. GPS depends on 24 satellites (courtesy of the US Department of Defense) that are strategically positioned in orbits such that, at any given time and place on earth, one can have a line of sight to at least four of these satellites. Using the techniques of "Trilateration" a user is able to access at least three satellite signals, each producing a surface that will overlap each other to locate the site in question, providing both longitude and latitude. A fourth satellite signal allows the altitude of the site to be calculated. Satellite signals can be clearly received even under inclement weather conditions. However, tall building, dense forests, mountains and hills and other such structures can obstruct clear signal communications. The GPS technology is used in everyday life as navigational aids, in automobiles, in tacking sites of breakdown, routing emergency response crew, and other applications. Individuals can purchase hand held units of varying resolution.

b) Geographical information system (GIS)

This is a computer based system for storing very large amounts of data (collected based on spatial location), retrieving, manipulating and displaying them for easy interpretation. The term "geographic" should be interpreted to mean "space" or "spatial". GIS has existed for over four decades, the concept of which dates back to the 1960's when computers were used for spatial analysis and quantative mapping. During the past two decades, GIS has been widely used in agricultural and natural resource mangagement. For crop production, some of the data of importance are soils, land use, vegetation, fertility hydrology and rainfall averages. Spatial analysis is concerned with analyzing data involved with changes with space on location within areas. GIS has the capability of linking multiple sets of data (in layers) to study relationships among various attributes and creating new relationships. GIS is an aid for decision making and can be used to answer the location question "what is there? " as well as to study trends by answering the question "what has changed since a certain point in time? " Another application of GIS is in the area of prediction of change or modeling. The success of GIS analysis depends on the availability of accurate and reliable data. These databases are created by various private or public entities and are available for a fee or freely accessible via the internet or other means. One of the leading suppliers of GIS products is the environmental systems research institute (ESRI), makers of Arcview® and Arcinfo® Software and various accessories. In view of the enormity of data routinely involved in GIS applications, one needs to have access to a computer with appropriate capabilities and speed, in order not to make use of the technology a drag.

c) **Yield monitors**

These are crop yield measuring devices installed on harvesting equipment. The yield data from the monitor is recorded and stored at regular intervals along with positional data received from the GPS unit. GIS software takes the yield data and produces yield maps.

d) **Variable rate technology**

It consists of farm/field equipment with the ability to precisely control the rate of application of crop inputs and tillage operations. Typical VRT system components include a computer controller, GPS receiver and GIS map database. The computer controller adjusts the equipment application rate of the crop input applied. The computer controller is integrated with the GIS database which contains the flow rate instructions for the application equipment. A GPS receiver is linked to the computer. The computer controller uses the location coordinates from the GPS unit to find the equipment location on the map provided by the GIS unit. The computer controller reads the instructions from the GIS system and varies the rate of the crop input being applied as the equipment crosses the field. The computer controller will record the actual input being applied as the equipment runs across the field. The computer controller will record the actual rates applied at each location in the field and store the information in the GIS system, thus maintaining precise field maps of materials applied. Although VRT can control inputs applied to crops, it cannot control factors such as soil type, weather, climate and topography that are fixed.

e) **Remote sensing**

It is the art and science of obtaining information from a distance *i.e.,* about objects or phenomena without being in physical contact with them. The science of remote sensing provides the instruments and theory to understand how objects and phenomena can be detected. The art of remote sensing is in the development and uses analysis techniques to generate useful information (Aronoff, 1995).

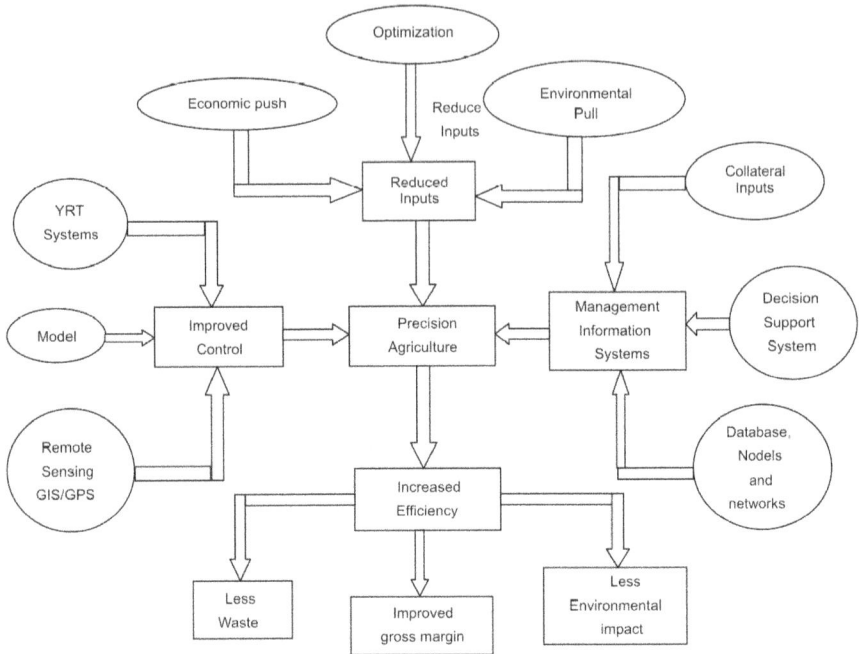

APPLICATION OF REMOTE SENSING IN PRECISION AGRICULTURE

1) Soil and drainage maps

a. Soil maps

The grid sampling technique takes separate soil samples from uniform sized grids laid out over the field. A problem with this type of sampling is the variability that can exist in soil types within each grid. This variability makes it much tougher to determine soil characteristics within the grid for crop input management purposes. To minimize this problem smaller grids are needed which then requires many more soil samples to be taken for a longer number of grids. An alternative to grid sampling is targeted or zone sampling. The soil samples are located in homogenous management zones instead of uniformly spaced grids. The zones are laid out using a process similar to computer based unsupervised image classification. Images obtained from multispectral remote sensors are taken of the vegetated areas of the field. The pixel digital numbers for each band are separated into statistically separable clusters that are classified into homogenous zones. This cuts down on the soil, terrain, plant growth, and other variability within each area to be managed; thus fewer soil samples are needed for each area (Anderson and yang 1996).

b. Drainage maps

Colour infrared (CIR) aerial photographs have been shown to be an effective tool in locating unknown subsurface tile lines. The image data is digitized for pre-processing and then geo-referenced using ground control points. The CIR photographs show different tones of gray depending on soil type and moisture. By filtering out spectral reflectance differences due to soil type, soil moisture content in dry soils that have a higher reflectance can be identified from lower reflectance wet soils. The resulting image shows where the tile lines are located and whether they are working properly (Verma and Vendte 1997). Normal colour aerial photographs can also be used to locate tile lines.

2) Monitor crop health

Remote sensing data and images provide farmers with the ability to monitor the health and condition of crops. Multispectral remote sensing can detect reflected light that is not visible to the naked eyes. The chlorophyll in the plant leaf reflects green light while absorbing most of the blue and red light waves emitted from the sun. Stressed plants reflect various wave lengths of light that are different from healthy plants. Healthy plants reflect more infrared energy from the spongy mesophyll plant leaf tissue than stressed plants. By being able to detect areas of plant stress before it becomes visible, farmers will have additional time to analyze the problem area and apply corrective treatment.

3) Water stress

The use of remote sensors to directly measure soil moisture has had very limited success. Synthetic Aperture Radar (SAR) sensors are sensitive to soil moisture and they have been used to directly measure soil moisture. SAR data requires extensive use of processing to remove surface induced noise such as soil surface roughness, vegetation and topography.

A crop evapo-transpiration rate decreases is an indicator of crop water stress or other crop problems such as plant disease or insect infestation. Remote sensing images have been combined with a crop water stress indeed Model to measure filed variations.

4) Weed management

Aerial remote sensing has not yet proved to be very useful in monitoring and locating dispersed weed populations. Some difficulties encountered are that weeds often will be dispersed throughout a crop that is spectrally similar, and very large-scale high resolution images will be needed for detection and identification.

The use of machine vision technology systems to detect and identify weeds places remote sensors directly on the sprayer equipment. Being close to the crop allows for very high spatial resolutions. Machine vision systems have the ability to be used in the field with the real time capabilities that are necessary to control sprayer equipment.

5) Insect detection

Aerial or satellite remote sensing has not been successfully used to identify and locate insects directly. Indirect detection of insects through the detection of plant stress has generally not been used in annual crops. The economic injury level for treatment is usually exceeded by the time plant stress is detected by remote sensing. Entomologists prefer to do direct in field scouting in order to detect insects in time for control treatments to be effective and economical.

6) Nutrient stress

Plant nitrogen stress areas can be located in the field using high resolution colour infrared aerial images. The reflectance of near infrared, visible red and visible green, wavelengths have a high correlation to the amount of applied nitrogen in the field. Canopy reflectance of red provides a good estimate of actual crop yields (Geopalapillai *et al.*, 1998).

7) Yield forecasting

Plant tissue absorbs much of the red light and is very reflective of energy in near infrared (NIR) wavebands. The ratio of these two bands is referred to as the vegetation index. The difference of red and NIR measurements divided by their sum is normalized difference VI (NDVI). For crops such as grain sorghum, production yields, leaf area index, crop height and biomass have been correlated with NDVI data obtained from multispectral images (Anderson *et al.*, 1996). In order to get reasonably accurate yield predictions this data must be combined with input from weather models during the growing seasons. (Moran *et al.*, 1997).

c. Decision support (management)

Just having information about variability within the field does not solve any problems unless there is some kind of decision support system in order to make VRT recommendations. Have suggested the following steps for a DSS: Identify environmental and biological states and processes in the field that can be monitored and manipulated for the betterment of crop production.

- Choose sensors and supporting equipment to record data on these states and processes.
- Collect, store and communicate the field recorded data.
- Process and manipulate the data into useful information and knowledge.
- Present the information and knowledge in a form that can be interpreted to make decisions.
- Choose an action associated with a decision to change the identified state or process in a way that makes it more favourable to profitable crop production.

COMPONENTS OF PRECISION FARMING

Precision agriculture basically, is characterized by reduced cost of cultivation (through optimization of inputs), improved control and increased resource use efficiency, through appropriate applications of management information system (MIS). While the reduced cost of cultivation is achieved through optimization of agricultural inputs taking into account economic push and environment pull related factors, the control mechanisms are introduced by the help of VRT systems, model out puts and conjunctives use of remote sensing, GIS and GPS. The MIS comprises decision support systems (DSS), collateral inputs and associated GIS databases on crops, soil and weather. Dynamic remote sensing inputs on in season's crop conditions, crop simulation model outputs on the potential production under the different constraining scenario and the networks of labs and farms form the essential ingredients of MIS. Increased efficiency does not employ only efficient resource use, but also reflects in terms of less waste generation, improved gross margins and reduced environmental impact. Precision agriculture thus calls for the use of appropriate tools and techniques, within a set of the framework as mentioned, to address the micro-level variations between crop requirements and application of agricultural inputs.

METHODS OF IMPLEMENTING A PRECISION FARMING STRATEGY

1. Map based technologies

This is most widely used method and may involve a GIS based method of presampling and mapping of the field. The computer generated maps are then converted into a form that can be used by the variable rate applicator. The applicators controller then calculates the desired amount of an agronomic input to apply at each moment in time as the equipment traverses the field. The farmer knows how much of the agronomic input is needed before starting the application. A Differential Global Positioning System (DGPS) must be used to constantly evaluate the location in the field with a coordinate on the map and the desired application rate for the coordinate.

2. Sensor-based technologies

This method offers real-time sensing and variable rate control. The strategies provide on-the-go sensing of field characteristics, thereby eliminating the need for a positioning system. The sensors must be mounted strategically (*e.g.*, at the front of the tractor) to allow the variable rate applicators controller adequate time to adjust the rate of the agronomic input accordingly, before it passes the sensed location.

Steps involved in precision agriculture adoption

1. Purchase a mapping programme

The first and most essential acquisition will be an entry level mapping programme, which will allow to import, overlay and manage spatial data. This software will enable to develop a geographically referenced database for the operation.

2. Collect spatial data

Look for existing geographically referenced data on the Internet. These and other layers of information will form the basis of the farms geographic database.

3. Map field boundaries

Service providers will map your fields for a nominal fee, or you can obtain a digital global positioning system (DGPS) receiver and a laptop or hand held PC to do it yourself.

4. Keep records

To the extent possible, these records should be archived electronically in the mapping system. Farm records should be organized on a field by field basis. These records should include: Historical yields (whole field), Field boundary locations, Soil test values and Management history, including past fertility, tillage, pest management practices and financial records.

5. Obtain remote images

Aerial or remotely sensed images of the fields should be obtained one or more times during each growing season. These images can be used to identify management problems that cannot be seen on the ground *e.g.*, we can locate areas where inputs such as nitrogen have been misapplied. It may also be possible to detect moisture or pest stresses. While it may be too late to correct some of these problems during the current growing season, the manager and employees can become mindful of the potential economic impact of these problems so they can be avoided in the future. Remotely sensed images can be obtained taking aerial pictures photographs with a 35 millimeter or digital camera.

6. Arrange a yield monitor

A yield monitor on the combine will determining total yields for fields or areas within fields without a weigh wagon or scales. This yield information will make it possible to do side by side comparisons of hybrids or management practices. In addition, the dynamic yield indicator in the combines cab will enable the combine operator to observe quantitative difference in yield throughout the field.

7. Take help of a DGPS receiver

We can greatly expand what a yield monitor can do if we add a differential global positioning system receiver. The yield monitor will record DGPS position along with yield, enabling to create yield maps and evaluate how the yields vary according to location.

8. Generate yield maps

With the yield maps monitor in place and a DGPS receiver connected, we can readily create yield maps. Most yield monitors have companion software that will create yield maps from field data. The mapping programme can also create yield maps.

9. Use yield maps for scouting

One of the fundamental uses of yield maps is to locate problems we may not have been aware of. Once we identify them, we can decide if they can be profitably corrected.

10. Generate profit maps

If we have been keeping good records, including accurate yield maps, we will then be ready to get the fields profitability. Use yield information along with the cost of all inputs and field treatments, such as tillage, to generate a profitability map of the field. Locate areas of high and low profitability.

11. Use yield and profit maps for land-use decisions

A land use decision is the first decision most farmers make about a tract of land. We can use PA technologies to identify certain areas of the field that are low yielding or unprofitable year after year and remove these areas from production and/or enroll them in government programmes, we can also combine profit maps with other field information, such soil types, to identify areas suitable for different crops or rotations.

12. Take site-specific soil fertility samples

Both grid and directed sampling are used to describe soil properties for management of variable rates of fertility. In either system, collect at least five or six subsamples from each cell or zone and mix them into one sample container. When soil fertility data is returned from the soil testing lab, we can enter the data into the mapping programme to create fertility maps.

13. Manage Subfields

Once fertility maps have been created, we can use them to make spot applications in parts of fields that are especially low or high and deficient in certain plant nutrients. This relatively crude application can be accomplished without special equipped machines. Simply look at the fertility map and try to make application in the deficient areas. By suing a DGPS receiver and computer, the accuracy of application can be increased by following the cursor on the field map.

14. Adopt variable rate technology (VRT)

Variable Rate management is the continuous adjustment of inputs to match local field conditions. Variable rate controllers are required on application equipments for automatic control of application or seeding rates in the field. They are used along with a task computer (a laptop PC) and DGPS receiver. In addition, the software must be able to generate application files to control the process.

PITFALL TO AVOID

- Don't expect large returns with minimal energy. PA technology will not make a mediocre farm manager a better manger. It simply provides tools to quantify spatial variation. Thus users must still follow fundamentally sound management practices.

- Expect a learning curve. Be prepared to spend extra time learning to use the technology's software and hardware.

- Be aware of software and hardware compatibility issues. Before purchasing any hardware or software, make sure that all components are compatible and that they are compatible with the systems your service providers may be using.

- Make user that the farm computer has the minimum requirements of random access memory and hard disk space.

- Be prepared to make management changes. Using Precision Agriculture technology won't increase your profits if all you do is monitor your existing management strategies. You have to make changes to see results.

• Don't expect results overnight. It may take several seasons to see and confirm positive results from using the technology. Have patience.

PRECISION AGRICULTURE: IS REALLY NEEDED IN INDIA'S CONTEXT?

• On the one hand there are depletion of ecological foundations of the agro-ecosystems, as reflected in terms of increasing land degradation, depletion of water resources and rising trends of floods, drought and crop pests and diseases. On the other hand, there is imperative socio-economic need to have enhanced productivity per units of land, water, time and manpower.

• At present, 3 ha of rainfed areas produce cereal grain equivalent to that produced in 1 ha of irrigated. Out of 142 m ha net sown areas, 92 m ha are under rainfed agriculture in the country.

• From equity point of view, even the record agricultural production to more than 200 Mt is unable to address food security issue. A close to 60 Mt food grains in the storehouses of Food Corporation of India (FCI) is beyond the affordability and access to the poor and marginalized in many pockets of the country.

• Globally, there are challenges arising from the globalization especially the impact of WTO regime on small and marginalized farmers.

• Some other unforeseen challenges could be anticipated. Global warming scenario and its possible impact on diverse agro-ecosystems in terms of alterations in traditional crop belts, micro-level perturbations in hydrologic-cycle and more uncertain crop weather interaction, etc.

At this stage, agriculture needs new paradigms to deal with the present situation. The strategy lies in integration of the dynamic information and scientific knowledge into the management of agro-ecosystems, and thereby optimizing the radiation, water and nutrient usages. Agriculture needs transition from high inputs material to the optimum input level, through the appropriate use of information, knowledge and strategies for efficient resources usages. In such cases, productivity of agriculture may not be the function of the quantum of agricultural input use alone, but will include information, knowledge and efficiency while managing the agricultural practices. However , a comprising close to precision farming status can be worked out to suit present circumstances confronting Indian agriculture in a bid to shift later, at the earliest possible, to most scientific precision agriculture.

TECHNOLOGICAL INTERVENTIONS IN PRECISION HORTICULTURE

Plant genetic resources are backbone of the crop improvement programmes of any country and their conservation is fundamental to their long term availability.

While systematic efforts to conserve such materials have been on, field conservation or conservation in a conventional gene bank is not adequate. There is a wide diversity of form in different horticultural crop and these require conservation approaches suited to them. One of the modern techniques for conservation is by the adoption of in vitro technique of tissue conservation. This has been successfully adopted in potato and other tuber crops, spices, many ornamental species, medicinal and aromatic plants. However, there are problems associated in management of large *in vitro* collections since these are labour intensive (in regular sub-culturing), carry risk of infection and breakdown of equipment. As such, need for safe long term conservation techniques have still been felt.

A comparatively more recent and hi-tech technique for germplasm conservation is cryopreservation which has come to stay as a method for storage of live material and ensures safety of rare or endangered plant seed and tissue, *e.g.*, meristem, embryo or recalcitrant seeds. Characteristics of genetic resources using molecular techniques assume paramount importance. While gene sequencing in horticultural crops may take many more years, right now various DNA fingerprinting techniques such as RFLP, AFLP etc. are being used to characterize germplasm.

1) Genetic engineering

As India marches ahead in the new millennium, the challenge of bringing in crop improvement at a much faster pace than ever before demands use of frontier technologies. In the conventional method of crop improvement, both desirable and detrimental genes get transferred from the parent population, the modern biotechnology enables us to insert a single two or three desired genes into the crop to give it new and advantageous characteristics. A number of varieties have already been released by various organizations and are in the commercial cultivation *e.g.* Flavr Savr and Endless Summer (tomato), New Leaf (potato) etc. Bt Brinjal like crops may also find place in our vegetation scenario, if cleared from socio-environmental objections hesitations.

2) Ensuring supply of planting material

a) Micropropagation

Micropropagation is perhaps the most popular and widely commercialized global application of plant biotechnology. It is a proven way of producing millions of identical plants. An added advantage is production of pathogen free planting material. Propagation of plants through tissue culture, inducing sophisticated techniques of meristem culture and molecular indexing for diseases, are of immense use in making available healthy propagules. Tissue culture is going to be of immense value in species where conventional propagation procedures are either slow or problematic or our requirement of planting material is very high.

b) Production of hybrid seed and seeding in vegetables

Cultivation of hybrid varieties of vegetables has already caught the large imagination of the Indian farmers. Considerably large area is being put under such varieties and these have contributed significantly to augment average productivity. However, several biotic and abiotic stresses limit hybrid seed production throughout the year in the open. Growing them under protected cultivation counters the ill effects of multiple stresses especially on solanaceous and cucurbitaceous vegetables. Requirement of optimum temperature, RH and pH range have already been standardized for this purpose.

The containerized hybrid seed production of high value vegetables like tomato, sweet pepper, muskmelon, watermelon, squash, cucumber and lettuce under greenhouse has become a service oriented industry in Europe and USA. In recent years, there has been a shift from growing vegetables transplanted on raised nursery beds in the open towards raising value added transplanting in especially designed containers such as pre-spaced peat pellets, strip peat pots, poly ethane foam flats, perforated plastic trays, plug trays, pruning planting trays and chemical (cupric hydroxide) transplant flats. A major advantage of this system is the significant reduction in transplanting shock resulting in improved crop establishment and yield especially in crops like cucurbits which do not withstand root disturbance.

Another innovation is Robot grafted seedling production. This is being practiced in Japan where production of grafted seedling in watermelon, cucumber and muskmelon have become a necessity due to endemic nature of several soil borne diseases. Grafted seedlings of the above mentioned crops account for 57 per cent production in the open and 70 per cent production in the greenhouses. Bottle gourd (*Lageneria ciceraria*) is used as a rootstock for grafting watermelon, while *Benicasa cerifera* and *Cucurbita ficifolio* are used for muskmelon. Rootstocks resistant to bacterial wilt and Fusarium wilt are used for grafting tomato and eggplant and is done by using specially produced robot G892 and 2. The successful union of the graft is evaluated by using a thermal camera.

3) Agro-techniques

a) High density planting

Introduction of high density plantings is one of the important technologies to achieve high productivity per unit area both in short duration crops and perennial crops. High density orchards not only provide higher yield and net economic return per unit area in the initial years but also facilitate efficient use of inputs and easy harvest distance between plants, rows and beds. The success of high density plantings also depends on optimum fertilizer and irrigation requirements. Efficient fertigation methods need to be standardized for such plantations.

b) Integrated nutrient management

Integrated nutrient management refers to maintenance of soil fertility and plant nutrient supply to an optimum level for sustaining the desired crop productivity through optimization of the benefits from all possible sources of plant nutrients in an integrated manner. Therefore, it is a holistic approach where we first know what exactly is required by plants for optimum level of production, in what different forms these nutrients can be applied in soil at what different timings in best possible method, and how best these forms can be integrated to obtain highest productivity with efficiency at economically acceptable limits in environment friendly way. Addition of inorganic fertilizer constitutes one of the most expensive inputs in agriculture. Simultaneously, their excessive and indiscriminate use in commercial horticultural crops has resulted in several problems. In a country like India, it is essential that such approaches are followed which do not adversely affect yield and quality and are simultaneously cost effective and eco-friendly. As such, it is necessary that fertilizer applications are made using both inorganic and organic manure including bio-fertilizers.

c) Bio-fertilizer

Biofertilizer offer an economically attractive and ecologically sound means of reducing external inputs and improving the quality and quantity of internal resources. These are inputs containing micro-organisms which are capable of mobilizing nutritive elements from non-usable form to usable form through biological processes. They are less expensive, eco-friendly and sustainable. *Azotobactor*, *Azospirillum*, PSB, Rhizobium, etc. are some of the popular examples used in vegetable crops.

d) Vermiculture

Harnessing earthworms as versatile natural bioreactors is vermiculture. The process of composting organic wastes through domesticated earthworms under controlled conditions is vermicomposting. Earthworms have tremendous ability to compost all biodegradable materials. Wastes subjected to earthworm consumption decompose 2-5 times faster than in conventional composting. These worms feed on partially decomposed organic matter and provide a rich / better compost.

e) Micro-irrigation and fertigation

Water is a critical input for production of crops and its efficient use is inevitable in context of achieving higher production from shrinking land resources. It is estimated that only 38 per cent of the area is irrigated and with harnessing of all the resources of ground and surface water with current efficiency of traditional irrigation, not more than 60 per cent of area could be irrigated. Micro-irrigation, popularly known as drip irrigation, is an efficient way of irrigation with high frequency application of water, where water is delivered in and around root zone of plant through a network of pipes provided with suitable emitting devices. It has been very successful for irrigating horticultural crops like mango, banana, grape, pomegranate, guava, citrus, brinjal, cucumber, okra, capsicum, coconut, cashew etc. Now the emphasis is on applying liquid fertilizers through micro-irrigation called fertigation. This results in saving of fertilizer upto 30 per cent, increase in yield upto 100 per cent, saving of water upto 70 per cent, prevention of weed growth, saving of energy and improvement in quality of the produce.

Table 22.1: Relative performance of crops with drip irrigation in comparison with that of traditional irrigation methods.

Crop	Location	Yield (q/ha)		Irrigation Water (cm)		WUE (a/ha/cm)		Advantage of Drip irrigation	
								Saving of % H$_2$O	Increase Yield %
		Surface	Drip	Surface	Drip	Surface	Drip		
(1)	(2)	(3)	(4)	(5)	(6)	(7)	(8)	(9)	(10)
Ashgourd	Jodhpur	108	120	84	74	1.3	1.6	12	11.1
Beet	Coimbatore	5.7	8.9	86	18	0.07	0.5	79.1	56.1
Bottlegourd	Jodhpur	380	558	84	74	4.5	7.5	12	47.0
Bittergourd	Chalakudy	32	43	76	33	0.42	1.3	56.6	34.4
Brinjal	Bathinda	217	333	159	45	1.36	7.4	71.6	54.0
	Akola	91	148	168	64	0.55	2.3	62.0	63.0
	Delhi	280	33.8	45	35	6.2	9.7	22.2	21.0
	NCPA	280	320	90	42	3.11	7.6	53.3	14.3
	Pune	225	245	78	51	2.9	4.8	34.6	9.0
	Rahuri	280	280	90	42	3.11	6.7	53.3	0.0
Broccoli	Delhi	140	195	70	60	2	3.25	14.3	39.3
Capsicum	Gayeshpur	22.6	37.9	40.0	28.0	1.77	9.90	30.0	68.0
Cauliflower	Akola	83.0	116.2	39.0	26	2.1	4.5	33.3	40.0
	Pantnagar	171.0	274.0	27.0	18.0	6.3	15.2	33.3	60.2
Chilli	NCPA	42.3	60.9	109	41.7	0.39	1.5	61.7	44.0
Cucumber	Pune	155	225	54	24	2.9	9.4	55.6	45.2
Curry leaf	Bhavanisagar	100	175	63.0	26.0	0.63	6.73	75.0	75.0
Okra	Coimbatore	100	113.1	53.5	8.6	1.87	13.2	84.0	13.0
	Delhi	360	480.0	42.0	26.0	8.6	18.5	38.1	33.3
	Rahuri	189	203.3	219.0	113.0	0.86	1.8	48.4	8.0
Onion	Sriganganagar	137	225.0	67.0	35.0	2.04	6.42	47.7	64.2
Potato	Delhi	284	342.0	52.0	26.0	5.5	13.2	50.0	20.4
	Hisar	93	112.0	60.0	45.0	1.6	2.5	25.0	20.4
	Delhi	172.0	291.0	60.0	27.5	2.9	10.6	54.2	69.2
Raddish	Hisar	235.7	344.2	20.0	20.0	11.8	17.2	0.0	46.0
	Parbhani	334.0	480.0	30.0	22.0	11.1	21.8	26.7	44.0
	Coimbatore	10.5	11.9	46	11	0.23	1.1	76.1	13.3
Sugarbeet	Hisar	418	489	50	37	8.4	13.2	26	17.0
Sweet Potato	Coimbatore	42.4	58.9	63	25	0.67	2.4	60.3	39.0

(Contd...)

Crop	Location	Yield (q/ha)		Irrigation Water (cm)		WUE (a/ha/cm)		Advantage of Drip irrigation	
								Saving of % H₂O	Increase Yield %
		Surface	Drip	Surface	Drip	Surface	Drip	Saving of % H₂O	Increase Yield %
(1)	(2)	(3)	(4)	(5)	(6)	(7)	(8)	(9)	(10)
Tomato	Chiplima	167.0	225	130.0	72.0	1.28	3.12	44.6	35.0
	Akola	45.0	58.0	102.0	77.0	0.44	0.75	24.5	29.0
	Coimbatore	61.8	88.7	49.8	10.7	1.24	8.2	78.5	44.0
	Delhi	257.0	396.0	47.0	25.0	5.5	15.8	46.8	54.1
	NCPA	320.0	480.0	30.0	19.0	10.7	25.3	36.7	50.0
	Pantnagar	104.0	137.0	22.0	14.0	4.7	9.8	36.4	32.0
	Pune	292.0	413.0	31.0	20.0	9.4	20.7	29.0	41.4
	Rahuri	16.4	17.2	29.7	20.8	0.6	0.82	30.0	5.0
	Udaipur	144.0	175.0	41.0	28.0	3.5	6.3	31.7	22.0

Source: International Conference on Plasticulture and Precision Farming, 2005

Crop	Locations	Yield (t/ha)	Optimum Levels		Savings (%) in	
			Irrigation (% PE)	Fertigation (% RDF)	Water	Fertilizer
Banana	Bhavanisagar	85	80	80*	20	15
	Rahuri	83	80	75	25	17
Brinjal	Gayeshpur	26	80	125	20	-
	Navsari	30	85	100	15	-
Cotton	Rahuri	3.7	75	100	25	-
	Parbhani	2.9	80	75	20	25
	Sriganganagar	3.4	60	80	37	20
	Bathinda	1.7	60	100	40	-
Dry chilli	Madurai	2.2	60	75	40	25
Sugarcane	Sivganga	137	80	125	20	-
	Rahuri	155	80	100	20	-
	Navsari	110	80	100	20	-
Tomato	Bhavanisagar	27	60	50	40	50
Average		51.3	71.9	91.2	26.3	25.3

RDF, Recommended Dose of Fertilizers
Source: AICRP (Water Management), Annual Report 2002-03, 2003-04, 2004-05

f) Organic farming

The use of organic farming is regarded as the best solution to restore our natural resources and to safeguard our environment. It is holistic production management system which promotes and enhances agro-ecosystems health including bio-diversity, biological cycles and soil biological activities. The farming system emphasizes upon management practices wherein agronomic, biological and mechanical methods are used for sustainable production avoiding the use of synthetic materials. With increasing health consciousness and concern for environment, organic farming system has been drawing attention all over the world. As a result, there is widespread organic movement. Demand for organic products especially in developed countries, has been increasing by leaps and bounds. Major components of organic farming are crop rotation, maintenance and enhancement of soil fertility through biological nitrogen fixation, addition of organic manures, biofertilizer and use of soil micro-organisms. Precision organic Agriculture through GIS fulfils the demands of design strategies and managerial activities in a continuing process. By implementing this combination, certified methods for defining the best policies, monitoring the results and the sustainability of the framework and generating a constructive dialogue for future improvement on environmental improvement and development and be developed.

4) Protected cultivation

India has varied climatic conditions in different regions, so the greenhouse and the supporting facilities have to be developed accordingly. Design of greenhouse depends upon the (i) ambient temperature (ii) relative humidity (RH) (iii) wind velocity (iv) latitude and snowfall and (v) the requirement of temperature and RH of the crop inside the greenhouse. The southern plateau requires only mild climate and coastal regions need only naturally ventilated polyhouses. The northern plains with extreme hot and cold climate need the cooling and heating facilities. The local climate conditions play a major role in design of the greenhouse structures and cladding materials. Under mild and hot climates, single film cladding of 200 micron thick UV stabilized LDPE film will be sufficient while for cold climates inflated double layer of UV stabilized LDPE film is required. In naturally ventilated greenhouses, the temperature is maintained at 13°C above ambient conditions due to wind and thermal buoyancy and these do not need any electric power or greenhouse for maintenance of temperature and RH. Under Indian conditions naturally ventilated greenhouses are most suitable due to low operating costs. These are used in the climates of southern peninsula. The naturally ventilated greenhouses are also economical for cultivation in northern hilly regions during summer. These vegetables and flower nursery, production of early vegetable crops, extension of growing season, vegetable production during frozen winters, protection of valuable germplasm and cultivation of cucurbits, brinjal, capsicum and certain flowering annuals.

5) Adopting Recent Techniques in Crop Protection

a) Integrated pest management.

Integrated pest management aims at a judicious use of cultural, biological, chemical host plant resistance tolerance, physical mechanical control and regulatory control methods. Currently the term Biointensive Integrated Pest Management is used to lay major emphasis on conservation and enhancement of natural enemies and utilization of all compatible methods for achieving effective, economical and safe pest suppression. The most common methods for the management of horticultural pests is the use of pesticides. Other methods used from time to time are sanitation practices, tillage operations, crop rotation, fertility management, manipulation of planting dates and crop duration, trap-cropping, destruction of alternate hosts, destruction of off types and volunteer plants, pruning and defoliation and water management.

In this new millennium, concerted efforts should be made to adopt biological suppression based pest management methods including conservation techniques for which research capabilities should be enhance and transfer of technology strengthened. For export oriented crops biological suppression alone may be used to produce pesticide residue free fruits, vegetables, spices and ornamentals.

b) Biopesticides

These pesticides of biological origin, *e.g.* viruses, bacteria, fungi, Bacloviruses *viz.* nuclear polyhedrosis virus (NPV) and grannulosis viruses (GV) are important and specific in infecting only certain closely related species of caterpillars attacking various crop plants *e.g.*, tomato, potato and brinjal like crops.

c) Pesticide residues

Horticultural crops though cultivated on 5 per cent of the cropped area of the country, consume 18 per cent of total pesticide used in India. The intensive and off-season cultivation of vegetables and flowers under controlled conditions in present scenario of hi-tech horticulture have further necessitated increased use of pesticides. Newer pesticides molecules are also being introduced with their high potency. These residues are not only harmful to consumers but are carcinogenic in several cases. These also result in bleaking our exports of various commodities. It is, therefore, necessary that only those plant protection schedules are recommended which ensure freedom from contaminants at harvest. This is possible by working out the safe waiting period based on the Maximum Residue Limits (MRL) permissible as prescribed by FAO/WHO and GOI to ensure safety of human health. Such build up of residues has also to be checked to ensure quality control.

d) Molecular diagnostics

Several pests cause serious constraints in production of horticultural crops. Crop protection through pathogen detection has become an important pre-requisite for their effective management. Considerable advances have been made and while serological methods are very important, they are not as precise and sensitive, as the latest nuclear probes and PCR (polymerase chain reaction) based technologies. A major breakthrough in the diagnosis of viral diseases came with the application of Enzyme Linked-Immunosorbent Assay (ELISA) for sensitive detection of viruses. Above all, direct electron microscopy and immune-electron microscopy are very useful for the diagnosis of viral infection.

6) Post-harvest technology

Efforts to increase production through enhanced area under cultivation, use of high yielding varieties, adoption of improved agro-techniques and plant protection measures will go waste unless post harvest losses are reduced to minimum level. It is very important in case of horticultural crops where the produce is mostly perishable in nature. Since the cost involved in preventing the losses is always lesser than cost of production, post harvest management attains greater significance. Therefore, there is, need for more developed marketing mechanism with improved grading, packing, storage and transport ensuring low spoilage and wastage. To improve the quantity of horticultural produce used in processing is also essential.

a) Grading

One of the important post harvest operation in relation to marketing is the grade of produce. At present, grade standards for all commodities are not available. There is an urgent need to develop and follow grade standards in different crops. Recent researches have succeeded in developing equipment for grading and commercial prototypes have been developed utilizing acoustic response ultrasonic photometry, light reflectance, delayed light emission, short wave radiation response and machine vision technique. Equipment has also been developed for orientation of fruits and vegetable to meet the needs of different types of grading machines.

b) Packaging and storage

Proper packaging and ideal storage of these perishable commodities are very important to ensure minimal losses and good quality. Significant advances have been made in recent years by introduction of cardboard packages, establishment of pre-cooling units and introduction of refrigerated vans. We have yet to make use of advances made in this field elsewhere. Therefore, there is, need for improving the design, operational cost and safety standards of cold storage. We have also to make use of hitech technologies in packaging and storage of fruits and vegetables for extending storage life. There is a need to standardize technologies for various commodities and a lot still needs to be done before these can be applied under our conditions.

c) Cold chain

Modern cold chain is a temperature management facility involving a number of equipment such as precooling units, cold storage, humidity controlled atmospheric storage, reefer-container, mobile cooler and integrated handling and storage systems. It provides ideal conditions for preserving perishable agro-commodities from their point of origin to the point of consumption. It consists of pre-cooling units, refrigerated transport, cold storage and refrigerated retail shops. The cool chain is vital for perishable horticulture commodities e.g., fruits, vegetables and flowers to keep commodities in prime condition and reduce their wastage. However, lack of infrastructure e.g. good roads, adequate precooling unit, refrigerated trucks and storage space with refrigeration facilities coupled with unreliable and insufficient power availability and low level of awareness are some major constraints in developing cool chain.

d) Irradiation

It refers to physical means of exposing fresh or pre- packaged food products or bulk food to gamma rays. Studies on various aspects i.e. wholesomeness, microbial safety, nutritional adequacy, lack of mutagenicity and toxicity have affirmed its safety. Experiments have conclusively shown that appropriate exposure to radiation such as gamma rays enhance the shelf life of processed fruit. Studies have shown that irradiation can delay ripening and senescence in fruits, inhibits sprouting of roots and tubers and can control fungal decay. In addition, application of irradiation technology for extending shelf life referred as cold storage has helped in maintenance of low temperature and extending the shelf-life. However, retention of quality during storage has assumed greater importance.

e) Processing

By adopting hi-tech processing methods and increasing the scale of operation quality of processed products can be improved and cost of product reduced. Asceptic packaging which is now commercially being utilized for several juices, requires high capital investment but enables continuous processing. Ohmic heating systems are now commercial in USA, Japan in aseptic processing plants. Commercial ohmic heating based plants are now available with fully automatic temperature control systems. Another process namely, osmotic process removes 30-50 per cent of water without heating and therefore such products are of superior quality. The use of oscillating magnetic field technology ensures no post processing contamination unlike in thermal processes. High intensity pulsed electric field technology relies on the lethal effect of electric field on micro-organisms. Foods are processed in a short period of time and energy lost due to heating is minimal. Over the years, a treasure of knowledge of food preservation methods which fulfills the requirement of being suitable for shelf stable products without refrigeration and are inexpensive but reliable have been achieved by a combination of several factors for which the term HURDLE has been introduced. By understanding the complex interactions responsible for the microbial stability of fruits, more than 50 hurdles have been identified. Defense Food Research Laboratory has introduced many products to special military missions through this technology. Further, exploration and application of hurdle technology is essential to pave the way for developing processes for preservation of many fruits and vegetables. Because of their fresh like qualities, stability at ambient temperature and suitability for direct consumption, it has become highly popular in advanced countries. Vegetable food products preserved by hurdle technology offer several advantages such as economy in processing, storage, transport and marketing convenience, quality and safety to consumer and scope for expanding commercial application.

CONCLUSION

- A truly comprehensive approach to precision agriculture must cover all phases of production from planning to post-harvest.
- Information, technology and management are combined into a production system that can increase productivity, improve product quality, allow more efficient chemical use, conserve energy and provide guidelines for soil and ground water protection.
- Technology and management practices such as field scouting, field mapping, variable rate control, yield, mapping and post harvest processing can readily be adopted to vegetable crop production. However, the technology related to precision farming needs refinement to realize benefits.

- Questions remains about cost effectiveness and the most effective ways and means to use the technological tools we now have but the concept of doing the right thing in the right place at the right time has a strong initiative appeal.

- Ultimately, the success of precision agriculture depend largely on how well and how quickly the knowledge needed to guide the new technologies can be found, disseminated and used .

Thus precision farming is a professional farming system, which aids in taking better management decisions in order to use input more efficiently and effectively for improving overall farm profitability. Ultimately the technology also helps in avoiding environment degradation and enhances sustainability. However, the constraints in the way are to be addressed for making it more and more popular.

A success story

A project on precision farming (PF) was implemented under the aegis of Tamil Nadu Agriculture University (TNAU) in two back ward districts *viz.* Dharmapuri and Krishnagiri in northern Tamil Nadu over three years (2004-2007). Dharmapuri considered to be the horticultural district of the state for largest producer of fruit crops like mangoes, banana, grapes, guava, papaya and sapota and vegetables such as tomato, brinjal, chillies, cabbage etc. About 400 farmers were selected in these two districts month wisely commencing from august 2004 till march 2007 and organized in clusters.

Table 22.2: Output and estimated gross income of a sample of PFP farmers

Crop	Sample of farmers	Average no. of harvests in sample	Average output/ crop for one season (t/ha).	Average of gross income/crop (Rs./ha).
Tomato	30	71	85	2,11,963
Brinjal	24	76	121	3,03,333
Chilli	8	42	25	1,52,273
Cabbage	8	4	56	1,68,109
Cassava	9	1	40	1,68,725
Okra	5	33	13	90,720
Muskmelon	6	3	34	2,03,000
Watermelon	5	3	33	97,860
Banana	24	2	79	6,12,842

The PF yield were found 3 to 12 times higher as compared to national average yield estimates of tomato (17.35t/ha), brinjal (10.46 t/ha) and banana (28.58 t/ha). The improvements in terms of lengthened crop duration, increased harvest periods, more picking and consequently, (i) By adopting PF package and (ii) Over traditional cultivation method.

Table 22.3: Farmers views about benefits derived from the PF techs.

Farmer	Crop	Remarks
Thangamani	Banana	Three fold increase In yield
Varadaraja	Tomato	-do-
Krishran	Tomato	150% or about 2.5 times increase in yield.
Pallani	Tomato	16% or about 2.6 times increase in yield.
Mahengran	Tomato	About five fold (26 t/ acre instead of 5-6 t/ acre) got Rs. 5/- more per kg.
Dorairaj	Tomato	Earned about Rs. 2.0 lakhs/acre. PF tomato weighed 28 kg/rate as against 26kg/rate of traditional crop.
	Sugarcane	Harvested in 8 months compared to 12 months.
Rajendran	Tomato	Earned about Rs. 3, 00,000/ha in first year.
Balasubramanium	Tomato	Got Rs. 20/- extra per crate for same volume or weight
Ambumani	Tomato	Twelve fold increase n in income under PF

The cost of converting a traditional non- precision farm to a precision farm is also important which includes irrigation/ fertigation equipment and the use of hi- tech inputs and other cultivation inputs. The total cost using class 3 cultivation cost (inputs consisting of hybrid seed, water soluble fertilizers and plant protection measures and preparation costs including labour cost for field preparation, transplantation, irrigation, harvesting and packing etc.) of about Rs. 50,000 per hectare, making a total raning between Rs. 1,35,000 to Rs. 1,50,000. However, the entire cost of such a conversion is possible to be recovered virtually in one season alone through the proceeds of precision farming, even if the amortization period of 10 years for equipments in not taken into consideration.

ECONOMICS AND CONSTRAINTS

Maheswari *et. al.* (2008) compared the economics of tomato and brinjal production under precision precision farming and found that the gross returns to be 166 per cent higher under precision farming than non- precision farming in case of tomato and 67 per cent in brinjal.

Table 22.4: Comparative economics of tomato and brinjal cultivation:

S. No.	Particulars	Difference (%) in PF over non PF system	
		Tomato	Brinjal
1.	Human labour	39.77	49.57
2.	Machine labour	14.28	12.00
3.	Seedlings	-10.52	-6.25
4.	Manures	82.30	0.93
5.	Plant protection chemical	-58.02	-123.64
6.	Fertilizers	297.73	151.66

(Contd...)

S. No.	Particulars	Difference (%) in PF over non PF system	
		Tomato	Brinjal
7.	Stacking	20.56	-
8.	Drip system	100.00	100.00
9.	Interest on work capital (7%)	42.15	32.92
10.	Total variable cost	42.15	25.47
11.	Main production (kg/ha)	80.16	34.00
12.	Difference in output (contribution)	63.86	28.14
13.	Gross income	120.20	51.32
14.	Gross margin	165.64	67.26

They also analyzed the reasons for non- adoption of PF technologies and found that the lack of difficulty in obtaining credit facilities was the major limiting factor, followed by drip irrigation/ fertigation equipments installation, lack of technical knowledge and labour shortage etc.

Table 22.5: Constraints to adoption of precision farming

Rank	Reasons	Mean Garrett's score
1	Lack of finance and credit facilities	73
2	Expensive drip/fertigation assemblies	65
3	Lack of technological knowledge	54
4	Labour scarcity	53
5	Farmer's perception on yield impact of low quantity of inputs	51
6	Lack of water availability and pumping efficiency	44
7	Lack of technical skills	42
8	Low price/ unprofitable negotiation	41
9	Inadequate extension/research activities	41
10	Inadequate land holding	27

REFERENCES

Asnderson G and Yang C. (1996). Determining with field management zones for grain sorghum using aerial videography. In: Proceeding of the 26th Symposium on Remote Sensing of Environment, Vancouver, B.C., March 25-29, 1996.

Aronoff S. (1995). In Geographic information systems: a management perspective, pp 62-63. WDL Publication, Ottawa, Ontario.

Arora S. (2005). Precision agriculture and sustainable development, *Kurkshetra*. **54**(2): 18-22.

Chadha KL.(2005). Precision farming in horticulture. Souveniri. International Conference of Plasticuture and precison farming (Nov. 17-21). Hotel The Ashok, Chanakyapuri, New Delhi, pp. 1-18.

Geopalapillai S, Tian L and Bea J. (1998). Detection of nitrogen stress in corn using digital aerial imaging. Presented at ASAE Annual international Meeting, July 12-16 paper No. 983030, ASAE, 2950 Niles Road, St. Joseph, MI 49085-9659, USA [cf: http:/ www.age.uiue.edu/faculty/lft/papers/rspaper.htm.

Kalia P. (2005). Precision vegetable farming. Compendium. Hi-Tech farming in vegetable crops. CASH ort. (Vegetables). Deptt. Of Vegetable Crops, Dr. Y.S.P.U.H.F., Nauni, Solan, H.P. pp. 16-19.[cf: http://www.bac.uky .edu/~precog/precisionAg/ Exten_pubs/pa-2.htm].

Maheswari R, Ashok KR and Prahadeeswaran, M. (2008). Precision farming technology, Adoption, Decision Productivity of vegetables in Resource poor environments. *Agriculture Economics Research* **21:**514-424.

Taylor J and Whelan B. (2005). A general introduction to precision agriculture [ef: http:// www.usyd.edu.au.su/agric/acpa].

Verma C and Wendte. (1997). Mapping subsurface drainage systems with color infrared aerial photographs. American Water Resource Association's 32nd Annual Conference and Symposium 'GIS and Water Resources'', September 22-26, Ft. Lauderdale, Florida [cf: http:///www.uwin.siu.edu/~awra/proceeding /gis32/florida/index.html].

23

CHAPTER

Rangelands : Concepts and Management

☞ R. Banyal[1] and ☞ Birbal[2]

INTRODUCTION

India, the seventh largest country in the world, lies between latitudes 8° 4' and 37° 6'N and longitudes 68° 7'and 97° 25'and occupies a geographical area of 3.29 M km². It measures about 3,214 km from north to south and about 2,993 km from east to west (Anonymous, 2011). The country exhibits great diversity in climate, topography, flora, fauna and land use. The precipitation ranges from 150 mm in western and north-western deserts to 3126 mm in north-eastern hills. The altitude varies from the coastline to the lofty, snow clad mountains of the Himalayas. The temperature ranges from sub-zero in the Himalaya to about 50°C in the central and western parts. India has common borders with Afghanistan, Bangladesh, Bhutan, China, Myanmar, Nepal and Pakistan (Anonymous, 1997).

Rangelands are vast natural landscapes in the form of grasslands, shrublands, woodlands, wetlands and deserts. Types of rangelands include tallgrass and short grass prairies, desert grasslands and shrublands, woodlands, savannas, chapparals, steppes and tundras. Rangelands occur around the globe dry in areas to or with soils and topography unsuitable for broad-acre farming but fertile and wet enough for pastoralism. Because, pastoralism may intensively utilize water and vegetation thereby some rangeland areas can become damaged. This damage means that pastoralism itself ultimately suffers and damage causes conflicts with other land uses such as conserving biodiversity, hunting and gathering bush foods or firewood. Agricultural lands have always preferential treatment as compared to both forests and rangelands. Moreover, grasses are considered to be a type of vegetation which comes up naturally, available to cattle grazing and vanishes. The increase in livestock population results in the major cause of deterioration. The productivity of such lands reduced and results in barren look to the areas. A deficit of 62.7 per cent in green and 23.4 per cent in dry fodder exists between demand and availability of the forage for feeding the livestock of the country. India with about 2 per cent of the total Worlds' geographical area sustains as much as 15 per cent of the Worlds' livestock population. Livestock plays an important role in countrys' rural economy and in meeting the demand for milk its products, manure, meat, hides and bones.

[1] Faculty of Forestry, Sher-e-Kashmir University of Agricultural Sciences & Technology of Kashmir, Shalimar, Srinagar (J&K) 191 121

[2] Sr. Scientist CAZRI, Regional Centre, Bikaner (Raj.)

Livestock rearing is an integral part of the various farming systems. Arable agriculture contributes a major fodder resource in the form of crop residues which are extensively fed to the animals. Wheat straw is transported from surplus areas such as Punjab and Haryana to deficit areas, mostly the Himalayan hills. Fodder crops like oats, Egyptian clover, fodder rape and chicory are grown during winter, while maize, pennisetum, sorghum and cowpeas are sown during the summer. Cultivation of forage crops is restricted to irrigated areas and land rich farmers. Sale of green fodder through retail outlets is a common practice. Cultivation of perennial grasses such as napier and napier × Bajra (*Pennisetum*) hybrids is becoming popular. Intensive fodder cultivation is restricted to States such as Punjab, Haryana, Uttar Pradesh, Madhya Pradesh, Gujarat, Maharashtra, Andhra Pradesh and Karnataka. The area under fodder cultivation amounts to 4 per cent of the total cultivable area. However, exclusive pastures and grasslands are widespread and are grazed by domestic animals. The total area of permanent pastures and grasslands is about 12.4 M ha or 3.9 per cent of the country's geographical area. An area of 15.6 M ha, classified as waste land, is also used for grazing. Forests and their associated grasslands and fodder trees are another major source of grazing and fodder collection.

Livestock are more significant for people living in drought-prone, hilly, tribal and other less favoured areas where crop production may not be certain. Animal rearing is a means of supporting the earning capacity of landless, marginal and small farmers. The importance of this sector can be estimated from the fact that the contribution of livestock sector to GDP is about 9.0 to 10 per cent. The country possesses 26 indigenous breeds of cattle and 6 breeds of buffalo. India has substantial imports of live cattle and dairy products, in particular dry milk from the neighbouring countries. The livestock population is increasing continuously on one hand and the land under permanent pastures is shrinking on the other hand. The emerging trend in livestock growth suggests an overall rising trend with an increase in small ruminants in coming years. This is likely to put greater pressure on natural fodder resources like forests and other grazing areas. To enhance farm income, incorporating fodder into joint forest management programmes is a welcome step but requires greater impetus. Optimisation of fodder production from arable lands and efficient utilisation of crop residues are expected to relieve at least some of the pressure on natural fodder resources. There is also an urgent need to adopt appropriate technologies backed by policy and institutional mechanisms to enhance rural incomes from livestock enterprises.

THE RANGELAND/PASTURE RESOURCES

The grazing of animals takes place on a variety of grazing lands. True pastures and grasslands are spread over an area of about 12.4 M ha. Other grazing lands are available under tree crops and groves (3.70 M ha), on wastelands (1.50 M ha) and on fallow lands (2.33 M ha). Pastures and grasslands have often resulted from degradation and destruction of forests until savannas are formed (Misra, 1983). True pastures as climax vegetation are found only in sub-alpine and alpine pastures in the higher altitudes of the Himalayas. Small areas of grazing lands are found in the states of Punjab (5.9 per cent) and Haryana (8.9 per cent), while 79.3 per cent and

81.7 per cent of grazing lands are found in Arunachal Pradesh and Nagaland, respectively. Punjab and Haryana states have large areas under intensive fodder production. One hectare of fodder cropped area supports 11 to 12 adult cattle units. In the states of Arunachal Pradesh and Nagaland, fodder is not cultivated (Singh and Misri, 1993).

The deterioration of Indian pastures, grasslands and other grazing lands may be ascribed to the large bovine population, free grazing practices, lack of management, and natural constraints like extremes of temperature, steepness of slopes, variable precipitation and scarcity of moisture in arid and semi-arid situations. The situation sin Himalayan pastures is even more alarming due to the severe pressure of the sedentary, semi-migratory and migratory graziers. Overgrazing has caused the nearly complete loss of edible species. Weeds such as *Stipa, Sambucus, Aconitum, Cincifuga, Adonis,* and *Sibbaldia* have heavily infested these pastures (Misri, 1995). Fodder cultivation has remained static at 4 per cent of the total cultivated area. Availability of fodder seed is another limiting factor. The annual requirement of fodder feed for 6.9 M ha area under fodder cultivation. Against this, the availability is of these crops very meagre to achieve the target.

WHAT IS RANGELAND?

Areas around the globe with arid, semi-arid and dry sub-humid climates and where topography and soils are unsuitable for broad-acre farming are generically referred to as rangelands because these areas are traditionally used for pastoralism (Harrington *et al.*, 1984). Roughly half of the globe's land surface is rangeland which is about 67 million km^2 (WRI, 1986). As a traditional and extensive land use, pastoralism tends to mask many other more local intensive uses of rangelands such as mining and tourism (Williams *et al.*, 1968). Protected areas within rangeland regions also serve to conserve biodiversity, and some areas are used for hunting and gathering of bush foods and firewood. Wildfires and intentional use of fire also affect rangelands. All these multiple uses affect rangelands and cause changes to various degrees and extents.

Rangelands may be defind as those systems having healthy (i) biophysical functions that include a high capacity to retain water, capture energy, produce biomass, cycle nutrients and provide habitats for diverse populations of native animals, plants and microorganisms, and (ii) socio-economic functions that adequately provide people with their material, cultural and spiritual needs. To maintain rangelands in good condition, these biophysical and socio-economic functions need to be measured and reported. This involves in development monitoring procedures and passing the information to stakeholders.

The term has been defined by different workers and some of the definitions are given here as under:

(i)   Rangeland is defined as land on which the potential plant community is composed periodically of native grass, forbs and shrubs for forage and sufficient quantity to justify grazing use.

OR

(ii) Rangelands refer to large, naturally vegetated mostly unfenced lands of low rainfall area that are grazed by domestic livestock and game mammals (Sampson)

The term 'condition' in standard dictionaries means "state of being" or "health". In human health terms, poor health is a 'state of being' in reference to good health which is typically assessed in terms of easily measured indicators such as body temperature, blood pressure and resting heart rate. Rangeland condition is analogous. It is a human perception of the state of health of a rangeland area in reference to an area perceived to be in a state of good health – a reference or benchmark site (Friedel *et al.*, 2000). The state of the benchmark site and other rangeland sites of interest, can be defined by a set of easily measured indicators related to, *for example*, production, conservation and aesthetic values (Keith and Gorrod, 2006). Given such indicators, the state or condition of the rangeland site is judged by people (stakeholders) to be in a given state of health relative to the benchmark site. This health analogy is widely used and it has proven useful for talking about the state of rangelands.

Types and distribution

The grazing of animals takes place on a variety of grazing lands. Dabadghao and Shankaranarayan (1973) have classified grasslands into five types.

(i) Sehima-Dichanthium Type

This type of grass cover is spread over the Central Indian plateau, Chota Nagpur plateau and Aravallis, covering an area of 1,74,000 km^2. The elevation ranges between 300-1200 m. There are 24 species of predominant perennial grasses, 89 species of annual grasses and 129 species of dicots including 56 legumes. Some of the important grasses are:

Perennial species

Aristida setacea, Arundinella mesophylla, Cynodon dactylon, Sehima nervosum, Heteropogon contortus, Dichanthium annulatum, Chrysopogen fulvus Themeda triandra and *Bothriochloa pertusa.*

Annual species

Apluda mutica, Aristida funiculate and *Chloris barbata.*

Management

The management of such grass cover requires adequate soil conservation measures like contour furrowing, trenching, etc.

(ii) Dichanthium - Cenchrus - Lasiurus Type

This type of grasslands are spread over an area of 436,000 km^2, including northern parts of Gujarat, Rajasthan, Aravali ranges, South-Western Uttar Pradesh, Delhi and Punjab. The elevation ranges between 150-300m. There are 11 perennial grass species, 43 annual grass species and 45 dicots with 19 legumes. The main grass species are:

Perennial species

Cenchrus ciliaris, Dichanthium annulatum, Cymbopogon spp., *Lasiurus sindicus, Cenchrus setigerus* and *Cynodon dactylon.*

Annual species

Apluda mutica, Aristida adsensionis, Chloris barbata, Eragrostis ciliaris and *Tragus biflours.*

Management

Grazing should be restricted in these areas as the grasses grown here are preferred by cattle. The reseeding is beneficial for rejuvenating the area.

(iii) Phragmites - Saccharum - Imperata Type

This type of grass cover spreads over an area of 2,80,000 km² in the Gangetic plains, the Brahamputra Valley and the plains of Punjab. The elevation ranges between 300-500 m. There are 10 perennial grasses, 26 annual grasses and 56 herbaceous species including 16 legumes. The predominant grass species are:

Perennial species

Imperata cylindrical, Saccharum spontaneum, S. arundinaceum, S. bengalense, Phragmites karka, Desmostachya bipinnata.

Annual species

Apluda mutica, Eragrastis uniolcides, Dactyloctenium aigyptium.

Management

There is no need for special management practices in such areas as the most of the grasses occupy the swampy areas. There is ample scope for the introduction of high yielding and palatable grasses.

(iv) Themeda - Arundinella Type

This type of grassland spreads over 230,400 km² and includes the States of Manipur, Assam, West Bengal, Uttar Pradesh, Himachal Pradesh and Jammu & Kashmir. The elevation ranges between 350-1200 m and there are 37 major perennial grasses, 32 annual grasses and 34 dicots with 9 legumes. The main grass species are:

Perennial species

Cynodon dactylon, Arundinella bengalensis, A. nepalensis, Euleliopisis binata, Heteropogon contortus, Bothriochloa intermedia, Chrysopogon fulvus.

Annual species

Apluda mutica, Arthraxen lancifolius, Eragrostis unioloides.

Management

Such areas are subjected to minimum biotic interference because stall feeding is preferred to grazing. Light grazing during monsoon and moderate grazing during post monsoon period is allowed. Adequate soil conservation measures like contour furrowing, contour terracing, gully plugging, manuring, etc. may be adopted for harnessing the full potential of such areas.

(v) Temperate - Alpine grasslands Type

The area lies across the altitudes higher than 2100 m and include the temperate and cold arid areas of Jammu and Kashmir, Himachal Pradesh, Uttar Pradesh, West Bengal and the north-eastern states. There are 47 perennial grasses, 5 annual grasses and 68 dicots including 6 legumes. The principal grass species are:

Perennial species

Agrostis filipes, Agrostis cannina, Poa pratensis, Agropyron canaliculatium, Chrosopogon gryllus, Phleum alpinum.

Annual species

Poa annua, Oryzopsis lateralis, Poa stewartiana.

Management

These are the highly nutritive forage areas for the grazing animals. The management interventions include fertilizer application for qualitative and quantitative improvement of the grasses. The other intensive management measures like reseeding, introduction of legumes may also be taken up at large scale for the overall improvement of the zone. Shankar and Gupta (1992) have classified the Indian grazing lands as fragile eco-systems and have ranked them as class IV and V in their land capability classification. The carrying capacity of these areas is 0.20 to 1.47 adult cattle units (ACU)/ha, but the present stocking rates are much higher. In semi-arid areas, the present stocking rates are 1 to 51 ACU/ha against the carrying capacity of 1 ACU/ha (Shankar and Gupta 1992) while in the arid areas, the stocking rates are 1 to 4 ACU/ha against the carrying capacity of 0.2-0.5 ACU/ha (Raheja, 1966).

EVALUATING THE CONDITION OF RANGELANDS AND CLASSIFICATION

Range condition of a range site is the present state of the vegetation compared with that of the climax for that range site. Issues on evaluating rangeland condition or health can be viewed as having two main components: first, how do individual stakeholders assess the information available on the functional state of the rangeland area of interest and judge its condition relative to their visions and values and second, if the area of rangeland is being evaluated by multiple stakeholders having different visions and values, how to best resolve any conflicting statements on the condition or health of the rangeland.

Assessing rangeland condition

An issue for information on the state of a rangeland is how to encourage stakeholders to broadly assess all the data available rather than narrowly focus on a few specific indicators of prime interest to them.

RESOLVING CONFLICTING EVALUATIONS OF RANGELAND CONDITION

The statements issued on the condition of rangelands are often conflicting and require resolution. In some cases, statements are strongly conflicting and resolution has been required at the highest levels of government. However, the site-based pasture monitoring was often conducted on more stable mid-slope areas, which usually miss those areas actively eroding or degrading. Pasture monitoring reported positive changes in perennial plants (Watson, 1998 and Watson and Thomas, 2003) whereas aircraft-based assessments reported areas of desiccation and vegetation change (Pringel *et al.,* 2006). Subsequent ground-based surveys of these latter areas suggested that landscape desiccation was due to hydrological changes at critical points in the landscape (*i.e.,* rill and gully cutting; referred to as 'nick-points') (Pringle and Tinley, 2003). The issue is to be best resolved through participative approach.

The purpose in the classification of condition of rangeland is to provide an approximate measure of any deterioration in the plant cover thus provide a basis for predicting the degree of improvement. There are four classes being commonly used for classification of the range condition: (Table 23.1)

Table 23.1: Four Classes commonly used for classification of the range condition

Range condition class	of climax vegetation present (%)
Excellent	76 to 100
Good	51 to 75
Fair	26 to 50
Poor	0 to 25

ASSESSMENT OF IMPROVEMENT AND/OR DETERIORATION OF RANGE CONDITIONS

The assessment of range condition set the road map for the proper management and improvement measures of the area. The response variables are as follows:

(a) Plant vigour

The size of the plant in relation to age and biomass of above and below ground portion reflect the plant vigour. The length of rhizomes and stolons is also considered as good indicator. The basal area of the clumps can be charted by field pantograph. The number of tillers and the height of tallest tiller can be sound base for assessing the range condition.

(b) Reproduction and regeneration

This is reflected by the preponderance of young seedlings and plants of various ages. These vary with the growth habits of the individual species that too with current growing conditions. Vigorous reproduction and growth of the species relished by livestock show that the range is improving. The density increases with the increase in reproduction and regeneration.

(c) Plant composition

The study of plant composition gives valuable information on plant succession (progressive or retrogressive) and an insight into the extent of improvement or deterioration in the condition of the range. If, there is deterioration, the causes can be analysed and the suitable measures may be sought as per the requirement of the site. Changes in plant composition may be determined by various methods but the most commonly used is transect laid at random in the site.

(d) Plant residue

The progressive accumulation of plant residues generally indicates the improvement in range condition. The extent of plant residues accumulates depends on:

    (i)   The amount of herbage the plant community can produce

    (ii)   The amount of herbage removed by grazing, haying, by wind, fire, water; and

    (iii) The amount decomposed in place

If the ranges are under stocked and much residue accumulates leads the appearance of desirable plants than the undesirable ones.

(e) Condition of the soil

The condition of the surface soil affects the range condition and the rate of improvement or deterioration of the range site. The lack of plant residues, forming a protective cover over the soil, exposes the soil to splash erosion and to surface crusting. Surface crusting, soil compaction caused by trampling, soil erosion and the increase in the area of bare ground lead to deterioration of the range site. This happens under over grazing.

(f) Forage yield

This is the most important consideration and ultimate test of the productivity of a range site. It depends upon climate (chiefly rainfall), soil, plant composition, plant vigour, reproduction, regeneration, etc. The forage production on a range site may be estimated by harvesting method. This is done by laying out quadrates of fixed area laid at random either permanent or temporarily in the sites closed for grazing and also under different management practices. The total forage production from all the vegetation and forage production from each species separately may be obtained on weight basis. The forage yield in terms of quantity or quality or both may be correlated with other observations over a desired length of time.

Delineation of different range sites determining the condition classes and initial assessment of the plant composition, vigour and forage yield are the preliminary operations to provide the bench mark of the initial range resources so indispensible for an effective planning of the rangelands for improvement and management.

RANGELANDS IMPROVEMENT

The systematic approach to scientific management of rangelands is based on ecological principles and knowledge of plant succession. A comparison of the existing vegetation and the climax vegetation is an indication of the condition of the rangeland. The condition of rangelands have been improving if the vegetation is progressing towards climax and deteriorating if the vegetation belongs to a stage lower in succession than the predominant vegetation. The extent of deterioration of rangelands depends on the cause of deterioration and period of misuse. Therefore, range manager should have more stress on the significance of changes in vegetation rather than manipulation of succession.

Rangelands get rehabilitated and improve with a closure of 1 to 5 years depending upon the class condition. It is very essential to protect the area from grazing by fencing and with scientific management such as deferred, rotational and deferred-rotational grazing. May be of different types like live hedge, ditch-corewall, corewall-thorn, stone posts with barbed wire, wooden posts with barbed wire and angle iron with barbed wire. Angle iron barbed wire is most effective and cheapest fencing in the long run. Since, the expenditure on fencing is very high, it is advisable to have the participatory mode of protection with the help local inhabitants. Regeneration is very slow even under closure in degraded sites and slow in the fair condition class. Majority of the rangelands in India fall under these condition classes. Moreover, succession may be erratic and not bringing the desired forage species to the desired standard quickly. Therefore, the other intensive measures needed for rangelands improvement are as:

(a) Soil and water conservation

(b) Weed control & eradication of obnoxious plants

(c) Raising of legume crops

(d) Application of fertilizer

(e) Re-seeding and sodding (planting)

(a) Soil and water conservation

The soil and water conservation measures are essential part of rangeland management because all the forage plants grow and derive nutrients and water only from soil. These measures depend upon the rainfall, topography, soil and condition of the class of the range. The objective behind such measures is to encourage the soil forming process, to improve the nutrient status and to conserve the moisture in the soil. In the arid and semi-arid regions light harrowing, contour furrowing at appropriate intervals are very effective for soil and water conservation. In gullied rangelands, the control measures adopted may be vegetative or mechanical or a combination of both depending on the stage of gully formation. The common measures are closure of the area for grazing, construction of contour & peripheral bunds, easing of the head of the gully and gully plugging. Small brush wood check dams have to be constructed for checking the erosion. The choice of the material to be used and size of gully plug depend on width, length & bed slope of the gully and anticipated run-off. Whatever may be the material used, the gully plugs are quite effective in improving the soil moisture regime and retarding the run-off. In arid and semi arid areas, the run off water is not only lost to use by forage but it may cause erosion. A permanent vegetative cover is the best mean to control the run-off. But, sometimes it has to be supplemented by construction of structures for the conservation of water. The measures normally adopted are pitting, basin listing, water spreading and irrigation of pasture lands. Efficient utilization from natural water courses by water spreading increases the productivity of the range. The most common type of spreader is small dam placed in a water course with gradient ditches or terraces leading the water out to gentle slopes where it is released. It should not be assumed that water erosion is insignificant in drought prone areas. Even rare high intensity storms in one year or in a period of two or three years are sufficient to cause serious erosion due to loosened soil and structural habit of forage plants. Hence, soil and water conservation measures and erosion control structures are utmost essential part for range improvement and efficient management.

(b) Weed control

The primary purpose of weeds control in rangelands is to increase the forage production by reducing the competition for available water, nutrients and light. In rangelands one plant weed for one category of livestock (cattle) may be forage for another category (sheep, goats, camel, etc.). It is therefore, advisable to study the forage value of all the species, their seasonal palatability and arrange them in such a way that the best utilization of rangelands would be possible in all the seasons by different livestock. Troublesome weeds should be grubbed out. The most common methods of weed control are:

    (i)   Mechanical method; and

    (ii)   Chemical method

The choice of methods depends on extent of area, resource available and economic factors. Sometimes, both the methods can be used in combination. The weeds can be eradicated manually by resorting to hand pulling, hoeing, tilling or by mowing. The use of such methods is restricted as it involves lot of physical energy. The method consists in use of implements which can also be attached to tractor. Some of commonly used implements are plough (disc mould board) harrow (disc, spike) and cultivators. Some other mechanical implements used for eradication are tree dozer, brush cutter, circular power saws and heavy duty mowers. The herbicides can be used in the vicinity of the plants which is not possible by mechanical methods. A herbicide is any chemical that kills the plants or inhibits their growth. The chemicals employed can be classified into 1) selective herbicides 2) Non-selective herbicides. Weedicides are efficient but costly and may have harmful effects on the grazing livestock. Some of the herbicides being commonly used are here as depicted in Table 23.2.

Table 23.2 : Commonly used Herbicides

S. No.	Chemical name	Chemical formula
1.	Atrazine	2-choloro-3 ethylamino-5 isopropylamine, 2,4,6-triazine
2.	Bromacil	5-bromo-3 (butan-2-yl)-6-methylpyrimidine-2, 4 dione
3.	Dalapan	2,2-dichloropropionic acid
4.	MCPB	4-(4-Chloro-2 methoxy) butanoic acid
5.	PCB	Pentachlorophenol
6.	Glycel	glyphosate
7.	Paraquat	1,1¹–Dimethyl-4, 4¹ bipyridinium dichloride
8.	2, 4-D	2, 4-diclorophenoxyacetic acid
9.	2, 4-DB	4-(2, 4-dichlorophenoxy) butyric acid
10.	2,4,5-T	2,4,5-trichlorophenoxyacetic acid
11.	TCA	Trichloroacetic acid]

To destroy the roots of perennial weeds translocated herbicides may be used. Some of the translocated herbicides are Carbyne, 2,4-DB, MCPB, 2,4,5-T, Glyphosate, etc. Herbicides may be select on the basis of nature of weeds, mode of action of herbicide and extent of weed control.

(c) Legume crops

Many legumes are well known for improving the fertility of the soil. These are available when the cattle have finished with the grasses by about September/October. Voluntary intake and digestibility of mature legumes are much higher than those of mature grasses. Some recommended legumes are *Atylosia scarabaeoides, Clitoria ternatea, Dolichos lablab, Rhyncosia minima, Pueraria hirsute, Alysicarpus vaginalis, Orotalaria medicaginea* var. *luxurians, Sesbania* spp., *Phaseolus trilobus*, etc. During *kharif* season, common pulse legumes may be grown in mixture with grasses. The best are *Phaseolus aureus, P. aconitifolius* and *Cyamopsis tetragonoloba*.

(d) Application of fertilizers

The application of fertilizers has been found to be more effective in areas of high rainfall such as meadows. The response is related to climate, soil and type of vegetation. The results of fertilizer application are not encouraging in arid lands. Application of farm yard manure has also become common practice. Fertilizers have an advantage over farm yard manure as it is smaller in bulk and is convenient to transport. Fertilizers are usually classified according to the particular plant food element which forms their principal constituents. The main plant food elements are nitrogen, phosphorus and potassium. 20 or 40 kgN/ha would boosted the growth of the perennial grasses. The study conducted at Jhansi on *Chrysopogen fulvus* revealed that the maximum forage production was obtained under 90kgN/ha and 40kgP/ha. The cost of fertilizers required for vast areas is very high that's why the use of fertilizers has to be restricted to selected sites where returns are assured.

(e) Reseeding and planting

The best way to rehabilitate and develop the range land is to manage it on ecological principles. The reseeding of range is resorted to only when the grass regeneration is inadequate, the native vegetation has disappeared and the range is required to be improved immediately. Poor rangelands can be stocked with selected high forage yielding perennial grasses by reseeding and planting. A good forage species must have quick growth, high yield, high nutritive value, high palatability for grazing and non toxicity at all feeding stages. Certain high forage yielding perennial grass species are identified after long research observations and evaluation as per the climatic and edaphic factors. *Lasiurus sindicus, Cenchrus ciliaris, C. satigerus, Panicum antidotale* are suitable for arid region experiencing less than 250 mm mean annual rainfall. In the year to reseeding or planting, a total rain of 100 to 150 mm, is reasonably sufficient for 75 per cent success but if the rain is below 100 mm or if there are prolonged dry spells then reseeding and planting would get failed. *Cenchrus ciliaris, C. satigerus* and *Panicum antidotale* are useful species for moderate sized sand dunes, light well drained soils and rainfall more than 250 mm. *Dichanthium annulatum* requires high rainfall (>250 mm), moderately heavy to heavy soils with good moisture regimes throughout the year. *Sehima nervosum* prefers hill slopes, rolling and undulating terrain with good soil and soil moisture. *Chrysopogen fulvus* is the characteristic of sub-montane tracts of Jammu & Kashmir, Himachal Pradesh and Uttarakhand.

Planting is always more successful than reseeding but costlier and requires more skilful operations. Reseeding and planting are done during monsoon period. The seed sowing is done in furrows of 75cm apart on contours on sloping lands just after one or two good showers. Three weedings are necessary within four months from July to October. Planting of sods is done at 50 cm or 75 cm spacing along the contour. On flat lands, planting is done at 50x50 cm or 75x75 cm spacing in lines. One or two weedings are necessary during August to October. The area which is reseeded or planted requires rest for at least two years. It is not possible to take up all the areas for reseeding or planting because the grazing livestock can't be left starving. A plan therefore, has to be devised whereby certain areas are open and certain areas are closed for holistic range improvement. Forage quality and production is greatly superior in the reserved areas to that badly damaged by heavy grazing.

OPPORTUNITIES FOR IMPROVEMENT OF RANGELANDS/ PASTURES

Despite various constraints on the productivity of pastures and grasslands, the development of grazing areas and fodder cultivation has tremendous potential in India. Research at the Indian Grassland and Fodder Research Institute, Jhansi (IGFRI) and at various other organizations, such as ICAR Institutes and State Agricultural Universities, has developed appropriate technologies for the improvement of these areas. Studies conducted at IGFRI, Jhansi have revealed that the initial protection from grazing of newly improved grasslands can lead to better establishment and higher biomass (3.31 tonnes/ha against 0.93 tonnes/ha without protection). Live-hedge fencing has been found to be economic and suitable. Extensive grazing studies have revealed that the appropriate stocking rates are 25-30, 20, 17, 12 and 6 ACU/100 ha for the management of excellent, good, fair, poor and very poor classes of rangelands, respectively. Basic moisture conservation techniques like contour furrowing, contour bunding and contour trenching can lead to increases in herbage yield (Ahuja, 1977). IGFRI studies, undertaken on natural pastures dominated by *Sehima nervosum, Heteropogon contortus* and *Iseilema laxum*, have revealed that their production can be increased from 4.1 to 7.6, from 3.4 to 5.6 and from 4.5 to 6.4 tonnes/ha/year by the application of nitrogen at a rate of 40 kg/ha (Shankar and Gupta, 1992). Kaul and Ganguli (1963) have recommended that pastures must have 14 per cent of the area under edible bushes to obtain best production results. Silvi-pasture systems on degraded grazing lands (Pathak and Roy 1995) have enhanced biomass by up to 7-15 tonnes/ha/year. Misri (1986) has reported an additional herbage availability of 35-48 tonnes/ha under horti-pasture systems. Singh and Hazra (1995) have suggested methods to substantially increase pasture seed production in India. A number of highly productive, disease resistant and area specific cultivars of various forage crops, and range species have been developed. Hazra (1995) has listed more than one hundred cultivars of fodder crops developed by various research institutes. There will be definite increase of biomass yield by many-fold when these cultivars will be adopted fully in Indian Pastures.

RANGE MANAGEMENT

All the improvement measures will be futile if the rangelands are not managed scientifically and sound management is not possible without improvement measures. Such is the intimate relationship between the two. Some of the important practices of rangeland management are:

(a) Balancing the number of animals to be grazed with carrying capacity or inconsistent with the forage supply

Balancing the number of animals with carrying capacity is the biggest and most serious problem of rangeland management. The lack of balance in various factors of management are scanty forage production, poor & low development of animals, catastrophic periodic starvation, losses livestock, soil erosion, etc.

Carrying capacity is defined as the ability of a grassland unit to give adequate support to a constant number of livestock for a stated period each year without deteriorating with respect to this and/or other proper land use.

This definition implies a harmonious balance among the land range resources and the livestock for maximum sustained production. Earlier, the measured criteria enumerated for better range management is the good health & increasing body weight and milk yield, etc. The efficiency of draught animals was also considered as the criteria for the status of the rangelands. The utilization of vegetation should not exceed from 60 to 75 per cent.

At present, on an average, the range condition is taking poor if the yield of air dry forage is 750 kg/ha which is sufficient for only 75 days for a cattle unit (equal to a cow of 300 kg body weight). The minimum requirement for a cattle unit is estimated at 10kg of air dry forage per day apart from the necessary supplemental feeds during the dry months (say 8) when the nutrient content of the range vegetation is low. At this rate, one cattle unit requires 5 ha for yearlong grazing and without supplements. The livestock population in India is so large that even if one can succeed with the best improvement and management practices to raise the carrying capacity to 1 CU/ ha, it would not be possible to balance the numbers of the livestock with carrying capacity. Therefore, it is imperative to regulate the livestock population by removing the un-productive animals from the competition. Reduction of livestock population may be achieved by preventing promiscuous breeding, fixing limits to the numbers of livestock owned by villagers, by heavy grazing fees and by restriction of grazing rights. Some of the methods for regulating the grazing incidence in the rangelands are as discussed below:

Controlled Grazing

The number of animals that are allowed to graze per unit area of rangeland is fixed in accordance with the carrying capacity of the rangeland is generally termed as controlled grazing. Moreover, it is regulation of grazing on the basis of carrying capacity for yearlong (continuous controlled grazing) or seasonal grazing or for deferred grazing or for rotational grazing or for deferred-cum-rotational grazing.

Early vs. Deferred grazing

The nutritive value of the range vegetation is highest during period of active growth from July to September. But, the grazing in this period is harmful because the grasses are grazed before flowering, fruiting and seed production which is essential for regeneration in the ranges. Deferred grazing is postponing or delaying grazing either completely for the entire growing season (complete deferment) or for the beginning of the growing season (early deferment) or for the later part of the growing season (late deferment) to enable the vegetation to grow well and production of abundant seed for regeneration of the rangeland. Grazing is allowed to after seeding but may be withdrawn to prevent the overgrazing and exposure of soil after few months. This system has a disadvantage that matured forage made available is less palatable and poor in nutritive value.

Rotational grazing

It is a system designed for yearlong grazing in blocks or compartments in rotation. The objective is to give rest to the rangeland and hence full opportunity for the vegetation to grow and develop well. The entire grazing land is divided into two blocks and to allow grazing in alternate years in each block. However, one year rest is not sufficient for the regeneration. Therefore, it is advisable to divide the area into three blocks to give rest for two years to each block. But, the blocks open for grazing would be overgrazed. Such situation has to be tolerated for two years. Later on, there will be improvement in all the three blocks.

Deferred rotational grazing

This is a combination of both the practices of deferred and rotational grazing. Here, all the three blocks are used each year in such a way that each block is grazed for 1/3rd year and protected for 2/3rd year. Thus, each block gets deferment once in three years with equal grazing stress in all the three blocks during the period of three years.

Table 23.3: Time bound deferred-rotational grazing

Year (s)	Block(s)		
	A	B	C
2008	Grazing in rains	Grazing in winter	Grazing in summer
2009	Grazing in summer	Grazing in rains	Grazing in winter
2010	Grazing in winter	Grazing in summer	Grazing in rains

However, it is better to have this system of grazing as per the development of the range vegetation rather than a fixed period of time.

Table 23.4: Plan for grazing where grassland is divided into four blocks and grazed in rotation

Block(s)			
A	B	C	D
Begin grazing on 15th July and continue until the height of grass is reduced to 10cm Then shift livestock to B	X	X	X
	Graze as in A and shift to C	Graze as in A & B and then shift to D	When area D is grazed to desi-red livestock are returned to A to begin the cycle again

The second year of grazing should start on areas having longest period of growth from previous season. The number of divisions would depend upon size and topography of the total area, number of livestock and availability of water. The value of rotational grazing is that it compels livestock to utilize with approximate equal intensity of the poor as well as the good plants. Besides, rotational grazing, there is need to develop several watering places distributed in the range uniformly, placing of salt licks and other mineral nutrients at scattered points equally distributed to attract the livestock in the areas which are un-grazed or less grazed. These practices should particularly be done in remote rangelands so that the pressure on the ranges near in habitat will be reduced.

Top feeds

Besides, grasses and legumes good top feed tree species should be incorporated in the ranges for green and nutritious fodder during winter. Some of the useful species are as follows:

Arid region	*Prosopis cineraria, P. juliflora, Acacia tortilis, Moringa pterygosperma, Albizzia lebbeck* and *Ailanthis excels.*
Semi-arid region	*Prosopis cineraria, P. juliflora, Acacia tortilis, Moringa pterygosperma, Albizzia lebbeck, Ailanthis excels, Acacia nilotica, Dalbergia sissoo, Azadirachta indica, Morus alba* and *Sygyzium cumini.*
Temperate region	*Robinia pseudoacacia, Celtis australis, Salix* spp., *Populus* spp., *Catalpa bignoindes, Ulmus wallichiana, Qurcus* spp., temperate bamboos.

Silage

It is a matter of common observation that forage production is quite high in good rain years and low in drought years. Good quantities of forage produced in good rain years may be suitably treated for preservation and use for the livestock in drought years.

Creation of experimental farms

Experimental farms as per the agro-climatic zones may be established to study the vegetation and different management practices. Such farms would prove to be milestone in educating the local masses and for holistic improvement in the rangelands.

Participatory management

The local inhabitants whose livestock utilize the common grazing lands should form the co-operative groups whereby the projects may be carried efficiently and effectively. It is true that without the local co-operation no one can achieve the measure of success.

Seed production areas

There is acute shortage of grass seeds in the Indian sub-continent. Therefore, it is the need of the hour to use some large experimental farms as seed production areas of good forage grasses.

Transhumant system of management

The transhumant system is practised in order to locate the best herbage resources from pastures and grasslands. There are also well recognized pastoral tribes who practise a complete transhumance, moving from one place to another on traditional migratory routes. The dates of migration have traditionally been fixed. Even, grazing rights rest with the migratory graziers by traditional usage, though they do not hold proprietary rights over the land. The transhumant system is prevalent in the Himalayan region in particular to Jammu & Kashmir. However, this system still exists in some states situated in the plains such as Rajasthan, Madhya Pradesh, Tamil Nadu, Gujarat and Uttar Pradesh.

In Jammu & Kashmir 15 lac livestock population migrate each summer from southern parts of the state to high altitude meadows and pasture. Besides, a sizeable population from adjoining areas of the Kashmir valley also use such grazing areas, with the result grass biomass is not produced even to cater seasonal requirements. Some of the important established bridal paths known as migratory routes in Jammu & Kashmir are as follows:

- Riasi-Margan-Krishnai route
- Poni-Barakh-Zaji Marg-Kolahoi route
- Lamberi-Budhal-Sonmarg route
- Chingas-Dubagan-Tilel route
- Poonchh-Khag-Gurez route
- Poonchh-Tangmarg-Lolab route
- Mansar-Doda-Marwah route

Unfortunately, most of the grazing resources fall within above mentioned migratory routes, therefore, results in over stocking of the areas. Hence, there is need for the regulation of the livestock to save the rich vegetation of such luxuriant meadows.

Livestock rearing is strongly integrated with various farming systems. Since, crop residues form the major portion of animal feed, the integration of livestock rearing in farming systems is common. Recently, hortipasture and silvipasture systems of fodder production have been adopted, thus integrating livestock husbandry with plantation crops as well. Under the on-going national programme on watershed development, these systems are being introduced to enhance biomass productivity of degraded lands which in turn helps in increasing the livestock production. Silvi-pasture system with trees such as *Acacia, Leucaena, Albizia, Melia*, and under-storey grasses *Cenchrus, Chrysopogon, Panicum, Pennisetum, Dichanthium* and legumes *Stylosanthes hamata* and *Macroptilium* are becoming very popular with the farmers.

The educational infrastructure for training and extension requires improvement. Pastures and grazing lands are commonly overgrazed and degraded. Therefore, improved management, better use of crop residues and production of dry fodder are required.

Methods for monitoring rangelands

Monitoring data on the functional state of rangeland systems can be acquired using a variety of methods, including ground-based and remote sensing-based approaches. The wider aim is to provide information so that the interests of a diverse group of stakeholders are met, which means that monitoring information must cover a broad range of spatial extents and time-frames. We recommend that rangeland information providers: use a combination of site-based and remote sensing-based data to report changes in the functional state of rangelands. In the future reporting will require a greater use of remote sensing-based information, especially as remote technologies improve and costs decline.

Research and development organizations at national level involved in improvement and/or management of rangelands

1. **Indian Grassland and Fodder Research Institute (IGFRI), Jhansi, Uttar Pradesh:** This national Institute has the mandate to carry out research on various aspects of fodder production, utilization and management. One hundred and sixty scientists are grouped in six scientific divisions and three regional research stations located at Avikanagar (Rajasthan), Dharwad (Karnatka) and Palampur (Himachal Pradesh).

2. **Regional Stations for Forage Production and Demonstration:** Seven stations have been established for the production of high quality forage seeds and demonstration of proven forage production technologies in various parts of the country. These are located at Hissar, Kalyani, Gandhinagar, Alamadi, Hyderabad, Suratgarh and Shehama.

3. **All India Co-ordinated Research Project (Forage Crops), IGFRI, Jhansi, Uttar Pradesh:** Responsible for national testing and release of forage crop varieties.

4. **Central Arid Zone Research Institute (CAZRI), Jodhpur, Rajasthan:** Responsible for development of range management technologies for arid areas. This station has been recently opted in Leh to look after the meadows of high attitude of (J&K) state.

5. **ICAR Research Complex for NEH Region, Barapani, Assam:** The work is focus on fodder tree management.

6. **National Research Centre for Agro-forestry, Jhansi, Uttar Pradesh:** Fodder tree production and management.

7. **BAIF Development Research Foundation, Pune, Maharashtra:** An NGO undertaking R&D activities on forage production and tree fodder research.

8. **G.B. Pant Institute of Himalayan Environment & Development, Kosi-Katarmal, Almora, Uttarakhand:** Fodder tree inventory, utilization and management.

Some of agricultural universities, research centres and state forest department are also working on need based aspects as per the region of rangelands/pastures for their improvement and efficient management.

SUMMARY

India exhibits great diversity in climate, topography, flora, fauna and land use. The total area under permanent pastures and grasslands is about 12.4 M ha or 3.9 per cent of the countrys' geographical area. The deterioration of Indian pastures, grasslands and other grazing lands may be ascribed to the large bovine population, free grazing practices, lack of management, and natural constraints like extremes of temperature, steepness of slopes, variable precipitation, and scarcity of moisture in arid and semi-arid situations. The situation in Himalayan pastures is even more alarming due to the severe pressure of the sedentary, semi-migratory and migratory graziers. The purpose in the classification of condition of rangeland is to provide an approximate measure of any deterioration in the plant cover thus provide a basis for predicting the degree of improvement. The assessment of range condition set the road for the proper management and improvement measures of the area. The systematic approach to scientific management of rangelands is based on ecological principles and knowledge of plant succession. A comparison of the existing vegetation and the climax vegetation is an indication of the condition of the rangeland. The condition of rangelands have been improving if the vegetation is progressing towards climax and deteriorating if the vegetation belongs to a stage lower in succession than the predominant vegetation. Rangelands get rehabilitated and improve with a closure of 1 to 5 years depending upon the class condition.

If, the improvement in the existing rangelands to is raise their carrying capacity to at least good class and succeed in even continuous controll then its' possible to achieve the self sufficiency in fodder production. Moreover, this will lead to the significant contribution to the prosperity of the country as whole.

REFERENCES

Ahuja LD. (1977). Improving rangeland productivity. In: *Desertification and its Control*. ICAR New Delhi, pp. 203-214.

Anonymous (1997). India 1996 - A Reference Manual. Ministry of Information and Broadcasting, Government of India, New Delhi, pp. 733.

Anonymous (2011). India State of Forest Report. Forest Survey of India, Dehradun, Ministry of Environment & Forests, Govt. of India.

Dabadghao PM and Shankarnarayan KA. (1973). The Grass Cover of India. ICAR, New Delhi.

Friedel MH, Laycock WA and Bastin GN. (2000). Assessing rangeland condition and trend. In: Mannetje L and Jones RM. Field and Laboratory Methods for Grassland and Animal Production Research. CABI, Wallingford, UK. pp.227-262.

Harrington GN, Wilson AD and Young MD. (1984). Management of rangeland ecosystems. In: GN Harrington A.D. Wilson & M.D. Young (editors) Management of Australia's Rangelands. CSIRO Publishing, East Melbourne, Australia. pp. 3-13.

Hazra CR. (1995). Improved Cultivars of Forage Crops for Different Agro- Environments. In: RP. Singh (ed.) *Forage Production and Utilization*. IGFRI, Jhansi (India), pp. 326-335.

Kaul RN and Ganguli BN. (1963). Fodder potential of *Zizyphus* in the shrub grazing lands of arid zones. *Indian Forester*, **(39)**: 623-630.

Misra R. (1983). Indian Savannas. In: F. Bourliere (ed.) *Tropical Savannas*. Elsevier, Amsterdam, pp. 155-166.

Misri B. (1986). Forage Production in the Kashmir Himalayas. In: P. Singh (ed.) *Forage Production in India*. RMSI, IGFRI, Jhansi (India), pp. 32-38.

Misri B. (1995). Range and Forest Grazing in the Himalaya. In: P. Singh (ed.) Workshop Proceedings. Temperate Asia Pasture and Fodder Sub-Regional Working Group. Kathmandu, Nepal pp. 28-33.

Pringle H and Tinley K. (2003). Are we overlooking critical geomorphic determinants of landscape change in Australian rangelands? Ecological Management & Restoration vol. 4, pp. 180-186.

Pringle HJ, Watson IW and Tinley KL. (2006). Landscape improvement, or ongoing degradation – reconciling apparent contradictions from the arid rangelands of western Australia. *Landscape Ecology* vol. **21**: pp. 1267-1279.

Raheja PC. (1966). Rajasthan desert can bloom with forage. *Indian Farming*, **15**: 47.

Shankar V and Gupta JN. (1992). Restoration of Degraded Rangelands. In: J. S. Singh (ed.). *Restoration of Degraded Lands-Concepts and Strategies*. Rastogi Publications, Meerut, India, pp. 115-155.

Singh P and Misri B. (1993). Rangeland Resources-Utilization and Management in India. Paper presented at International Symposium on Grassland Resources held at Huehot, Inner Mangolia, China, August 16-20.

Singh RP and Hazra CR. (1995). Forage Seed Production-Status and Strategy. In: R.P. Singh (ed.), *Forage Production and Utilization*. IGFRI, Jhansi (India), pp. 309-323.

Watson IW and Thomas PW. (2003). Monitoring shows improvement in the Gascoyne–Murchison rangelands. Range Management Newsletter No. 1, pp. 11–14.

Watson IW. (1998). Monitoring West Australian shrublands: what are the expectations of change? Range Management Newsletter No. 98/2, pp. 1-5.

Williams RE, Allred BW, Denio RM and Paulsen HA. (1968). Conservation, development, and use of the world's rangelands. *Journal of Range Management* vol. 21, pp. 355-360.

WRI (World Resources Institute). (1986). World Resources 1986: An Assessment of the Resource Base that Supports the Global Economy. Basic Books, New York, USA.

■

24

CHAPTER

Geoinformatics in Forest Mapping

☞ Sandeep Arya₁, ☞ V.S. Arya², ☞ R.S. Hooda³,
☞ K.S. Bangarwa⁴ and ☞ N.K. Goyal⁵

INTRODUCTION

Forest plays a vital role in economic, social and cultural development of any country. In addition to our essential requirements such as fuelwood, timber, shelter etc. forest plays an important role in carbon sequestration. In several region of India people depend on forest and their products. Therefore, our forests are continuously disappearing at alarming rate. It is well known that the decrease in forest cover will create a number of serious environmental problems like increase in CO_2 level, soil erosion, floods and adverse affects on biodiversity. Thus, it is necessary to take care of our renewable resources for sustainable development. Geoinformatics is the convergence of various technologies like Remote Sensing (RS), Geographical Information System (GIS), Global Positioning System (GPS) and Information and Communication Technologies (ICT) which help in gathering information about the earth and its resources. It has been a valuable source of information over the course of the past decade in mapping and monitoring forest activities. It is a powerful aspect of a technology to detect the changes accurately. Earlier, for the purpose of consistent and repeated monitoring of forests over larger areas, several types of remote sensing data, including aerial photography, multi-spectral scanner (MSS), radar (Radio Detection and Ranging), LIDAR (Light Detection and Ranging) laser and videography data have been used by forest agencies to detect, identify, classify, evaluate and measure various forest cover types and their changes. Gradually other types of remote sensing tools were developed with which forest object properties were registered from the air or from space. Nowadays aerial photography has been replaced by the satellite imagery. It is a cost effective and appropriate technique of data interpretation. For management purpose Geographical Information System (GIS) is a technique of hardware and software designed to store, retrieve and analyze the spatial and non-spatial data, whether generated using remote sensing or other means, and to generate the information in desired forms. Satellite images provide a synoptic view of a large area; it also captures bio-physical properties of the land

¹ KVK Yamunanagar, CCS HAU, Hisar-125 004
² Haryana Space Applications Centre, Hisar-125 004
³ Haryana Space Applications Centre, Hisar-125 004
⁴ Department of Forestry, CCS HAU, Hisar-125 004
⁵ Department of Forestry, CCS HAU, Hisar-125 004

feature through the reflected electro-magnetic radiations, often called signature in the remote sensing parlance.

INDIAN SATELLITE REMOTE SENSING PROGRAMME

Indian satellite remote sensing programme dawned a new era in March 1988, when Indian Remote sensing satellite (IRS-1A) was launched. Before that USA made series of Landsat satellites data was used. In India NRSA, now NRSC, used Landsat MSS false colour composites on 1:1 m scale and mapped nationwide forests for the first time for 1972-75 periods. Khuswaha *et. al.* (1988) evaluated the Landsat-MSS, TM, spot-PAN and MSS, IRS LISS-I and LISS-II data and found that IRS LISS-I data was better than Landsat –MSS, IRS-1 C, IRS-1 D and IRS-1-P series of satellites and provide further improvised sensor data of 23 m and 5.8 m resolutions along with a new Wide Field Sensor (WiFS) data having 188 spatial resolutions. WiFS data facilitated the assessment of land features and phenomena encompassing large areas because of its wide swath of about 750 kms.

In India, the first attempt to classify the forest cover types by digital classification of satellite data was made in 1978 in Nagaland (Madhavan, 1979) delineating temperate evergreen, tropical evergreen, tropical semi-evergreen, Sal forest, degraded forests and current shifting cultivation. Micro level planning in forestry is required to show the details of different forest patches, block plantations, forest blanks, degradation etc. IRS LISS-I (70 m) and IRS-LISS-II (36 m) data could be used for mapping at 1:2,50,000 and 1:50,000 scale, respectively. IRS LISS-III (23 m) and PAN (5m) data together support preparation of 1:25,000 scale maps. Experience has shown that a merging of LISS-III and PAN data is advantageous compared to any one of them alone. IRS PAN data has also been used for the preparation of maps on large scale than a 1:25,000 scale and have been found useful in delineation of smaller forest patches, strip plantation, etc. Mapping on a 1:25,000 scale can be taken as optimal, for planning at forest subdivision, considering various aspects of forest resources survey and mapping tasks. IKONOS/CARTOSAT data (about 2.5 m ground resolution and better) could be used to prepare maps of scale larger than 1:12,500. Such maps will be highly useful for growing stock estimation (using stratified random sampling) and revision/updation of detailed stock maps. (Joseph, 2007). Table-24.1 gives details of various satellites and their sensors whereas the description of high resolution satellites with sensors is given in Table-24.2 (*Geospatial Today, July-August, 2002*).

Table 24.1: Remote sensing satellites and sensors

Satellites	Sensors	No. of bands	Wavelength (μm)	Spatial Resolution (m)	Swath (km)
Landsat 1-5 (U.S.A.)	MSS	4	05 – 0.6, 0.6 – 0.7, 0.7 – 0.8, 0.8 – 1.1	80	185

(Contd...)

Satellites	Sensors	No. of bands	Wavelength (µm)	Spatial Resolution (m)	Swath (km)
Landsat 4/5	TM	7	0.45 – 0.52 0.52 – 0.60 0.60 – 0.69 0.76 – 0.90 1.55 – 1.75 2.08 – 2.35 10.4 – 12.5	30 120	185
SPOT-1 (France)	XS	3	0.50 – 0.59 0.61 – 0.68	20	117
	PAN	1	0.79 – 0.89 0.51 – 0.73	10	
SPOT-2/3	XS	1	-do-	20	
	PAN	3	-do-	10	
IRS-1A/1B (India)	LISS-I & II	4	0.45-0.52 0.52-0.59 0.62-0.68 0.77 – 0.86	72.5 (LISS-I) 36.25 (LISS-II)	148 (LISS-I) 2x74 (LISS-II)
IRS-1C	LISS-III	4	0.52-0.59 0.62-0.68 0.77-0.86 1.55-1.70	23.5	141
	PAN	1	0.5 – 0.75	5.8	70.5
	WiFS	2	0.62-0.68 0.77-0.86	188	810
IRS-1D	LISS-III PAN & WiFS	412	0.62-0.68 0.77-0.86 1.55-1.700.5 – 0.75-0.62-0.68 0.77-0.86	23.5 5.2-5.8 3.6-70.5 169-188	141 133-148 63-70.5 728-812
CartoSat-1	Panchromatic	1	0.50- 0.85	2.5	30
CartoSat-2	Panchromatic	1	0.50- 0.85	<1	9.6
Resource Sat-1 (IRS P6)	LISS-III	4	0.52-0.59 0.62-0.68 0.77-0.86 1.55-1.70	23.5	141
	LISS-IV		0.52-0.59 0.62-0.68 0.77-0.86	5.8	23.9 (Mx)

(Contd...)

Satellites	Sensors	No. of bands	Wavelength (μm)	Spatial Resolution (m)	Swath (km)
	AWiS		0.52-0.59 0.62-0.68 0.77-0.86 1.55-1.70	56	740
Resource Sat-2	LISS-III	4	0.52-0.59 0.62-0.68 0.77-0.86 1.55-1.70	23.5	140
	LISS-IV		0.52-0.59 0.62-0.68 0.77-0.86	5.8	70/23 (Mono & MX)
	AWiFS		0.52-0.59 0.62-0.68 0.77-0.86 1.55-1.70	56	740
ERS-1(EU)	SAR	1	C (5.3 GHz)	30	100
ERS-2(EU)		2	C (5.3 GHz)	30	100
JERS-1 (JAPAN)	SAR	1	L (1.275 GHz)	18	75
RADARSAT-1 (CANADA)		1	C (5.3 GHz)	10x50	100 165 150 45
IKONOS (USA)	PAN	1	0.45-0.90	1	11
	MSS	4	0.45-0.52 0.52-0.60 0.63-0.69 0.76-0.90	4	

Table 24.2: High resolution satellites and their specifications

Organisation	Digital Globe (EarthWatch) USA	OBRIMAGE, USA	Space Imaging, USA	Images at International (West Indian Space), Israel	ISRO, India	CNES, France
System	Quickbird 1&2	OrbView 3&4	IKONOS 1&2	EROS-A&B	Cartosat 1&2	SPOT 5A
On-Orbit Date	QB 1-Nov 2000 QB 2-Oct 2001	OV 3-End 1999 OV 4-Sep 2000	I-1-Apr 1999 I-2-Sep 2000	A:#1-Dec 2000 #2-Sep 2000 B:#3-#8 Dec'02-Dec'04	#1-2003-04 #2-2004	May 2002

(Contd...)

Organisation	Digital Globe (EarthWatch) USA	OBRIMAGE, USA	Space Imaging, USA	Images at International (West Indian Space), Israel	ISRO, India	CNES, France
Spatial	0.61 PAN	OV-3&4 1.0 PAN	1.0 PAN	#A 1.8 PAN	#C1 - 2.50 PAN	2.5 PAN
Resolution (m)	2.50 MS	OV-3&4 4.0 MS OV-4 8.0 HS	2.0 4.0 MS	#B 0.82 PAN 3.28 MS	#C2-<1.0 PAN	20 SWIR 10 HRG
Revisit Interval	1-3.5 days	Less than 3 days	1-3 days	3 days	#1-5 days #2-4 days	5 days
Altitude (km)	600	470	680	480 & 600	#2-600	824
Spectral Band Width (μ)	0.45 - 0.90 PAN 0.45-0.52 0.52-0.60 0.63-0.69 0.76-0.89 MS	0.45-0.90 0.45 - 0.90 2.50 3.0-5.0 200 bands 80 bands	0.45-0.90 0.45 - 0.52 0.52 - 0.60 0.63 0.69 0.76 - 0.89 MS	A&B PAN 0.5-0.9 #B-MS 3 bands or more	PAN 0.5-0.7	PAN 0.51-0.730
Imaging System	Pushbroom	Pushbroom	Pushbroom	Pushbroom	CCD	NA
Swath Width (km)	16.5	8.0 PAN 5.0 HS	11-11.0 12-12.6	EROS A-12.6 EROS B-16.4	#C1-30 #C2-10	60
System Life	5 yrs	5 yrs	7 yrs	#A-4 yrs #B-6 yrs	5 yrs	5 yrs
Stereo	Along & Across : ±38° & ±30°	Along & Across: ±45° & ±45°	Along & Across: ±45° & ±45°	Along : ±45°	#1-Along: ±26° & -5° #2-along & Across : ±450°	Across : ±26°

For monitoring and management of forest resources, we need to have accurate information of these resources. Broadly Remote Sensing techniques can be used in forestry as given below:

Forest cover mapping

Forest cover mapping includes forest type and forest density. Champion and Seth (1968) defined forest type as a unit of vegetation which possesses (broad) characteristics in physiognomy and structure sufficiency pronounced to permit its differentiation from other such units. Management prescriptions in any region are based upon the existing forest type in that region. The National Forest Policy of India, 1952 set a goal of bringing one third of the country under forest covering plains and sixty per cent in the hilly regions. The policy goal gives the direction for conservation, afforestation and tree planting outside the traditional forest areas. Realizing the need for a close and scientific monitoring of forest cover, India started the regular programme for the same with the use of remote sensing technology way back in early 1980. In the report of Forest Survey of India (FSI, 2009) country have 21.02 per cent forest area. The forests of India have been divided into 16 types as given in Table 24.3.

Table 24.3: Forests type with Forests area of India

Sr. No.	Major groups	Forest Area
1.	Tropical Wet Evergreen Forest	8.75
2.	Tropical Semi-Evergreen Forest	3.35
3.	Tropical Moist Deciduous Forest	33.92
4.	Tropical Littoral and Swamp Forest	0.38
5.	Tropical Dry Deciduous Forest	30.16
6.	Tropical Thorn Forest	5.11
7.	Tropical Dry Evergreen Forest	0.29
8.	Subtropical Broad Leaved Hill Forest	0.38
9.	Subtropical Pine Forest	5.99
10	Subtropical Dry Evergreen Forest	0.36
11.	Maotane Wet Temperate Forests	3.45
12.	Himalayan moist temperate Forest	3.79
13.	Himalayan dry temperate Forest	0.28
14.	Sub-Alpine Forests	3.79
15.	Moist Alpine	
16.	Dry Alpine	
		100 %

Source: FSI, 2009 report

Forest Survey of India carries out forest density mapping of India biannually using LISS-III data at a 1:50,000 scale. Forest maps are being prepared from satellite imagery using either visual interpretation or digital classification. Forest density refers to the canopy density classes- canopy density >70 per cent is very dense forest, canopy density 40 to 70 per cent is called moderately dense forest, canopy density 10 to 40 per cent is open forest and <10 per cent is called degraded forest. The best time for taking the satellite imageries during the year is when maximum difference occurs due to phenological changes such as leaf fall, flowering, etc. improves the capability of satellite data in forest type delineation. Table 26.4 shows the appropriate season for selecting the satellite data of different forest region of India.

Table 24.4: Appropriate season for aerial/satellite data acquisition for forest mapping

Sr.No.	Region/Vegetation Zone/Geographical Set-up	Proper season
1.	Humid/moist evergreen and semi-evergreenforests of western Ghats and eastern Ghats	January-February
2.	Humid and moist evergreen and semi-evergreen forests of north-east India and Andaman and Nicobar Islands	February-March

(Contd...)

Sr.No.	Region/Vegetation Zone/Geographical Set-up	Proper season
3.	Tropical moist deciduous forests of northern and central India	December-January
4.	Temperate evergreen forests of western Himalayas	March-May
5.	Temperate, sub-alpine, alpine evergreen, deciduous forests of Jammu and Kashmir	September-October
6.	Arid and semi-arid dry deciduous and scrub forest	October-December
7.	Mangrove forests	February-March

(Source: Ranganath et.al., 2000)

Monitoring/Change detection

Looking at the change in spatial structure that can be monitored using remote sensing information and landscape metrics, natural disturbances can be analyzed using remote sensing imagery, such as forest damage, defoliation and fire. Change in a forest as a result of many imperceptible and yet powerful forces may be apparent only after long periods of time. Many forests are slow-growing and relatively long-lived. Change is a defining characteristic of forests. An example might be the creation of a soil horizon layer in a conifer forest, predictable by considering the climate conditions, litter fall and microbial activities. Successional changes, growth changes, changes as a result of structural and age processes, all accrue slowly and with generally small daily, weekly, monthly and even annual variability. Change can also be rapid and transformative (For example, leaves can change colour and cell structure overnight). Powerful, even cataclysmic, forces can arrive with little or no warning. Examples might include wildfire, insect outbreak, wind throw, harvesting operation, or a prescribed burn (Franklin, 2001). Detection and monitoring of such forest changes across large areas are of most important tasks that remote sensing can accomplish.

Another practical example is the encroachment studies carried out in the Sanjay Gandhi National Park, Borivali, Maharastra and at a few other places (Jadhav, 1995). Judicial courts have found geoinformatics technique reliable and three cases have been disposed off based on the findings prepared by satellite data. Change detection technique also helps in assessment of forest degradation, deforestation, afforestation and Joint Forest Management (JFM) related forest cover change monitoring.

Forest Modelling

Forest models represent a key piece of infrastructure required in support of sustainable forest management. Models allow generalizations from sites to regions and can be used to predict, investigate or simulate effects over a wide range of conditions and scales. Geoinformatics is very useful and fast tool for making the models for wildlife habitat suitability, timber volume, fire risk zone, etc. Ecological models have been developed "as tools for projecting the consequences of observations or theories about how ecosystems may change over time". A wide variety of forest models exist, ranging from the individual tree growth and mortality models, to gap or stand models of competition and structure, to global models of productivity (Shugart, 1998). Substitute "stands" for "ecosystems" and the value of this new tool is quite apparent under any forest management strategy. The sustainable forest management with pressing need to better understand ecosystem models may be an indispensible information resource. Models facilitate experimental design and interpretation of results, the testing of current hypotheses and the generation of new ones; models form a framework around which empirical observations can be organized (Lauenroth *et al.*, 1998). By recognizing the cultural aspects of data management and modeling, a three-way relationship designed to alleviate the problems that flow from the enormous accumulation of scientific data is emerging between. (i) Empirical data collection, (ii) Multidisciplinary data analysis and (iii) Computer modeling (Olson *et al.*, 1999).

Obviously, GIS and remote sensing are wonderful ways of accumulating enormous collections of empirical observations. This creates the need for better, more powerful tools to help for making the sense of these data. Models represent one such powerful tool. The proliferation of models threatens to overwhelm their promising role as a helpful tool in forest management. *For example*, Landsberg and Coops (1999) list three types of models that have been developed to deal with aspects of, or approach to, forest productivity: (i) standard growth and yield models, (ii) gap models, and (iii) carbon balance or biomass models. Battaglia and Sands (1998) and Shugart (1998) provide more comprehensive listings of models and few of them are expected to emerge as bonafide management tools. Models are reaching new levels of sophistication at the same time that they are increasingly able to answer questions posed by managers (Battaglia and Sands, 1998). Here, the promise appears to be in those carbon balance or ecosystem process models; at least in some forests, such models appear to have a greater likelihood of current or near-future use as tools by managers. Their uses in operational settings have been made more likely by virtue of the wider use of remote sensing and, especially, GIS technology (Bateman and Lorett, 1998). Increased interest in the results of forest ecosystem (Leblon *et al.*, 1993) and grassland modeling (Burke *et al.*, 1990) has spurred wider availability of various types of GIS data, biogeographical data, DEMs, forest inventory cover type maps, spatially explicit meteorological data and finally, biophysical remote sensing information. Modeling technique has been used to estimate Net Primary Productivity (NPP) and Carbon Fixation in Forest-Ecosystems. In the ecosystem process models, such as BEPS and BGC ++, a daily time step procedure is used to determine the rates of photosynthesis, autotrophic respiration (sum of growth and maintenance respiration), and nitrogen transformation. These depend heavily on LAI and forest cover type assumptions. LAI is one of the most important variables in many process models; step back, and consider that "The terrestrial biosphere is like a chlorophyll sponge blanketing the Earth with a thickness proportional to LAI" (Running, 1994). With GIS and modeling approaches, it is possible to simulate empirical or natural history and to devise experimental and comparative ecosystem studies (Likens, 1998). Hooda *et. al.* (2002) estimated agricultural and non-agricultural carbon fixation over India using remote sensing data in the PEM. Arya *et. al.*, 2010 studied the estimation of forest biomass of Haryana at block level with the help of equations by using the geoinformatics. The availability of GIS data and ecological models has created a number of new analytical possibilities, including a new emphasis on ecological impact assessment (Treweek, 1999). On the other hand, database development to serve forest modeling applications has stimulated progress in using and refining forest models.

A CASE STUDY

The study on "Mapping of Forest Crown Density and Tree outside Forest (TOF) in Barwala block, Panchkula" of Haryana state was carried out to demarcate the forest cover and analyze crown density of forests sites (Kumar et al., 2011). The study area consists of 235.37 sq.km area, where tropical dry deciduous forests and sub-tropical forests are found majorly. This study shows the utility of satellite remote sensing technique and GIS for preparation of more consistent and accurate information of different forest categories and Tree outside Forest (TOF). Cartosat-I Panchromatic data was used and based upon the standard image characteristics, the visual interpretation of satellite data was carried out. Doubtful areas were checked

Tree outside Forest (TOF) has done on 1:10,000 scale. A Geo-database was created to manage the spatial data and its attributes. As per National Remote Sensing Centre (NRSC), The forest area was classified into five categories based on density of trees 0-10, 10-20, 20-40, 40-60 and 60-80 per cent and the forest area marked under these five categories was 15.60, 14.38, 17.93, 21.50 and 0.77 sq.km, respectively. The total forest area of Barwala block is 70.18 sq.km which constitutes 29.82 per cent of the total geographical area. Total area under Tree outside Forest (TOF) is 93.02 sq.km which constitutes 39.56 per cent of the total geographical area. Total length of Tree outside Forest (TOF) along Canal, Road, River and Farm Bunds is 282.13 km. The study helped in preparing latest crown density maps of the forest area in digital form which can be used for forest management. The spatial information generated on forests cover and Tree outside Forest (TOF) on 1:10,000 can be utilized for various reclamation measures and other uses for the district level planning in the Department of Forest.

SUMMARY

The satellite applications for effective forest mapping on a more scientific way with the priorities set at Macro and Micro levels studies is requirement of the era. The GIS for data base creation and requirement of forest resources information system involving effective inventory data analysis packages supporting volume yield and cull factor analysis are effective part of forest mapping. The scientific and effective forest management information could be operationalised using the GIS approach as a functional unit. The organization of data feasibility adopting micro scale to macro scale may be a viable approach rather than adopting top to bottom approach starting with coarse resolution data which may fail to reveal any meaningful outputs required at the working level. The organization of data base, data inputting, updating, retrieval and analysis for specific purposes and to obtain outputs apparently is feasible cost effective.

REFERENCES

Arya S and Hooda RS. (2010). Estimation of Agricultural and Forest Biomass in Haryana. Report by Dept. of Science & Technology, Govt. of Haryana HAESAC/TR/22/2010 pp. 233.

Bateman I and Lorett A. (1998). Using a GIS and large area databases to predict yield class: a study of Sitka spruce in Wales. *Forestry*. **71**: 147–168.

Battaglia M and Sands PJ. (1998). Process-based forest productivity models and their application in forest management. For. Ecol. Manage. **102**: 13–32.

Burke IC, Schimel DS, Yonker CM, Parton WJ, Joyce LA and Lauenroth WK. (1990). Regional modeling of grassland biogeochemistry using GIS. Landscape Ecol., **4**: 45–54.

Champion HG and Seth SK (1968). Revised Survey of Forest types of India, Manager of publications, Govt. of India, New Delhi.

Forest Survey of India (2009). State of Forest report 2009. Published by Ministry of Environment & Forests, Govt. of India.

Franklin Steven E. (2001). Remote sensing for sustainable forest management. Lewis Publishers CRC Press LLC USA. pp. 393

Geospatial Today (2002). Spatial networks private limited. Khairatabad, Hyderabad.

Hooda RS, Dye DG and Shibasaki R (2002). Evaluating agricultural and non-agricultural carbon fixation over India using remote sensing data. In Owe, M. *et. al.* (Eds.) Remote Sensing for agricultural, Ecosystems and Hydrology, IV: 108-113.

Jadhav RN. (1995). Photonirvachak, *Journal of Indian Society of Remote Sensing*, **23**, pp. 87-88.

Joseph G. (2007). Fundamental of Remote Sensing, Second Edition. University press (India) Pvt. Ltd. Hyderabad, India.

Khuswaha SPS, Murthinaidu KS and Madhavan Unni NV. (1988). Comparative visual evaluation of Landsat, SPOT and IRS-1A data for land features differentiation and mapping. Proc. IRS data Evaluation Seminar, December 23, NRSA, Hyderabad.

Kumar M, Hooda R S and Arya S. (2011). Mapping of Forest crown density and tree outside forest (TOF) in Barwala block, Panchkula using Geo-informatics. Thesis of Haryana Space Applications Centre HARSAC. pp-46.

Lauenroth WK, Canham CC, Kinzig AP, Poiani KA, Kemp M and Running SW. (1998). Simulation modeling in ecosystem science. pp. 404–414 in Pace ML and. Groffman PM, Eds. Successes, Limitations, and Frontiers in Ecosystem Science. Springer-Verlag, New York.

Madhavan Unni. (1979). Satellite remote sensing survey of natural resources of Nagaland. Report, National Remote Sensing Agency, Hyderabad.

Olson RJ, Briggs JM, Porter JH, Mark GR and Stafford SG. (1999). Managing data from multiple disciplines, scales, and sites to support synthesis and modelling. *Rem. Sensing Environ.* **70**: 99–107.

Ranganath B.K, Roy P.S, Dutt CBS and Divakar. (2000). Use of modern technologies and information systems for sustainable forest management: Status Report, ISRO, Department of Space.

Running S W. (1994). Testing FOREST-BGC ecosystem process simulations across a climatic gradient in Oregon. *Ecol. Appl.* **4**: 238–247.

Shugart H.H. (1998). Terrestrial Ecosystems in Changing Environments. Cambridge University Press, Cambridge, U.K.

Treweek J. (1999). Ecological Impact Assessment. Blackwell Scientific, Oxford, U.K.

■

www.ingramcontent.com/pod-product-compliance
Lightning Source LLC
Chambersburg PA
CBHW020216290326
41948CB00001B/65